(New $285-) $195—
 OF-119

MECHANICS OF OIL SHALE

MECHANICS OF OIL SHALE

Edited by

KEN P. CHONG

Professor of Civil Engineering, University of Wyoming, Laramie, Wyoming, USA

and

JOHN WARD SMITH

Synthetic Fuels Consultant, Laramie, Wyoming, USA

ELSEVIER APPLIED SCIENCE PUBLISHERS
LONDON and NEW YORK

ELSEVIER APPLIED SCIENCE PUBLISHERS LTD
Ripple Road, Barking, Essex, England

Sole Distributor in the USA and Canada
ELSEVIER SCIENCE PUBLISHING CO., INC.
52 Vanderbilt Avenue, New York, NY 10017, USA

British Library Cataloguing in Publication Data

Mechanics of oil shale.
1. Oil-shale industry
I. Chong, Ken P.   II. Smith, John Ward
333.8'2   HD9571.5

ISBN 0-85334-273-3

WITH 49 TABLES AND 232 ILLUSTRATIONS

© ELSEVIER APPLIED SCIENCE PUBLISHERS LTD 1984

The selection and presentation of material and the opinions expressed in this publication
are the sole responsibility of the authors concerned.

All rights reserved. No part of this publication may be reproduced, stored in a retrieval
system, or transmitted in any form or by any means, electronic, mechanical, photocopy-
ing, recording, or otherwise, without the prior written permission of the copyright owner,
Elsevier Applied Science Publishers Ltd, Ripple Road, Barking, Essex, England

Printed in Northern Ireland by The Universities Press (Belfast) Ltd

DEDICATED

TO

Mr & Mrs Chih-Hwan Chong
Mr & Mrs John H. Smith
Shuang-Ling Chong
Harriet Smith

Foreword

Oil shale is but one of several unconventional sources of feedstock for production of transportation fuel that will be developed as the disequilibrium in the supply/demand situation for those commodities develops toward the end of this century. Whatever proportions are assumed by the energy mix that includes coal, tar sands, heavy oil, and enhanced recovery, oil shale will share in some significant measure. The difficulties in turning away from petroleum based transportation fuels to electrical power, methanol, hydrogen and the like range from formidable to totally impractical. The problems in the development of oil shale to help fill the inevitable petroleum supply shortfall appear immensely difficult only until compared with those attendant to development of its alternatives.

In the United States, the roller coaster performance in the development of the Green River oil shales has been sufficiently discouraging to foster some predictions to the effect that commercial production from them is impractical. The reasoning is illogical in the light of the slow development of some other highly beneficial technologies and belied toward the end of 1983 when the first commercial plant in the USA came on stream. The progression from exploitation of small very rich deposits to that of very large, low grade deposits has been the norm for virtually all of the minerals used by man. We are simply looking to the certain development of this low grade ore of oil.

The competing specific technologies for production of shale oil, broadly divided into mining with surface retorting and *in situ* methods, have been demonstrated to widely varying degrees. The largest tests have produced a few hundred thousand barrels and the least tested schemes are still on paper. As to which of these technologies will ultimately prove the most successful, the jury is still out. Nonetheless, practitioners and proponents of all of these different approaches vitally need precise data and a broad understanding of a wide variety of physical properties and mechanical behaviors of oil shale, both as a process feedstock and as the envelope within which the miner works. This collection of writings on the mechanics of oil shale is the only single source to which one can turn presently to find needed information in a single volume. The editors have skillfully assembled the contributions of a group of distinguished scientists and engineers to produce a practical tool for the people who are molding a major new industry.

The various competing concepts for the production of salable products from these unusual rocks are straightforward, simple, and in very large part, technologically feasible. The immense scale of the required investment with its consequent high cost of technologic or economic error together with the large benefits to be gained from subtle optimization, render the final design of these facilities difficult and complex. The underground miner wants his rooms as large and his pillars as small as is compatible with safety. The surface miner would like his final pit slope as steep as possible without risking slope failure. The modified *in situ* designer strives for the fewest blastholes and smallest powder charge as will produce that ideal rubble that upon retorting yields maximum production in the shortest time with the least grief. All of these people and others connected with yet other facets of oil shale plant design require information on the physical behavior of oil shale under a wide variety of conditions. If the complete solution to your problem is not within the covers of this book, information and insight sufficient to enable one reasonably skilled in the art to find the answer, is.

EUGENE L. GROSSMAN
Oil Shale Mining Consultant,
Englewood, Colorado, USA

roleum industry's monomania for retorting with the statement 'If the petroleum industry got into the lumbering business, they'd try to drill a tree down from the top.' Dr Ertl knew the value of working with the resource that actually exists and of maximizing the recovery of oil shale from Colorado's Piceance Creek Basin. After participating in design and development of the US Bureau of Mines room and pillar demonstration mine at Anvil Points, Colorado, Dr Ertl rebelled against the waste represented by high-grade mining of oil shale. In 1948 he proposed to open pit mine the entire 1200 mile^2 (3100 km^2) Green River oil shale deposit in Colorado's Piceance Creek Basin. This is a powerful concept with many real advantages. Two advantages are that virtually 100% resource recovery could be achieved and that the huge resource available could be developed as necessary to supply the entire United States oil consumption for several hundred years. Dr Ertl amplified this proposal in 1965, and it has been echoed several times since. Although the scope of such a project is so vast that it almost requires national effort, the potential rewards to the United States are spectacular. Provided, of course, that the nature of the rock is respected in the development. As Harry K. Savage says 'There is an enormous field and a great need for creative oil shale thinking.'

Although this book's articles tend to center on development of the Green River Formation oil shales, the editors have attempted to indicate how at least part of the basic information included can be extrapolated to other deposits. Although each deposit is an individual, the oil shale families have some common factors permitting this extrapolation.

Technical units used in oil shale studies across the world tend to be confusing. The mechanical property studies appear usually in the unit system in which they were developed. Oil shale grade and resource units are also confusingly diverse. To try to provide some magic route among all of the diverse units the editors have compiled a table of unit conversions.

The editors deeply appreciate the willingness of all of the chapter authors to contribute to this book. They have participated in a pioneering venture.

KEN P. CHONG
JOHN WARD SMITH

Contents

Foreword vii

Preface ix

List of Contributors xv

Common Conversion Units xix

1. Introduction to Mechanics of Oil Shale 1
   JOHN WARD SMITH AND KEN P. CHONG

2. Historical Perspective 43
   STEPHEN UTTER

3. The Continuum Theory of Rock Mechanics . . . 71
   W. F. CHEN

4. Sampling, Logging and *in situ* Stresses of Oil Shale . . 127
   GEORGE F. DANA

5. Mechanical Characterization of Oil Shale 165

KEN P. CHONG AND JOHN WARD SMITH

6. Statistical Analysis and Modeling of Physical, Mechanical and Strength Properties of Oil Shale 229

NIEN-YIN CHANG AND ELLEN J. BONDURANT

7. Stratigraphic Variations in Fracture Properties 291

CHAPMAN YOUNG, BRUCE C. TRENT and NANCY C. PATTI

8. Model Studies of Fragmentation 337

W. L. FOURNEY, D. C. HOLLOWAY AND D. B. BARKER

9. Surface Uplift Blasting for Oil Shale Production . . 389

KEITH BRITTON

10. A Strain-Rate Sensitive Rock Fragmentation Model. . 423

E. P. CHEN, MARLIN E. KIPP AND D. E. GRADY

11. Mining and Fragmentation Oil Shale Research . . . 457

WILLIAM HUSTRULID, ROGER HOLMBERG AND ERNESTO PESCE

12. Temperature Effects 523

JOEL DuBow

Index 579

List of Contributors

D. B. BARKER

Mechanical Engineering Department, University of Maryland, College Park, Maryland 20742, USA

ELLEN J. BONDURANT

Engineer, Structure Stability Branch, Pacific Northwest Region, US Bureau of Reclamation, Boise, Idaho 83724, USA

KEITH BRITTON

President, Blast Masters Inc., 325 Beverly Avenue, San Leandro, California 94577, USA

NIEN-YIN CHANG

Associate Professor, Department of Civil and Urban Engineering, University of Colorado at Denver, 1100 Fourteenth Street, Denver, Colorado 80202, USA

E. P. CHEN

Applied Mechanics Division II, Sandia National Laboratories, Albuquerque, New Mexico 87185, USA.

W. F. CHEN

Professor and Head of Structural Engineering, School of Civil

Engineering, Purdue University, West Lafayette, Indiana 47907, USA

KEN P. CHONG

Professor of Civil Engineering, University of Wyoming, Laramie, Wyoming 82071, USA

GEORGE F. DANA

Manager, Geology and Geochemistry Research Group, Western Research Institute, Laramie, Wyoming 82071, USA

JOEL DUBOW

Professor, Electrical and Computer Engineering Department, Boston University, 110 Cummington Street, Boston, Massachusetts 02215, USA

W. L. FOURNEY

Professor and Chairman, Mechanical Engineering Department, University of Maryland, College Park, Maryland 20742, USA

D. E. GRADY

Thermomechanical and Physical Division, Sandia National Laboratories, Albuquerque, New Mexico 87185, USA

D. C. HOLLOWAY

Mechanical Engineering Department, University of Maryland, College Park, Maryland 20742, USA

ROGER HOLMBERG

Swedish Detonic Research Foundation, Tellusborgsvagen 34–36, PO Box 32058, S-126 11 Stockholm, Sweden

WILLIAM HUSTRULID

Professor, Department of Mining, Colorado School of Mines, Golden, Colorado 80401, USA

MARLIN E. KIPP

Computational Physics and Mechanics II Division, Sandia National Laboratories, Albuquerque, New Mexico 87185, USA

NANCY C. PATTI

Science Applications Inc., Steamboat Springs, Colorado 80477, USA

ERNESTO PESCE

Palermo 5560, Apto 003, Montevideo, Uruguay

JOHN WARD SMITH

Synthetic Fuels Consultant, 1742 North Fifth Street, Laramie, Wyoming 82070, USA

BRUCE C. TRENT

Department of Civil and Mineral Engineering, University of Minnesota, Minneapolis, Minnesota 55455, USA

STEPHEN UTTER

Senior Staff Engineer, Bureau of Mines, US Department of the Interior, PO Box 25086, Denver Federal Center, Colorado 80225, USA

CHAPMAN YOUNG

Sunburst Recovery Inc., PO Box 1173, Steamboat Springs, Colorado 80477, USA

Common Unit Conversions

US customary unit	SI/metric unit	US customary unit	
1 in.	2·54 cm	1 cm	0·394 in.
1 ft	0·305 m	1 m	3·28 ft
1 mile	1·61 km	1 km	0·62 miles
1 lb (weight)	0·453 kg	1 kg	2·21 lb
1 lbf (force)	4·45 N	1 N	0·225 lbf
1 stone	6·35 kg	1 kg	0·157 stone
1 psi	$6·9 \text{ kN m}^{-2}$	1 kN m^{-2}	0·145 psi
1 psi	$0·069 \text{ bar (dynes cm}^{-2})$	$1 \text{ bar (dyne cm}^{-2})$	14·50 psi
1 ksi	6·9 MPa	1 MPa	0·145 ksi
1 ksi	0·069 kbar	1 kbar	14·49 ksi
1 ksi	0·0069 GPa	1 GPa	144·93 ksi
1 gallon	$0·0038 \text{ m}^3$	1 m^3	263·2 gallons
1 gallon	3·8 liters	1 liter	0·26 gallons
1 gallon ton^{-1} (gpt)	$4·17 \text{ liters tonne}^{-1}$	$1 \text{ liter tonne}^{-1}$	0·24 gallons ton^{-1} (gpt)
1 gallon ton^{-1} (gpt)	$4·17 \text{ cm}^3 \text{ kg}^{-1}$	$1 \text{ cm}^3 \text{ kg}^{-1}$	0·24 gallons ton^{-1}
1 barrel	0·1596 kliter	1 kliter	6·3 barrels
1 barrel	0·145 tonne	1 tonne	6·9 barrels
1 yard^3	$0·765 \text{ m}^3$	1 m^3	$1·307 \text{ yard}^3$
1 horsepower	745·7 W	1 W	0·00134 horsepower
1 ft.lb	1·356 J	1 J	0·737 ft.lb
1 Btu	1055 J	1 J	0·00095 Btu
1 Btu	252 cal	1 cal	0·004 Btu

Chapter 1

Introduction to Mechanics of Oil Shale

JOHN WARD SMITH
Consultant, Laramie, Wyoming, USA

and

KEN P. CHONG
Professor of Civil Engineering, University of Wyoming, Laramie, USA

SUMMARY

This chapter serves to introduce the subject of oil shale and to provide background for study of the mechanics of oil shale. Included in this background is what oil shale is, how oil shale formed, and how this affects any investigation of the mechanics of oil shale. The effects include the fact that most oil shale is bedded and isotropic in both directions along the bedding planes but anisotropic perpendicular to the bedding planes. Tensile strength perpendicular to the bedding planes is a strictly local effect linked to the anisotropy, but most of the mechanical properties are bulk effects, linearly reflecting the relative volumes of the organic matter and the mineral matter which have significantly different mechanical properties and their relative volumes control the contribution of each fraction. Relationships are presented to permit determining the volume of organic matter. How organic matter influences oil shale's thermal behavior is also described. How a relatively few oil shale deposits can represent most oil shales of the world is discussed as is how deposit joint systems affect oil shale development. The problem of hydraulic fracturing oil shale is reviewed. The chapter also looks briefly at why and when oil shale will be developed to provide oil, how much shale there is, where it is, where it will be developed and why and how the mechanics of oil shale must contribute to this development.

1.1. GENERAL INTRODUCTION

Oil shale is a source of oil, and its development as a source of oil must come from a knowledge of its mechanical properties. This book attempts to assemble in one volume a record of the research on the mechanics of oil shale currently available. The present chapter introduces the subject under the following headings: 'Why Oil Shale Will be Developed', 'Where Oil Shale Will be Developed', 'How Oil Shale Will be Developed', 'What Oil Shale is to Rock Mechanics' and 'Mechanical Properties of Oil Shale Deposits'. These sections present an oil shale background important to understanding the mechanics of oil shales. The discussion leads to the conclusion that oil shale will be developed and that the mechanics of oil shale will be vitally important to this development.

1.2. WHY OIL SHALE WILL BE DEVELOPED

Oil shale is sedimentary rock containing solid, combustible organic matter in a mineral matrix. One particular property of this organic matter distinguishes oil shale—it can yield oil. This organic matter, largely insoluble in petroleum solvents, decomposes to produce oil when heated (Smith, 1982). Most sedimentary rocks contain some organic matter, but only those capable of yielding oil on heating are classed as oil shales. No real minimum oil yield can be established to define oil shale. Any attempt to set a minimum oil-yield limit imposes an indefinite economic condition on the definition of rock. Obviously, however, those rocks yielding the most oil attract the most interest.

Oil from oil shale is often referred to as a 'synthetic' fuel, but the organic matter in oil shale is a naturally occurring fossil fuel. In fact oil shale's organic matter is the world's largest supply of fossil fuel. Williams (1983) estimated that the world's sediments contain about 1×10^{16} tonnes of this organic matter. In the world of petroleum exploration oil shale deposits are called 'potential source rocks' containing 'immature' organic matter. Since the potential oil in oil shale doesn't flow, it is ignored in petroleum geochemical exploration to search for what are termed 'source rocks'. Source rocks have already yielded their oil by natural processes involving deep burial, the earth's heat, and geological time. If oil shale is subjected to these conditions, it will eventually yield oil. The objective of oil shale development is to replace these slow natural processes with procedures having more immediate and practical results in producing oil from the 'immature' organic matter.

Oil has come to dominate the world's energy economy. Current world economic patterns depend on oil as an inexpensive, highly concentrated, easily mobile, easily stored and extremely convenient energy source. While oil's use as a raw material in chemical synthesis is impressive, most oil used is simply burned. Much of this burning is conducted in ways that require petroleum products. Transportation demands have proved particularly inflexible. In the United States, for example, over 50% of the 1983 oil consumption was due to gasoline. Gasoline is almost totally consumed for transportation. The other transportation fuels—diesel and jet fuels—increase the US 1983 transportation fraction of petroleum consumption to perhaps 70%. In 1980 transportation fuel cost the US $164 000 000 000 according to the US Department of Energy, and transportation represented 42% of the US *total* energy consumption. Virtually all of this transportation was provided by petroleum. A significant part of the remaining 30% of petroleum consumed in the US in 1983 went to providing fuel oil for space heating. Transportation and space heating are only the primary two of a substantial number of specific petroleum applications.

These petroleum users cannot quickly adapt to other energy forms. For example, the National Safety Council reports that in 1982 the US had 165 700 000 registered motor vehicles. These all depend on oil; the number operating on anything else is miniscule. Substituting another energy supply (methanol, for example) for automotive use requires modifying at least the engines of every one of these vehicles. Methanol also requires modifying carburetors and fuel systems to prevent corrosion. Accompanying this massive modification must be new engineering, new development and testing, new manufacturing systems, new parts supply, new repair and maintenance training and manuals, etc., just to be able to use methanol as the primary transportation fuel. This is in addition to the immense problems and costs inherent in developing both the manufacturing and the distribution systems at the monstrous scale necessary for comprehensive use of methanol (or any other chosen fuel!). Space heating consumption of fuel oil is similarly specialized. Millions of furnaces individually designed to burn fuel oil do not adapt cheaply to burning coal (or anything else!). And methanol doesn't pave roads.

We depend on oil. Your morning coffee, orange juice, eggs, bacon, toast and newspaper rode on oil from their diverse origins to your breakfast table. Given time and economic push, we can develop and instrument replacement forms of energy. Do we have time? The world's oil supply is finite. Hubbert and Root's 1981 projection of the world's petroleum supply indicates that by about 2027 (given orderly development) 90% of all the petroleum the world will ever produce will already be gone. M. K. Hubbert's projections have proven disconcertingly accurate. The US, the world's most prodigious energy consumer, will reach the 90% point in its domestic production around about 1997, given orderly development. However, the continued increase in petroleum well drilling from 1970 to 1979 plus the sharply accelerated rate of US exploration drilling in 1980 and 1981 and the first half of 1982 was not orderly development. This may have advanced to the mid-1980s the break-even point—where on average each well drilled costs as much oil to drill as it will produce (Hall and Cleveland, 1981). Perhaps the abrupt decline in US drilling in 1982, usually credited to the world wide oil 'glut', included petroleum accountants' recognition of uneconomic production results. In 1982 the US petroleum reserve resumed its long-term drop (down 5·3%, according to the US Department of Energy) after two relatively level years. The level years resulted from the federally inspired massive drilling effort in 1980–81, which temporarily slowed the decline in reserves but which hastened the arrival of the break-even point (Hall and Cleveland, 1981). When this break-even point is reached, the US 8-year reserve of proven petroleum will dwindle more rapidly because exploration drilling is discouraged. Domestic oil will become more expensive as time passes.

Time is a stringent limitation. The US has relatively few years to convert its transportation system to another energy source. But do we have the economic push to make the conversion? Major fuel substitution for transportation will begin in the US *only* when the substitution becomes economically advantageous. Once found, oil can be produced cheaply. We will continue to use already discovered natural oil because it is cheapest. As this is depleted the US will buy more and more imported oil, a move loaded with economic, political and military risks. Only as these risks are translated into costs will the price of oil in the US rise enough to provide the economic push. The comprehensive replacement of petroleum for transportation in the US and possibly the world just will not be accomplished in time to permit an orderly transition to non-petroleum fuels. Without an orderly replacement of

Introduction to mechanics of oil shale 5

petroleum fuels for transportation, US society will slide downhill, not abruptly perhaps, but definitely. Conservation merely delays the agony slightly. This same problem is faced in the other specialized applications of natural oil.

Oil from secondary sources may provide additional time to make the required adjustment (Smith, 1982). Only two secondary sources actually yield oil at prices nearly competitive with petroleum—oil shale and tar sands (Smith, 1982). Oil shale is distributed worldwide and has a long history of development. Some of these developments, like the 100-year oil production from the Lothian shales in Scotland, and the South African Torbanite Mining and Refining Co., actually operated successfully in direct competition with natural oil (Smith, 1984). From about 1860 until about 1900 when gas mantles for lighting became common, Australia marketed, worldwide, a kerosene shale oil fraction as lamp oil with superior luminosity under the name 'Bottled Sunshine of Australia' (Ferguson, 1980). Development of oil shale as the next most nearly competitive source of oil is highly probable in order to buy the time needed to change from our ingrained petroleum consumption patterns.

1.3. WHERE OIL SHALE WILL BE DEVELOPED

Rocks meeting the oil shale definition are relatively common throughout the world. As an example Fig. 1.1 presents the distribution of oil shale deposits in the US (Smith, 1980). Oil shale lies under 20% of the US land area. World distribution is not so readily presented. Matveyev (1974) compiled and published a comprehensive survey of the world's oil shale deposits outside the USSR. Although this rather brief book has faults arising primarily from the scope of the task it undertook (Smith, 1977), it presents a broad picture of the world's oil shale resources. Matveyev describes oil shales on six continents, demonstrating how much oil shale is known. However his descriptions could only touch the known major deposits. For example three oil shale deposits are described in Morocco, but by diligent coring this industrious little country has raised its total to at least 28 deposits scattered across the nation.

Knowledge of oil shale resources is by no means complete. Most sequences of sedimentary rocks contain materials which could be classed as oil shale. Duncan and Swanson (1965) summed *known* oil

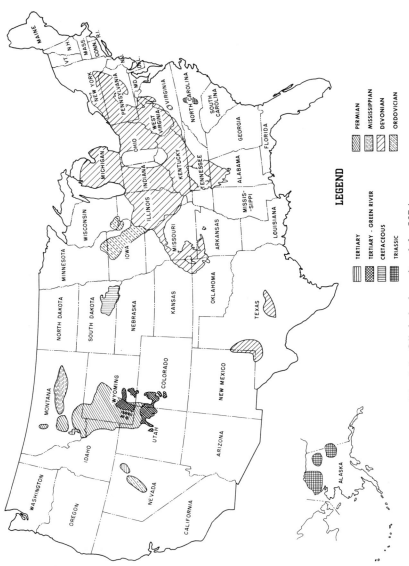

Fig. 1.1. Oil shale deposits of the USA.

Introduction to mechanics of oil shale 7

resources in relatively rich shale (25–100 gallons ton^{-1}; 104–417 liters tonne^{-1}) as representing 910×10^9 barrels (132×10^9 tonnes) of potential shale oil. They then applied a statistical approach to estimate the order of magnitude of the world's total oil shale resources by considering the total amount of sedimentary rock in the world and its probable organic content. Their estimate of the *total* rich or high grade shale comparable to the known value above was 17×10^{12} barrels ($2 \cdot 5 \times 10^{12}$ tonnes) of potential shale oil. Their estimate indicates that 95% of the world's better grade oil shales remain to be discovered! Even smaller fractions of lower grade shales were classed as known. The rich resource estimate given above makes up less than 1% of their estimated total resource, a good indication of the distribution of richness of oil shale. However, the world's crude oil reserves at the end of 1979 (Doscher, 1981) represent only 4% of the projected total resource of the rich oil shale alone.

Comparison of the values for known and estimated oil shale resources given above emphasizes how little is known of the world's oil shale resources. Much less effort has been invested in finding oil shales than in finding oil and gas hydrocarbons. Oil shale deposits have not been deliberately sought ordinarily. They have become known from outcrops, from unrelated geological investigations, or by accident. Outcrops generally have been detected, but subsurface deposits must be located by drilling. Records of discoveries are difficult to compile because these discoveries attract very limited interest. Discoveries, especially of small deposits, are usually noted only in local geological records. For example, Smith (1980) briefly describes 17 additional US oil shale deposits *not* shown in Fig.1.1.

The units used above to describe and quantify oil shale resources warrant comment. Oil shale resources are evaluated in terms of the oil they can generate potentially. The Fischer assay procedure, now standardized by the American Society for Testing and Materials (ASTM, 1980), provides the analytical basis for resource evaluation. It measures the amount of oil an oil shale sample will generate when heated under standard conditions approximating those met in production. While oil yield in weight percent is also determined by Fischer assay, the US expresses oil yield in the volume unit of gallons per ton. This unit is converted from weight percent oil yield by multiplying by a conversion factor, $2 \cdot 4$, and dividing the result by absolute oil specific gravity determined at $15 \cdot 6°C$. The metric volume unit, liters tonne^{-1}, corresponds to gallons ton^{-1}. 1 gallon ton^{-1} is equivalent to

4·17 liters tonne^{-1}. Oil shale resource values are expressed in barrels of oil in the US. A barrel of oil, the 42-gallon volume unit used in Western petroleum commerce, has no direct metric equivalent. It represents 0·159 kliters, or m^3. Specification of oil density, a variable, is necessary to convert the volume unit, barrels, into the metric weight trade unit, the tonne. The 1975 World Energy Conference agreed to define a barrel of oil as 0·145 tonnes, and 1 gallon ton$^{-1} \times 0·29$ as 1 kg tonne^{-1}, approximations that ignore density variations in oil.

Basic to the development and use of any oil shale deposit as an energy source is knowledge of the resource itself. Information essential to this development includes such things as the total resource, where it is, how it varies, where and how good the best parts are, and what can be done with the raw material. This is an extremely tall order. Oil shale is a complex material which at best qualifies only as a lean ore. Producing any appreciable quantity of energy requires processing huge volumes of rock. Consequently all kinds of questions require answers in planning development—mechanical properties of the rock and its deposit; nature and amount of its energy component, the organic matter; heating value of the rock and its components; mineralogy; heat requirements for oil production; thermal behavior of the rock and its components; properties of the residue after energy production; and dozens of other fundamental properties. These properties may be determined on individual 1 lb samples but how can the results be meaningful to a production system required by economics to process the equivalent of 150 000 000 1-lb samples every day? And can such hugh production even be planned? Postulated mining plans assuming huge thicknesses of high-grade oil shale make nice engineering exercises, but production can only work on what is actually in place. Resource evaluation provides the required answers—defining the resource and correlating the composition and properties of its rocks and how they vary. Smith (1981) describes a pattern for accomplishing and coordinating resource characterization. Chapter 4 in this book includes a description of some on-site techniques for determining deposit characteristics.

Nearly all of the world's oil shale deposits remain to be evaluated. Only when production of energy from oil shale becomes economic in competition with petroleum will most deposits, particularly the smaller ones, attract detailed investigation.

Development will occur on the best known deposits, the richest deposits and the most economically and politically practical deposits.

Introduction to mechanics of oil shale

These factors all work together to decide if, when and how development of any particular deposit will occur. The importance of the first two is readily apparent, but the third factor covers a lot of facets.

Oil shale economics as modified by political and social influences can best be illustrated by histories of a few developments. Yes, the world has already seen quite a number. Estonia (now Estonia SSR) began massive oil shale development about 1920 to supply energy in this fuel-short area. Oil or coal was difficult to ship in during winter, just when energy was needed most. This oil shale development continues although oil and gas can now be piped into the area. Estimating economics is difficult in the USSR, but production from these very rich, easily-mined shale beds continues without apparent national subsidy (Smith, 1984). Across the Baltic Sea, Sweden, once an enthusiastic oil shale developer, no longer produces energy from its somewhat leaner shales. Scotland produced shale oil as an economic business for over 100 years. The grade of their resource gradually decreased (Smith, 1984), but competition from cheap near-east oil finally killed production. Only in its last years (after Second World War) did the Scottish industry receive government support. The South African Torbanite Mining and Refining Co. operated commercially from 1935 to 1962, marketing oil shale products in competition with imported crude oil. This operation had two advantages—a cheap and inexhaustible labor supply and the high cost of importing petroleum into the interior of the country. This company did not fail—it mined itself out of existence. Australia was producing some shale oil commercially in the middle 1850s and commercial production continued into the 1900s. Additional shale oil production was carried on sporadically in Australia, usually with governmental support, until after the Second World War (Ferguson, 1980). These successful developments have a couple of common factors—lack of other fuels and geographic or political isolation. World War II created a burst of oil shale development activity by disrupting petroleum supply patterns. These World War II developments did not survive postwar petroleum developments in the near east.

The largest deposit of rich oil shale in the world provides an example of how the factors combined to decide against production. This resource is the Green River Formation in the western US. It is also the best evaluated of US deposits. Figure 1.2 outlines the 16 500 square mile (43 000 km^2) Eocene Green River Formation in adjoining corners of Utah, Colorado and Wyoming. Smith (1980) recently estimated that

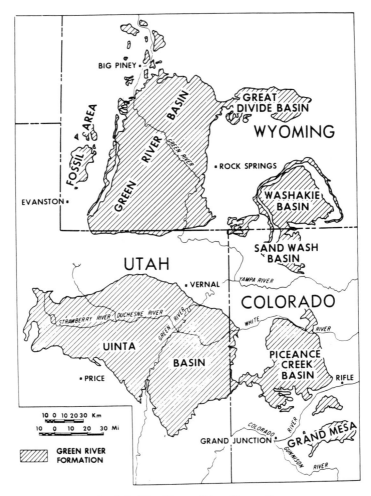

Fig. 1.2. The Green River Formation.

the source of potential oil represented by the Green River oil shale of all grades was $1{\cdot}8 \times 10^{12}$ barrels (256×10^9 tonnes). Little of this massive resource is in shales leaner than 15 gallons per ton (63 liters tonne^{-1}), yet it has not been developed. More than 60 years of violent ups and downs in shale oil prospects has yet to produce an

Introduction to mechanics of oil shale 11

industry. Only now (1983) does the first major production appear imminent. With federal price guarantees as support, Union Oil is nearing production of 10 000 barrels per day (1500 tonnes per day) in Colorado. Native Coloradans say they'll believe it when they see it (they've been there before).

What factors have deterred development of Green River Formation oil shale? It is the largest deposit in the world and has much rich oil shale in it. It is probably the world's best defined undeveloped oil shale deposit. Two or three negative factors have contributed. The first is that the US has been a massive producer of domestic oil. Development of the East Texas oil field killed a Green River boom about 1930, for example. The second is that the US Government has used their ownership of 80% of the Green River Formation oil shale resource to prevent or direct oil shale development. This 80% of the Green River Formation was withdrawn from 'leasing or other disposal pending evaluation of the resource' since President Herbert Hoover signed Executive Order 5327 in April 1930. Although the Executive Order withdrew oil shale, this order has only been applied to the Green River Formation. Figure 1.1 indicates how myopic the application is. This executive order, drafted by and presented to Hoover by Ray Wilbur, then Secretary of the Interior, reflects the authoritarian attitude towards oil shale that has persisted in the US Department of Interior. Experimental development of a new resource requires two land positions—a resource experimental site to test an idea, and a resource position to develop if the testing is successful. These have not been available under the withdrawal order. The US Department of Interior has exploited the withdrawal order to limit, direct, and control the nature of oil shale experimentation. While this was done with the best of intentions, it was also done with a marked lack of understanding of how needed oil shale development could be achieved. Political decisions on research programs have a very low probability of encouraging successful development. US Government agencies have made a special case of Green River Formation oil shale and are managing it in ways very unlike those on any other federal land.

Political factors in the US Government have been a major blockage to Green River Formation development. Recent additions to the government's battery of impediments include environmental factors. By insisting on absolute answers to everything, environmental groups have raised development costs markedly, adding on both the direct cost of environmental studies and the inflation incurred during delays while the required studies are conducted. Environmental concerns should be incorporated into overall development strategies rather than become a political tool. A pioneering industry cannot provide absolute answers which will be available only after production. If every possible answer is required before development begins, development will never begin. Also regulatory contests between the US Federal Government and the State of Colorado have impeded development by creating an unstable atmosphere. Estimated development costs for Green River Formation projects have escalated into the billion (10^9) dollar range. Any unstable political atmosphere around oil shale makes potential investors and developers very nervous.

The privately held Green River Formation land has largely been accumulated by major oil companies. These companies feel that Green River oil shale will be developed, but they can afford to wait. The resource will stay. It is often said that the major companies wish to be third in oil shale development, beginning only after all the pioneering costs are paid and all the pioneering problems solved. Potential developers generating production cost estimates for Green River sites using tested methods have consistently obtained shale oil prices somewhat higher than imported petroleum. Development of privately held land will occur only when shale oil appears cheaper than the natural oil.

The question of where oil shale will be developed can only be answered by time. Known oil shale resources are required. Many deposits are known already. The world is full of oil shale resources waiting to be discovered and become better known. A lack of oil is a prerequisite. This is the situation now in Morocco, Brazil and Australia, for example. All have limited supplies of domestic petroleum and all are investigating production of oil from shale. A lack of oil exists today in the developing countries where local oil shale might be developed on a scale suitable to provide local industry and local products. There is a lack of oil now in the US, and it will become more acute. Production of shale oil at prices competitive with petroleum is required. This will occur as petroleum supplies dwindle. For the US and for other potential production sites, political recognition of the lack of oil is required. This recognition must be followed by creation of a political environment conducive to development. A growing lack of oil may accelerate political recognition of the problem.

1.4. HOW OIL SHALE WILL BE DEVELOPED

Oil shale has a long history of development. The only general way yet found to produce oil from oil shale economically is to mine the shale, break the resulting ore up into sizes appropriate for a particular processing system and then to heat the ore in that processing system. The processing systems, usually called retorts, take two general forms depending on how the required heat is generated. Figure 1.3 illustrates these forms. In the internally heated system the oil shale supplies its own heat by burning part of its organic matter or its products inside the retort. Heat supplied internally may be generated by burning raw oil shale, the residual carbon, the gas evolved, part of the oil, or some combination of these. In the externally heated system, heat generation is carried out separately in the section labeled 'furnace'. The fuel for this furnace has usually been retort off-gas but has included the carbon on spent shale, part of the product oil, locally available coal or natural gas and even heat from nuclear power plants or the sun. The heat generated or collected is then transported to the retort chamber. In one externally heated form the retort shell itself is used as a heat carrier. In other forms hot heat-carrier material from the furnace joins the raw shale to transfer heat for retorting. Proposed or tested heat carrier materials have included ceramic balls, recycle gas, sand, spent shale and molten metal.

In meeting the challenge of oil shale development, human ingenuity has produced a remarkably diverse set of interpretations of the

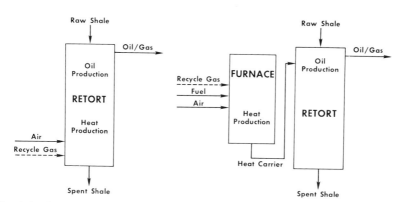

Fig. 1.3. Generalized oil shale retorting systems: left, internally heated; right, externally heated.

generalizations presented in Fig. 1.3. Several published demonstrations of this diversity exist. Smith (1984) reviews this diversity, Russell (1980) its history in the western US and Allred (1982) an assortment of the best tested current retorting systems. In a series of US Bureau of Mines bulletins, Klosky (1949, 1958) surveyed oil shale and shale oil patents across the world. Klosky reports on perhaps 1500 different retort patents. Every one describes a device capable of producing oil from oil shale. Most of the differences in retorting systems are intended to provide or improve process efficiency and heat management. Inventors have enthusiastically attacked these problems by designing new retorts, but oil production costs differ little among all the conventional systems.

Few systems directly attack the basic problem common to the surface processing of oil shale—materials handling. The major costs in conventional processing are attributable to the fact that even the best oil shales are low grade starting materials. For example, 1 ton oil shale rock yielding 25 gallons of oil (104 liters per tonne), a relatively good grade, must be handled five times to produce 0·6 barrels (0·09 tonnes) of oil. The rock must be mined, gathered up and hauled, crushed and heated to produce oil. Then the residue consisting of most of the original rock must be disposed of. That is a lot of materials handling for the price of the product and is where shale oil's production costs lie. *Every one of these materials handling steps can benefit from an understanding of the mechanics of oil shale.*

Oil shale will have to be processed in huge volumes in order to contribute significantly to meeting world's oil demands. Large volume production of oil from oil shale rewards (and may even require) improvement in efficiency or cost reduction which are achievable from a knowledge of the mechanics of the rock and the deposit. Some factors in achieving this production by mining include blast design, blast breakage, product particle size control, and the influence of joints on these factors. If the mine is subsurface, as most must be, the factors added include roof support, mine stability and longevity, recovery factors, mechanization possibilities and subsurface safety. By adding hauling, crushing and spent shale disposal to the mining package, a realistic picture of cost distribution in oil shale processing by conventional techniques is obtained. The retorting (heating) of oil shale is required, but it is not the cost dominator all the retort patenting would lead us to expect. Materials handling from mining to disposal offers a much wider, more realistic range for cost control. Crushing may furnish an example. In dedicating the Utah Mining and Minerals Resources Research Institute establishèd in 1983 at the University of Utah, Professor John J. Herbst stated that ore crushing was a neglected area of research, 'yet the United States currently uses about 36 billion kWh of energy just to break rock'. Oil shale processing in the US will increase that huge figure markedly. Matzick *et al.* (1960) provide the only basic data used in making cost estimates for crushing Green River Formation oil shale. Certainly this is a major area of research for cost saving in oil shale processing. Allred (1982) points out that processing technologies developed for one oil shale resource do not automatically apply to all others. Implicit in his technology discussion is the concept that the entire oil shale processing system must be integrated and designed completely as a site-specific package. The mechanics of oil shale has a vital role to play.

The problems of the mechanics of oil shale have been solved locally and empirically any number of times. For example, oil shale mining was practised in Scotland for over 100 years, but relatively little of the Scots' empirically gained experience is recorded. Their underground oil shale mines demonstrated mine stability, absence of gas and minimum explosion hazard, and impermeability of the oil shale beds to water (Gavin, 1922). Many underground mines were generated and they still have spent shale dumps—mounds called bings—scattered over the production area. This oil shale area is heavily faulted (Smith, 1984) but no detailed record of how this was accommodated or how it affected their costs exists. Many places in the world have had mining experience on oil shale; several are listed below with the type of mining conducted (Guthrie and Klosky, 1951; East and Gardner, 1964; Ferguson, 1980; Smith, 1984).

Country	Type of mining
Australia	Subsurface
Brazil	Open pit
China	Open pit
France	Open pit, subsurface
Germany	Open pit
Morocco	Subsurface
Sweden	Open pit
USA	Subsurface
South Africa	Sursurface
USSR	Open pit, subsurface

This represents a substantial amount of empirically developed experience on the mining and handling of oil shale. None of the above seem to have generated a comprehensive report on their mining and rock mechanics knowledge. One objective of this book is to compile in one place records of the oil shale mechanics data currently being generated on Green River Formation oil shales. We are also attempting to build-up information and methods which can be extrapolated to other oil shales and oil shale deposits.

Recognition of the materials handling problems inherent in conventional surface processing of oil shale generates a search for methods of avoiding them. An obvious first step, enrichment of the organic matter from mined oil shale, may save on the amount of material which must be heated, but no materials handling is eliminated by enrichment. An enrichment process on kukersite is being practised in Estonia (Smith, 1984), but it was installed to minimize environmental effects and to cut down on fouling of the furnace by mineral residue during direct combustion. Although investigations continue, no other enrichment process is incorporated in a production process.

Production of oil from oil shale in place is the only real way to avoid costly materials handling. However, production in place faces a different problem—oil shales in place tend to be impermeable. Research and development efforts toward processing oil shale in place have concentrated on generating the permeability necessary for creating or transferring heat into oil shales and for moving the products out. No in-place processes have reached commercial production yet (1983), but two have been tested intensely in Green River Formation oil shales. Both produce oil successfully and perhaps economically. These are the Geokinetics process which heaves the surface up with explosives to obtain permeability, and the Occidental process which mines out void space in an oil shale deposit and then distributes this void space to create permeability by shattering the oil shale with explosives. Both generate the heat required by burning the now-permeable oil shale. They are both discussed in this book. *These and most other proposed in place techniques can only succeed with understanding of the mechanical and thermal properties of the rocks they treat.*

1.5. WHAT OIL SHALE IS TO ROCK MECHANICS

Oil shale as defined in Section 1.2 is a sedimentary rock consisting of a mineral matrix containing solid combustible organic matter insoluble in

Introduction to mechanics of oil shale 17

petroleum solvents but yielding oil on heating. This is a good general definition of oil shale which includes all oil shales. Some oil shales consist of virtually nothing but organic matter, and some oil shales contain only trace amounts of organic matter. The ability of the natural rock to yield oil on heating is the key characteristic. The bitumen deposits, like tar sands or gilsonite, do not qualify as oil shales because their organic matter is soluble in petroleum solvents. However, like all definitions attempting to put the complete realm of nature in a small bag, this definition has one indistinct boundary. Coals usually distinguish themselves from oil shales by yielding little or no oil. On heating, however, some coals yield oil, usually high molecular weight tar with a distinctive characteristic odor, which differs markedly from the odor of shale oil. Because of this yield of oil (tar) from coal, the oil shale definition does not precisely distinguish coals from oil shales—neither does Mother Nature. Some organic-rich sediments meeting the oil shale definition occur in company with coals. These materials, often called cannel coals, are oil shales forming a transition series from oil shales to coals (Ashley, 1918). The cannel coal forms a relatively small fraction of the world's oil shales.

Oil shales have some characteristics in common in spite of the diversity of materials included in their definition. The oil-yielding character of the organic matter is one. Another is that oil shales developed from subaqueous sediments deposited under geochemical conditions which preserved organic matter. Both ocean and lake (more formally marine and lacustrine) sediments have formed oil shale, but active ocean or lake sediments oxidize organic matter very readily. The chemical conditions required to preserve organic matter in water were strikingly described by Garrels and Christ (1965) in terms of the acid–base balance (pH) and the oxidation–reduction potential (E_h). While specific values for these properties are particularly useful in describing the chemistry behind the creation of an organic-rich deposit (Smith, 1974), Garrels and Christ (1965) point out that these conditions arise *only* in sediments isolated from the atmosphere. This is a fact important to the mechanics of oil shales. Sedimentary conditions capable of achieving the required isolation have specific consequences to the nature of the rock formed when these sediments lithify to form oil shale. Provo (1977) summarizes the general condition as follows:

'Deposition of organic-rich sediments depends on (production and) preservation of organic matter. . . . .The water mass must be stratified. . . . . Such stratification inhibits turnover or vertical mixing of the water mass, thus favoring a low oxygen content and creating a reducing environment.'

Provo postulates several depositional arrangements in oceans which can give rise to this stratification, usually maintained by density differences in the water. Smith (1974, 1983) described a chemically generated density difference which produced and maintained the lacustrine stratification which created Green River Formation oil shales.

The consequences of deposition under stratified water with limited circulation to the mechanics of oil shale are that the sediments deposited are fine grained, laminated, uniform laterally and rather uniform in composition. Organic-rich oil shales are not usually created by massive production and deposition of organic matter, but by stringent limitation of mineral deposition in relation to the amount of organic matter being deposited and preserved. The tiny particles of mineral and organic matter are intimately mixed. Very slow deposition under stable conditions with limited transport of mineral matter are natural consequences of persisting water stratification.

The rock formed reflects the characteristics of the sediment. Oil shale is virtually always fine-grained. In the petrographic examinations Smith et al. (1959) conducted on 25 oil shale samples from deposits in five countries (Australia, New Zealand, Brazil, France, South Africa) the specimens were always described as 'fine grained'. Several descriptions preceded this with 'very'. The Devonian black shales in the US (Fig. 1.1) are uniformly described as fine grained. Tisot and Murphy (1960) measured the mineral particle size distribution in Green River oil shale, determining the mean particle size by weight to be 5 μm. Because particle weight varies as the third power of the particle diameter, less than 1% of the particles are larger than 5 μm.

Laminations are characteristic of oil shales. Smith et al. (1959) commented on the fine laminations visible in specimens from several countries. The Green River Formation shares this characteristic. Figure 1.4 shows the laminations in a 2-in. (5-cm) oil shale block from the Formation's Mahogany Zone in Colorado.

Oil shale tends to be uniform laterally. The regular laminations visible in Fig. 1.4 indicate this. Bradley (1929) discussed detection of very small layers, alternating darker and lighter, in Green River oil shale. These layers, apparently seasonal, he called varves. The varves are much too small—perhaps 30 μm for a complete pair (Smith, 1974)—to be reproduced in Fig. 1.4, but they could not even be

Introduction to mechanics of oil shale 19

Fig. 1.4. Laminations in a 2-in (5-cm) long block of Green River Formation oil shale.

detected if they were not laterally persistent. Lateral uniformity on the scale of the laminations shown in Fig. 1.4 has been shown to be usual in the Green River Formation oil shales. Trudell *et al.* (1970) published a classic photograph demonstrating precise correspondence and correlation of markings similar to those in Fig. 1.4 across a 65-mile (105-km) span.

Oil shales in a deposit created by persistence of deposition under stratified water tend to be uniform in composition. The conditions continue to generate and deposit the same organic matter. The limited influx of mineral matter tends to select the same incoming minerals. The chemistry of the water over and in the sediment tends to create the same mineral and organic matter modifications. Consequently, the composition of the resulting oil shale also tends to be uniform. However, because oil shale is deposited slowly, the composition controls generated by the stratified water can be overridden temporarily by cataclysmic events such as falls of volcanic ash. The markedly white band across the oil shale block in Fig. 1.4 was created by such an ash fall. This band consists primarily of analcime formed by chemical alteration of the ash in the sediment. Analcime is not a mineral found in Green River oil shale itself (Robb *et al.*, 1978). However, stratification and organic matter deposition persisted through the ash fall. Oil shale deposition resumed as soon as the ash stopped falling and settling (Fig. 1.4).

The physical nature of oil shale rock can now be evaluated. The minerals which predominate in most oil shales are quartz and clays, particularly illite, smectite (formerly montmorillonite) and mixed-layer combinations of these two clay families. While pyrite (and/or marcasite) formed in the sediment is ubiquitous in oil shales, it usually constitutes a small volume fraction which has little effect on mechanical properties of oil shale rock. The quartz is fine-grained and occurs as uncemented discrete particles. In this somewhat simplified oil shale picture, oil shale structure is provided by the clay layers and by the organic matter.

Some oil shales incorporate carbonate minerals—calcite and aragonite, for example. Depending on how they are formed, these minerals may be small, discrete particles, lending little structural strength to the oil shale rock. If they are interconnected or cemented together, however, they become structural members lending strength to the rock. Some oil shales incorporate dolomite as a structural member in the resulting rock. This is characteristic of much of the Green River

Formation oil shales. The other minerals in Green River oil shales—quartz, albite, and K-feldspar—form in-place and are not interconnected. Consequently the Green River Formation's oil shale structure is controlled by its dolomite structure, by its clay structure when clay is a predominant mineral instead of dolomite, and by the volume of organic matter in the rock. In very lean oil shales the Green River rock is a structurally sound laminated dolomite (clay content effects are to be discussed shortly). As the relative volume of the organic matter increases (as the shale gets richer), the proportion of cemented dolomite decreases, and the rock displays more and more the physical character of the organic matter itself. If the oil shale is rich enough, the organic matter becomes the continuous phase in the rock, and the mineral matter, including the dolomite, behaves as isolated aggregate particles suspended in organic matter.

In the clay-rich oil shales not containing substantial amounts of cemented carbonates, the clay is the primary structural member in lean shales. Clay structure is significantly weaker than the structure of cemented carbonates. As the high clay shales become richer, containing larger volumes of organic matter, the rock's strength changes faster toward the strength of the organic matter.

The physical characteristics of the organic materials in the world's oil shales are significant factors in the mechanics of oil shale rocks. As indicated above, if the shales become richer in organic matter, the organic characteristics influence and ultimately dominate the physical characteristics of the rock. The hydrogen content of the organic matter and how this hydrogen is bonded influences the organic matter's mechanical properties. Organic matter containing a large proportion of aliphatically associated hydrogen, like Green River Formation oil shales (Miknis and Smith, 1982), tends to have a low density and to deform plastically under load. Organic matter containing less hydrogen has a higher density and less plasticity. This mechanical property variation with hydrogen content parallels a decrease in the proportion of total organic matter converted to oil on heating with decreasing available hydrogen (Smith *et al.*, 1959).

Deposition of the oil shale sediment as described above has consequences for the testing of the mechanical properties of the oil shale rock. As was pointed out earlier, oil shale tends to be fine grained, laminated, uniform laterally and rather uniform in composition. The three factors—fine grained, uniform laterally and uniform in composition—tend to make easy the selection of representative materials

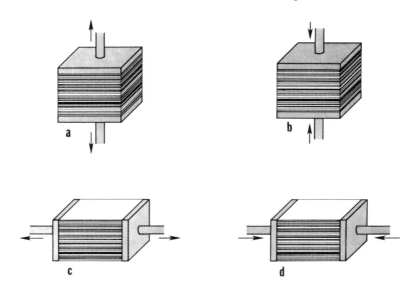

Fig. 1.5. Testing modes in laminated oil shale: (a) tensile strength perpendicular to bedding; (b) compressive strength perpendicular to bedding; (c) tensile strength parallel to bedding; (d) compressive strength parallel to bedding.

for testing. Figure 1.4 indicates the regular nature of the material from which samples may be chosen. This figure also emphasizes the consequences of bedding plane laminations—namely that oil shale is isotropic in the two directions along the bedding planes but is anisotropic in the direction perpendicular to the bedding planes. This produces the four testing modes for oil shales illustrated in Fig. 1.5. The two compressive strengths, perpendicular to the bedding (b) and parallel to the bedding (d), and the tensile strength parallel to the bedding (c) are all bulk characteristics of the rock. Testing in these modes produces average values based on the strength characteristics and the relative volumes of the oil shale mineral matrix and the organic matter in the rock. Tensile strength testing perpendicular to the bedding (Fig. 1.5(a)) tends to generate bedding plane failures which express instantaneous events during deposition—formation of a clay layer or a crystal layer, abrupt change in materials deposited like appearance of ash falls, etc. Testing the tensile strength perpendicular to the bedding will have only a limited relation to the primary rock variables.

Introduction to mechanics of oil shale 23

The ordinary fracture toughness testing has limited application to laminated oil shale. Fracture toughness can be oriented: parallel to the bedding where the test tries to pry apart just one bed of many; perpendicular to the face of the bedding where the test tries to break whichever bed the testing notch happens to stop on; and perpendicular to all of the beds. Only this last test might have a general and useful meaning.

Sample selection for mechanical property testing requires choosing a specimen representative of a huge mass of material. Not all oil shale specimens are as tidy as that shown in Fig. 1.4, and even that block shows a small crack. Desiccation cracks from drying after formation sampling and blemishes from sample preparation must be avoided. In addition there are a number of minor lithologic horrors which do not affect bulk properties but may well destroy representation of selected samples. Figure 1.6 presents three such minor 'horrors' collected from the Green River Formation. Figure 1.6(A) shows a small slump fault offset at the bottom but not at the top; Fig. 1.6(B) shows natural tension cracks in very lean shale which fail to penetrate the organic rich darker layers; Fig. 1.6(C) shows irregular bedding generated around pyrite and chert nodules which sometimes form in most oil shales. In Section 1.6 another minor 'horror' is described which occurs in bulk in Colorado's oil shale—nahcolite nodes. Incorporating these or other anomalous lithologic structures in test specimens fails to represent the bulk properties of the rock.

The volume of organic matter in oil shale rocks is the oil shale variable controlling the bulk mechanical properties of the rock illustrated in Fig. 1.5. Low-grade shales have the mechanical properties of the mineral matrix. As the volume fraction of organic matter increases in the oil shales of a particular deposit, the mineral matter exerts less and less influence on the rock's mechanical properties. These properties come more and more to resemble the properties of the organic matter. In very rich oil shales the organic matter is the continuous phase, and the mineral matter behaves as a dispersed and discontinuous aggregate. The change in physical properties is gradational and occurs across the usual grade of oil shales found in most deposits. The relationship between volume of organic matter and the mechanical properties of the rock through this range is shown to be linear in Chapter 5. Chong *et al.* (1979*a,b*, 1980, 1982) showed that the volume of organic matter in an oil shale could be used efficiently to estimate an oil shale's mechanical properties.

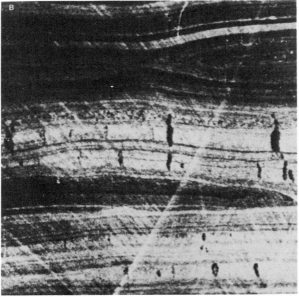

Fig. 1.6. Minor lithologic horrors for mechanical property samples: (A) slump fault; (B) tension cracks.

Introduction to mechanics of oil shale 25

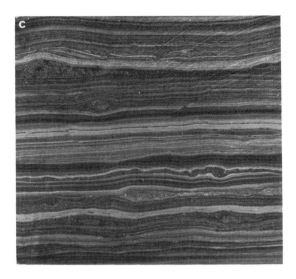

Fig. 1.6.—contd. (C) Chert and pyrite lumps.

Measuring the volume of organic matter in an oil shale is not a straightforward analytical procedure, but approaches for estimating values for organic volume exist. In 1976 Smith developed a theoretical equation relating volume of organic matter to the absolute density of an oil shale rock:

$$V_O = \frac{100(D_B - D_T)}{D_B - D_A} \qquad (1.1)$$

where V_O = organic volume in volume percent of the rock; D_A = average density (g cm^{-3}) of organic matter; D_B = average density (g cm^{-3}) of mineral matter; D_T = rock density (g cm^{-3}) of oil-shale specimen. This is a general equation valid for any oil shale. The rock density, D_T, is the absolute density without porosity, not the bulk density which includes porosity. Most oil shales have little porosity. The absolute density, D_T, can be measured directly on a specimen using an air comparison pycnometer. Average values for the densities of the organic matter (D_A) and the mineral matter (D_B) are functions of the specific deposit. Because of the way oil shale forms, however, each deposit will probably have uniformity sufficient to permit determination of D_A and D_B values typical of the deposit. The density of organic matter in the world's oil shales lies between about $1\cdot0$ and $1\cdot3$ g cm^{-3}. The more hydrogen-rich the organic matter is, the lighter it is. The average density of the mineral matter in the world's oil shales is around $2\cdot6$–$2\cdot8$ g cm^{-3}. Smith (1969) tested the influence on the average mineral density, D_B, of variations in amounts of constituent minerals. He concluded that only a few minerals, like pyrite with a density near 5 g cm^{-3}, can appreciably affect the average mineral density.

Few oil shale deposit values for D_A and D_B are on record. However, for Green River Formation, Smith (1976) reports $D_A = 1\cdot07$ g cm^{-3} and $D_B = 2\cdot72$ g cm^{-3}. Obviously the shale containing larger amounts of organic matter is lighter. Substituting these values in eqn 1.1 gives the following linear relationship

$$V_O = 164\cdot8 - 60\cdot6 D_T \qquad (1.2)$$

where V_O is the volume percent organic matter in a Green River Formation oil shale and D_T is the absolute rock density (g cm^{-3}). This simple linear relationship can be used to estimate organic volume from the rather easily measured rock density for Green River oil shale. Realistic values for D_A and D_B can be developed for most oil shale deposits because the organic matter and the mineralogy tend to be uniform in density.

An even more readily applicable method for estimating organic volume has been developed. Oil yield by Fischer assay (ASTM, 1980) or organic carbon in weight percent raw shale are frequently available for oil shale specimens. Both are used frequently in evaluating oil shale deposits and both are determined by standard laboratory procedures. However, both are weight measures rather than volume measures. Oil yield particularly has been used to relate to physical property measurements. Unfortunately, oil yield is not linearly related to organic volume. Consequently many oil shale physical property studies show curved results (e.g. Sellers *et al.*, 1972), usually with no attempt at explaining why the relationships curve. The following relationship derived by Smith (1969) permits an estimation of organic content by volume from the organic content by weight:

$$V_O = \frac{100 A D_B}{A(D_B - D_A) + D_A} \qquad (1.3)$$

where A is the weight fraction of organic matter in the rock and the other symbols have the same meaning as in eqn 1.1. In this equation A

Introduction to mechanics of oil shale 27

is limited to values from 0 to 1 in order to represent oil shales. If the proportion of organic carbon in the organic matter is known, together with D_A and D_B, organic volume can be calculated directly from organic carbon. Analysis for organic carbon is relatively simple and accurate. Typical proportions of carbon in organic matter have been determined for some deposits. For example, Matveyev (1974) reports several, Smith (1961) reports the ultimate composition of Green River organic matter, and Smith and Young (1967) report average elemental composition for organic matter in Devonian black shales of Kentucky. For the Green River Formation oil shales of the Mahogany Zone (described in Section 1.6), Smith (1961) reports that carbon averages 80·5 weight percent of the organic matter. If this is used to calculate the organic content, A, from organic carbon determined on a sample, the volume of organic matter in an oil shale sample can be calculated from eqn 1.3.

If the proportion of organic matter converted to oil by Fischer assay is known, organic volume can be estimated from the oil yield of a specimen using eqn 1.3. This conversion proportion is usually rather uniform in a deposit when assay is done under standard conditions (e.g. ASTM, 1980). If oil yield is reported in weight percent, calculation of A, the weight fraction of organic matter in the rock, is direct and simple if the conversion factor is known. If oil yield is reported in volume units like gallons per ton, back calculation is possible. However, direct conversion factors can be developed statistically. Smith (1966) reported and evaluated the following conversion constant for Green River Formation oil shales of the Mahogany Zone

$$\frac{\text{Organic matter (weight percent raw shale)}}{\text{Oil yield (gallons per ton)}} = 0 \cdot 580$$

If this is applied in eqn 1.3 to calculate organic matter content, A, and the values for D_A and D_B given above are entered, the following expression results:

$$V_O = \frac{164 \cdot 9M}{M + 111 \cdot 8} \tag{1.4}$$

where M = oil yield in gallons per ton. Straightforward conversion factors can convert this equation to metric units. In applying the equation Chapter 5 uses the symbol O_c for V_O, the organic volume as a percent of total rock volume. Equation 1.4 permits calculation of organic volume directly from oil yield in gallons per ton for Green

River Formation oil shales whose conversion to oil is similar to that of the Mahogany Zone. Most Green River oil shales meet these specifications. Similar simple equations for estimating organic volume from oil yield in any units can be developed for any oil shale deposit once the deposit constants are known. The form of eqn 1.4, a hyperbola, removes the curvature inherent in correlating mechanical properties to oil yield. The mechanical properties are related linearly to organic volume. Estimating mechanical properties is much easier using a linear relationship.

Yet another route exists for estimating organic volume from oil yield. The absolute density of Green River Formation oil shale has been related empirically to oil yield for Green River oil shale (Smith, 1976) and also for Devonian black shales of Kentucky (Smith and Young, 1964). The empirical relationship for Green River Formation oil shale is

$$M = 31 \cdot 563(D_\mathrm{T})^2 - 205 \cdot 998 D_\mathrm{T} + 326 \cdot 624 \qquad (1.5)$$

where M = oil yield (gallons ton^{-1}) and D_T = rock density (g cm^{-3}). Equation 1.5 was developed as a fitted parabolic function by regression techniques although the theoretical equation is hyperbolic. Correlation between these variables is so high ($R^2 > 0 \cdot 99$) that eqn 1.5 may be used to estimate rock density from oil yield as well as oil yield from rock density. If D_T estimated from oil yield using eqn 1.5 is inserted for D_T in eqn 1.1, a corresponding value for V_O is obtained. Smith (1976) exploited these relationships to compare in table form the rock density (D_T), the corresponding organic volume (V_O) and the oil yield (M) (actually the symbol OY is used) for Green River Formation oil shales. While all of the rock density and organic volume values are comparable and valid in that tabulation, Smith (1976) extrapolated the empirical eqn 1.5 far beyond its limits of applicability. Consequently, oil yields entered in Smith's (1976) table above 70 gallons ton^{-1} (290 liters tonne^{-1}) are incorrect with the error progressively increasing.

Although rock density is an efficient estimator of organic volume, two problems affect its use. The first arises because the organic matter density (D_A) and the mineral density (D_B) incorporated in the equations are average values representing an entire deposit. Smith (1969) evaluated the effect of variation in mineral amounts on rock density, showing that for many of the normal oil shale minerals even large variations in quantity have minor effects on rock density. How-

Introduction to mechanics of oil shale 29

ever, some minerals occurring in Green River oil shale may have more substantial effects. These include pyrite, analcime, dawsonite, nahcolite and shortite. The first three of these occur in the shale's mineral matrix and are not readily seen, while nahcolite and shortite are usually detectable. If significant amounts of these minerals occur in an oil shale sample, density correction may be necessary to estimate organic volume accurately. Analytical techniques now available can determine the quantity of these minerals and the analytical results can be used to correct rock density as described by Smith (1969). Organic matter density is ordinarily uniform in a deposit because of persistent depositional conditions. If part of a deposit has been buried deeply enough to undergo natural decomposition to form oil, as in the Phosphoria Formation of Idaho and Montana (Miknis *et al.*, 1982) the organic matter density may change in a deposit. Geologic knowledge can detect such events, however.

The second problem centers around possible porosity in an oil shale specimen. The relationships presented require that only the absolute density, not the bulk density, of the rock be considered. Most oil shale and most Green River oil shale is non-porous, but some layers, particularly those associated with analcimized tuffs, may contain appreciable void space. Permeability of these layers is usually very low, making determination of density by immersion techniques impractical. The best means of solving this problem is to measure the specimen density (D_T) by gas displacement techniques. An apparatus known commercially as an air comparison pycnometer will eliminate porosity as an analytical problem in the determination of rock density.

The different thermal properties of the organic matter and the mineral fractions of oil shales also affect the thermal behavior of oil shale rocks. Oil shale must be heated to make oil, so the thermal effects on mechanical properties influence this process. Chapter 12 evaluates thermal changes in mechanical properties together with the difficult problem of evaluating these subsurface. Several effects occur. These are functions both of the rock temperature and of the nature of the organic and mineral fractions, so the effects will be described briefly in relation to increasing temperature.

The first effect to appear as oil shale is heated is that the organic material expands faster (usually) than the mineral matter. Hydrogen-rich, low-density organic matter like that found in the Green River Formation expands perhaps 20 times as much as its surrounding minerals for the same temperature change. At the other extreme hydrogen-poor, high density organic matter expands at rates similar to the minerals. One consequence of the thermal expansion of hydrogen-rich organic matter is a parting of oil shale layers along bedding planes as stresses develop between layers of differing organic content (Smith and Johnson, 1976).

As the organic matter reaches retorting temperature it begins to undergo decomposition and evolve oil and gas. In low grade shales the rock structure remains intact as the organic matter is volatilized. If organic matter is the continuous phase as in richer shales, the rock structure disintegrates as the organic matter disappears. In oil shales containing hydrogen-rich organic matter much of the organic matter volume disappears. The Green River Formation volatilizes about 90% of its organic volume during oil evolution (Smith and Young, 1975). The soundness of the remaining rock depends to a large degree on the volume of organic matter volatilized. The less organic matter present in the original rock, the more stable the spent shale is. The organic matter leaves coke behind as it decomposes to yield oil and gas. This coke is distributed across all the mineral surfaces. If the organic matter's volume were a large fraction of the rock's total volume, the decomposing organic matter may fuse onto itself, forming a coke which is pliable and sticky when hot and physically strong when cold.

One thermal characteristic of Green River oil shale is that the organic matter loses some of its strength before oil evolution occurs. After conducting a series of tests by heating oil shale at constant load, Tisot and Sohns (1971) concluded that richer oil shale could collapse before retorting and plug the large undergound retorts being proposed at that time. However, rich oil shales do not collapse under overburden load during heating. When oil shale is broken up for processing in-place, the loads in the rubblized sections are point loads. If the organic matter begins to lose strength in a rich shale, the area of these points rapidly becomes larger. This easily compensates for the loss of strength of the organic matter before retorting. The widely publicized experiments misinterpreted by Tisot and Sohns (1971) were conducted under constant pressure. This constant pressure condition is not maintained in a column of broken shale (Smith et al., 1978b).

As oil shale temperatures go above about 500°C the minerals become thermally active. Mechanical property conditions vary with temperature, atmosphere and mineral composition. These all will be site- and process-specific and can't be covered here. Smith et al. (1978b) reviewed some of the high temperature reactions and effects in Green

Introduction to mechanics of oil shale 31

River Formation oil shale. Miller *et al.* (1978) reported on some of the mechanical and thermal properties at these elevated temperatures. Temperature effects in retorts processing Green River Formation oil shale are discussed in Chapter 12.

1.6. MECHANICAL PROPERTIES OF OIL SHALE DEPOSITS

The mechanical properties of most of the world's oil shales and how these properties vary with volume of organic matter can be illustrated by specific examples of the clay–quartz and the carbonate-cemented oil shales. In this book the clay–quartz oil shales are represented by the Devonian black shales of Kentucky and Tennessee (Fig. 1.1) and by the clay-rich, fissile oil shales of the Tipton Member of Wyoming's Green River Formation. The Tipton Member is the bottom member stratigraphically of Wyoming's Green River oil shales (Culbertson *et al.*, 1980). The Tipton Member oil shales have the same fissile, clay-rich character as oil shales of the clay zone in Colorado (Smith, 1980), which lies under the dolomite-rich oil shales of the Parachute Creek Member. Robb *et al.* (1978) illustrate the marked mineral change between these zones in Colorado, showing that clay decreases from more than 60% of the oil shale's mineral matter to less than 10% moving up across this transition. Dolomite increases correspondingly. The chemical mechanisms which produced this mineral change have been outlined (Smith, 1974). While no detailed study of mechanical properties of shales in the clay zone of Colorado are included here, the Tipton Member in Wyoming represents them very well. The mechanical characteristics of other clay-rich, carbonate-poor oil shales of the world will tend to resemble the characteristics presented for the Devonian black shales and for the clay-rich Tipton Member oil shales of Wyoming's Green River Formation. Differences in mechanical characteristics between these two high clay oil shales may be typical of differences between physical properties of the hydrogen-poor, high density organic matter in the Devonian black shale (Smith and Young, 1967) and the hydrogen-rich, low density organic matter in the Green River Formation oil shales.

Most of the Green River Formation oil shales proposed for development are dolomite-cemented and relatively poor in clay materials contributing to structure. The total thickness of Colorado's Green

River oil shales ranges up to 2100 ft (640 m) and incorporates four different lithologic sections. Figure 1.7 shows the location in Colorado's Piceance Creek Basin of seven oil yield histograms correlated in Fig. 1.8 to show the four primary lithologic zones in the complete oil shale section.

Smith (1980) points out the differences in development techniques required by the nature of the oil shale in these four zones. The clay zone characteristics were described earlier. The saline mineral zone is characterized by occurrence of nahcolite ($NaHCO_3$) nodules of assorted sizes incorporated in oil shales distorted by their occurrence. The dark blobs in the oil shale of about 10 ft of quartered core from the saline zone (Fig. 1.9) illustrate how nahcolite occurs. Although nahcolite is competent in the formation under compressive load, it is brittle. Any mechanical property testing done on oil shales containing this material has trouble representing the rock in the formation. Some experimental mining has been done in the saline zone oil shales (Utter and Hawkins, 1978; Cox, 1979).

In the leached zone the nahcolite (Fig. 1.9) has been extracted by ground water, leaving behind solution cavities, some of which have collapsed. These water-filled cavities are interconnected along the bedding, but not vertically. Efforts to exploit this natural permeability to generate oil from the shales in-place, in the leached zone, successfully produced oil (Dougan and Dockter, 1981), but apparently could not access enough oil shale for economic recovery. Evaluating and then mining in or through this water-loaded, somewhat less competent oil shale section presents some special mechanics problems.

The Mahogany Zone and its overyling oil shales have been the primary target for development efforts. Chapter 2 summarizes much of this development history. Most mechanical property and mining data available on Green River Formation oil shale refers to Mahogany Zone oil shale. Toward the south end of the Piceance Creek Basin (Fig. 1.7) up to 600 ft (180 m) continuous oil shale lies atop the Mahogany Zone (Smith *et al.*, 1978*a*). This overlying oil shale together with the Mahogany Zone oil shale was the target of the underground combustion process involving creation of gigantic chambers filled with broken oil shale, a process pioneered by Occidental Oil Shale Corp. The mining and mechanics of this process are discussed in this book.

The Mahogany Zone exists across all of the area outlined in Fig. 1.2 for the Green River Formation in Colorado and for the primary oil shale resource in Utah (Trudell *et al.*, 1983). The Mahogany Zone and

Introduction to mechanics of oil shale 33

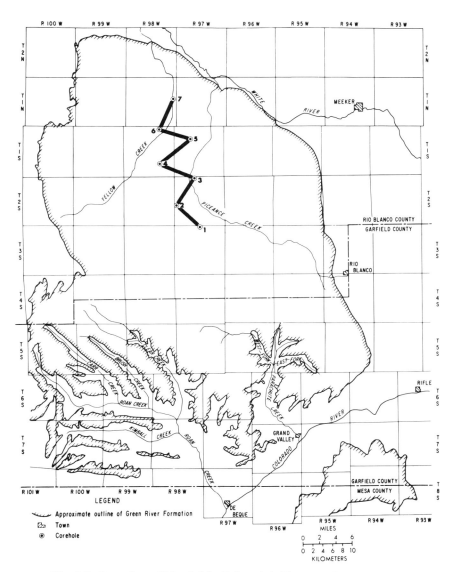

Fig. 1.7. Location of Fig. 1.8 in Colorado's Piceance Creek Basin.

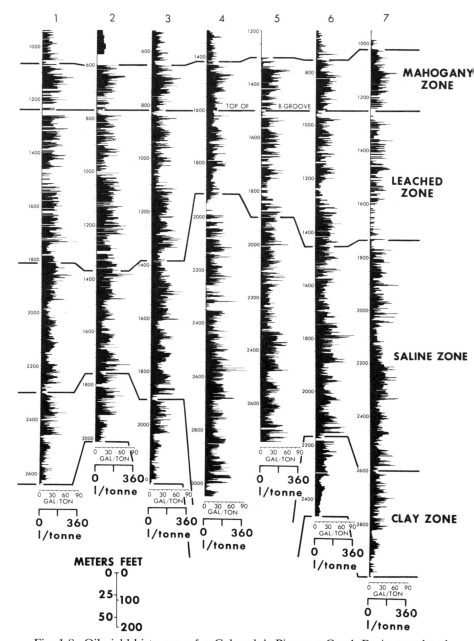

Fig. 1.8. Oil yield histograms for Colorado's Piceance Creek Basin correlated to show lithologic sections in the Green River Formation.

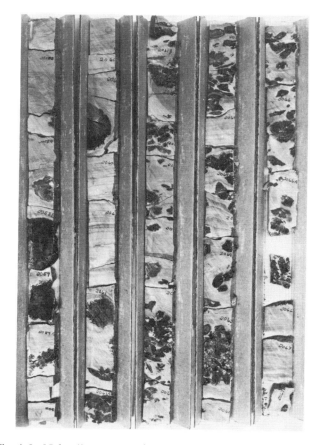

Fig. 1.9. Nahcolite nodules (dark blobs) in saline zone oil shale.

its overlying oil shales are dolomite-cemented. Because they are laterally uniform, they all tend to have similar properties which represent the properties of carbonate-cemented oil shales across the world.

One area where mechanical properties of the Mahogany Zone differ from those in the rest of the deposit is in the southern end of the Uinta Basin in Utah (Fig. 1.2). The only property which differs is the deposit's tensile strength perpendicular to the bedding (Fig. 1.5(a)). In the Mahogany Zone shale of the southern Uinta Basin the tensile strengths begin at levels of only a small fraction of those found in the rest of the Mahogany Zone. This weakness arose because the oil shales in this area were deposited in relatively shallow water. Depositional variations in the shallow water generated numerous variations in bonding between the bedding planes. The resulting deposit's tensile strength perpendicular to the bedding resembles that found for the clay-rich Tipton Member shales in Wyoming rather than that found for the balance of the Mahogany Zone. This effect is discussed in Chapter 7.

Formation joints and their spacing constitute one deposit-specific and site-specific factor particularly important to oil shale development. In a formation formed of rocks as strong and laterally continuous as most oil shales, joints tend to decide what can be accomplished efficiently in explosive treatment of the rock in place. Mine initiation, size distribution of rubble from blasting, mine layout, mining direction, etc., are influenced by joint patterns and joint distribution. Processes planned to generate permeability in oil shale deposits may find strong directional effects produced by joint systems. Joints may control water transmission in otherwise impermeable oil shale deposits.

Detailed descriptions of joint systems for oil shale deposits are not common. However, Verbeek and Grout (1983) are studying fracture history in the Piceance Creek Basin including its oil shales (Fig. 1.7). In this study they report that the character and spacing of fractures formed in the Green River Formation deposits are influenced strongly by organic content. This observation helps explain the discontinuous nature of some of the jointing in Mahogany Zone oil shales of Colorado and Utah (Fig. 1.2), where joints occur in specific beds and not in adjacent beds. Verbeek and Grout (1984) describe the dependence of joint spacings on bed thickness in sedimentary rocks, illustrating their relationship with Green River Formation examples. The joints are farther apart in thicker beds of similar lithology. This is particularly useful in interpreting joints and their distribution in oil shale. Site-specific studies of joint sytems are essential to oil shale development programs.

Hydraulic fracturing offers an interesting possibility of generating permeability in an oil shale formation. If hydraulic fracturing can be conducted to separate oil shale along its bedding planes, controlled access to an oil shale formation may be produced. Hydraulic fracturing techniques have become widely used in petroleum production in the US. Availability of this technology has led to tests on Green River Formation oil shales. Shell Oil Co. tested hydraulic fracturing at a site near Piceance Creek in Colorado's oil shale (Fig. 1.7). Results of these

Introduction to mechanics of oil shale 37

tests are not available in detail, but apparently Shell produced directionally oriented vertical fractures from tests at depths greater than 1000 ft (300 m). While at Shell Development in Houston, Hubbert and Willis (1957) produced a classic report on hydraulic fracturing. Conclusions incorporated in this study include the following statements applicable to oil shale:

> 'Hydraulically induced fractures should be formed approximately perpendicular to the least principal stress. Therefore in tectonically relaxed areas they (the fractures) should be vertical, while in tectonically compressed areas they should be horizontal.'

Further

> 'In geologically simple and tectonically relaxed areas, not only should the fractures in a single field be vertical but they also should have roughly the same direction of strike.'

Shell's hydraulic fracture experience in Piceance Creek Basin oil shales seems to conform to the Hubbert and Willis (1957) specifications for a 'relaxed area'. However, horizontal fractures along the bedding planes of oil shale have been created hydraulically in Wyoming's Green River Formation where bedding planes dip 1° or less. Wyoming's Green River Basin (Fig. 1.2) is even more relaxed tectonically than Colorado's Piceance Creek Basin. Thomas *et al.* (1972) produced a horizontal fracture at a depth of about 400 ft (122 m) near Rock Springs (Fig. 1.2) in the clay-rich Tipton Member described earlier. They report that the general stress condition at the test site was probably hydrostatic, indicating that oriented vertical fractures would be expected. Their experiment was conducted between straddle packers set about 4·5 ft (1·4 m) apart in a smooth-walled impermeable section of the test borehole. The final lines of their abstract read:

> 'A corehole drilled 100 ft (30 m) from the fractured well intersected the fracture and confirmed its horizontal orientation. Because a horizontal fracture was successfully initiated, it is inferred that the creation of multiple, closely spaced horizontal fractures may be feasible in Green River Basin oil shale.'

A subsequent borehole drilled 200 ft (60 m) out from the fractured hole at this site also encountered the fracture at the same stratigraphic position. A hydraulic fracturing experiment was also conducted at a site 25 miles (40 km) north of the initial experiment. A horizontal bedding plane fracture was successfully created in the Green River Formation's Wilkins Peak Member at a depth of about 920 ft (280 m). Before funding ran out on this test, the bedding plane fracture was verified by intersecting it at the same stratigraphic position in a borehole offset 50 ft (15 m) from the fractured borehole.

Multiple hydraulic fractures designed to carry liquid explosive into the formation were the basis for an experiment conducted by Talley Energy Systems in Tipton Member oil shale in Wyoming. Jensen (1979) reports that four hydrofractures were created, all approximately parallel and horizontal. Some directional differences were found in permeability along these fractures. Jensen (1979) concludes that the four hydraulically induced fractures were adequately sized and appropriately configured to allow injection and detonation of liquid explosive. The planned explosion was carried out, but the combustion experiment to produce oil was never conducted.

Hubbert and Willis (1957) indicate that such horizontal fracturing by hydraulic pressure is unlikely—obviously 'unlikely' does not mean 'impossible'. Hubbert subsequently (1971) indicated that 'Hydraulic-fracture orientation may ... be influenced by anisotropy..... such as bedding. .....hydraulic fractures may follow such a zone of weakness'. The anisotropy perpendicular to the bedding planes in Green River oil shales of Wyoming permitted horizontal fracturing to occur.

CONCLUSION

Oil shale will be developed and oil shale mechanics will be vitally important to this development.

REFERENCES

Allred, V. D. (1982). *Oil Shale Processing Technology*, Center for Professional Advancement, East Brunswick, New Jersey.

Ashley, G. H. (1918). Cannel coal in the United States, *US Geological Survey Bulletin 659*.

ASTM (1980). *Standard Test Method for Oil from Oil Shale (Resource Evaluation by the USBM Fischer Assay Procedure) Method D3904-80*, American Society for Testing and Materials, Philadelphia.

Bradley, W. H. (1929). The varves and climate of the Green River epoch, *US Geol. Survey Prof. Paper 158-E*, pp. 87–110.

Britton, K. C. and Lekas, M. A. (1979). Process for producing an underground zone of fragmented and pervious material, *US Patent No. 4175490*.

Chong, K. P., Smith, J. W., Chang, B. and Roine, S. (1979*a*). Oil shale properties by split cylinder method, *J. the Geotechnical Eng. Div., ASCE*, **105** (GT5), 595–611. (Proc. Paper 14567.)

Chong, K. P., Uenishi, K. and Smith, J. W. (1979*b*). Complete elastic constants and stiffness coefficients for oil shale, *US DOE/LETC Rept. Invest. RI-79/8*.

Chong, K. P., Uenishi, K., Smith, J. W. and Munari, A. C. (1980). Non-linear three dimensional mechanical characterization of Colorado oil shale, *Int. J. Rock Sci. & Geomech. Abstr.*, **17**, 339–47.

Chong, K. P., Smith, J. W. and Borgman, E. S. (1982). Tensile strengths of Colorado and Utah oil shales, *J. Energy, AIAA*, **6**(2), 81–5.

Cox, K. C. (1979). Experience in equipping a 96-inch diameter shaft and mining bulk samples of oil shale, dawsonite and nahcolite, in: *Twelfth Oil Shale Symposium Proceedings*, ed. J. H. Gary, Colorado School of Mines Press, Golden, pp. 179–83.

Culbertson, W. C., Smith, J. W. and Trudell, L. G. (1980). Oil shale resources and geology of the Green River Formation in the Green River Basin, Wyoming, *US DOE/LETC Rept. Invest. 80/6*.

Doscher, T. (1981). Petroleum reserves, in: *Encyclopedia of Energy*, 2nd edn, ed. S. Parker, McGraw-Hill, New York, p. 529.

Dougan, P. M. and Dockter, L. (1981). BX *in-situ* oil shale project, in: *Fourteenth Oil Shale Symposium*, ed. J. H. Gary, Colorado School of Mines Press, Golden, pp. 118–27.

Duncan, D. C. and Swanson, V. E. (1965). Organic rich shale of the United States and world land areas, *US Geological Survey Information Circular 523*, Washington DC.

East, J. H., Jr and Gardner, E. D. (1964). Oil shale mining, Rifle, Colorado, 1944–56, *US Bureau Mines Bulletin 611*.

Ferguson, P. (1980). History of the development of the shale oil industry in Australia, *Southern Pacific Petroleum News Letter*, Sydney.

Garrels, R. M. and Christ, C. L. (1965). *Solutions, Minerals and Equilibria*, Freeman Cooper and Co., San Francisco (figure 11.2).

Gavin, M. J. (1922). Oil shale. An historical, technical, and economic study, *US Bureau of Mines Bulletin 210*.

Guthrie, B. and Klosky, S. (1951). The oil shale industries of Europe, *US Bureau of Mines Rept. Invest. 4776*.

Hall, C. A. S. and Cleveland, C. J. (1981). Petroleum drilling and production in the United States: yield per effort and net energy analysis, *Science*, **211**(4482), 576–9 (6 February).

Hubbert, M. K. (1971). Natural and induced fracture orientation, *Bull. Am. Assoc. Petrol. Geol.*, **55**(11), 2086–7.

Hubbert, M. K. and Root, D. H. (1981). Outlook for fuel reserves, in: *Encyclopedia of Energy*, 2nd edn, ed. S. Parker, McGraw-Hill, New York, pp. 43–56.

Hubbert, M. K. and Willis, D. G. (1957). Mechanics of hydraulic fracturing, *AIME Petroleum Transactions*, V. **210,** (T. P. 4597), 153–6.

40 John Ward Smith and Ken P. Chong

Jensen, H. B. (1979). Oil shale *in situ* research and development, *Talley Energy Systems Final Report, US Department of Energy DOE/LC/01791-T1*.

Klosky, S. (1949). An index of oil shale patents, *US Bureau of Mines Bulletin 468*.

Klosky, S. (1958). Index of oil shale and shale oil patents, 1946–1956; part I, US patents; part II, UK patents; part III, European patents and classification. *US Bureau of Mines Bulletin 574*.

Matveyev, A. K. (ed.) (1974). *Deposits of Fossil Fuels: Oil Shales Outside the Soviet Union*, G. K. Hall, Boston. (Authorized English translation.)

Matzick, A., Dannenberg, R. O. and Guthrie, B. (1960). Experiments in crushing Green River oil shale, *US Bureau of Mines Rept. Invest. 5563*.

Miknis, F. P. and Smith, J. W. (1982). An NMR survey of United States oil shales, in: *Proceedings Fifteenth Oil Shale Symposium*, ed. J. H. Gary, Colorado School of Mines Press, Golden, pp. 50–62.

Miknis, F. P., Smith, J. W., Maughan, E. K. and Maciel, G. E. (1982). Nuclear magnetic resonance: a technique for direct nondestructive evaluation of source rock potential, *Bull. Am. Assoc. Petrol. Geol.*, **66**(9), 1396–401.

Miller, R., Wang, F. D. and DuBow, J. (1978). Mechanical and thermal properties of oil shale at elevated temperatures, *Eleventh Oil Shale Symposium Proceedings*, ed. J. H. Gary, Colorado School of Mines Press, Golden, pp. 135–46.

Provo, L. J. (1977). Stratigraphy and sedimentology of radioactive Devonian-Mississippian shales of the central Appalachian Basin. PhD Thesis, Department of Geology, University of Cincinnati.

Robb, W. A., Smith, J. W. and Trudell, L. G. (1978). Mineral and organic distribution and relationships across the Green River Formation's saline deposition center, Piceance Creek Basin, Colorado, *US DOE/LETC Rept. Invest. RI-78/6*.

Russell, P. L. (1980). *History of Western Oil Shale*, Center for Professional Advancement, East Brunswick, New Jersey.

Sellers, J. B., Haworth, G. R. and Zambas, P. G. (1972). Rock mechanics research in oil shale mining, *Transactions AIME*, **252**, 222–32.

Smith, J. W. (1961). Ultimate composition of organic material in Green River oil shale, *US Bureau of Mines Rept. Invest. 5725*.

Smith, J. W. (1966). Conversion constants for Mahogany Zone oil shale, *Bull. Amer. Assoc. Petrol. Geol.*, **50**(1), 167–70.

Smith, J. W. (1969). Theoretical relationship between density and oil yield for oil shales, *US Bureau Mines Rept. Invest. 7248*.

Smith, J. W. (1974). Geochemistry of oil shale genesis in Colorado's Piceance Creek Basin, in: *Guidebook to the Energy Resources of the Piceance Creek Basin*, ed. D. K. Murray, Rocky Mountain Association of Geologists, Denver, pp. 71–9.

Smith, J. W. (1976). Relationship between rock density and volume of organic matter in oil shales, *US ERDA/LERC/Rept. Invest. 76/6*.

Smith, J. W. (1977). Review of 'Deposits of fossil fuels: oil shales outside the Soviet Union', *Sedimentary Geology*, **19**, 75–80.

Smith, J. W. (1980). Oil shale resources of the United States, *Mineral and Energy Resources*, **23**(6).

Smith, J. W. (1981). Oil shale resource evaluation, in: *Symposium Papers: Synthetic Fuels from Oil Shale II, Nashville, Tennessee*, Institute of Gas Technology, Chicago, pp. 123–42.

Smith, J. W. (1982). Synfuels: oil shale and tar sands, in: *Perspectives on Energy*, 3rd edn, eds L. C. Ruedisili and M. W. Firebaugh, Oxford University Press, New York, pp. 225–49.

Smith, J. W. (1983). The chemistry which formed Green River Formation oil shale, in: *Geochemistry and Chemistry of Oil Shales*, Am. Chem. Soc. Series 230, Washington DC, pp. 225–48.

Smith, J. W. (1984). Additional oil shale technologies, in: *Handbook of Synfuels Technologies*, ed. R. A. Meyers, McGraw-Hill, New York, chapters 4-7, pp. 4-159–4-203.

Smith, J. W. and Johnson, D. R. (1976). Mechanisms helping to heat oil shale blocks, *Am. Chem. Soc., Division of Fuel Chemistry Preprints*, **21**(6), 23–35.

Smith, J. W. and Young, N. B. (1964). Specific gravity of oil yield relationships for black shales of Kentucky's New Albany Formation, *US Bureau Mines Rept. Invest. 6531*.

Smith, J. W. and Young, N. B. (1967). Organic composition of Kentucky's New Albany shale: determination and uses, *Chem. Geol.*, **2**, 157–70.

Smith, J. W. and Young, N. B. (1975). Dawsonite: Its geochemistry, thermal behavior, and extraction from Green River oil shale, *Colorado School of Mines Quarterly*, **70**(3), 69–93.

Smith, H. N., Smith, J. W. and Kommes, W. C. (1959). Petrographic examination and chemical analyses of several foreign oil shales, *US Bureau of Mines Rept. Invest. 5504*.

Smith, J. W., Beard, T. N. and Trudell, L. G. (1978a). Colorado's primary oil shale resource for vertical modified *in situ* processes, *US DOE/LETC Rept. Invest. 78/2*.

Smith, J. W., Robb, W. A. and Young, N. B. (1978b). High temperature reactions of oil shale minerals and their benefit to oil shale processing in place, in: *Eleventh Oil Shale Symposium Proceedings*, ed. J. H. Gary, Colorado School of Mines Press, Golden, pp. 100–12.

Thomas, H. E., Carpenter, H. C. and Sterner, T. E. (1972). Hydraulic fracturing of Wyoming Green River oil shale: field experiments, phase I, *US Bureau of Mines Rept. Invest. 7596*.

Tisot, P. R. and Murphy, W. I. R. (1960). Physicochemical properties of Green River oil shale: particle size and particle size distribution of inorganic constituents, *J. Chem. Eng. Data*, **5**, 558–62.

Tisot, P. R. and Sohns, H. W. (1971). Structural deformation of Green River oil shale as it relates to *in situ* processing, *US Bureau of Mines Rept. Invest. 7576*.

Trudell, L. G., Beard, T. N. and Smith, J. W. (1970). Green River Formation lithology and oil shale correlations in the Piceance Creek Basin, Colorado, *US Bureau of Mines Rept. Invest. 7357*.

Trudell, L. G., Smith, J. W., Beard, T. N. and Mason, G. M. (1983). Primary oil shale resources of the Green River Formation in the eastern Uinta Basin, Utah, *US DOE/LC Rept. Invest. 82/4*.

Utter, S. and Hawkins, J. E. (1978). Drilling and casing a large diameter shaft in the Piceance Creek Basin, in: *Eleventh Oil Shale Symposium Proceedings*, ed. J. H. Gary, Colorado School of Mines Press, Golden, pp. 292–310.

Verbeek, E. R. and Grout, M. A. (1983). Fracture history of the northern Piceance Creek Basin, northwestern Colorado, in: *Sixteenth Oil Shale Symposium Proceedings*, ed. J. H. Gary, Colorado School of Mines Press, Golden, pp. 26–44.

Verbeek, E. R. and Grout, M. A. (1984). Fracture studies in Cretaceous and Paleocene strata in and around the Piceance Basin, Colorado: preliminary results and their bearing on a fracture-controlled natural gas reservoir at the MWX site, *US Geological Survey Open-File Report 84–156*.

Williams, P. V. F. (1983). Oil shales and their analysis, *Fuel*, **83**, 756–71.

Chapter 2

Historical Perspective

STEPHEN UTTER

Senior Staff Engineer, Bureau of Mines, Denver, Colorado, USA

SUMMARY

This chapter reviews the past 40 years or so of research in the mechanics of the Green River oil shales. Although the first shale oil was produced in Colorado in 1917, it was not until the Anvil Points project of 1944 that research in the mechanics of oil shale probably began. Because of its importance to the safety and economy of any oil shale operations, special emphasis is directed to the mining aspects of rock mechanics.

Later experimental oil shale mines, some patterned on the Anvil Points work, are described and true and modified in situ *experiments are summarized. Plans for the Colony Development Operation and the Union Parachute Creek Shale Oil Program are presented.*

Since the scope of this brief overview is mainly limited to the major mining and in situ *projects, several oil shale studies on other aspects of the subject are not included.*

2.1. INTRODUCTION

During the past 70 years, interest in oil shale as an energy source in the US has changed as the cost and availability of conventional petroleum has varied. At present, oil shale activities appear to have once again reached a plateau. However, one fact is evident—this tremendous energy resource will eventually be tapped.

In this chapter, a brief overview of the history of oil shale mechanics and past research into this fascinating subject is presented.

Oil shales are fine-grained sedimentary rocks that contain a solid hydrocarbon called kerogen. Kerogen yields oil, gas and a carbon residue when heated (retorted) at temperatures ranging from 480 to 650°C. Attention in this country has focused on the rich, thick, deposits of the Green River Formation in northwestern Colorado, northeastern Utah and southwestern Wyoming. This formation, which covers an area of about $44\,000\,\text{km}^2$, contains an estimated $290 \times 10^9\,\text{m}^3$ of shale oil. Of this total, $95 \times 10^9\,\text{m}^3$ are considered to be of current commercial interest. Figure 2.1 shows the location of the Green River Formation.

Retorting methods to produce shale oil from shale rock are divided into surface and *in situ*. For surface methods, the shale is mined, crushed and sized before retorting in surface vessels. For *in situ* methods, the shale is heated underground, in place. True *in situ* processes involve the drilling of wells, fracturing or otherwise treating the shale to increase permeability and then retorting by underground combustion, heated gas, superheated steam or other agents. Modified *in situ* processes involve mining part of the shale to provide space for expansion. The remaining shale is drilled and blasted and then retorted by underground combustion or other methods.

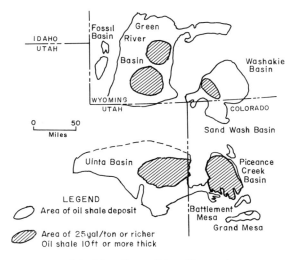

Fig. 2.1. The Green River Formation.

Historical perspective 45

The design of oil shale mine structures is important because it affects the safety, economy and environment of the mining operation. Rock fragmentation (drilling, blasting, crushing) is part of the overall excavation and processing system. Mine design and well fragmentation can be categorized as that branch of science known as rock mechanics. Rock mechanics is the theoretical and applied science of the mechanical behavior of rock and rock masses: it is that branch of mechanics concerned with the response of rock and rock masses to the force fields of their mechanical environment (Panel on Rock Mechanics Research Requirements, 1981).

2.2. BACKGROUND

Interest in Green River oil shale began in the early 1900s. The first oil shale producing retort in Colorado was built in 1917 (Russell, 1980). By 1918, several companies were mining and retorting oil shale in that state (East and Gardner, 1964). From 1926 to 1928 the US Bureau of Mines conducted intermittent mining and experimental retorting operations near Cottonwood Point, between the towns of Rifle and Parachute (Grand Valley), Colorado (East and Gardner, 1964).

After this flurry of activity, interest in oil shale lapsed as petroleum became cheaper and more plentiful. It is doubtful that any of these early oil shale operations were commercial. During Second World War, increasing oil consumption coupled with decreasing domestic reserves once again revived interest in oil shale as a source of synthetic liquid fuels.

Studies of the mechanics of oil shale probably started with the Synthetic Liquid Fuels Act of 1944 which authorized the construction and operation of plants to produce synthetic fuels. Under this Act, the Bureau of Mines constructed and operated an experimental mine and synthetic fuels plant from 1944 to 1956 on US Naval Oil Shale Reserves Nos 1 and 3 at Anvil Points, near Rifle, Colorado. Research at Anvil Points and at the Bureau's Laramie Petroleum Research Center, Laramie, Wyoming, greatly expanded the existing technical information about oil shale geology, mineralogy, mechanics, mining and processing.

After Anvil Points was closed, the Laramie Petroleum Research Center continued research on oil shale and shale oil. The Union Oil Co. constructed a mine and pilot plant on Parachute Creek, Colorado.

The Sinclair Oil and Gas Co. conducted true *in situ* retorting tests in wells drilled on Haystack Mountain also on Parachute Creek. The Colony Development Corp. developed a pilot mine and semiworks plant on upper Parachute Creek. The Colorado School of Mines Research Foundation leased the Anvil Points Facility from the US Department of the Interior and acted as a contractor for a consortium of oil companies. This consortium conducted retorting tests at the Facility and mining and rock mechanics studies in an experimental mine adjoining the Anvil Points Mine. Equity Oil Co. conducted true *in situ* retorting tests in wells drilled on Black Sulphur Creek, Colorado. Later, the Atlantic Richfield Co. conducted *in situ* field tests on this property. A nuclear *in situ* test, Project Bronco, was proposed as a joint experiment by industry and government but was never started. The Shell Oil Co. conducted true *in situ* retorting tests on Piceance Creek, Colorado.

From 1964 to 1981 the Bureau of Mines conducted rock mechanics studies under cooperative research agreements with companies engaged in oil shale research and also developed a pilot shaft and research facility at Horse Draw in Rio Blanco County, Colorado.

The oil embargo and price increases of 1973–1974 together with the Department of the Interior's Prototype Oil Shale Leasing Program stimulated a wide range of research activities in oil shale mechanics, mining, retorting and refining.

The Rio Blanco Oil Shale Co., which leased Prototype Oil Shale Tract C-a in Colorado, sank a shaft and developed two experimental modified *in situ* retorts. The Cathedral Bluffs Project sank three shafts on Prototype Oil Shale Tract C-b in Colorado in a program to conduct modified *in situ* retorting. The White River Oil Shale Corp. planned a mine on Prototype Oil Shale Tracts U-a and U-b in Utah. The Occidental Oil Shale Corp. developed a mine and conducted modified *in situ* retorting tests at Logan Wash near De Beque, Colorado.

The Department of Energy (formerly the Energy Research and Development Agency), created by Congress in 1977, included an organization intended to conduct research programs in fossil energy. The Laramie Petroleum Research Center was transferred to the Department of Energy and renamed the Laramie Energy Technology Center. Here, studies continued on oil shale geology and mineralogy, true *in situ* retorting tests at Rock Springs, Wyoming, and batch type, surface retorting tests to simulate *in situ* retorting conditions.

Other studies have been or are being conducted by the Colorado

School of Mines, Denver University, University of Wyoming, Geokinetics Inc., Equity Oil Co., Illinois Institute of Technology, Institute of Gas Technology, Dow Chemical Co., National Bureau of Standards, Lawrence Livermore National Laboratory, Sandia National Laboratory, Los Alamos Scientific Laboratory, Multi Mineral Corp. and others. Some of this work is described in the sections that follow.

2.3. ANVIL POINTS

2.3.1. Mine Design

From 1944 to 1956, the Bureau of Mines constructed and operated a selective mine, a demonstration mine and a processing plant at Anvil Points, Colorado. The purpose of the project was to mine oil shale for process testing, to conduct the testing and to develop a method for mining the Mahogany Zone. The Mahogany Zone at the site is about 22 m thick and averages 117 liter tonne^{-1}.

The selective mine was developed by driving adits above and below the Mahogany Zone, developing levels in the various oil shale beds, and connecting the levels with raises. Stopes about 12 m wide were mined by standard drilling and blasting methods. Shale was mucked by small scrapers to raises and loaded into diesel-powered trucks on the lower level.

Core holes were drilled to determine stratigraphy, grade and thickness of the oil shale. Physical properties of the shale were determined and a raise was driven to select a roof horizon for the demonstration mine.

Test results from three laboratories, which were in general agreement, showed that the shale could be classified as a moderately strong rock with an average compressive strength of 110·3 MPa and a modulus of rupture of 27·6 MPa. Variations in sample size, location, shale grade, direction of loading with respect to bedding planes and the experimental error that occurs in all measurements contributed to differences in the test results. Physical properties are summarized in Table 2.1 (Merrill, 1954).

A room-and-pillar mining method was selected for the demonstration mine. Safe roof spans and pillar sizes were calculated from theory and from model tests. Results from theory showed that a single-member roof 2·4 m thick could span 22 m with a safety factor of 8. Results from model tests showed that a three-member roof with the

TABLE 2.1
Average Physical Properties of Oil Shale

Property	Roof cored parallel to bedding	Roof cored perpendicular to bedding	Pillar cored perpendicular to bedding
Apparent specific gravity	2·16	2·25	—
Compressive strength (MPa)	—	114·4	156·5
Modulus of rupture (MPa)	20·7[a]	2·5[b]	—
Young's modulus (MPa)	21 374	12 411	—
Poisson's ratio	0·58	−0·10	—

[a] Load applied perpendicular to bedding.
[b] Loading applied parallel to bedding.

lowest member 2·4 m thick could span 26 m with a safety factor of 5·6. An 18 m span, which was well within safety limits, was chosen. For a 75% extraction ratio it was found that pillars 18 m square would provide support with safety factors ranging from 3 to 13 depending on depth of overburden (East and Gardner, 1964).

Calculations were verified by developing an instrumented test room in the selective mine. The room was excavated 15 m wide by 30·5 m long directly under the selected roof strata. This room was gradually enlarged to a final dimension of 24×61 m, well past the design limits established for the production mine. Two months later, a layer of roof 51 cm thick fell. Drill holes in the roof were examined for separations and fractures with a stratascope, roof sag was measured and microseismic tests were conducted. It was found that roof sag was greater than that calculated from theory. Separations were observed at various depths in the roof strata (Merrill, 1954).

The demonstration mine was developed by driving an adit on each of two upper levels. A third adit, which was driven to develop the lower level, was not completed. The mine layout consisted of pillars in staggered rows. Pillars were 18 m wide. Three levels were initially planned, a top heading 8·2 m high and two benches 7 m high. Later the plan was changed to a two level system with a top heading 12 m high and a bench 10·4 m high. This plan provided shale of similar grade from each level and was more cost-effective than the three-level system.

Fig. 2.2. Anvil Points Mine. (From US Bureau of Mines.)

Standard drilling and blasting methods were used. Special equipment was built for drilling and scaling the high faces. Shale was loaded with a 3 m^3 electric shovel and hauled with a 16·3 tonne diesel truck as shown in Fig. 2.2.

Rock performance in the demonstration mine was observed continuously for several years. Separations in the roof were observed in the observation holes drilled in the roof of the demonstration mine. In 1953, a roof fall occurred in a cross-cut that had been opened for $4\frac{1}{2}$ years. The fall ranged from 15 to 61 cm thick and covered about 700 m^2. Roofbolts 25 mm in diameter and 1·8 m long were installed on 3 m centers. In 1955, a second roof fall occurred in a cross-cut that had been opened for about 2 years. The fall ranged from about 2·4 to 4·6 m thick and covered about 650 m^2. About 9000 tonne of rock fell. The roof had been bolted. Figures 2.3 and 2.4 show the roof fall.

Two sets of vertical joints were revealed by the falls. It was concluded that (1) 'time sag', a plastic deformation of the shale, occurred that led to a progressive deterioration of the roof; (2) two and possibly three horizontal planes of weakness were present at depth in the roof; and (3) the second roof fall probably started at the intersection of two vertical joints (Obert and Merrill, 1958). Although these failures

Fig. 2.3. Roof fall in Anvil Points Mine. (From US Bureau of Mines.)

indicated failure processes continuing to occur in the roof virtually all of the area mined out by the Bureau of Mines still (1982) stand open.

2.3.2. Rotary Drilling Experiments

Two electric-hydraulic rotary test drills were designed and built to test rotary drilling of blast holes. Results showed that rotary drilling of oil shale was found to be significantly faster and more economic than standard percussion drilling (East and Gardner, 1964).

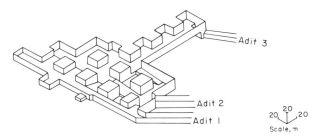

Fig. 2.4. Schematic view of Anvil Points demonstration mine in 1955. Roof falls occurred in the first cross-cut to the right of adit 2 and in the long cross-cut to the left of adit 3. (From US Bureau of Mines.)

2.3.3. Blasting Experiments

An extensive series of blasting tests were conducted using various types of drill rounds and the standard explosives and delays available at that time. ANFO (ammonium nitrate/fuel oil) was not in use. Best results were obtained with 45% semi-gelatin dynamite and millisecond-delay electric detonators. A typical heading round consisted of 112 drill holes in a V-cut pattern. This round consistently pulled 4·6 m in the 12×18 m face and broke about 1800 tonnes of shale with 475 kg of explosive. Two miners could drill out a round with the percussion-drill heading jumbo in less than one 8 h shift.

2.3.4. Crushing Experiments

Crushing tests were conducted with conventional jaw, gyratory, impact and roll crushers on mine-run shale. Grade ranged from 75 to 263 liters tonne^{-1} and averaged 117 liters tonne^{-1}. The crushers were usually set to produce a maximum particle size of 76 mm with a minimum of fines (6 mm and smaller) for discard. Although these tests were not complete, results showed that Mahogany Zone shale is heterogeneous, tough and resilient. High-grade shale is slippery and breaks into large slabs (Matzick *et al.*, 1960).

2.4. UNION MINE

From 1955 to 1958, Union Oil Co. constructed and operated a demonstration retort (retort A) and an underground mine on

Parachute Creek about 19 km north of Parachute, Colorado. Early production was from a quarry. Later, two 9×9 m adits were driven to develop a room-and-pillar mine in the Mahogany Zone. Rooms 18 m wide and pillars 18 m square were planned. Conventional drilling and blasting methods were used. Shale was hauled in 18-tonne trucks to an 107 cm gyratory crusher where it was reduced to 15 cm size. A jig-back aerial tram transported the shale to a stockpile in the valley near the retort. When the operation closed, one adit had been driven 122 m where it divided into two branches each 33·5 m long. One room had been started.

The retort, which had a design capacity of 227 tonnes day^{-1}, had reached a maximum rate of 1090 tonnes day^{-1} and produced up to 127 m^3 day^{-1} of shale oil. Over 2000 m^3 of shale oil was processed into gasoline and other products (Carver, 1964).

The use of tunnel-boring machines for mining oil shale was investigated. Preliminary results indicated that mechanical excavation of oil shale was feasible (Carver, 1965; Hamilton, 1965).

2.5. MOBIL OIL

In 1964, the Department of the Interior leased the Anvil Points facility to the Colorado School of Mines Research Foundation. The Foundation acted as a research contractor for a consortium of six petroleum companies: Mobil Oil Corp. (Project Manager), Continental Oil Co., Humble Oil and Refining Co., Pan American Petroleum Co., Phillips Petroleum Co. and Sinclair Research Inc. After a small amount of shale had been mined from the demonstration mine, it was decided to develop a new experimental mine on the adjacent Mobil property because of safety and ventilation requirements.

A ventilation drift was driven from the demonstration mine and a 12 m high adit was driven to develop a two-level room-and-pillar mine in the Mahogany Zone. A 6×6 m decline was driven about midway in the adit to provide access to the bottom level. A short winze was sunk and a $2·4 \times 2·4$ m instrumentation drift was developed below the location of the projected center pillar row. Borehole extensometers were installed in advance of mining to measure subsequent pillar deformations.

The mine was laid out for two levels, a 12 m top heading followed by a 11·6 m bench. Room spans were 18 m. Four parallel rooms were mined leaving 12 m wide rib pillars. Crosscuts were blasted in the final stage, leaving 12 m square pillars (Zambas *et al.*, 1972).

2.5.1. Mine Design

An extensive rock mechanics program was conducted. It included *in situ* stress measurements conducted under a cooperative research agreement with the Bureau of Mines, pillar deformation measurements, physical property determinations, finite element analyses by the Bureau of Mines, pillar stress change measurements with Bureau of Mines' hydraulic borehole pressure cells, photoelastic stressmeters, multiple point borehole extensometers, roof sag pins and roof bolt load cells.

Results showed that an 80% extraction ratio was feasible. Final calculated stresses in the pillars, which were 12 m square and 24 m high, were about 13·8 MPa. The minimum compressive strength of the shale was about 55·2 MPa. Roof sag was most pronounced at the intersection of the rooms and cross-cuts. However, measurements at nine sag pin stations showed little or no sag in strata that were over 4·6 m above the roof.

A section of unbolted mine roof about 51 cm thick and measuring 15×17 m fell at an intersection with the adit. Sag pin measurements had shown accelerated movement at two roof horizons before the fall. This section was stabilized by installing high-strength rock bolts 7·6 m long. It was concluded that a system of long rib pillars with a minimum of cross-cuts would improve roof stability, because by reducing the number of intersections the incidence of long roof spans would also be reduced (Sellers *et al.*, 1972).

2.5.2. Drilling Experiments

Benches were drilled with a rotary down-the-hole drill. Penetration rate averaged $1·2$ m min^{-1} at 100 rpm and a thrust of 5400 kg.

2.5.3. Blasting Experiments

A top heading round consisting of 28 drill holes in a V-cut pattern was developed from drilling and blasting tests. Holes were pneumatically loaded with ANFO and bottom-primed with one stick of 70% dynamite. Millisecond delay electric detonators were used. This round pulled 7·6 m and broke about 3600 tonnes of shale. The powder factor was $0·25–0·29$ kg tonne^{-1}.

A typical bench round consisted of 20 drill holes with a spacing of

4·3 m and a burden of 4·6 m. ANFO was used. The powder factor was $0·14$ kg tonne^{-1}. It was found that this round caused excessive damage to the pillars. After an extensive seismic program which involved measurements of particle displacement, velocity and acceleration, pre-split heading and bench rounds were developed which reduced pillar damage and scaling requirements.

Over 454 000 tonnes of shale were mined during 13 months of operation. The maximum production rate was 3629 tonnes day^{-1} (Zambas *et al.*, 1972).

2.6. COLONY MINE

From 1964 to 1972, the Colony Development Co. developed and operated a room-and-pillar pilot mine and a prototype TOSCO retort on the Middle Fork of Parachute Creek. The Company began as a joint venture of The Oil Shale Corp. (TOSCO), Standard Oil Co. of Ohio and Cleveland-Cliffs Iron Co. In 1969, The Atlantic Richfield Co. joined the venture and the name was changed to the Colony Development Operation. Cleveland-Cliffs Iron Co. and Standard Oil Co. of Ohio withdrew from the venture in 1971.

The mine was developed by an adit driven into a Mahogany Zone section 18 m high which averages 146 liters tonne^{-1}. Overburden ranges from about 183 to 259 m. The mine was laid out for two levels, a 9 m high heading followed by a 9 m high bench. Roof span averaged 18 m. Pillars were about 18 m square. A full-seam mining experiment was also conducted in which the 18 m high headings were mined in one stage. Conventional drilling and blasting methods were used (Marshall, 1974).

2.6.1. Mine Design

Approximately 1 million tonnes of oil shale were mined during an extensive research program on mining operations, equipment and rock mechanics. Under a cooperative agreement, the Bureau of Mines conducted laboratory tests to determine the effect of joints on uniaxial and shear strength. *In situ* stresses were determined. Measurements were taken to determine pillar loads and roof sag. The *in situ* modulus of elasticity was determined.

Results showed that the average uniaxial compressive strength of the shale ranged from 87·6 MPa for solid shale to 23·4 MPa for shale with a place of weakness inclined at an angle of approximately 30° to the vertical (Agapito, 1974).

One pillar failed during benching operations and a section of adjacent roof fell because of the increased roof span. This event provided an opportunity for studying the sequence of pillar failure and an assessment of *in situ* pillar strength. A finite element computer model was developed and used to simulate load transfer during pillar failure and stress distribution in pillars near the rib walls. Model and experimental results were found to agree.

After the engineering test programs were completed in 1972, a commercial operation was designed.

2.7. OCCIDENTAL LOGAN WASH MINE

From 1972 to date, the Occidental Petroleum Corp. conducted field tests to develop a modified *in situ* retorting process at Logan Wash near De Beque, Colorado. The Garrett Research and Development Co., a subsidiary of Occidental, began the experimental program in 1972.

The site consists of medium-grade deposits of oil shale 76 m thick and averaging 70 liters tonne^{-1}. An adit was driven and three small experimental retorts, 1E, 2E and 3E, were developed, rubblized and burned. The objective of the program was to test the feasibility of the method. Particle size, permeability, pressure drop, ignition, retorting and oil recovery were studied (Ridley, 1978).

After these tests, Occidental Oil Shale Inc. was organized in 1974 and began an experimental program to develop larger retorts of a commercial size. To date, eight retorts have been tested as summarized in Table 2.2.

The process consists of developing the retort by excavating three void levels, an upper, an intermediate and a lower level, and drilling and blasting the intervening shale. From 15 to 25% of the shale on the total shale column is excavated from the void levels. The rubblized shale is then retorted by igniting the top of the shale rubble. Air and steam are pumped into the top of the retort to continue the retorting process. As the burn front progresses downward, four zones occur: (1) the hot spent zone at the top where incoming air, gas and steam are preheated; (2) the combustion zone where residual carbon and some oil and gas are burned; (3) the retorting zone below the combustion

TABLE 2.2
Occidental *in situ* Retorts

Retort number	Size, m	Oil production, m^3
1E	$9.4 \times 9.4 \times 21.9$	190
2E	$9.8 \times 9.8 \times 28.7$	220
3E	$9.8 \times 9.8 \times 34.4$	250
4	$36.6 \times 36.6 \times 78.3$	5100
5	$36.0 \times 36.0 \times 50.9$	1800
6[a]	$49.4 \times 49.4 \times 82.0$	8700
7[b]	$50.3 \times 50.3 \times 75.0$	—
8[b]	$50.3 \times 50.3 \times 75.0$	—

[a] Retorting upset.
[b] Total production from both retorts by November 1982 was over 30 210 m^3.

zone where gas and oil are produced; and (4) the condensation zone, where the oil, water and gases are cooled. Oil and water flow to a sump at the bottom for recovery.

In 1976, a two-phase cooperative agreement was signed with the Department of Energy. The first phase was the engineering development of two vertical modified *in situ* retort designs (retorts 5 and 6). The second phase was demonstration of the technical feasibility of the design selected (retorts 7 and 8). Occidental has also conducted an extensive rock fragmentation test program in connection with retort rubblization. Under a cooperative research program, the Bureau of Mines conducted stress measurements and seismic tests to determine pillar stability (Romig, 1981).

2.8. FEDERAL LEASE TRACT C-a

Tract C-a is one of the two tracts in Colorado leased by the US Department of the Interior under the Prototype Oil Shale Leasing Program. In 1974, Gulf Oil Corp. and Standard Oil Co. of Indiana were successful with a bid of $210·3 million (Rutledge, 1982). They organized the Rio Blanco Oil Shale Co. to develop the tract.

The tract is on Yellow Creek in the Piceance Creek Basin of Colorado. Although originally selected for an open pit mining operation, that method could not be used because additional Federal land

TABLE 2.3
Rio Blanco *in situ* Retorts

Retort number	Size, m	Oil production, m^3
0	9·1× 9·1× 50·6	298
1	18·3×18·3×121·9	3887

was not available for off-tract disposal of overburden and spent shale. A modified *in situ* retorting method with surface retorting of development shale was then selected.

A service and production shaft 4·6 m in diameter, was sunk by conventional methods to a depth of 296 m. Three levels were developed and an exhaust shaft and a service/escape shaft were raise bored. Two modified *in situ* retorts were developed, rubblized and burned. Oil production results are summarized in Table 2.3 (Berry *et al.*, 1982).

Five dewatering wells, drilled before shaft sinking, produced about 189 liters s^{-1}. As draw-down continued, water production dropped to about 63 liters s^{-1}.

The mine operated under the Mine Safety and Health Administration (MSHA) gassy mine classification. Methane and hydrogen sulfide

Fig. 2.5. Aerial view of Rio Blanco Oil Shale Co. Tract C-a. (Courtesy of Minerals Management Service.)

gases were diluted and removed by the ventilation system. The retorts were operated under a vacuum and retort gases were not a problem. Instrumentation was installed to measure temperature, pressure and rock movement. Under a cooperative research agreement with the Bureau of Mines, a method for backfilling *in situ* retort chambers with spent shale was investigated. Such a method would permit disposal of part of the spent shale produced from surface retorts and also help to stabilize the mine workings (Watson *et al.*, 1982).

The company is constructing a Lurgi-type pilot retort in Pennsylvania and conducting engineering and planning studies for an open pit operation. Congressional legislation, required before the Department of the Interior can lease federal land for off-tract disposal, has recently been passed. An aerial view of the surface plant is shown in Fig. 2.5.

2.9. FEDERAL LEASE TRACT C-b

Tract C-b is the second tract leased in Colorado by the Department of the Interior. In 1974, the original partners, Atlantic Richfield Co., The Oil Shale Corp. (TOSCO), Ashland Oil Co. and Shell Oil Co. were successful with a bid of $117·8 million. All of the original partners withdrew from the project at various times. The present partners are Occidental Oil Shale Inc. and Tenneco. Known at various times as C-b Oil Shale Project, Roxana Shale Oil Co. and C-b Shale Oil Venture, the operation is now called the Cathedral Bluffs Shale Oil Co. (Office of Shale Resource Applications, 1981).

The tract is in the Piceance Creek Basin (T.3S., R96 and 97W, Garfield County, Colorado) just south of Piceance Creek. A room-and-pillar system was first proposed to mine a 23 m thick section of the Mahogany Zone which has an average grade of 146 liters tonne^{-1}. The shale would be processed in a TOSCO II surface retort. Later, a modified detailed development plan was approved by the Department of the Interior for a 9063 m^3 day^{-1} operation using the Occidental modified *in situ* process. Plans were to extract 191 million m^3 of oil from a 94·5 m thick interval of oil shale that contains about 477 million m^3 of shale oil in place.

Three shafts were sunk by conventional methods. The service shaft is 10·4 m in diameter, concrete lined and 537 m deep. The concrete head frame, which was slip-formed in 10 days, is 54·2 m high. The tower-mounted friction hoist has a capacity of 27 tonnes. The produc-

Fig. 2.6. Aerial view of Cathedral Bluffs Shale Oil Co. Tract C-b. (Courtesy of Minerals Management Service.)

tion shaft is 6·8 m in diameter, concrete lined and 569 m deep. The concrete head frame, which was slip-formed in 26 days, is 95·4 m high. Two tower-mounted friction hoists will each handle two 47·6 tonne skips in balance. The ventilation–escape shaft is 4·6 m in diameter, concrete lined and 493 m deep. It has a structural steel head frame 44·2 m high and two 7·3 tonne skips (McDermott, 1980). Grouting was required to control the inflow of ground water. Maximum inflow was about 63 liters s^{-1}.

Stations were cut on five levels in the service and production shafts and 6×9 m drifts were driven to connect the shafts. The mine operated under the MSHA gassy mine classification (Stellavato, 1982).

In December 1981, the company announced that development would be halted while project plans were re-evaluated. An aerial view of the surface plant is shown in Fig. 2.6 (Rutledge, 1982).

2.10. FEDERAL LEASE TRACTS U-a AND U-b

Tracts U-a and U-b are adjoining tracts in Utah leased by the US Department of the Interior under the Prototype Oil Shale Leasing

TABLE 2.4
White River Shale Project Production Schedule

Phase	Tonnes mined[a]	Oil production, m^3
I	24 794	2 360
II	84 790	9 045
III	160 340	16 900

[a] Per stream day.

Program. In 1974, Sun Oil Co. (now Sunedco) and Phillips Petroleum Co. were successful bidders for Tract U-a with a bonus bid of $76·6 million. White River Shale Oil Corp. (Sunedco, Phillips, and Standard Oil Co. of Ohio) were successful bidders for adjoining Tract U-b with a bonus bid of $45·1 million. Rights were assigned to the Sohio Shale Oil Co. Joint development of the tracts, known as the White River Shale Project, is planned.

The tracts are near the White River in the eastern part of the Uinta Basin of Utah. Development was delayed because of legal questions about property titles. In March 1982, a court injunction order, which had suspended the lease terms, was lifted and the detailed development plan was approved by the Department of the Interior (Office of Shale Resource Applications, 1981).

A room-and-pillar mining method is planned to mine oil shale from a 17 m thick section of the Mahogany Zone which has an average grade of 117 liters tonne^{-1}. Development will be by a vertical shaft and a decline. Production ore will be transported by conveyor up the decline to a surface plant. Three phases are planned to reach commercial production of 16 900 m^3 per stream day by 1996 as shown in Table 2.4.

Plans are to use a Union B retort in Phase I. TOSCO II and/or Superior-type retorts may also be used in the later phases.

2.11. PARAHO DEVELOPMENT CORP.

From 1973 to 1976, the Paraho Development Corp. conducted the Paraho Oil Shale Demonstration at Anvil Points under lease from the Federal Government. The project was funded by a consortium of 17 companies which received a preferential license to use the Paraho process. Two Paraho retorts, a pilot retort and a semi works retort, were constructed and operated. The demonstration mine was re-opened to supply oil shale.

From 1976 to 1978, Paraho contracted with the Department of the Navy and the Department of Energy for continued research and development and for the production of 15 900 m³ of shale oil. About 91 000 tonnes of shale were mined (Office of Shale Resource Applications, 1981). A study of engineering properties of spent shale was conducted by Woodward-Clyde Consultants under a cost-sharing agreement funded by Paraho and the Bureau of Mines. Results showed that Paraho retorted shale has properties similar to a low-grade cement. It can be compacted to construct strong impervious structures. Permeability rates were less than $1 \times 10^{-6} \, \text{cm s}^{-1}$.

In 1980, Paraho and a consortium of 14 companies signed a cooperative agreement with the Department of Energy to design a 1590 m³ per stream day Paraho retort module. This design was completed.

In 1982, Paraho Development Corp. and Davy McKee Corp. submitted a proposal to the US Synthetic Fuels Corp. for a loan guarantee and possible price guarantees. The plant for the project (Paraho-Ute Project) would be on Paraho's State Lease in the Uinta Basin, Utah.

It is designed to produce 6725 m³ per stream day of hydrotreated shale oil (Anon., 1982).

2.12. COLONY DEVELOPMENT OPERATION

After completion of tests with the TOSCO semi-works plant, The Atlantic Richfield Co. (ARCO) and TOSCO continued to conduct environmental and engineering studies at the site of the Colony Mine. In 1974, an environmental impact analysis was released for a commercial shale oil complex to mine and process 59 875 tonnes day⁻¹ of oil shale and produce 7500 m³ day⁻¹ of low sulfur fuel oil and other products (Marshall, 1974). A room-and-pillar underground mine was planned in a section of the Mahogany Zone 18 m thick with an average grade of 146 liters tonne⁻¹.

A bench would be constructed in the Middle Fork Canyon of Parachute Creek and six 9 m square adits would be developed. Mine layout would be similar to that previously tested in the Colony pilot mine. A two-level system, top heading and a bench, was planned. The top heading would be 9 m high and 18 m wide. It would be followed by a 9 m bench. Pillars would be 18 m square.

A V-cut heading round would be drilled 9 m deep with a twin-boom, rubber-tired drill jumbo. Rounds would be loaded with ANFO and detonated by non-electric blasting caps. Front-end loaders of 7·6–11·5 m^3 capacity would load the shale into 72·6 tonne diesel-powered trucks. Rotary or rotary–percussive drills would be used. Bench rounds would be drilled with quarry-type rotary drills. The extraction ratio would be 60% (Trepp, 1975).

Shale would be retorted in six individual TOSCO II retorting trains. Upgrading units would process the raw pyrolysis gas and oil into low sulfur fuel oil, LPG, sulfur, ammonia, coke and other products (Office of Oil Shale Resource Applications, 1981).

In late 1974, Colony announced that construction plans were suspended. One of the reasons for the suspension was a 40% increase in construction cost estimates in just 6 months because of inflation. In 1980, Exxon Co. (USA) acquired ARCO's interest in the operation and started construction. The Department of Energy granted TOSCO a $1·1 billion loan guarantee in 1981. In 1982, Exxon announced a decision to discontinue funding for its share of the project. TOSCO exercised an option to sell its interest to Exxon. Escalating costs and the poor state of the economy were blamed for stopping the operation. At present Exxon is continuing some reclamation work and environmental monitoring.

2.13. GEOKINETICS INC.

In 1975, Geokinetics started field tests in the Uinta Basin, Utah to develop a method for true *in situ* retorting of thin beds of oil shale under shallow overburden. An ERDA (later DOE) cooperative agreement was signed in 1977. A series of experimental retorts were drilled, blasted and burned in the Mahogany Zone which is about 30 m thick and averages 60 liters kg^{-1}. The rich zone, which outcrops on site, reaches a maximum depth of 33·5 m. 27 retorts were drilled and blasted. 16 of these were burned. Total oil production was 7000 m^3 (Lekas *et al.*, 1982). A view of the test site is shown in Fig. 2.7.

In this method, the oil shale is fractured by drilling and blasting a pattern of holes from the surface. Air injection wells are drilled at one end of the retort and offgas and production wells are drilled at the other end. The oil shale is ignited at the injection end and the burn

Fig. 2.7. Aerial view of Geokinetics Inc. test site. (Courtesy of Minerals Management Service.)

front moves horizontally through the rubblized shale, heating and retorting the oil shale. The oil, which drains into the bottom of the retort, is pumped from the production wells (Mason and Sinks, 1982).

Retort 23 was an *in situ* retorting experiment sponsored by DOE and conducted by Geokinetics Inc. and the Sandia National Laboratories. It was 30 m long, 15 m wide and 7·3 m high. The overburden was 12 m thick. Shale grade averaged 108 liters tonne^{-1}. The retort contained 5830 tonnes of shale and 630 m^3 of oil. The retort was rubbled in September 1979, ignited on 15 March 1981 and shut-in on 30 June 1981. It was heavily instrumented and several operating conditions were tested. Results showed an average void of 12% and a retorting efficiency of 58% of Fischer assay (Lekas *et al.*, 1982).

Geokinetics expects to complete its research program in 1982 and then to design and construct a commercial unit (Office of Oil Shale Resource Applications, 1981).

2.14. LARAMIE ENERGY TECHNOLOGY CENTER

In 1924, the Bureau of Mines established a station in Laramie, Wyoming, to conduct petroleum research. In 1935, the station was named US Bureau of Mines Petroleum Experiment Station. When the Anvil Points project began in 1944, this station's research was expanded into oil shale, and its name was changed to the Petroleum and Oil Shale Experiment Station. After the close of Anvil Points in 1956, the name of the Station was changed again, to the Laramie Petroleum Research Center. Basic research continued on the lithology and oil yields of Green River oil shales, the composition of kerogen and shale oil and chemical reactions and conversion methods (Madonia, 1982).

In 1964, a series of true *in situ* retorting tests were begun. Laboratory and field studies were conducted to evaluate the fracturing of oil shale by electricity. Explosive-fracturing experiments were conducted near Rock Springs, Wyoming. Hydraulic fracturing treatments and combinations of hydraulic and explosive fracturing were tested.

In situ retorting experiments were conducted. These were based on the concept of drilling an injection well from the surface and a pattern of production wells ringing the injection well. After fracturing to increase permeability, the shale was ignited in the injection well and the retorting front progressed toward the production wells where the gases and liquid products were removed. Two surface retorts, one of 9 tonnes capacity, the other of 150 tonnes capacity, were constructed and operated at Laramie to study the operating variable of the *in situ* method. The *in situ* research program was continued when the Center was transferred to the Energy Research and Development Agency (ERDA) in 1974 and later to the Department of Energy (DOE) in 1977. The name was changed to the Laramie Energy Technology Center.

Rock Springs site 9 is an example of the *in situ* retorting experiments. A zone of oil shale 12·2 m thick in the Tipton Member of the Green River Formation was tested. The depth was 41·8 m. The shale grade was 97 liters tonne^{-1}. A square pattern, 21·4 m on a side, was drilled with eight production wells on the periphery, an injection well in the center of the square and four monitoring wells. On 5 April 1976 a slurried explosive was charged to the injection well and detonated. The oil shale was ignited and the advance of the combustion front was monitored as air was injected. After 150 days of operation 11·3 m^3 of shale oil was produced. This represented about 1% of the shale oil in place. Oil migration beyond the periphery of the pattern was noted (Long *et al.*, 1977).

In 1979, the center was assigned the role as lead laboratory for the Fossil Energy Oil Shale Program. In addition to in-house research it

Historical perspective 65

managed government–industry cooperative agreements with Occidental Oil Shale Inc., Geokinetics Inc., Talley Energy Systems Inc., Equity Oil Co., Illinois Institute of Technology Research Institute, Institute of Gas Technology, Dow Chemical Co., the National Bureau of Standards and others (Laramie Energy Technology Center, 1979).

2.15. UNIVERSITY OF WYOMING

The University of Wyoming, in cooperation with the Laramie Energy Technology Center and the Office of the Naval Petroleum and Oil Shale Reserve, is conducting a comprehensive mechanical characterization program on Green River oil shales. Tests include uniaxial compression, creep and relaxation behavior, tensile strength, strain rate effects, fracture strengths, fatigue resistance and three-dimensional non-linear properties (Chong *et al.*, 1980*a*).

Results from these experiments are used in deriving prediction equations for the three-dimensional non-linear stiffness coefficients based on organic volume and stress level as independent variables. These coefficients are used in mine design fragmentation models and methods for increasing the permeability of shale in-place for *in situ* retorting.

It was found that compressive fracturing strength is strongly dependent upon organic volume and on strain rate. A linear relationship was observed between the ultimate stress, organic volume and the logarithmic strain rate (Chong *et al.*, 1980*b*).

2.16. UNION PARACHUTE CREEK SHALE OIL PROGRAM

In 1978, the Union Oil Co. announced plans to develop a commercial oil shale plant on its Parachute Creek property. Phase I consists of an underground room-and-pillar mine, a Union B retort and an upgrading facility. Mine production of 11 350 tonnes day^{-1} would be processed to produce 1590 m^3 day^{-1} of high quality syncrude.

The first mine is located near the mining operations of 1955–1958. The original bench was enlarged and adits developed in the Mahogany Zone.

A two-level system, top heading and a bench, would be used to mine the 18·3 m thick zone. Rooms would be 16·8 m wide and pillars would be 30·5 m long and 15·2 m wide. The extraction ratio would be 69%.

Conventional drilling and blasting methods would be used. Broken shale would be loaded by 11.5 m^3 front-end loaders into 45-m^3 dump trucks. Primary and secondary crushers would be underground. Shale would be crushed to a size range of 1·27–5 cm. It would then be transported by conveyer belt to the retort (Randle and McGunegle, 1982).

In 1980, under a cooperative research program with the Bureau of Mines, an incline was driven from the old adit and geotechnical and rock mechanics tests were conducted to determine the physical properties of the rock and to select a roof horizon for the mine.

Construction mining was completed in 1981. Underground space for facilities has been excavated and construction of the retort is underway.

In 1981, the DOE awarded Union a $400 million purchase agreement for delivery of $480 \text{ m}^3 \text{ day}^{-1}$ of military aircraft turbine fuel and $1100 \text{ m}^3 \text{ day}^{-1}$ of diesel to the Department of Defense when actual production begins.

Phase I is scheduled for completion in 1983. Phase II is scheduled to increase production to $7950 \text{ m}^3 \text{ day}^{-1}$ by 1990 and to $14\,300 \text{ m}^3 \text{ day}^{-1}$ by 1993.

2.17. US BUREAU OF MINES HORSE DRAW SHAFT

From 1977 to 1978, the Bureau of Mines, under a cooperative research agreement with the Bureau of Land Management, drilled and cased a 3 m diameter shaft to a depth of 723 m on Horse Draw in the Piceance Creek Basin. The lower 488 m of the shaft penetrates virtually the full thickness of oil shale and accessory saline minerals (nahcolite and dawsonite) in the Parachute Creek member of the Green River Formation. The purpose of the project was to conduct mining environmental research, prototype mining tests and to obtain bulk samples of oil shale, nahcolite and dawsonite for processing experiments.

The shaft was equipped, two stations were cut and one level was developed. With the establishment of the DOE, plans for prototype mining tests were cancelled and research focused on environmental research. The mine was classified gassy by MSHA. In 1979, a cooperative research agreement was signed with the Multi Mineral Corp. for a two-phase research program of mining, environmental and geotechnical tests. Multi Mineral developed three levels, a stope, three raises and two sub-levels. All development was in the saline zone at depths ranging from 560 to 695 m.

2.17.1. Mining Research

Mine development drifts ranged in size from $3 \times 3 \cdot 6$ m to $2 \cdot 1 \times 7 \cdot 3$ m. A total of 17 780 tonnes was mined. Standard mining practices were followed. A typical round on the upper level consisted of a 20 hole V-cut pattern in a $2 \cdot 1$ m high by $7 \cdot 3$ m wide drift. Average advance was 2 m. The powder factor was $0 \cdot 72$ kg tonne^{-1}. A typical round on the lower level consisted of a 40 hole parallel hole cut pattern with four unloaded drill holes in the center. The face was 3 m high and $3 \cdot 6$ m wide. Advance was $1 \cdot 9$ m. The powder factor was $1 \cdot 9$ kg tonne^{-1}. The vertical crater retreat method was used to mine the stope on the lower level. Stope block dimensions were $19 \cdot 8$ m long by $11 \cdot 9$ m wide by $26 \cdot 9$ m high. Total stope tonnage was 15 979 tonnes. The powder factor was $0 \cdot 74$ kg tonne^{-1}. Fragmentation ranged from 80% minus $50 \cdot 8$ cm to 80% minus $35 \cdot 6$ cm. Large slabs, which fell from the stope back after blasting, required secondary blasting in the draw points.

2.17.2. Rock Mechanics Research

Core holes were drilled to determine stratigraphy, assay values, mechanical properties of the shale, and methane gas content. Mechanical properties determined for oil shale from two of the levels and for nahcolite are summarized in Table 2.5. On the upper level, four rooms 3 m high and ranging in width from $3 \cdot 6$ to $9 \cdot 1$ m were mined around a

TABLE 2.5
Average Mechanical Properties of Nahcolite and Oil Shale

Property	L4B shale	L4B nahcolite	R3 shale
Tensile strength (MPa)	—	—	$5 \cdot 4$
Compressive strength (MPa)	$56 \cdot 1$	$96 \cdot 0$	$46 \cdot 1$
Modulus of elasticity (MPa)	9 929	36 337	3 516
Poisson's ratio	$0 \cdot 76$	$0 \cdot 31$	$0 \cdot 33$

central pillar 18·3 m square. *In situ* stresses and physical properties were determined and deformations of the pillar and roof were measured. After the mining program was completed in 1982, the project was placed on a stand-by basis. The Bureau of Mines is continuing to monitor methane gas at the shaft.

2.18. EQUITY PROJECT

From 1962 to 1971, the Equity Oil Co. conducted laboratory studies and field tests of a true *in situ* method for extraction of shale oil from the Leached Zone of the Green River Formation. The field tests were conducted at a site on Black Sulfur Creek in the Piceance Creek Basin.

The BX *in situ* Project is designed to extract shale oil from the leached zone which is located between the Mahogany Zone and the underlying saline zone. The strata in the leached zone which is permeable and porous, contain highly saline water. The leached zone is 165 m thick with an average grade of 100 liters tonne^{-1}. Injection and production wells were drilled and heated natural gas was injected. Saturated steam injection was also tested (McCoy, 1979).

From 1979 to 1981, Equity conducted field tests under a cost-sharing contract with the DOE. Eight injection wells, five production wells and three temperature monitoring wells were drilled. Between September 1979 and February 1981, 99 400 m^3 of water as steam were injected at an average wellhead temperature of 318°C and pressure of 9053 kPa. During this period, 83 400 m^3 of fluid was produced. The first retorted oil was produced in October 1980. Through February 1981, a total of 7·3 m^3 of oil was produced (Dougan and Dockter, 1981).

REFERENCES

Agapito, J. F. T. (1974). Rock mechanics applications to the design of oil shale pillars, *SME/AIME Preprint No. 74-AIME-26*.

Anon. (1982). Paraho proceeds with the shale oil facility, *Mining Record*, 9 June.

Berry, K. L., Hutson, R. L., Sterrett, J. S. and Knepper, J. C. (1982). Modified *in situ* retorting results of two field retorts, in: *Fifteenth Oil Shale Symposium Proceedings*, ed. J. H. Gary, Colorado School of Mines Press, Golden, pp. 385–96.

Carver, H. E. (1964). Conversion of oil shale to refined products, in: *First Symposium on Oil Shale*, Vol. 59, Colorado School of Mines, Golden, pp. 19–38.

Carver, H. E. (1965). Oil shale mining: a new possibility for mechanization, in: *Second Symposium on Oil Shale*, Vol. 60, ed. S. W. Spear, Colorado School of Mines, Golden, pp. 215–34.

Chong, K. P., Uenishi, K., Smith, J. W. and Munari, A. C. (1980*a*). Non-linear three dimensional mechanical characterization of Colorado oil shale, *Int. J. Rock Mech. Sci. & Geomech. Abstr.*, **17,** 339–47.

Chong, K. P., Hoyt, P. M., Smith, J. W. and Paulsen, B. Y. (1980*b*). Effects of strain rate on oil shale fracturing, *Int. J. Rock Mech. Sci. & Geomech. Abstr.*, **17,** 35–43.

Dougan, P. M. and Dockter, L. (1981). BX *in situ* Oil Shale Project, in: *Fourteenth Oil Shale Symposium Proceedings*, ed. J. H. Gary, Colorado School of Mines Press, Golden, pp. 118–27.

East, J. H. and Gardner, E. D. (1964). Oil shale mining, Rifle, Colorado, 1944–56, *US Bureau of Mines Bulletin 611.*

Hamilton, W. H. (1965). Preliminary design and evaluation of an alkirk oil shale miner, in: *Second Symposium on Oil Shale*, Vol. 60, ed. S. W. Spear, Colorado School of Mines, Golden, pp. 235–65.

Laramie Energy Technology Center (1979). *US Department of Energy Annual Report, 1979*, pp. 38–9.

Lekas, J. M., Tyner, C. E., Parrish, R. L. and Major, B. H. (1982). Sandia/Geokinetics retort 23: a horizontal *in situ* retorting experiment, in: *Fifteenth Oil Shale Symposium Proceedings*, ed. J. H. Gary, Colorado School of Mines Press, Golden, pp. 370–84.

Long, Jr., A., Merriam, N. W. and Mones, C. G. (1977). Evaluation of Rock Springs site 9 *in situ* oil shale retorting experiment, in: *Tenth Oil Shale Symposium Proceedings*, ed. J. H. Gary, Colorado School of Mines Press, Golden, pp. 120–35.

Madonia, S. T. (1982). At the Laramie Center, change is the only certainty, *Shale Country*, **4,** 12–14.

Marshall, P. W. (1974). Colony development operation room-and-pillar oil shale mining, in: *Seventh Oil Shale Symposium*, Vol. 69, ed. J. H. Gary, Colorado School of Mines, Golden, pp. 171–84.

Mason, G. M. and Sinks, D. J. (1982). Results of post burn caving at Geokinetics retort 16, Uinta County, Utah, in: *Fifteenth Oil Shale Symposium Proceedings*, ed. J. H. Gary, Colorado School of Mines Press, Golden, pp. 361–9.

Matzick, A., Dannenberg, R. O. and Guthrie, B. (1960). Experiments in crushing Green River oil shale, *US Bureau of Mines Report of Investigation 5563.*

McCoy, D. E. (1979). Oil shale, *Synthetic Fuels*, **16,** 2-6–2-9.

McDermott, W. F. (1980). Shale oil—a mining frontier, in: *1980 Mining Convention Session Papers*, American Mining Congress, Washington DC.

Merrill, R. H. (1954). Design of underground mine openings, oil shale mine, Rifle, Colo., *US Bureau of Mines Report of Investigations 5089.*

70 Stephen Utter

Obert, L. and Merrill, R. H. (1958). Oil shale mine, Rifle, Colo.: a review of design factors, *US Bureau of Mines Report of Investigations 5429.*

Office of Shale Resource Applications (1981). *Oil Shale Projects,* US Department of Energy.

Panel on Rock Mechanics Research Requirements (1981). (US National Committee of Rock Mechanics, Assembly of Mathematical and Physical Sciences, National Research Council.) *Rock Mechanics Research Requirements for Resource Recovery, Construction and Earthquake Hazard Reduction,* National Academy Press, Washington DC, p. 1.

Randle, A. C. and McGunegle, B. F. (1982). Union Oil Company's Parachute Creek shale oil program, in: *Fifteenth Oil Shale Symposium Proceedings,* ed. J. H. Gary, Colorado School of Mines Press, Golden, pp. 224–30.

Ridley, R. D. (1978). Progress in Occidental's shale oil activities, in: *Eleventh Oil Shale Symposium Proceedings,* ed. J. H. Gary, Colorado School of Mines Press, Golden, pp. 169–75.

Romig, B. A. (1981). Progress of Phase II of the DOE/OOSI cooperative agreement, in: *Fourteenth Oil Shale Symposium Proceedings,* ed. J. H. Gary, Colorado School of Mines Press, Golden, pp. 91–117.

Russell, P. L. (1980). *History of Western Oil Shale,* Center for Professional Advancement, East Brunswick, New Jersey.

Rutledge, P. A. (1982). The prototype oil shale program—an update, in: *Fifteenth Oil Shale Symposium Proceedings,* ed. J. H. Gary, Colorado School of Mines Press, Golden, pp. 210–23.

Sellers, J. B., Haworth, G. R. and Zambas, P. G. (1972). Rock mechanics research on oil shale mining, *Trans. SME/AIME,* **252,** 222–32.

Stellavato, N. (1982). Results of the geologic mapping program during shaft sinking and subsequent station development at C-b Tract, in: *Fifteenth Oil Shale Symposium Proceedings,* ed. J. H. Gary, Colorado School of Mines Press, Golden, pp. 115–35.

Trepp, D. W. (1975). Mining of oil shale commercially by the room-and-pillar method, *SME/AIME Preprint No. 75-Au-326.*

Watson, G. H., Ziemba, E. A., Bissery, P., Namy, D., Griffis, R. L. and Nicholson, D. E. (1982). The filling of oil shale mines with spent shale ash, ash characteristics and grout development, in: *Fifteenth Oil Shale Symposium Proceedings,* ed. J. H. Gary, Colorado School of Mines Press, Golden, pp. 397–410.

Zambas, P. G., Haworth, G. R., Brackenbusch, F. W. and Sellers, J. B. (1972). Large scale experimentation in oil shale, *Trans. SME/AIME,* **252,** 283–9.

Chapter 3

The Continuum Theory of Rock Mechanics

W. F. Chen

Professor and Head of Structural Engineering, Purdue University, West Lafayette, Indiana, USA

SUMMARY

This chapter presents a simple, concise and reasonably comprehensive introduction to the mechanics of rock that sets the stage for discussions of the subsequent chapters. The subject matter of this chapter has been divided into two major parts. The first part includes definitions and concepts of stress and strain, and basic equations of mechanics of solids. The second part includes the general technique used in the constitutive modeling of rocks in general and oil shales in particular, together with a discussion of the properties of the materials involved and the way in which they are idealized to form a basis for the mathematical theories of elasticity, plasticity and viscosity. Here, our emphasis is placed on the assumptions involved in the theory and the way in which they affect the solutions, rather than in the study of special problems. Special treatments of problems and material models are given in details in the subsequent chapters.

3.1. INTRODUCTION

3.1.1. Rock Mechanics

The mechanics of oil shale is a branch of the science of rock mechanics. The word mechanics implies a mathematical formulation of the problem and of the basic equations to be used in its solution. In the continuum theory of rock mechanics that includes the mathematical theories of elasticity, plasticity and viscosity, the basic sets of equations are (1) equations of equilibrium or motion; (2) conditions of geometry or compatibility of strains and displacements; (3) material constitutive laws or stress–strain relations.

Clearly, both the equations of equilibrium and the equations of compatibility are independent of the characteristics of the material. They are valid for metals, soils as well as rock or oil shale materials. The differentiating feature of various material behaviors is accounted for in the material *constitutive relationships* which idealize the behavior of actual materials. Once the material stress–strain relationship is known, equations of equilibrium and of compatibility are used to determine the state of stress or strain when an idealized body is subjected to prescribed forces.

Although the art of rock engineering is old and in many respects well established (Farmer, 1968), the science of rock mechanics is relatively recent (Jaeger, 1969; Asszonyi and Richter, 1979). At present, the practical engineer bases his design primarily on experience and case history, while the academic engineer bases his design on the assumption of rock as a simple brittle elastic solid, adjusting the final design empirically to take account of the actual field conditions that will affect the rock properties in a specific state.

The earliest studies in rock mechanics were based on classical soil mechanics (Asszonyi and Richter, 1979). This approach was found unsuitable for a proper interpretation of the true causes of phenomena in some rock mechanics problems. They could not, for instance, account for time-dependent rock behavior. A substantial advance in the mathematical description of rock mechanics phenomena as well as in identifying the causes for them was made possible in recent years by the application of the mathematical theories of elasticity and viscosity. At present, linear elastic or viscoelastic analysis is used widely for the interpretation of different phenomena involving rocks. This theoretical mechanics approach has partly clarified a number of controversial problems but the linear elastic assumption has a limited scope of applications.

With the present developments in computational techniques like the finite element method, a more general theory of *continuum mechanics*, like hyper- or hypo-elasticity, plasticity and viscoplasticity, must be developed to describe the very complex behavior of rocks involving phenomena like inelasticity, cracking, time dependency and discontinuity.

3.1.2. Scope

The development of material models for oil shales is a particularly challenging field in rock mechanics (Reed, 1966; Sellers *et al.*, 1972). The special properties of oil shales are nonlinear functions of organic volume and stress levels among others and they must be considered for all stages of the computational process, including static stress analysis for mining design, fracture propagation, temperature and pressure effects, strain-rate effects and explosive dynamic stress analysis for *in situ* fragmentation. All these aspects will be discussed in the chapters that follow. In this chapter, we confine our presentation to the most fundamental aspects of rock mechanics, that is, the general technique used in the discussion of stress–strain laws based on the theories of elasticity, plasticity and viscosity leading from the modeling of rocks to modeling the layered transversely isotropic oil shales.

This chapter presents a simple, concise and reasonably comprehensive introduction to the mechanics of rock that will set the stage for the subsequent chapters. The reader is assumed to be familiar with the more elementary aspects of stress analysis and some basic concepts of elasticity, viscosity and plasticity. A recent comprehensive book entitled *Constitutive Equations for Engineering Materials* by Chen and Saleeb (1982, 1985) may prove helpful in this respect as an introduction to the constitutive modeling of engineering materials. Although the basic concepts of stress analysis and strain analysis can be found in a number of standard books, for completeness, some of the developments involving stress and strain transformations in three dimensions are collected here in a form which is keyed directly to the main exposition of the present chapter and the chapters that follow.

The subject matter of this chapter has been divided into two major parts. The first part includes definitions and concepts of stress and strain and basic equations of mechanics of solids. The second part includes the general technique used in the constitutive modeling of rocks in general and oil shales in particular, together with a discussion of the properties of the materials involved and the way in which they are idealized to form a basis for the mathematical theories of elasticity, plasticity and viscosity. Here, our emphasis is placed on the assumptions involved in the theory and the way in which they affect the solutions, rather than on the study of special problems. Special treatments of problems and material models are given in detail in the subsequent chapters.

3.1.3. Notations

For purposes of generalization, symbolic forms of the equations, using index notation and summation convention, have been used. The notations used in the text are those conventionally used in continuum mechanics.

For computer programming purposes, matrix notations are most convenient. Thus, for specific material models, the numerical procedures for a solution have been illustrated in matrix notations in a cartesian coordinate system.

In this book, we restrict ourselves to a *right-handed cartesian coordinate* system with a set of three mutually orthogonal x, y and z axes. For future convenience, the axes are more conveniently named as x_1, x_2 and x_3 for a general discussion, rather than the more familiar x, y and z for a specific engineering application. Herein, they will be used interchangeably.

In continuum mechanics, it is conventional to use tensile stress as a positive quantity and compressive stress as a negative quantity. Problems of engineering analysis and design in rock, however, are generally concerned with compressive stresses in most cases. For convenience, therefore, the continuum mechanics sign convention is followed for general discussion but is reversed in the text when a specific application is made.

Notations and symbols in the text are explained when they first occur. Detailed discussion of index notation and summation convention can be found in the book by Chen and Saleeb (1982).

3.2. STRESSES IN THREE DIMENSIONS

In rock mechanics theory, the rocks or oil shales are regarded as *continua* as a rule. This permits the use of the notions of stress and strain. The relationship between stress and strain in an idealized material forms the basis of the mathematical theories of elasticity, plasticity and viscosity which can in turn be applied to actual oil shale materials to estimate stress or strain in a specified force field.

An understanding of stress and strain and principles of stress and strain analysis is therefore essential to the engineer modeling the behavior of, and designing structures in rock in general and oil shale in particular. These principles are summarized briefly here and in the following section. Details of this development are given elsewhere (Chen and Saleeb, 1982).

3.2.1. Definitions and Notations

The analysis of stress is essentially a branch of statics which is concerned with the detailed description of the way in which the stress at a point of a body varies. In two dimensions, this involves only elementary trigonometry and the use of Mohr's circle is found to be most convenient. In three dimensions, however, index notation is preferred for the calculation of stresses across any plane at the point. Herein, only the three-dimensional case will be worked out.

The stress at a point P in a solid body may be obtained by considering a small plane area δA at random orientation with a unit normal vector n_i originating at P (Fig. 3.1). Then, if δF_i is the resultant of all the forces exerted on δA, the limit of the ratio $\delta F_i/\delta A$ as δA tends to zero is called the *stress vector* T_i at the point P across the plane whose

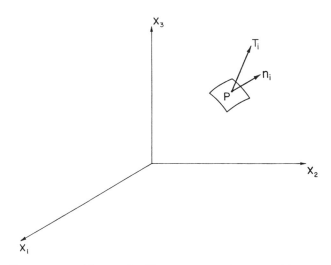

Fig. 3.1. Stress vector T_i at point P on an area element with a unit normal vector n_i.

76   W. F. Chen

normal vector is n_i, that is

$$T_i^{(n)} = \lim_{\delta A \to 0} \frac{\delta F_i}{\delta A} \tag{3.1}$$

The dimensions of $T_i^{(n)}$ are force per unit area.

The *state of stress* at the point P is completely specified or defined if we know all the values of $T_i^{(n)}$ corresponding to various n_i. Since there are an infinite number of n_i through the point P, we shall have an infinite number of values of $T_i^{(n)}$ which, in general, differ from each other. Thus, the infinite number of values of $T_i^{(n)}$ are needed in order to characterize the state of stress at the point. It turns out, however, that these stress vectors are related to each other through Newton's law of motion or equilibrium. In fact, the value of $T_i^{(n)}$ for any n_i can be calculated once the stress vectors $T_i^{(1)}$, $T_i^{(2)}$ and $T_i^{(3)}$ are known for the three mutually perpendicular area elements whose normals are in the direction of the coordinate axes x, y, z or equivalently x_1, x_2 and x_3, respectively. This is known as *Cauchy's formula*; it has the simple form

$$T_i^{(n)} = T_i^{(1)} n_1 + T_i^{(2)} n_2 + T_i^{(3)} n_3 \tag{3.2}$$

where the stress vector $T_i^{(n)}$ at the point P with a unit normal $n_i = (n_1, n_2, n_3) = (l, m, n)$ is expressed as a linear combination of the three stress vectors on the plane-area elements perpendicular to the three coordinates at the point. Therefore, it is clear that the three stress vectors $T_i^{(1)}$, $T_i^{(2)}$, $T_i^{(3)}$ define the state of stress at the point completely.

Since T_i is a vector quantity, it can be more conveniently represented by three components: A normal stress component and two tangential stress components. For example, the stress vector $T_i^{(2)}$ associated with the coordinate plane area y or x_2 has three components: normal stress, σ_y or σ_{22}, and shear stresses, τ_{yx} and τ_{yz} or σ_{21} and σ_{23}, in the direction of the three coordinate axes y, x, and z or x_2, x_1 and x_3, respectively, as shown in Fig. 3.2, or

$$T_i^{(2)} = (\sigma_{21}, \sigma_{22}, \sigma_{23}) = (\tau_{yx}, \sigma_y, \tau_{yz}) \tag{3.3}$$

in which the index notation of σ_{22} for normal stress and σ_{21}, σ_{23} for sharing stresses, and the engineering notation of σ_y for normal stress and τ_{yx}, τ_{yz} for shearing stresses will always be used interchangeably. In the former case, the first suffix denotes the direction of the normal of the small area δA and the second suffix the direction in which the

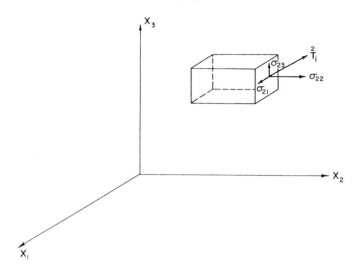

Fig. 3.2. Typical components of stress tensor σ_{ij}.

component acts. In the latter case, only one suffix is used for the normal stress since the direction of the component is the same as that of the normal to the surface. If the normal component of the stress across a surface is positive, it is called a *tensile* stress and if it is negative it is called a *compressive* stress.

In the same way the stress at P across a plane whose normal is in the direction of x_1 or x will have components

$$T_i^{(1)} = (\sigma_{11}, \sigma_{12}, \sigma_{13}) = (\sigma_x, \tau_{xy}, \tau_{xz}) \tag{3.4}$$

and that across a plane whose normal is in the direction x_3 or z will have components

$$T_i^{(3)} = (\sigma_{31}, \sigma_{32}, \sigma_{33}) = (\tau_{zx}, \tau_{zy}, \sigma_z) \tag{3.5}$$

The nine quantities in eqns 3.3–3.5 which define the three stress vectors $T_i^{(1)}$, $T_i^{(2)}$, and $T_i^{(3)}$, are called the components of the tensor σ_{ij} at the point P, which are collected in eqn 3.6:

$$\sigma_{ij} = \begin{bmatrix} T_i^{(1)} \\ T_i^{(2)} \\ T_i^{(3)} \end{bmatrix} = \begin{bmatrix} \sigma_{11} & \sigma_{12} & \sigma_{13} \\ \sigma_{21} & \sigma_{22} & \sigma_{23} \\ \sigma_{31} & \sigma_{32} & \sigma_{33} \end{bmatrix} = \begin{bmatrix} \sigma_x & \tau_{xy} & \tau_{xz} \\ \tau_{yx} & \sigma_y & \tau_{yz} \\ \tau_{zx} & \tau_{zy} & \sigma_z \end{bmatrix} \tag{3.6}$$

The components of σ_{ij} are shown in the positive directions in Fig. 3.2 referred to the x_i coordinate system. The dual notation in eqn 3.6, where σ represents a normal component of stress and τ a shearing component of stress, known as von Karman's notation, is used widely in practice as well as in the well known book by Timoshenko and Goodier (1951).

The nine components in eqn 3.6 are the components of a mathematical entity called a *second-order tensor*. A second-order tensor is defined completely by three vectors just as a vector is defined completely by three scalars. A vector is therefore called a *first-order tensor*. Tensor analysis is much used in developing the higher parts of the continuum theory. The main change, from the present point of view, is that the notation x, y, z for the coordinates is replaced by x_1, x_2, x_3, so that they are specified by the numbers 1, 2, 3. The nine components of the *stress tensor* is denoted by σ_{ij}, with values of i and j running from 1 to 3.

3.2.2. Cauchy's Formulas, Index Notation and Summation Convention

Unlike the two-dimensional stress analysis which can be conveniently described by the simple geometrical construction of Mohr's circle, many of the difficulties of the three-dimensional stress analysis are caused by the complication of three-dimensional geometry. This can best be handled by the use of index notation and summation convention. These relations are summarized briefly in the following.

From the consideration of equilibrium of forces acting on an arbitrary small volume of material, it can be shown that the components of the stress vector $T_i^{(n)}$, or simply T_i, in the x_i coordinate direction can be written as

$$T_i = \sum_{j=1}^{3} \sigma_{ji} n_j \tag{3.7}$$

where we take it for granted that, since the subscript i is unspecified, the equation must hold for each of the three possible values of this subscript. The index i is therefore called a *free index*. The index j in eqn 3.7 is, however, a *dummy* subscript because of the fact that the particular letter used in this subscript is not important; thus,

$$T_i = \sum_{j=1}^{3} \sigma_{ji} n_j = \sum_{m=1}^{3} \sigma_{mi} n_m \tag{3.8}$$

The continuum theory of rock mechanics 79

To allow for further brevity, we adopt the following *summation convention*: when a subscript occurs twice in the same term, it will be understood that the subscript is to be summed from 1 to 3. Thus, eqn 3.8 can be abbreviated as

$$T_i = \sigma_{ji}n_j = \sigma_{mi}n_m \tag{3.9}$$

where the free index must occur precisely once in each term of the expression or equation. It represents a vector with three components, $i = 1, 2, 3$, respectively. The dummy index j or m is to be summed from 1 to 3. The dummy index may or may not occur precisely twice in any other term. Note that if a subscript occurs more than twice in one term of an expression or equation, it is a mistake in the use of the index notation and summation convention.

From the consideration of equilibrium of moments of a material element, it can be shown that the stress tensor σ_{ij} is symmetric, i.e.

$$\sigma_{ij} = \sigma_{ji} \tag{3.10}$$

Thus, eqn 3.7 can be rewritten somewhat more conveniently and conventionally as

$$T_i = \sigma_{ij}n_j \tag{3.11}$$

and it therefore follows that T_i for any n_i can be calculated from a knowledge of the nine basic quantities as given in eqn 3.6.

The stress vector T_i acting on the area element n_i at P can be resolved into normal stress components $\sigma_n n_i$ and shear-stress component S_i, as shown in Fig. 3.3.

The magnitude of the normal stress σ_n is clearly $T_i n_i$, so that the normal-stress component can be expressed in terms of the stress tensor σ_{ij} through eqn 3.11 as

$$\sigma_n = T_i n_i = \sigma_{ij}n_i n_j \tag{3.12}$$

The magnitude of the shearing stress component S_n is given by

$$S_n^2 = S_i S_i = T_i T_i - \sigma_n^2 \tag{3.13}$$

Using eqns 3.11 and 3.12, the above equation becomes

$$S_n^2 = \sigma_{ij}\sigma_{ik}n_j n_k - (\sigma_{ij}n_i n_j)^2 \tag{3.14}$$

Equations 3.12 and 3.14 allow the determination of the normal and shearing components of the stress vector T_i acting on an arbitrary plane n_i (Fig. 3.3). These equations are called *Cauchy's formulas*. The

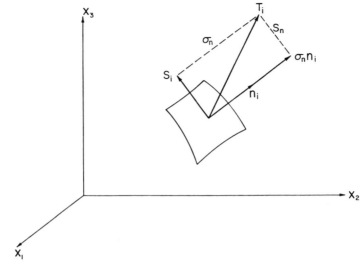

Fig. 3.3. Normal and shearing stress components of stress vector T_i acting on an arbitrary area element n_i.

vector $\sigma_n n_i$ is in the direction of the normal n_i, and the vector S_i lies in the plane formed by the two vectors T_i and n_i, or

$$S_i = T_i - \sigma_n n_i \tag{3.15}$$

3.2.3. Principal Axes of Stresses

Cauchy's formulas, eqns 3.12 and 3.14, give a complete picture of the variation of stress at P with direction, showing that both σ_n and S_n are related to the direction cosines $n_i = (n_1, n_2, n_3)$. If σ_n is differentiated with respect to n_i it can be shown that stationary values of normal stress σ_n occur when $S_n = 0$. This condition implies that the stress vector T_i at P is perpendicular to the area element with n_i, that is, $T_i = \sigma n_i$, or using eqn 3.11, we have

$$\sigma_{ij} n_j = \sigma n_i \tag{3.16}$$

which can be written

$$(\sigma_{ij} - \sigma \delta_{ij}) n_j = 0 \tag{3.17}$$

where δ_{ij} is the Kronecker delta; it has the value 1 if $i = j$ and 0 if $i \neq j$.

The above are a set of three linear homogeneous equations for (n_1, n_2, n_3). This set possesses solutions if and only if the determinant of the coefficient vanishes

$$\det |\sigma_{ij} - \sigma\, \delta_{ij}| = 0 \tag{3.18}$$

Expanding eqn 3.18 leads to the *characteristic equation*

$$\sigma_3 - I_1\sigma^2 + I_2\sigma - I_3 = 0 \tag{3.19}$$

where I_1, the sum of the diagonal terms of σ_{ij}, is

$$I_1 = \sigma_{ii} = \sigma_{11} + \sigma_{22} + \sigma_{33} \tag{3.20}$$

I_2, the sum of the principal two-rowed minors of the determinant of σ_{ij}, is

$$I_2 = \tfrac{1}{2}(I_1^2 - \sigma_{ij}\sigma_{ji}) = \begin{vmatrix} \sigma_{22} & \sigma_{23} \\ \sigma_{32} & \sigma_{33} \end{vmatrix} + \begin{vmatrix} \sigma_{11} & \sigma_{13} \\ \sigma_{31} & \sigma_{33} \end{vmatrix} + \begin{vmatrix} \sigma_{11} & \sigma_{12} \\ \sigma_{21} & \sigma_{22} \end{vmatrix} \tag{3.21}$$

and I_3, the determinant of σ_{ij}, is

$$I_3 = \tfrac{1}{6}(2\sigma_{ij}\sigma_{jk}\sigma_{ki} - 3I_1\sigma_{ij}\sigma_{ji} + I_1^3) = \det |\sigma_{ij}| \tag{3.22}$$

The quantities I_1, I_2, I_3 are called the *invariants of the stress tensor* σ_{ij}. As their name implies, these invariants are unaffected by rotation of the coordinate axes.

Equation 3.19 has three roots, $\sigma_1, \sigma_2, \sigma_3$. Using these roots, eqn 3.17 can be solved to give three directions of n_i at right angles to each other on which act maximum, intermediate and minimum normal stress and zero shear stress. These three directions are called *principal directions* at P, and the corresponding normal stresses are called *principal stresses*. A right-hand coordinate system can be oriented to line up with the principal directions at P; such a coordinate system is called *principal axes* for the stress state at P.

Once this principal orientation is known, it is convenient to use the principal axes as the axes of reference. Thus, if in Cauchy formulas 3.12 and 3.14, the principal stresses $\sigma_1, \sigma_2, \sigma_3$ are substituted for the stress tensor σ_{ij}, the normal and shear stress in the direction n_i have the simple form:

$$\begin{aligned} \sigma_n &= \sigma_{ij}n_in_j = \sigma_1 n_1^2 + \sigma_2 n_2^2 + \sigma_3 n_3^2 \\ S_n^2 &= (\sigma_1 n_1)^2 + (\sigma_2 n_2)^2 + (\sigma_3 n_3)^2 - (\sigma_1 n_1^2 + \sigma_2 n_2^2 + \sigma_3 n_3^2)^2 \end{aligned} \tag{3.23}$$

An *octahedral plane* is a plane whose normal makes equal angles with each of the principal axes of stress; it follows that the normal and shear stresses on the octahedral plane, whose normal n_i is defined by

$$n_i = (n_1, n_2, n_3) = \frac{1}{\sqrt{3}}(1, 1, 1) \qquad (3.24)$$

are given by Cauchy's formulas, eqns 3.23 and 3.24, in the form

$$\sigma_{\text{oct}} = \tfrac{1}{3}(\sigma_1 + \sigma_2 + \sigma_3) = \tfrac{1}{3}I_1$$
$$\tau_{\text{oct}}^2 = \tfrac{1}{9}[(\sigma_1 - \sigma_2)^2 + (\sigma_2 - \sigma_3)^2 + (\sigma_3 - \sigma_1)^2] \qquad (3.25)$$

Once the principal values of stress are known, the Mohr circle construction can proceed in the usual manner. That is, on the normal stress axis, lengths are marked off to represent σ_1 and σ_2, σ_2 and σ_3, σ_3 and σ_1, respectively. From this construction of Mohr's circles, it follows that there are three *principal shear stresses*, τ_1, τ_2, τ_3, corresponding to the radii of the three Mohr's circles with

$$\tau_1 = \tfrac{1}{2}|\sigma_2 - \sigma_3|, \qquad \tau_2 = \tfrac{1}{2}|\sigma_1 - \sigma_3|, \qquad \tau_3 = \tfrac{1}{2}|\sigma_1 - \sigma_2| \qquad (3.26)$$

each of which occurs on an area element whose unit normal makes an angle of 45° with each of the corresponding principal axes. The largest numerical value of the principal shears is called the *maximum shear stress*

$$\tau_{\text{max}} = \tfrac{1}{2}|\sigma_1 - \sigma_3| \quad \text{or} \quad \tau_{\text{max}} = \max(\tau_1, \tau_2, \tau_3) \qquad (3.27)$$

for $\sigma_1 > \sigma_2 > \sigma_3$. The maximum shear stress is half the difference of the major (greatest) and minor (least) principal stresses and occurs on a plane whose normal bisects the major and minor principal axes. Thus, if a material such as rock is known to fail in shear, it will obviously tend to fail in the plane of maximum shear. It is a relatively easy matter to define the likely direction and magnitude of failure from a given state of stress at a point. The importance of this will be seen in Section 3.6, when the failure criteria of rocks are given, and later in Chapter 5, when the tendency of oil shale to fail in shear is discussed. Further discussion on this can be found in the book by Jaeger (1969).

3.2.4. Deviatoric Stress

The stress state σ_{ij} can be decomposed into two stress states, one of which represents *pure shear* s_{ij} and the other *hydrostatic tension* $\sigma_m \delta_{ij}$.

The continuum theory of rock mechanics 83

Alternatively, the deviation stress tensor s_{ij} is defined by

$$s_{ij} = \sigma_{ij} - \sigma_m \delta_{ij} \qquad (3.28)$$

which is a state of pure shear.

Since subtracting a constant normal stress in all directions will not change the principal directions, s_{ij} and σ_{ij} have the same principal axes. The principal values of the deviatoric stress tensor s_1, s_2, s_3 can be found just as one finds the principal values of the stress tensor, σ_1, σ_2, σ_3, i.e.

$$\det |s_{ij} - s\,\delta_{ij}| = 0 \qquad (3.29)$$

or

$$s^3 - J_1 s^2 - J_2 s - J_3 = 0 \qquad (3.30)$$

where

$$J_1 = s_{ii} = s_{11} + s_{22} + s_{33} = s_1 + s_2 + s_3 = 0 \qquad (3.31)$$

$$J_2 = \tfrac{1}{2} s_{ij} s_{ji} = \tfrac{1}{2}(s_1^2 + s_2^2 + s_3^2) \qquad (3.32)$$

$$= \tfrac{1}{6}[(\sigma_x - \sigma_y)^2 + (\sigma_y - \sigma_z)^2 + (\sigma_z - \sigma_x)^2] + \tau_{xy}^2 + \tau_{yz}^2 + \tau_{zx}^2$$

$$J_3 = \tfrac{1}{3} s_{ij} s_{jk} s_{ki} = \det |s_{ij}| = s_1 s_2 s_3 \qquad (3.33)$$

The quantities, J_1, J_2, J_3, defined analogously to the invariants of the stress tensor I_1, I_2, I_3, are called the *invariants of the deviatoric stress tensor*. A single subscript indicates a principal value. The deviatoric stress tensor and their invariants play a very important role in the stress–strain modeling of both elastic and plastic materials. A simple geometric interpretation of these stress invariants will be given in Section 3.6 when failure criteria of isotropic materials based upon invariant functions of the state of stress are discussed. Physical interpretations of these stress invariants can be found elsewhere (e.g. Chen, 1982). Note that τ_{oct} and J_2 are related by

$$\tau_{\text{oct}} = \sqrt{\tfrac{2}{3} J_2} \qquad (3.34)$$

3.3. STRAINS IN THREE DIMENSIONS

The analysis of strain is essentially a branch of geometry which deals with the deformation of an assemblage of particles. For the present chapter, only the case of infinitesimal strain will be considered. This development is formally very similar to that of stress. Thus, there is no need to write all the details explicitly here because almost every step would be identical after the substitution of components of strain, ε_{ij}, for components of stress, σ_{ij}, of components of strain deviation, e_{ij}, for stress deviation, s_{ij}, and the associated word changes. Here, as before, only the three-dimensional case will be discussed.

3.3.1. Definitions and Notations

In the analysis of stress, the stress at a point is defined by the process of cutting a small area element at the point and taking limits of force divided by area (Fig. 3.1). The strain at a point, however, is defined by the process of drawing line elements through the point and taking the limits of the change in length of the line element divided by its original length and also comparing the change in angle between any two line elements radiating from this point. Thus, rigid-body rotation or translation produces no change in length or angle and therefore causes no strain.

Denote the changes in length per unit of length in the directions of the coordinate axes by ε_{11}, ε_{22}, ε_{33}, and the decreases in angle between the positive directions of the two coordinate line elements by $\gamma_{12} = \gamma_{21}$, $\gamma_{23} = \gamma_{32}$, and $\gamma_{13} = \gamma_{31}$. Then, for any line element at the point P in any direction n_i having the normal strain ε_n in the direction of n_i and shear strain ε_{nt} between the direction n_i and a perpendicular direction t_i, the corresponding Cauchy's formulas for strain have the form

$$\varepsilon_n = \varepsilon_{ij} n_i n_j \tag{3.35}$$

$$\varepsilon_{nt} = \varepsilon_{ij} n_j t_i \tag{3.36}$$

where the *strain tensor* ε_{ij} is defined as

$$\varepsilon_{ij} = \begin{bmatrix} \varepsilon_{11} & \varepsilon_{12} & \varepsilon_{13} \\ \varepsilon_{21} & \varepsilon_{22} & \varepsilon_{23} \\ \varepsilon_{31} & \varepsilon_{32} & \varepsilon_{33} \end{bmatrix} = \begin{bmatrix} \varepsilon_{11} & \dfrac{\gamma_{12}}{2} & \dfrac{\gamma_{13}}{2} \\ \dfrac{\gamma_{21}}{2} & \varepsilon_{22} & \dfrac{\gamma_{23}}{2} \\ \dfrac{\gamma_{31}}{2} & \dfrac{\gamma_{32}}{2} & \varepsilon_{33} \end{bmatrix} \tag{3.37}$$

Note an important difference between the results for stress and strain; for example, $\varepsilon_{12} = \gamma_{12}/2$ or $\varepsilon_{xy} = \gamma_{xy}/2$ appears in eqn 3.37 in place of γ_{12} or γ_{xy}, which would correspond to τ_{xy} in the engineering sense. The shear–strain components ε_{12}, ε_{23}, ε_{13} are called the *tensorial shear strains*, while the shear strains γ_{12}, γ_{23}, γ_{13} are called *engineering*

The continuum theory of rock mechanics

shear strains. When the components of the strain array are chosen appropriately, (ε_{11}, $\varepsilon_{12} = \gamma_{12}/2$, etc.) as in eqn 3.37, the array has a second-order tensorial character and indicates the existence of Mohr's circle construction for coordinate transformation.

The *state of strain* at a point is defined as the totality of *all* the changes in length of line elements divided by their original length which pass through the point and also the totality of all the changes in the angle between any pair of line elements radiating from this point. Here, as in stress, these changes can be determined by Cauchy's formulas, eqns 3.35 and 3.36, from the nine quantities ε_{ij}. Thus, the strain tensor ε_{ij} defines the state of strain at the point completely.

3.3.2. Deviatoric Strain

Like the stress tensor, the strain tensor can be decomposed into two parts: a spherical part associated with change in *volume* and a deviatoric part associated with a change in *shape* (distortion)

$$\varepsilon_{ij} = e_{ij} + \tfrac{1}{3}\varepsilon_v \, \delta_{ij} \tag{3.38}$$

where e_{ij} is defined here as the *deviatoric strain tensor* which is a pure shear state, and

$$\varepsilon_v = \varepsilon_{kk} = \varepsilon_{11} + \varepsilon_{22} + \varepsilon_{33} = \varepsilon_x + \varepsilon_y + \varepsilon_z \tag{3.39}$$

is the volume change per unit volume, or *dilatation*. Note that e_{ij} and ε_{ij} have the same principal axes.

3.3.3. Octahedral Strains and Principal Shear Strains

An octahedral fiber is a fiber of material which before deformation is equally inclined to the three principal strain axes 1, 2 and 3. The corresponding *octahedral normal strain* is given by

$$\varepsilon_{\text{oct}} = \tfrac{1}{3}(\varepsilon_1 + \varepsilon_2 + \varepsilon_3) = \tfrac{1}{3}\varepsilon_v \tag{3.40}$$

which represents the mean of the three principal strains, and the *engineering octahedral shear strain* is

$$\begin{aligned} \gamma_{\text{oct}} = \tfrac{2}{3}[(\varepsilon_x - \varepsilon_y)^2 + (\varepsilon_y - \varepsilon_z)^2 + (\varepsilon_z - \varepsilon_x)^2 \\ + 6(\varepsilon_{xy}^2 + \varepsilon_{yz}^2 + \varepsilon_{zx}^2)]^{1/2} \end{aligned} \tag{3.41}$$

Similarly, the *engineering principal shear strains* are

$$\gamma_1 = |\varepsilon_2 - \varepsilon_3|, \qquad \gamma_2 = |\varepsilon_1 - \varepsilon_3|, \qquad \gamma_3 = |\varepsilon_1 - \varepsilon_2| \tag{3.42}$$

86   *W. F. Chen*

and the *maximum shear strain* is the largest value of the principal shear strains,

$$\gamma_{max} = \max(\gamma_1, \gamma_2, \gamma_3) \quad \text{or} \quad \gamma_{max} = |\varepsilon_1 - \varepsilon_3| \qquad (3.43)$$

for $\varepsilon_1 > \varepsilon_2 > \varepsilon_3$.

3.4. EQUATIONS OF SOLID MECHANICS

In an outline form, the solution of a *solid mechanics problem* at each instant of time must satisfy the following three conditions:

1. Equations of equilibrium or of motion for a static or dynamic analysis, respectively.
2. Conditions of geometry or the compatibility of strains and displacements.
3. Material constitutive laws or stress–strain relations.

The initial boundary conditions on forces and displacements which also must be satisfied in a particular problem are included for brevity under (1) and (2). A brief discussion of each of the items listed above is given in what follows.

3.4.1. Equations of Equilibrium (or Motion)

From considerations of statics (or dynamics), one can relate the components of the stress field, σ_{ij}, in a body to the components of body forces, F_i (forces per unit volume), and external surface forces, T_i (forces per unit area), acting on the boundaries of the body. Stress fields which satisfy these statical (or dynamical) conditions are said to be *statically* (or *dynamically*) *admissible*. For example, in the static analysis of the body shown in Fig. 3.4(a), a statically admissible set of stresses σ_{ij}, and external surface and body forces, T_i and F_i, respectively, must satisfy the following equilibrium equations:

At surface points:

$$T_i = \sigma_{ji} n_j \qquad (3.44)$$

At interior points:

$$\sigma_{ji,j} + F_i = 0 \qquad (3.45)$$

$$\sigma_{ji} = \sigma_{ij} \qquad (3.46)$$

where n_j is the outward unit normal vector to a surface element on

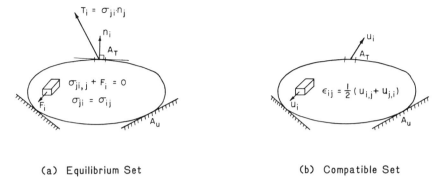

(a) Equilibrium Set (b) Compatible Set

Fig. 3.4. Conditions of equilibrium and compatibility (geometry) in the static analysis of a solid mechanics problem.

which T_i is acting, as shown in Fig. 3.4(a). In the above equations, summation convention for repeated indices is employed, and a comma in the term $\sigma_{ji,j}$ denotes partial differentiation with respect to the coordinate axes (space variables), x_j, with $j = 1, 2, 3$.

As can easily be seen from eqn 3.45, only three equations of equilibrium (or motion) are obtained at any point in the body for prescribed body forces F_i. Thus, we have three equilibrium equations with six unknowns, namely the stress components σ_{ij} at a given point in the body. Hence, an equilibrium set is merely a set and by no means a unique one. In general, an infinite number of stress states can be found which will satisfy the stress boundary conditions of eqn 3.44 and the equilibrium eqns 3.45 and 3.46.

3.4.2. Geometry (Compatibility) Conditions

Compatibility or geometry conditions are derived from kinematic considerations which relate the components of a strain field, ε_{ij}, to the components of a displacement field, u_i. It may also be necessary to impose the conditions of compatibility (integrability) of strains and displacements in order to ensure that these strain-displacement relations are integrable for a prescribed strain field. A set of displacements, u_i, and strains, ε_{ij}, which satisfies these geometry conditions in addition to the imposed displacement boundary conditions is called a *kinematically admissible set* or simply a *compatible set*. Referring to Fig. 3.4(b), kinematic considerations lead to the following conditions for small deformations.

88 W. F. Chen

Strain–displacement relations:

$$\varepsilon_{ij} = \tfrac{1}{2}(u_{i,j} + u_{j,i}) \tag{3.47}$$

Compatibility (integrability) conditions:

$$\varepsilon_{ij,kl} + \varepsilon_{kl,ij} - \varepsilon_{ik,jl} - \varepsilon_{jl,ik} = 0 \tag{3.48}$$

Therefore, a compatible set of displacements, u_i, and strains, ε_{ij}, must satisfy eqns 3.47 and 3.48 together with the prescribed displacement boundary conditions. Further, for an assumed displacement field, u_i (which may not be the actual displacement field induced by the prescribed distribution of body forces, F_i, and surface forces, T_i), the corresponding compatible strain components, ε_{ij}, can be derived directly from eqns 3.47. This compatible set of strains and displacements is, of course, only a set among many other possible sets of strain and displacement fields.

It is important to note that the strain integrability conditions given in eqns 3.48, are needed only when the displacements, u_i, are not explicitly retained as unknowns in the formulation of a problem. For instance, for solutions in the classical theory of elasticity, stress functions are frequently introduced as the only unknown functions (e.g. Airy's stress function in two dimensional elasticity). In such cases, eqn 3.48 must be imposed on the strain field to ensure the existence of a continuous single-valued displacement field. In most practical problems, the displacements are generally taken explicitly as unknowns in the formulation (e.g. in the finite element technique for numerical solutions). Then, the integrability conditions of eqn 3.48 are not needed, and only eqn 3.47 is used to derive the strains from the displacements. In such cases, there are nine independent unknowns (namely, six stress components, σ_{ij}, plus three displacement components, u_i, while strains are expressed in terms of the displacements). On the other hand, only three equations of equilibrium (or motion) are available (for example, eqn 3.45 in the static analysis). Thus, six additional equations are needed to complete the formulation of the problem. These additional equations are furnished by the constitutive or stress–strain relations of the material.

3.4.3. Constitutive Relations

Since both static (dynamic) and kinematic (or geometric) conditions are treated independently and no relating conditions are introduced, they are valid for elastic as well as inelastic or plastic materials. The differentiating feature of various material behaviors is accounted for in the material constitutive laws. These laws give the relations between stress components σ_{ij} and strain components ε_{ij} at any point in the body. They may be simple or extremely complex depending upon the material of the body and the conditions to which it has been subjected.

Once the material constitutive law is established, the general formulation for the solution of a solid mechanics problem can be completed. The interrelationships of variables (F_i, T_i, σ_{ij}, ε_{ij} and u_i) encountered in a general formulation are shown schematically in Fig. 3.5 in the case of static analysis.

The constitutive relations for a particular material are determined experimentally and they may involve measurable physical quantities other than stresses and strains, such as temperature and time, or internal parameters which cannot be measured directly. The effects of such internal parameters on the stress–strain behavior of the material can often be more conveniently expressed in terms of history of stress and strain, or memory of past mechanical events inherent in the material.

The constitutive relationship for a material depends on many factors, including the homogeneity, isotropy and continuity of the body material, its reaction to loading over a period and the rate and magnitude of loading. The behavior of an actual material varies enormously with temperature, confining pressure and rate of strain, among other factors. However, under certain limited conditions, it is possible to

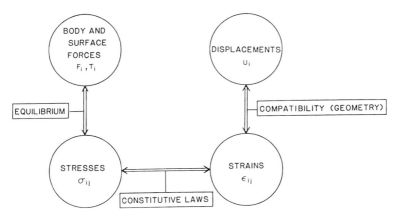

Fig. 3.5. Interrelationships of variables in the solution of a static solid mechanics problem.

idealize the rocks as elastic, plastic, or viscous materials for the purpose of stress and strain analysis. The general techniques to develop these relationships for rocks in general and oil shale in particular, based on an elastic-plastic or viscous model or a combination of the two are discussed in the following sections.

3.5. CONSTITUTIVE MODELING OF ROCKS

3.5.1. Rock as a Continuum

Unlike a steel which can be refined to consistent internal state before use, a rock is a naturally occurring material and must be used in its natural state. All rocks consist of mineral particles, air voids and water. The proportion of each mineral in the rock, together with the granular structure and the texture, serves as a basis for geological classification. For example, *shale* is classified as a *compressed clay* or *mudstone* which differs from clay in that it will not lose all strength when wet and failure is normally due to the laminated structure of the rock and the presence of 'slippery' minerals such as kaolinite. Therefore, the mathematical characterization of rock behavior should ideally be based on a consideration of the behavior of individual mineral grains or particles in a rock mass and their interaction so that the 'macroscopic' continuum stress–strain–time relations are derived from the more basic 'microscopic' interactions of many particles including the use of probabilistic and statistical theories to handle the probabilistic nature of the interparticle contact relationships.

However, such an approach in modeling rock behavior can be rather complex and would not be particularly fruitful in practical applications. For most rock engineering problems, the scale of the geometry of interest is very large. At this scale, therefore, the 'microscopic' or 'discontinuous' effects can be averaged and a continuum view of rock is necessary for progress. The mechanical behavior of the rock and its mathematical characterization can then be studied based upon the principles of continuum mechanics; i.e. an entirely phenomenological approach on a macroscopic level. This latter approach forms the basis of most of the commonly used constitutive models for rocks. The main purpose of this chapter is to discuss and evaluate various types of these models.

Within the framework of continuum mechanics, the behavior of real materials is generally idealized as time-independent or time-dependent. In the time-independent idealization, such as elastic and elastic-plastic models, time effects are neglected. Time does not appear explicitly as a variable in the constitutive relations, and phenomena like rate sensitivity, aging effects and creep are not included in these material models. Further, for an ideal elastic model, the behavior is reversible and independent of the loading path, while it is irreversible and load path dependent in a plasticity-based model. On the other hand, in the time-dependent material idealization, such as the visco-elastic and viscoplastic models, time effects are considered and, therefore, they are generally capable of describing rate- and history-dependent behavior.

In rock mechanics constitutive modeling of rocks is of central importance to the analyses and engineering design of mining systems. Elastic modeling has been used most widely and is well understood, but irreversible deformation is not. This is an area of great importance to rock in general and to oil shale in particular. Directions of current research on this subject will be emphasized.

It must be emphasized here that the previous idealizations and subsequent classification of the constitutive models are only for mathematical convenience in describing the actual complex behavior of real materials. Nothing can compel the material to behave according to any of these idealized models. Indeed, for rocks in general and oil shales in particular, the actual material response will exhibit the behavior characteristics of most of these models under certain conditions of stresses, temperatures, vibration and strain rates. Therefore, in any practical problem, it is essential that we determine the limits and conditions under which the material can sensibly be assumed to exhibit the dominant characteristics of a particular type of the idealized models. Furthermore, since any idealized model has its own shortcomings, all the results obtained must be interpreted carefully in terms of these shortcomings.

3.5.2. General Approaches
There are different approaches to formulating a constitutive model of rocks. In general, these can be considered in terms of two classifications:

1. Finite material characterizations in the form of secant (total) stress–strain models. Included in this class of models are those based on nonlinear elasticity and deformation theory of plasticity.

2. Incremental (differential or rate) material descriptions in the form of tangential stress–strain models. The most prominent models of this category are those based on hypoelasticity and flow theory of plasticity.

In what follows, the underlying concepts and the general characteristics of various types of elasticity-, plasticity- and viscosity-based models are briefly summarized. A comprehensive review of the presently available rock models related to oil shale applications is given in subsequent chapters.

3.6. FAILURE CRITERIA OF ROCKS AND OIL SHALES

3.6.1. Uniaxial Strength

The best known material property for rocks appears to be that of uniaxial compressive and tensile strengths. Strength values depend on the type of test. The compressive strength, σ_u, (or crushing strength) is normally obtained by crushing a cylindrical rock sample unconfined at its sides. Its value depends on the direction in which the sample is taken. For example, the compressive strengths perpendicular (z axis) and parallel (x and y axes) to the bedding planes of Colorado oil shale are significantly different. The compressive strengths in the x direction, σ_{ux}, (horizontal) and the z direction, σ_{uz}, (vertical) have the following mean values (Chong *et al.*, 1980*a*):

$$\sigma_{ux} = 88 \cdot 0 \text{ MPa} (12 \cdot 8 \text{ ksi})$$

$$\sigma_{uz} = 104 \text{ MPa} (15 \cdot 1 \text{ ksi})$$

The tensile strength, σ_{ut}, is usually determined from the split cylinder test, among others (Dismuke *et al.*, 1972). For example, by orienting the test's failure plane perpendicular to bedding planes in the Colorado oil shale, the tensile strength is found to have the mean value of $\sigma_{ut} = 11 \cdot 5 \text{ MPa} (1 \cdot 7 \text{ ksi})$.

Since most problems in rock design involve compressive stresses, the use of tensile strength in rocks is normally made indirectly. Note that the tensile strength is important in the Griffith failure criterion as well as in the dynamic fracturing of rocks by explosive action.

There appears to be a general linear relationship between tensile or compressive strength of oil shale and the organic content O_c, taking an approximate form:

$$\sigma_u = a - bO_c \tag{3.49}$$

where a, b and O_c are material constants depending on the type of oil shale. For Green River Formation oil shale, for example, the organic content in percentages by volume (O_c) can be calculated from the oil yield in gallons per ton ($=4\cdot19$ cm^3 kg^{-1}), M, using the following relationship (Smith, 1969)

$$O_c = \frac{1\cdot578M}{0\cdot957M + 107} \tag{3.50}$$

Typical compressive, σ_u, and tensile strength, σ_{ut}, values in MPa for Colorado oil shale have the following material constants:

$$\sigma_{ux} = 127\cdot73 - 1\cdot1215O_c$$
$$\sigma_{uz} = 161\cdot60 - 1\cdot5415O_c \tag{3.51}$$
$$\sigma_{ut} = 14\cdot78 - 0\cdot0928O_c$$

Further discussions on typical values of oil shales and the type of testing and its effect on strength magnitude are given in Chapter 5.

3.6.2. Strength Models in Three Dimensions

The failure of rock in mass in a three-dimensional state of stress is extremely complicated. Numerous criteria have been devised to explain the conditions for failure of a material under such a loading state. These models can be classified as one-parameter models including the Rankine or Griffith criterion of tensile failure and the Coulomb criterion of failure at maximum shear stress, and two-parameter models including the well-known Mohr or Mohr–Navier criterion of shear failure.

In all these strength models, two basic postulates are adopted: isotropy, and convexity in the principal stress space. The first assumption is mainly introduced because of the inherent simplification of the failure model. It is certainly true some rocks exhibit significant anisotropy with respect to their strength which requires the formulation of the failure surface in the six-dimensional stress space instead of the three-dimensional space of principal stresses. For many rocks, however, the assumption of isotropy is reasonable. On the other hand, convexity is an assumption which is supported by global stability arguments in plasticity (Chen and Saleeb, 1982). Clearly, there are some questions on the validity of this postulate, and in fact, there is a strong indication that the failure envelope for rocks over a wide range of hydrostatic (confining) pressures may be non-convex with respect to the hydrostatic axis.

3.6.2.1. Coulomb Criterion (Tresca)

The type of stress leading to failure (compressive or tensile) is particularly important to rock material. The simplest of the compressive type of strength criteria is that based on the *maximum shear stress criterion* of Coulomb. In applied mechanics, this criterion is widely called the *Tresca criterion* which states that failure occurs at a point in a material when the maximum shear stress τ_{max} is equal to the shear strength of the material and that failure occurs on a plane bisecting the angle between the two principal stresses, i.e.

$$\tau_{max} = \tfrac{1}{2}(\sigma_1 - \sigma_3) = k \tag{3.52}$$

or in terms of stress invariants, it has the general form

$$\sqrt{J_2} \sin{(\theta + \tfrac{1}{3}\pi)} - k = 0 \tag{3.53}$$

where k is the yield or failure stress in *pure shear* which is equal to one-half the uniaxial tensile strength, σ_{ut}, and θ is the *angle of similarity* defined by

$$\cos{3\theta} = \frac{3\sqrt{3}}{2} \frac{J_3}{J_2^{3/2}} = \frac{\sqrt{2}J_3}{\tau_{oct}^3} \tag{3.54}$$

in which J_2, J_3 and τ_{oct} are invariants defined by eqns 3.32, 3.33 and 3.34 respectively.

Since the effect of confinement or hydrostatic pressure on the yield or failure criterion is not considered in this model, it follows that eqn 3.53 represents a cylindrical surface in principal stress space whose generator is parallel to the hydrostatic axis. The failure surface has a *regular* hexagon cross section on a deviatoric plane (Chen and Saleeb, 1982).

3.6.2.2. Coulomb–Navier Criterion (Mohr–Coulomb)

The Coulomb criterion can be extended to the general form

$$\tau = f(\sigma) \tag{3.55}$$

where the limiting shearing stress τ in a plane is dependent only on the normal stress σ in the same plane at a point, and where eqn 3.55 is the

The continuum theory of rock mechanics 95

failure envelope for the corresponding Mohr circles. The envelope $f(\sigma)$ is an experimentally determined function. This is known as the Mohr criterion. According to this criterion, failure of material will occur for all states of stress for which the largest of Mohr's circles is just tangent to the envelope. Here, as in the Coulomb criterion, the intermediate principal stress, σ_2, has no influence on the failure.

The simplest form of Mohr envelope is the straight line or a linear relation between τ and σ

$$|\tau| = c - \sigma \tan \phi$$
$$= c - \mu \sigma \tag{3.56}$$

This equation is known as *Mohr–Coulomb criterion* in civil engineering, but called *Coulomb–Navier criterion* in rock mechanics. The constants c, ϕ and $\mu = \tan \phi$ are known as cohesion, internal friction angle and coefficient of internal friction, respectively. In the special case of frictionless materials for which $\phi = 0$, eqn 3.56 reduces to the maximum-shear-stress criterion of Coulomb or Tresca, $\tau = c = k$. From eqn 3.56 it can be shown that eqn 3.56 is identical with ($\sigma_1 > \sigma_2 > \sigma_3$)

$$\frac{\sigma_1}{\sigma_u} - \frac{\sigma_3}{\sigma_{ut}} = 1 \tag{3.57}$$

where

$$\sigma_u = \frac{2c \cos \phi}{1 - \sin \phi}, \qquad \sigma_{ut} = \frac{2c \cos \phi}{1 + \sin \phi} \tag{3.58}$$

or identically in terms of the stress invariants I_1, J_2, θ,

$$\tfrac{1}{3}I_1 \sin \phi + \sqrt{J_2} \sin \left(\theta + \frac{\pi}{3} \right)$$
$$+ \frac{\sqrt{J_2}}{\sqrt{3}} \cos \left(\theta + \frac{\pi}{3} \right) \sin \phi - c \cos \phi = 0 \tag{3.59}$$

In the $\sigma_1, \sigma_2, \sigma_3$ coordinate system, eqn 3.59 represents an irregular hexagonal pyramid.

The material constants c, ϕ and μ may be computed directly from the uniaxial strengths σ_u and σ_{ut}. For example, for shale materials, typical values are $c = 30–300 \text{ kg cm}^{-2}$, $\phi = 15°$ and $\mu = 0{\cdot}25–0{\cdot}6$. In fact, an empirical relationship of the form $c = 2\sigma_{ut}$ for competent rocks has been suggested (Farmer, 1968) for practical use.

As a further example, some statistics on the Mohr–Coulomb strength parameters of oil shale from the Colorado C-b Tract have been reported recently by Chang and Bondurant (1979). Both the peak (or ultimate) and the residual strength of oil shale are given. Typical mean values are (1) ultimate strength, $c_u = 4 \cdot 1$ ksi (28·3 MPa) and $\phi_u = 24 \cdot 8°$ and (2) apparent residual strength, $c_r = 0 \cdot 78$ ksi (5·4 MPa) and $\phi_r = 37 \cdot 1°$.

3.6.2.3. Rankine Criterion

Although used extensively to predict failure in rocks, the Coulomb–Navier criterion does not represent a good failure surface for rocks under tensile type of loading, which tends to fail by a *cleavage type of brittle fracture* instead of shear type of compression failure. The *maximum tensile stress criterion* of *Rankine* is generally accepted today for determining whether a tensile or a compressive type of failure has occurred for rock.

According to this criterion, *brittle fracture* of rock takes place when the maximum tensile strength of the material reaches a value equal to the tensile strength σ_{ut} of the material found in a simple tension test, regardless of the normal or shearing stresses that occur on other planes through the point. *Rankine's fracture surface* is defined by

$$\sigma_1 = \sigma_{ut}, \qquad \sigma_2 = \sigma_{ut}, \qquad \sigma_3 = \sigma_{ut} \tag{3.60}$$

which result in three planes perpendicular to the $\sigma_1, \sigma_2, \sigma_3$ axes, respectively. This surface will be referred to as the *fracture cutoff surface* or *tension failure surface* or simply *tension cutoff*.

In terms of stress invariants, this criterion can be written as

$$2\sqrt{3}\,\sqrt{J_2}\cos\theta + I_1 - 3\sigma_{ut} = 0 \tag{3.61}$$

To obtain a better approximation for rock failure under both tensile and compressive types of loading, it is commonly the practice to combine the Coulomb–Navier criterion with the maximum tensile strength cutoff (Chen, 1982). This combined criterion is a *three-parameter criterion* (two states determine the values of c and ϕ, and one stress state determines the maximum tensile stress).

3.6.2.4. Griffith Criterion

The *Griffith criterion* for brittle failure explains the condition for propagation of a crack. This criterion has wide applications in rock engineering. The basic assumption used in developing this criterion is that fracture will be initiated in a brittle material by tensile failure around the tips of microcracks and flaws present in the material. In

The continuum theory of rock mechanics

the solution for a crack in a stress field, it assumes that the crack is elliptical and flat, that maximum tensile stress criterion controls the tensile failure and that the material is linearly elastic.

In a homogeneous, isotropic material, the Griffith fracture criterion has the simple form, assuming random orientation of cracks ($\sigma_1 > \sigma_2 > \sigma_3$):

$$\sigma_{ut} = \frac{(\sigma_1 - \sigma_3)^2}{8(\sigma_1 + \sigma_3)} \tag{3.62}$$

which predicts that when $\sigma_3 = 0$, the uniaxial compressive strength represented by σ_1 is equal to $8\sigma_{ut}$. This value is quite near the experimental relationship obtained from the Coulomb–Navier criterion.

3.7. LINEAR ELASTICITY OF ROCKS

3.7.1. General

The simplest way of relating stress and strain is by direct *linearity* for an ideal *elastic isotropic* medium in which all strain is instantaneously and totally recoverable on the removal of the stress (elasticity), and all the grains and particles are assumed in random orientation so the material point has no directional preference (isotropy). A linear elastic isotropic medium is an *idealization* of actual material properties. For some fine-grained metamorphic rocks, they do approximate in varying degrees to the ideal material, particularly under low deforming loads.

In the linear elastic range, the constitutive equations are embodied in Hooke's law. The mechanics of rock in this range are well understood. Predictions of acceptable accuracy can be found in many mining operations (Pariseau, 1977). The basic relations between stress and strain are described completely by two elastic constants, Poisson's ratio, ν, and Young's modulus, E, in the usual form

$$\varepsilon_{ij} = \frac{1+\nu}{E} \sigma_{ij} - \frac{\nu}{E} \sigma_{kk} \delta_{ij} \tag{3.63}$$

$$\sigma_{ij} = \frac{E}{1+\nu} \varepsilon_{ij} + \frac{\nu E}{(1+\nu)(1-2\nu)} \varepsilon_{kk} \delta_{ij} \tag{3.64}$$

A neat and logical separation exists between the mean response and the deviatoric response, which is exhibited very clearly by subscript notation in the above equations. For example, substituting $s_{ij} + \sigma_{kk}\,\delta_{ij}/3$ for σ_{ij} and $e_{ij} + \varepsilon_{kk}\,\delta_{ij}/3$ for ε_{ij} into eqn 3.63, we find

$$s_{ij} = \frac{E}{1+\nu}\,e_{ij} = 2Ge_{ij} \tag{3.65}$$

$$p = \tfrac{1}{3}\sigma_{kk} = \frac{E}{3(1-2\nu)}\,\varepsilon_{kk} = K\varepsilon_{kk} = K\varepsilon_\nu \tag{3.66}$$

where

$$K = \frac{E}{3(1-2\nu)} \quad\text{and}\quad G = \frac{E}{2(1+\nu)} \tag{3.67}$$

are the *bulk* and *shear* modulus respectively. Volume change $\varepsilon_{kk} = \varepsilon_\nu$ is produced by the pressure or mean normal stress $p = \sigma_{kk}/3$; and distortion or shear deformation e_{ij} is produced by the shear stress or the stress deviator s_{ij}. Each is independent of the other. A generalization of eqns 3.65 and 3.66 for defining isotropic nonlinear stress–strain behavior of rock will be given later when the nonlinear elasticity of rocks are presented.

To define the stress–strain relationships elastically, two elastic constants are required from the four available (E, ν, K, G). In practical applications, E and ν are the most commonly used. For elastic analysis of rocks, a value of $0\cdot25$ for Poisson's ratio is usually assumed. This suggests the approximate relation $E = 2\cdot5G$ from eqn 3.67. As for E, the rule-of-thumb formula

$$E = 0\cdot9(\rho - 2\cdot1) \times 10^6 \quad (\text{kg cm}^{-2}) \tag{3.68}$$

may be used, where ρ is the *apparent density* of a material. The accuracy of this formula is about $\pm20\%$ (Farmer, 1968). Alternatively, the moduli of elasticity may be estimated from the approximate form $E = 350\sigma_u$, $G = 140\sigma_u$ where σ_u is rock compressive strength.

3.7.2. Transversely Isotropic Elastic Model

It has been shown that some oil shales can be adequately characterized as a *transversely isotropic elastic material* (Chong et al., 1979a,b, 1980a). In this case, the material exhibits a rotational elastic symmetry about one of the coordinates (z-axis) which is perpendicular to the bedding planes. The horizontal x–y plane is the plane of isotropy as shown in Fig. 3.6.

The stress–strain relationships for a transversely isotropic elastic material are completely described by five constants (Chen and Saleeb,

The continuum theory of rock mechanics

1982):

E_x, E_z = Young moduli in the plane of isotropy and perpendicular to it, respectively;

$G_{xy} = E_x/2\,(1+\nu_{xy})$ = shear modulus for the plane of isotropy;

G_{xz} = Shear modulus for a plane normal to the plane of isotropy;

ν_{xy} = Poisson's ratio that characterizes the transverse strain reduction in the plane of isotropy due to a tensile in the same plane;

ν_{zx} = Poisson's ratio which characterizes the transverse strain reduction in the plane of isotropy due to tensile stress in a direction normal to it.

Note that the five independent constants are chosen to be E_x, E_z, ν_{xy}, ν_{zx}, G_{xz} ($G_{xy} = E_x/2\,(1+\nu_{xy})$ is not independent). Using the engineering matrix notation,

$$\{\sigma\} = \{\sigma_x,\ \sigma_y,\ \sigma_z,\ \tau_{xy},\ \tau_{yz},\ \tau_{zx}\} \tag{3.69}$$

$$\{\varepsilon\} = \{\varepsilon_x,\ \varepsilon_y,\ \varepsilon_z,\ \gamma_{xy},\ \gamma_{yz},\ \gamma_{zx}\} \tag{3.70}$$

The stress–strain relationship for a transversely isotropic linear elastic material is

$$\{\varepsilon\} = [D]\{\sigma\} \tag{3.71}$$

where the *compliance matrix* takes the following form

$$[D] = \begin{bmatrix}
\dfrac{1}{E_x} & -\dfrac{\nu_{xy}}{E_x} & -\dfrac{\nu_{zx}}{E_z} & 0 & 0 & 0 \\[2ex]
-\dfrac{\nu_{xy}}{E_x} & \dfrac{1}{E_x} & -\dfrac{\nu_{zx}}{E_z} & 0 & 0 & 0 \\[2ex]
-\dfrac{\nu_{zx}}{E_z} & -\dfrac{\nu_{zx}}{E_z} & \dfrac{1}{E_z} & 0 & 0 & 0 \\[2ex]
0 & 0 & 0 & \dfrac{1}{G_{xy}} & 0 & 0 \\[2ex]
0 & 0 & 0 & 0 & \dfrac{1}{G_{xz}} & 0 \\[2ex]
0 & 0 & 0 & 0 & 0 & \dfrac{1}{G_{xz}}
\end{bmatrix} \tag{3.72}$$

Note that eqn 3.72 is symmetric because of the requirement for

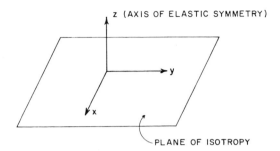

Fig. 3.6. Coordinate axes for a transversely isotropic material.

Green hyperelastic materials. This will be discussed further in the following section.

These relationships and their applications to Green River Formation oil shales are discussed in detail in Chapter 5. From their experimental results on these oil shales, Chong *et al.* (1980a) found that G_{xz} can be expressed approximately in terms of E_x, E_z, ν_{xy} and ν_{zx} in the form

$$G_{xz} = \frac{E_x E_z}{E_x + (1 + 2\nu_{zx})E_x} \qquad (3.73)$$

if the *degree of anisotropy* is slight (E_x/E_z is less than about 1·8). These four elastic constants E_x, E_z, ν_{xy} and ν_{zx} can be obtained by conducting uniaxial compressive tests on samples taken from the same horizon parallel and perpendicular to bedding planes (Fig. 3.6). A list of typical E and ν values for oil shales are quoted in Chapter 5. A more accurate method of determining G_{xz} was presented by Chong *et al.* (1980b) and will be discussed in Chapter 5.

3.8. NONLINEAR ELASTICITY OF ROCKS

3.8.1. Cauchy Elastic Material

Linear elasticity for isotropic and transversely isotropic materials constitutes the oldest and simplest approach to modeling the stress–strain behavior of rocks under low deforming loads. However, for rocks with large pore space such as the weaker sedimentary rocks and oil shales, the stress–strain curve is generally nonlinear, and any analysis based on linear elasticity would be dangerous. Such rocks should be characterized by variable stress–strain moduli.

The simplest approach to formulate such nonlinear models is to simply replace the elastic constants in the linear stress–strain relations with tangent moduli dependent on the stress and/or strain invariants. Nonlinear models of this type have been discussed in the paper by Chong et al. (1980a). For oil shales, for example, the four constants, E_x, E_z, ν_{xy} and ν_{zx} in the transversely isotropic elastic compliance matrix $[D]$ (eqn 3.72) can be assumed to be dependent on the stress levels applied and on the organic volume O_c (eqn 3.50). These elastic constants are obtained either incrementally in the form of *tangent moduli* such as K_t and G_t, using piecewise linear models, or in the form of *secant moduli* (K_s and G_s) expressed in terms of octahedral normal and shear stress–strain curves as shown in Fig. 3.7.

These models are mathematically and conceptually very simple. The models account for two of the main characteristics of rock behavior: nonlinearity and the dependence on the hydrostatic stress.

The main disadvantage of the models is that they describe path-independent behavior. Therefore, their application is primarily directed towards monotonic or proportional loading regimes. For arbitrary assumed functions for the secant moduli, there is no guarantee that the strain and complementary energy functions W and Ω will be path-independent and energy generation may be indicated in certain stress cycles, which is physically not acceptable. This type of engineering approach described above is called the *Cauchy elastic formulation*. The material is called the *Cauchy elastic medium*.

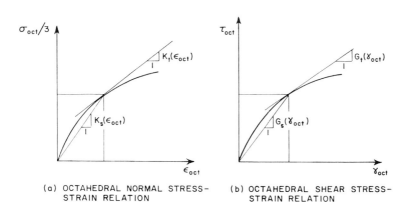

(a) OCTAHEDRAL NORMAL STRESS–STRAIN RELATION

(b) OCTAHEDRAL SHEAR STRESS–STRAIN RELATION

Fig. 3.7. Octahedral normal and shear stress–strain relations.

3.8.2. Hyperelastic Material

A more rational approach in formulating secant stress–strain models for rocks can be developed on the basis of *hyperelasticity* theory. Here the constitutive relations are based on the assumption of the existence of a strain energy function, W, or a complementary energy function, Ω, such that

$$\sigma_{ij} = \frac{\partial W}{\partial \varepsilon_{ij}}; \qquad \varepsilon_{ij} = \frac{\partial \Omega}{\partial \sigma_{ij}} \qquad (3.74)$$

in which W and Ω are functions of the current components of the strain or stress tensors, respectively. Equation 3.74 yields a one-to-one relation between actual states of stress and strain. In addition to the reversibility and path independence of stresses and strains in the hyperelastic type of elastic models, thermodynamic laws are always satisfied, and no energy can be generated through load cycles.

For an initially isotropic elastic material, W and Ω are expressed in terms of any three independent invariants of strain tensor, ε_{ij}, or stress tensor, σ_{ij}, respectively. Based on an assumed functional relation of W in terms of the strain invariants, or Ω in terms of the stress invariants, eqn 3.74 can be used to obtain various nonlinear elastic stress–strain relations in the form of secant formulation.

From Fig. 3.8, it is observed that tangent moduli are identical for loading and unloading. Thus, the hyperelastic model yields a constitutive relation which is incapable of describing load history dependence and rate-dependence. Hyperelasticity exhibits strain induced anisotropy in the material.

The hyperelastic formulation can be quite accurate for rocks straining in proportional loading. Moreover, use of these models in such cases satisfies the rigorous theoretical requirements of continuity, stability, uniqueness and an energy consideration of the continuum mechanics. However, as noted previously, models of the hyperelastic type fail to identify the inelastic character of rock deformations, a shortcoming that becomes apparent when the material experiences unloading.

The main objection to the hyperelastic formulation is the complications involved with the material constants. Even when initial isotropy is assumed, a nonlinear hyperelastic model often contains too many material parameters. For instance, a third-order model requires 9 constants; while 14 constants are needed for a fifth-order hyperelastic model. A large number of tests is generally required to determine

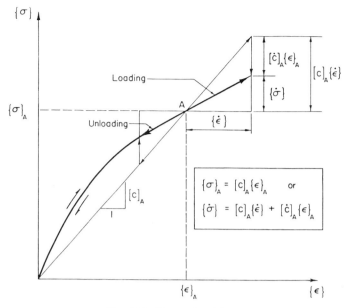

Fig. 3.8. Hyperelasticity.

these constants, which limits the practical usefulness of the models. Thus, the linear generalized Hooke's law as described above is most commonly used.

3.8.3. Hypoelastic Material

An obvious shortcoming in both of the previous types of non-linear elasticity models is the path-independent behavior implied in the secant stress–strain formulation, which is certainly not true for rocks in general. A more improved description of the rock behavior is provided by the hypoelastic formulation in which the incremental stress and strain tensors are linearly related through variable material response moduli that are functions of the stress or strain state:

$$\dot{\sigma}_{ij} = C_{ijkl}(\sigma_{mn})\dot{\varepsilon}_{kl}$$

or

$$\{\dot{\sigma}\} = [C]\{\dot{\varepsilon}\} \qquad (3.75)$$

in which the material tangential response function C_{ijkl} describes the instantaneous behavior directly in terms of the time rates of stress $\dot{\sigma}_{ij}$

and strain $\dot{\varepsilon}_{ij}$. These incremental stress–strain relations provide a natural mathematical model for materials with limited memory. This can be seen by an integration of eqn 3.75

$$\sigma_{ij} = \int_0^t C_{ijkl}(\sigma_{mn}) \frac{\partial \varepsilon_{kl}}{\partial \tau} d\tau + \sigma_{ij}^0 \qquad (3.76)$$

The integral expression clearly indicates the path-dependency and irreversibility of the process. The hypoelastic response is therefore stress history (path) dependent. In the linear case for which $C_{ijkl}(\sigma_{mn})$ is a constant, the hypoelasticity degenerates to hyperelasticity, which corresponds to the history independent secant modulus formulation. The integration in eqn 3.76 can be carried out explicitly and leads to the hyperelastic formulation.

As observed from Fig. 3.9, the tangential stiffnesses $[C]$ are identical in loading and unloading. This reversibility requirement only in the

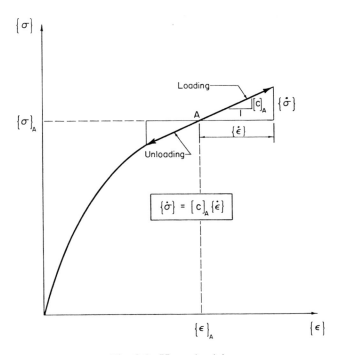

Fig. 3.9. Hypoelasticity.

infinitesimal (or incremental) sense justifies the use of the term hypoelastic or *minimum elastic*.

There are two problems associated with hypoelasticity modeling. The first problem is that, in the nonlinear range, the hypoelasticity-based models exhibit stress induced anisotropy. This anisotropy implies that the principal axes of stress and strain are different, introducing coupling effect between normal stresses and shear strains. As a result, a total of 21 material moduli for general triaxial conditions have to be defined for every point of the material loading path. This is a difficult task for practical application.

The second problem is that under the uniaxial stress condition, the definition of loading and unloading is clear. However, under multiaxial stress conditions, the hypoelastic formulation provides no clear criterion for loading or unloading. Thus, a loading in shear may be accompanied by an unloading in some of the normal stress components. Therefore, assumptions are needed for defining loading–unloading criteria.

In the simplest class of hypoelastic models, the incremental stress–strain relations are formulated directly as a simple extension of the isotropic linear elastic model with the elastic constants replaced by variable tangential moduli which are taken to be functions of the stress and/or strain invariants. This approach has been discussed previously in Section 3.8.1.

Models of this type are attractive from both computational and practical viewpoints. They are well suited for the implementation of finite element computer codes. The material parameters involved in the models can be easily determined from standard laboratory tests using well defined procedures; and many of these parameters have broad data base.

The application of this type of hypoelastic models should be confined to monotonic loading situations which do not basically differ from the experimental tests from which the material constants were determined or curve fitted. Thus, the isotropic models should not be used in cases such as non-homogeneous stress states, non-proportional loading paths or cyclic loadings. Also, as has been shown in the book by Chen and Saleeb (1982), the material tangent stiffness matrix for a hypoelastic model is generally unsymmetric which results in a considerable increase in both storage and computational time. Further, in such cases, uniqueness of the solution of boundary value problems cannot generally be assured.

106 W. F. Chen

3.9. DEFORMATION PLASTICITY OF ROCKS

3.9.1. General

Constitutive equations for rock deformed beyond the elastic range represent an area of great importance to rock mechanics. This is because deformation of rock in the elastic range (which is usually quite small) does not represent a realistic mining problem. In general, as a practical matter, it is the irreversible, inelastic range of deformation that is the main concern. This is an area, at present, we know very little about despite the fact that this is precisely the area where genuine rock mechanic problems arise.

Classical plasticity theory has been extended successfully in recent years to soil media and has been used to solve problems in geotechnical engineering (Chen, 1984a,b; Chen and Mizuno, 1985). However, little has been done on rock mechanics. In order to calculate the distribution of stress and the progress of post-elastic deformation as mining proceeds, stress–strain models that reflect *strain hardening* in the post-elastic range and *strain softening* in the post-failure range are required. This calls for the development of elastic-plastic-fracture models for rocks. A great deal of research needs to be done in this area. We attempt to characterize this time-independent inelastic behavior of rocks in terms of *plasticity* and to approach the analysis of stress in a rock in terms of the fundamental concept of plasticity. Note that plasticity, like elasticity, is based on the concept of an ideal material. Thus, the validity of such an approach must be justified by comparing the actual properties of a real material with an idealized plastic body. However, at present, there has never been a single experiment done that is intended to substantiate the basic plasticity assumptions of 'normality', and associated flow rules for rock.

3.9.2. Deformation Theory of Plasticity

The fundamental difference between elasticity and plasticity models lies in the distinction in the treatment of loading and unloading in plasticity theories. This is achieved by introducing the concept of a loading function. In addition, the total deformations ε_{ij} are decomposed into elastic and plastic components ε_{ij}^e and ε_{ij}^p by simple superposition:

$$\varepsilon_{ij} = \varepsilon_{ij}^e + \varepsilon_{ij}^p \tag{3.77}$$

The plastic strain is obtained from

$$\varepsilon_{ij}^p = \phi \frac{\partial F}{\partial \sigma_{ij}} \tag{3.78}$$

where ϕ is a scalar function relating to a one-dimensional test curve, positive during loading and zero during unloading, and F is a scalar function of the stress state and possible also for some hardening parameters.

In the deformation theories of plasticity for work-hardening materials, it is postulated that the state of stress determines the state of strain uniquely as long as plastic deformation continues. Thus, they are identical with nonlinear elastic stress–strain relations of the secant type as long as unloading does not occur.

If any of the elasticity models described above is to be used to describe rock behavior under general loading conditions involving loading and unloading, it must therefore be accompanied by special unloading treatment based on a criterion defining loading–unloading.

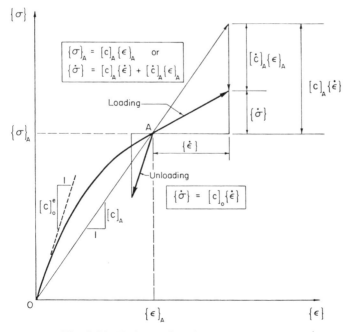

Fig. 3.10. Deformation theory of plasticity.

Such a formulation is closely related to the deformation theory of plasticity. This is illustrated graphically in Fig. 3.10: during loading it is identical with nonlinear elasticity, while during unloading, it follows linear elasticity.

3.9.3. Deformational Plastic Model

An example of a deformational plasticity formulation has been given in the book by Chen and Saleeb (1982) where the third-order hyperelastic model was augmented by a loading–unloading criterion. This criterion is expressed in terms of the complementary energy function Ω which is an invariant of the stress tensor σ_{ij} with respect to coordinate transformation. Unloading is indicated by the condition $\dot{\Omega} < 0$, where $\dot{\Omega} = \varepsilon_{ij} \, d\sigma_{ij}$ is the incremental change in Ω; the condition $\dot{\Omega} > 0$ indicates loading. Reloading is defined by the condition $\dot{\Omega} > 0$ and $\Omega < \Omega_{max}$, where Ω_{max} is the maximum previous value of Ω at the material point. Mathematically, these general conditions may be written as

$$\text{loading:} \quad \text{when } \Omega = \Omega_{max} \text{ and } \dot{\Omega} > 0$$
$$\text{unloading:} \quad \text{when } \Omega \leqslant \Omega_{max} \text{ and } \dot{\Omega} < 0 \qquad (3.79)$$
$$\text{reloading:} \quad \text{when } \Omega < \Omega_{max} \text{ and } \dot{\Omega} > 0$$

For the cases of unloading or reloading the initial tangential moduli may be applied whereas for loading, the following general form of deformation theory for an isotropic material may be used.

$$\varepsilon_{ij}^{\text{p}} = P \, \delta_{ij} + Q s_{ij} + R t_{ij} \qquad (3.80)$$

where $t_{ij} = s_{ik} s_{kj} - \frac{2}{3} J_2 \, \delta_{ij}$, s_{ij} is the stress deviation, $J_2 = \frac{1}{2} s_{ij} s_{ij}$ and $\varepsilon_{ij}^{\text{p}}$ is the plastic strain component. In general, the scalar functions P, Q, and R depend on the three invariants of the stress tensor. (For $P = R = 0$, eqn 3.80 reduces to Hencky's relations used for metals.)

The only objection to the present definition of loading and unloading in eqn 3.79 is the ambiguity encountered at the *neutral loading* condition $\dot{\Omega} = 0$ where one may arbitrarily assign either value of the loading or unloading moduli. The result is that infinitesimal stress changes near neutral loading may produce finite strain changes, and the *continuity condition* may be violated. This is not physically acceptable.

Considering the loading criterion of eqn 3.79, it seems that apart from severe multi-dimensional loading conditions, neutral loading paths are not likely to occur in many practical situations where

The continuum theory of rock mechanics 109

moderate loading conditions are generally encountered. However, the validity of such a statement, and the consequences, cannot be ascertained unless numerical studies of practical problems are performed. Such numerical studies are not presently available for the third-order hyperelastic models combined with a loading criterion such as the complementary energy function, Ω.

In general, it has been clearly demonstrated that, except for certain special cases of loading (e.g. increasing proportional loading), the deformation type of theories cannot lead to meaningful results, and sometimes they lead to contradictions. As mentioned previously, these types of models do not satisfy the *continuity* requirement for loading conditions near or at neutral loading. Basically, the difficulty lies in the fact that the deformation theory and the existence of the loading function, f, even in the most limited sense, are incompatible. This has led naturally to the consideration of the second type of formulation based on an incremental theory of plasticity. This theory is based on three fundamental assumptions: the shape of an *initial yield surface*, the evolution of subsequent *loading surfaces* (*hardening rule*), and the formulation of an appropriate *flow rule*. In addition, the total strain increments, $\dot{\varepsilon}_{ij}$, are assumed to be the sum of the elastic and plastic strain components $\dot{\varepsilon}_{ij}^e$ and $\dot{\varepsilon}_{ij}^p$, respectively. Different constitutive models based on the incremental theory of plasticity are described in the following section.

3.9.4. Variable Moduli Models

A generalization of the deformation theory of plasticity for the case of incremental stress–strain models is that now known as the *variable moduli models* (Chen and Saleeb, 1982). In these later models, different forms for the material response functions apply in initial loading and in subsequent unloading and reloading, i.e. the models are generally irreversible, even for incremental loading.

The mathematical description of the variable moduli model is given in terms of the incremental stress–strain relations

$$\dot{p} = K\dot{\varepsilon}_{kk}; \qquad \dot{s}_{ij} = 2G\dot{e}_{ij} \qquad (3.81)$$

where $\dot{p}$ and $\dot{\varepsilon}_{kk}$ are the mean hydrostatic stress and the volumetric strain increments, respectively, and $\dot{s}_{ij}$ and $\dot{e}_{ij}$ are the deviatoric stress and strain increments. Generally, different functions for shear modulus G and bulk modulus K apply in initial loading, and in subsequent

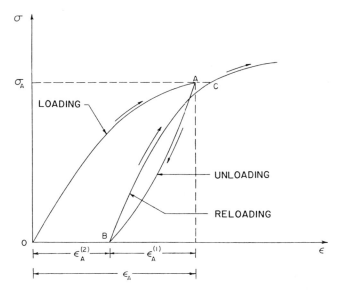

Fig. 3.11. Typical uniaxial stress–strain relation for a variable moduli model.

loading, unloading and reloading. This is illustrated in Fig. 3.11. These relations are incrementally isotropic.

Variable moduli models have many advantages. They can give a good overall fit to the full set of tests available, and they are capable of fitting repeated hysteretic data in cyclic loading. In addition, they are computationally simple and relatively easy to fit to data. However, phenomena such as dilatation, cross effects and noncoincidence of principal axes of stress and strain increments cannot be described by such relations as eqns 3.81. The other problem associated with this type of formulation is that the model may not satisfy all rigorous theoretical requirements for all stress histories. For instance, at or near neutral loading conditions in shear, the model fails to satisfy the continuity condition (Chen and Saleeb, 1982).

3.10. FLOW PLASTICITY OF ROCKS

In the flow theory of plasticity for strain or work hardening materials, the development of the incremental stress–strain relation is based on the

The continuum theory of rock mechanics 111

following three fundamental assumptions (Chen and Saleeb, 1985):

1. The existence of initial and subsequent yield (loading) surfaces.
2. The formulation of an appropriate hardening (softening) rule that describes the evolution of subsequent loading surfaces.
3. A flow rule which specifies the general form of the stress–strain relationship.

In addition, during plastic flow, the total incremental strains, $\dot{\varepsilon}_{ij}$, are assumed to be divisible into elastic, $\dot{\varepsilon}^e_{ij}$, and plastic, $\dot{\varepsilon}^p_{ij}$, components, i.e.

$$\dot{\varepsilon}_{ij} = \dot{\varepsilon}^e_{ij} + \dot{\varepsilon}^p_{ij} \tag{3.82}$$

The elastic response is normally assumed to be governed by an incremental form of the generalized Hooke's law, although any other isotropic or anisotropic nonlinear elastic model may be used as well. Therefore, one may write the elastic strain increments, $\dot{\varepsilon}^e_{ij}$, in terms of the stress increment, $\dot{\sigma}_{kl}$, as

$$\dot{\varepsilon}^e_{ij} = D^e_{ijkl} \dot{\sigma}_{kl} \tag{3.83}$$

where D^e_{ijkl} is the elastic tangential compliance tensor. Similarly, in the flow theory of plasticity, the plastic strain increments, $\dot{\varepsilon}^p_{ij}$, are assumed to be linearly related to the stress increment tensor, i.e.

$$\dot{\varepsilon}^p_{ij} = D^p_{ijkl} \dot{\sigma}_{kl} \tag{3.84}$$

where, as for D^e_{ijkl}, D^p_{ijkl} is the *plastic tangential compliance* tensor, which may depend on the stress and strain states and the loading history.

Substituting eqns 3.83 and 3.84 into eqn 3.82, the total incremental strains, ε_{ij}, can be expressed in terms of $\dot{\sigma}_{kl}$ as

$$\dot{\varepsilon}_{ij} = (D^e_{ijkl} + D^p_{ijkl}) \dot{\sigma}_{kl} \tag{3.85}$$

or

$$\dot{\varepsilon}_{ij} = D^{ep}_{ijkl} \dot{\sigma}_{kl} \tag{3.86}$$

in which D^{ep}_{ijkl} is the elastic-plastic tangential compliance tensor.

In general, it is always possible to obtain the inverse relations for eqns 3.85 and 3.86, and these relations are often written in the following forms

$$\dot{\sigma}_{ij} = (C^e_{ijkl} - C^p_{ijkl}) \dot{\varepsilon}_{kl} \tag{3.87}$$

or

$$\dot{\sigma}_{ij} = C^{ep}_{ijkl} \dot{\varepsilon}_{kl} \tag{3.88}$$

where C_{ijkl}^e, C_{ijkl}^p and C_{ijkl}^{ep} are the elastic, plastic and elastic-plastic tangential stiffness tensors, respectively. The matrix forms corresponding to the incremental stress–strain relations of eqns 3.87 and 3.88 can be expressed as

$$\{\dot{\sigma}\} = ([C]^e - [C]^p)\{\dot{\varepsilon}\} \tag{3.89}$$

or

$$\{\dot{\sigma}\} = [C]^{ep}\{\dot{\varepsilon}\} \tag{3.90}$$

Figure 3.12 depicts a general loading or unloading situation shown for two corresponding stress and strain components. Note the similarity between flow plasticity and hypoelasticity, and the difference in unloading. The flow plasticity exhibits both stress and strain history dependence and stress induced anisotropy.

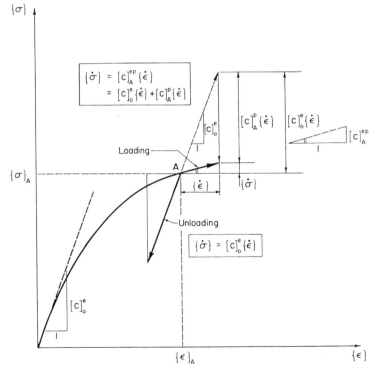

Fig. 3.12. Flow theory of plasticity.

3.11. PROGRESSIVELY FRACTURING THEORY OF ROCKS

Classical plasticity theory is not well adapted to the model softening behavior of rocks beyond the post-failure range. The progressive loss of strength with increasing strain in the post-peak strength range is called *strain softening*. The only physical mechanism which can explain strain-softening is microcracking. For this type of material, Dougill (1976) developed a theory for an ideal *progressively fracturing solid*. This theory is completely analogous to incremental plasticity described in the previous section.

Instead of loading surface in stress space, he assumes the existence of a *fracturing surface* $\Phi(\varepsilon_{ij}, H_k) = 0$ in strain space. This surface is used to distinguish loading and unloading. The surface can translate and change its shape. During unloading, the stress–strain relation is assumed to be linearly elastic. During loading, it is assumed that the rate

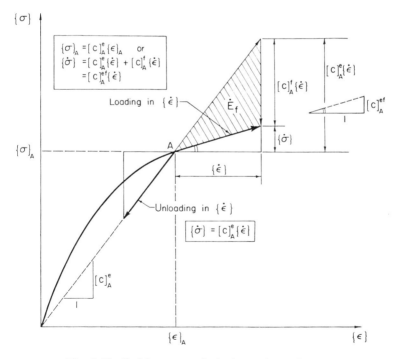

Fig. 3.13. Stable progressively fracturing solids.

of *fracturing stress* $\dot{\sigma}_{ij}^f$ can be obtained from the gradients of a fracturing potential $\Psi(\varepsilon_{ij}, H_k)$,

$$\dot{\sigma}_{ij}^f = \dot{\lambda} \frac{\partial \Psi}{\partial \varepsilon_{ij}} \tag{3.91}$$

An associated fracturing rule is obtained if $\Psi = \Phi$.

Following the similar procedures as those used for eqns 3.82–3.90, and using the consistency condition $\dot{\Phi} = 0$, an explicit progressively fracturing stress–strain relation can be written, in the usual matrix form

$$\{\dot{\sigma}\} = [C]^e \{\dot{\varepsilon}\} + [C]^f \{\dot{\varepsilon}\} = [C]^{ef} \{\dot{\varepsilon}\} \tag{3.92}$$

where $[C]^e$, $[C]^f$ and $[C]^{ef}$ are the elastic, fracturing and elastic-fracturing tangential stiffness matrices. The rate of dissipated energy during loading is given by

$$\dot{E}_f = \tfrac{1}{2} \{\dot{\sigma}\}^f \{\varepsilon\} \tag{3.93}$$

This elastic-fracturing material model is illustrated graphically in Fig. 3.13. The fracturing model exhibits strain induced anisotropy and strain history dependence. Bazant (1978) has extended this theory to plastic-fracturing solids. Good modeling of the behavior of concrete material under increasing loads has been reported.

3.12. LINEAR VISCOELASTICITY OF ROCKS

The elastic and plastic characterizations of materials are essentially time-independent. However, engineering materials in general and rocks in particular are subject to time-dependent deformation when loaded over a finite time interval. To account for the time-dependent behavior, a rock may be symbolized as deforming not only as an elastic, plastic solid, but also as a viscous fluid. This section attempts to characterize *time-dependent flow* in rocks in terms of *viscosity* as well as elasticity.

The basic elements that may be used to form a series of 'linear' models are represented by (1) a *spring* (elastic deformation—*Hookean* substance); (2), a *dashpot* (viscous deformation—*Newtonian* substance). A typical example is the *Maxwell model*, consisting of a spring and dashpot in series and used to describe a perfect viscoelastic material. The total strain of this model is given by

$$\varepsilon = \frac{\sigma}{E} + \frac{1}{\eta} \int \sigma \, dt \tag{3.94}$$

where η is the *coefficient of viscosity*. Equation 3.94 can be expressed in the differential form

$$\frac{\sigma}{\eta} + \frac{1}{E} \frac{d\sigma}{dt} = \frac{d\varepsilon}{dt} \tag{3.95}$$

In the case of *creep* which is under a constant stress condition ($d\sigma/dt = 0$, $\sigma = \sigma_0$), eqn 3.95 reduces to

$$\frac{d\varepsilon}{dt} = \frac{\sigma_0}{\eta} \tag{3.96}$$

which indicates that the strain increases linearly with time.

In the case of *stress relaxation* which is under a constant strain condition ($d\varepsilon/dt = 0$), eqn 3.95 reduces to

$$\frac{\sigma}{\eta} + \frac{1}{E} \frac{d\sigma}{dt} = 0 \tag{3.97}$$

which solves to give the exponential form

$$\sigma = \sigma_0 \exp\left(-\frac{Et}{\eta}\right) \tag{3.98}$$

This demonstrates the exponential process of stress relaxation at constant strain.

Another example is the *Kelvin–Voight* model comprising a spring and dashpot in parallel. The total stress of this model is given by

$$\eta \frac{d\varepsilon}{dt} + E\varepsilon = \sigma \tag{3.99}$$

which under constant stress (σ_0) conditions solves to give

$$\varepsilon = \frac{\sigma_0}{E} \left[1 - \exp\left(-\frac{Et}{\eta}\right) \right] \tag{3.100}$$

representing an exponentially increasing creep strain to a maximum value σ_0/E and exponentially decreasing strain leading to recovery.

A three-parameter solid model, generalized from the Kelvin–Voight model, has been used successfully to predict the creep behavior of oil shale. The following creep function of the generalized Kelvin–Voight model has been proposed by Chu and Chang (1980).

$$\varepsilon = \frac{\sigma_0}{k_2} + \frac{\sigma_0}{k_1}\left[1 - \exp\left(\frac{k_1 t}{\eta_1}\right)\right] \tag{3.101}$$

where k_1 and k_2 are spring constants and η_1 is the coefficient of viscosity. These three material constants depend on temperature, stress level and oil (or organic) content (O_c). They are obtained by curve-fitting with creep tests. It was found that the three-parameter solid model can be used to calculate *primary* and *secondary* creep of oil shale under various stress and temperature conditions.

The effects of stress level, organic and mineral contents on the creep and relaxation behaviors of oil shale and rheological models for predicting creep and the relaxation of oil shale have been discussed by Chong *et al.* (1978), Olsson (1980), among others. As an example, the following linear viscoelastic constitutive equation is found to describe the stress–relaxation behavior of oil shale very well (Olsson, 1980)

$$\sigma(t) = \int_{-\infty}^{t} R(t-\tau)\dot{\varepsilon}(\tau)\,\mathrm{d}\tau + E_e\varepsilon(t) \tag{3.102}$$

The function R is referred to as the *relaxation function*, t is the current time, τ is variable time and E_e is the equilibrium modulus. If $E_e \neq 0$, the equation describes the stress response of a viscoelastic solid and if $E_e = 0$ the material is a viscoelastic liquid.

The linear viscoelastic constitutive law, eqn 3.102, can be generalized to the form

$$\sigma_{ij}(t) = \sigma_{ij}^0 + \int_{0}^{t} R_{ijkl}(t,\tau)\dot{\varepsilon}_{kl}(\tau)\,\mathrm{d}\tau \tag{3.103}$$

where R_{ijkl} denotes the *relaxation tensor* and σ_{ij}^0 denotes the initial stress. Equation 3.103 expresses a *superposition principle* of stress. To facilitate more effective computer analysis, the relaxation tensor R_{ijkl} can be approximated as an exponential series, and eqn 3.103 can be written in the matrix form

$$\{\sigma(t)\} = \{\sigma\}^0 + \sum_{i=1}^{N} \{\sigma(t)\}_i \tag{3.104}$$

where $\{\sigma(t)\}_i$ can be otained from

$$\{\dot{\sigma}(t)\}_i = [R(t)]_i\{\dot{\varepsilon}(t)\} - \frac{1}{\tau_i}\{\sigma(t)\}_i = \{\dot{\sigma}\}_i^e - \{\dot{\sigma}\}_i^r \tag{3.105}$$

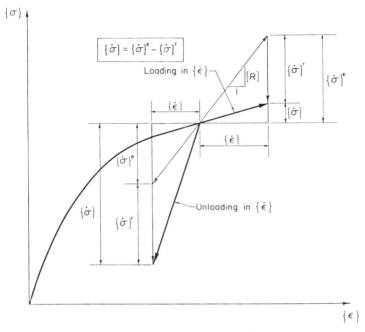

Fig. 3.14. Linear viscoelasticity.

where τ_i may be interpreted as the discrete relaxation time, and N is the number of elements in the series. Equation 3.102 corresponds to a generalized Maxwell chain model with age-dependent stiffnesses $[R]$ and viscosities $\tau_i[R]_i$.

Figure 3.14 illustrates the response of the viscoelastic model to an increment in strain. It can be seen that the tangential stiffness is age and strain history dependent, and an unloading results in a changed tangential stiffness. Due to the relaxation stress, the viscoelastic model exhibits rate-dependence.

3.13. ELASTOVISCOPLASTICITY OF ROCKS

This is a theory for rate-dependent plastic materials. The total strain rate $\dot{\varepsilon}_{ij}$ is decomposed into a linear elastic strain rate $\dot{\varepsilon}_{ij}^e$ and a combined viscous-plastic (or simply viscoplastic) strain rate $\dot{\varepsilon}_{ij}^{vp}$

$$\dot{\varepsilon}_{ij} = \dot{\varepsilon}_{ij}^e + \dot{\varepsilon}_{ij}^{vp} \tag{3.106}$$

where the viscoplastic strain rate is obtained from the *viscoplastic potential function* $G(\sigma_{ij}, H_k)$ and the loading function $F(\sigma_{ij}, H_k)$ in analogy with the flow theory of plasticity

$$\dot{\varepsilon}_{ij}^{vp} = \Phi(F) \frac{\partial G}{\partial \sigma_{ij}} \qquad (3.107)$$

where $\Phi(F) = 0$ if $F \leq 0$, and if $F > 0$ the function $1/\Phi$ may be regarded as a viscosity parameter which may be chosen to best fit experimental results. By analogy with the flow theory of plasticity, an associated flow rule is obtained if $F = G$. From eqns 3.106 and 3.107, the classical viscoplastic constitutive relation can be expressed in the matrix form

$$\{\dot{\sigma}\} = [C]^e (\{\dot{\varepsilon}\} - \{\dot{\varepsilon}\}^{vp}) = \{\dot{\sigma}\}^e - \{\dot{\sigma}\}^{vp} \qquad (3.108)$$

Equation 3.108 may be considered as a generalization of the flow

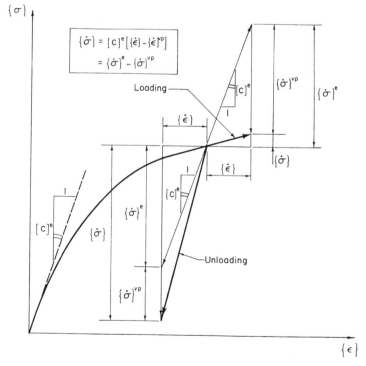

Fig. 3.15. Elasto-viscoplasticity.

The continuum theory of rock mechanics   119

theory of plasticity. Thus, the modeling techniques used in plasticity can be adopted in the theory of viscoplasticity.

Figure 3.15 illustrates the response of the viscoplastic model to an increment in strain. It is observed that the unloading from a viscoplastic state of stress can manifest itself either as an elastic or an elastoviscoplastic one. The elastoviscoplastic model is capable of describing age, rate and strain history dependence. For rock materials, it offers therefore the potential of an accurate modeling technique for a wide range of loading conditions. Despite this, little application to rocks in general and oil shale in particular has been reported in the literature.

3.14. ENDOCHRONIC PLASTICITY OF ROCKS

Endochronic theory is best approached as a special case of viscoplasticity (Bazant, 1978). By introducing a variable ζ, called the *intrinsic time measure*, defined as

$$\dot{\zeta}^2 = P_{ijkl} \, d\varepsilon_{ij} \, d\varepsilon_{kl} \tag{3.109}$$

which is a measure of the deformation history or path length, the *intrinsic time scale z* is then defined by the equation

$$\dot{z} = \frac{\dot{\zeta}}{f(\zeta)}, \qquad \frac{dz}{d\zeta} > 0, \qquad z \geqslant 0 \tag{3.110}$$

where P and f are material functions and therefore, scale z becomes a unique property of a material.

The incremental stress in endochronic theory is obtained from

$$\{\dot{\sigma}\} = [C]^e (\{\dot{\varepsilon}\} - \{\dot{\varepsilon}\}^{en}) = \{\dot{\sigma}\}^e - \{\dot{\sigma}\}^{en} \tag{3.111}$$

where the *endochronic strain increment* $\{\dot{\varepsilon}\}^{en}$ is given by

$$\dot{\varepsilon}_{ij}^{en} = \frac{\partial F}{\partial \sigma_{ij}} \dot{z} \tag{3.112}$$

and $F(\sigma_{ij}, \varepsilon_{ij})$ is an *endochronic loading function*. Note the similarity of eqn 3.112 with the viscoplastic relation eqn 3.107, but also note the main difference. No distinction is made between loading and unloading for an ordinary endochronic formulation. The 'trick' that makes this possible is illustrated in Fig. 3.16, in which the stress increment $\{\dot{\sigma}\}$ corresponding to given $\{\dot{\varepsilon}\}$ values is shown as a sum of the elastic increment $\{\dot{\sigma}\}^e$ and the inelastic decrement $\{\dot{\sigma}\}^{en}$. When the sign of $\{\dot{\varepsilon}\}$

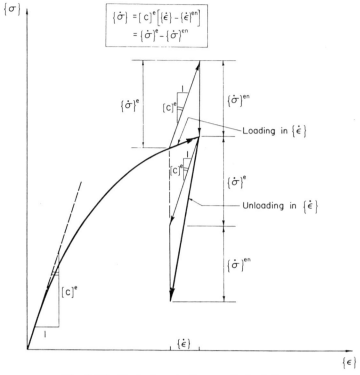

Fig. 3.16. Endochronic theory of plasticity.

is reversed, elastic increment $\{\dot{\sigma}\}^e$ changes sign but $\{\dot{\sigma}\}^{en}$ does not, which automatically makes the unloading slope steeper than the loading slope. Thus, the irreversibility of inelastic behavior of a material is built in *a priori*, without any inequalities.

Here, as in viscoplasticity, the endochronic theory has great potential and offers versatility in modeling the response of rocks to repeated and cyclic loads, but at present none has been reported in the literature for rocks.

3.15. ENDOCHRONIC VISCOPLASTICITY OF ROCKS

This is a direct generalization of previous endochronic plasticity theory. The time scale $\dot{z}$ defined in eqn 3.110 can now be extended to the new time scale $\dot{\xi}$ in the form

$$\dot{\xi}^2 = \alpha^2(\dot{z}^2 + \beta^2 \dot{t}^2) \tag{3.113}$$

where α depends on $\dot{z}$ and β is a material parameter.

The general behavior of this model is illustrated in Fig. 3.17. In many respects it is very close to the viscoplastic model. The theory of endochronic viscoplasticity has been applied to concrete material with good success (Bazant, 1978). This theory may be combined with the fracturing solid theory of Dougill (1976) described previously in Section 3.11 to model the softening behavior of rocks.

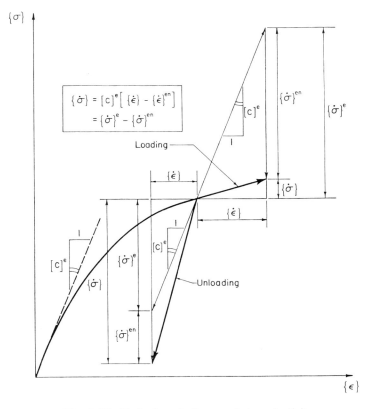

Fig. 3.17. Endochronic theory of viscoplasticity.

3.16. CONCLUSIONS

The state-of-the-art in the application of solid mechanics to the mining problems related to rocks in general and oil shales in particular has been considered. Various analytical approaches to the estimation of stress, strain and displacement are available, but the finite element method is certainly the most versatile. With the present state of development of finite-element computer programs, the problem of modeling the mechanical behavior of rocks for use in analytical studies of mining problems remains one of the most difficult challenges in the field of rock mechanics. Herein, special emphasis has been placed on constitutive modelings, and the difficulties of characterizing rock behavior in the field in a realistic way, within the framework of continuum mechanics in general and the theory of elasticity, viscosity and plasticity in particular. A unified treatment of available mathematical models of rocks as commonly used in rocks and oil shales and also those offering a future possibility has been presented.

The mechanics of rock in the elastic range are well understood. Predictions of acceptable accuracy for some mining problems can be made. However, elastic analyses do not account for such important phenomena as roof falls, pillar spalls, creep and irreversible deformations. In general, it is this inelastic, irreversible range of deformation that is of great concern in a mining problem.

In order to describe the behavior of rock beyond the elastic range, an elastic-plastic-viscous approach is advocated in which yield is pressure-dependent and time-dependent. To this end, the general techniques used in the modeling of rocks in general and oil shales in particular for elastic, plastic, viscous materials are discussed in some detail. In this brief chapter, one can only address the basic concepts and general approaches. Fortunately, it is from these areas that the fundamental research questions of rock mechanics are taken.

All rock is discontinuous to some extent, but it becomes irrelevant at some scale of aggregation and a *continuum* view is necessary for progress. The mathematical theories of elasticity, viscosity and plasticity all follow the same course. Firstly, the notions of stress and strain are developed; secondly, strain equations describing the geometry of deformation of a continuum, and stress equations expressing the basic physical principles of equilibrium or of motion are set up but in order to arrive at a system of equations which enable the state of stress and strain to be calculated, stress–strain–time relations must be obtained,

The continuum theory of rock mechanics 123

which idealize the behavior of actual materials. The form of such relations are not entirely arbitrary. They must satisfy the basic principles of continuum mechanics including the restrictions from thermodynamic laws. It is, of course, the simple expressions on rocks that determine the actual relations between stress and strain.

In the linear elastic range, the stress–strain relations are embodied in *Hooke's law*. The linear model can be significantly improved by assuming bilinear or higher polynomial fits for the nonlinear elastic stress–strain relationship, the range of reversible deformation. The mechanics in this range are well developed. The elastic models must of course be combined with criteria defining '*failure*' of the material.

Constitutive equations for rock deformed beyond the elastic range represent an area of great importance to rock mechanics. It is this irreversible, inelastic range of deformation that is of major concern and that is precisely the area where genuine rock mechanics problems arise.

The elastic formulations fail to identify *inelastic* deformations, a shortcoming that becomes apparent when the material experiences *unloading*. This can to some extent be rectified by introducing unloadings as in the *deformation theory of plasticity*. However, the deformation theory can not account for load *history* effects.

The *flow theory of plasticity* using pressure dependent yield and *associated flow rules* enables one to go beyond the elastic range in a time-independent but theoretically consistent way, because the theory satisfies the conditions of uniqueness, continuity, stability and thermodynamic laws. Consistency is admirable and the practicability of the theoretical situation is something that cannot be avoided. Thus, where it is possible, efforts should be made to introduce theoretical concepts and mathematics to rocks. If this is not possible, an empirical and engineering treatment may be introduced. However, experiments for determining the stress–strain diagrams beyond the elastic range for two- and three-dimensional general states of stress are difficult to perform. Little has been done in this area in rock mechanics in general and oil shale in particular.

The question of modeling *strain softening* using plasticity is highly controversial. This can to some extent be rectified by introducing the concept of *progressively fracturing solids* (Section 3.11). The formulation of a purely fracturing material is completely analogous to flow plasticity.

The presence of *viscous* mechanism of deformation raises the possibility of time-dependent deformation. Although much has been done about *creep* and *stress relaxation* in rock and applications of *viscoelasticity* and *viscoplasticity* have been reported in the literature, oil shale and tar sand deposits have shown significant time effects at ambient as well as elevated temperatures. The role of time, in general, does not appear to be well understood at ambient or elevated temperatures.

As apparent from the preceding discussion, it is necessary to include many complex effects in the material model in order to describe the behavior of rock properly. An important step in the direction of developing a more unified and comprehensive material model for rock is the generalization of the theory of viscoplasticity by introducing the measure of *intrinsic time* (Valanis, 1971). This is known as the *endochronic theory of viscoplasticity*. This theory offers the possibility of accurately modeling of rock behavior in a wide range of loading conditions. Despite this, there seems to be no application to rock yet either in static or dynamic loading conditions. Note that the theory has been adopted with good success to concrete materials. It should be emphasized here that this modeling requires the determination of a large set of functions. Thus, it is expected to demand excessive programming and computer effort.

ACKNOWLEDGMENT

The preparation of this material is supported in part by the National Science Foundation under grant no. PFR-7809326 to Purdue University.

REFERENCES

Asszonyi, C. S. and Richter, R. (1979). *The Continuum Theory of Rock Mechanics*, Trans Tech Publications, Rockport, MA.

Bazant, Z. P. (1978). Inelasticity and failure of concrete: a survey of recent progress, *Proc. Spec. Sem. Anal. Reinforced Concr. Struct. Means Finite Element Method*, Milan, pp. 5–59.

Chang, N. Y. and Bondurant, E. J. (1979). Oil shale strength characterization through multiple stage triaxial tests, *Proc. 20th US Symp. on Rock Mech.*, Austin, Texas, 4–6 June, pp. 393–401.

Chen, W. F. (1975). *Limit Analysis and Soil Plasticity*, Elsevier, Amsterdam.

Chen, W. F. (1982). *Plasticity in Reinforced Concrete*, McGraw-Hill, New York.

Chen, W. F. (1984a). Soil mechanics, plasticity and landslides. In: *Mechanics of Inelastic Materials*, eds G. J. Dvorak and R. T. Shields, Elsevier, Amsterdam, pp. 31–58.

Chen, W. F. (1984b). Constitutive modeling in soil mechanics. In: *Constitutive Laws for Engineering Materials: Theory and Applications*, eds C. S. Desai and R. H. Gallagher, John Wiley, London.

Chen, W. F. and Mizuno, E. (1985). *Plasticity Cap Models in Geotechnical Engineering*, Elsevier, Amsterdam.

Chen, W. F. and Saleeb, A. F. (1982). *Constitutive Equations for Engineering Materials*, Vol. 1—Elasticity and Modeling, Wiley-Interscience, New York.

Chen, W. F. and Saleeb, A. F. (1985). *Constitutive Equations for Engineering Materials*, Vol. 2—Plasticity and Modeling, Wiley-Interscience, New York.

Chong, K. P., Smith, J. W. and Khaliki, R. A. (1978). Creep and relaxation of oil shales, *Proc. 19th US Symp. on Rock Mech.* Reno, Nevada, pp. 414–21.

Chong, K. P., Smith, J. W., Chang, B. and Roine, S. (1979a). Oil shale properties by split cylinder method, *J. of the Geotech. Engrg. Division*, ASCE, **105**(GT5), 595–611.

Chong, K. P., Uenishi, K., Munari, A. C. and Smith, J. W., (1979b). Three dimensional characterization of the mechanical properties of Colorado oil shale, *Proc. of the 20th US Symposium on Rock Mechanics*, Austin TX, 4–6 June.

Chong, K. P., Uenishi, K., Smith, J. W. and Munari, A. C. (1980a). Non-linear three dimensional mechanical characterization of Colorado oil shale, *Int. J. Rock Mech. Min. Sci. & Geomech.*, **17**, 339–47.

Chong, K. P., Chen, J. L., Uenishi, K. and Smith, J. W. (1980b). A new method to determine the independent shear moduli of transversely isotropic materials, *Proc. 4th Int. Congr. on Experimental Mechanics*, Boston, Mass.

Chong, K. P., Smith, J. W. and Borgman, E. S. (1982). Tensile strengths of Colorado and Utah oil shales, *J. of Energy* **6**(2), 81–5.

Chu, M. S. and Chang, N. Y. (1980). Uniaxial creep of oil shale under elevated temperatures, *Proc. of the 21st Symposium on Rock Mechanics*, Univ. of Missouri, Rolla, Mo., 28–30 May, pp. 207–16.

Dismuke, T. D., Chen, W. F. and Fang, H. Y. (1972). Tensile strength of rock by the double-punch method. *Rock Mech.* **4**, 79–87.

Dougill, J. W. (1976). On stable progressively fracturing solids, *ZAMP* **27**(4), 423–37.

Farmer, I. W. (1968). *Engineering Properties of Rocks*, E. & F. N. Spon, London.

Jaeger, J. C. (1969). *Elasticity, Fracture and Flow*, 3rd Edn, Methuen, London.

Olsson, W. A. (1980). Stress-relaxation in oil shale, *Proc. of the 21st Symposium on Rock Mechanics*, Univ. of Missouri, Rolla, Mo., 28–30 May, pp. 517–22.

Pariseau, W. G. (1977). Mechanics considerations in coal mine design, *Proceedings of a NSF Workshop on Mechanics Problems Associated with Mining and Processing of Energy Related Minerals*, Asilomar, California, pp. 120–42.

Reed, J. J. (1966). Rock mechanics applied to oil shale mining, *Colorado School Mines Quart.* **61**(3), 45–53.

Sellers, J. B., Haworth, G. R. and Zambas, P. G. (1972). Rock mechanics research in oil shale mining, *Soc. Min. Eng. AIME, Trans.*, **252**(2), 222–32.

Smith, J. W. (1969). Theoretical relationship between density and oil yield for oil shales. *US Bureau of Mines Report of Investigations* 7248, 14pp.

Timoshenko, S. P. and Goodier, J. N. (1951). *Theory of Elasticity*, McGraw-Hill, New York.

Valanis, K. C. (1971). A theory of viscoplasticity without a yield surface, *Archiwum Mechaniki Stossowanej* (*Archives of Mechanics*), **23**, 517–51. (In Polish.)

Chapter 4

Sampling, Logging and *in situ* Stresses of Oil Shale

GEORGE F. DANA

Manager, Geology and Geochemistry Research Group,
Western Research Institute, Laramie, Wyoming, USA

SUMMARY

Any evaluation of oil shale from its in-place oil yield to its final form of spent shale requires sampling of the strata as it exists in place. The sampling should take into account weathering, vertical depth from surface, horizontal distance from outcrop, related fractures and joints, accessory minerals, both primary and secondary, and fluids, both migratory and static. Some mechanical and physical characterization research has been accomplished on cores from oil shale, but very little research has been done in boreholes or coreholes penetrating oil shale strata. This chapter will discuss the sampling of oil shale, some common types of analyses, the determination of fractures and joints, and their relationship to in situ *stresses.*

4.1. SAMPLING

4.1.1. General

The best method of obtaining oil shale samples from oil shale strata is to cut core samples in a manner providing material from which maximum scientific determinations may be made. The larger the diameter of the core, the more complete the information will be, especially the tectonic or fracture/joint data. Smith *et al.* (1963) discussed the relative merits of core versus drill cuttings samples for oil

128 *George F. Dana*

yield determinations under both air-drilling and mud-drilling conditions. The conclusions reached were that coring produced the most reliable samples, that air/foam cuttings, when carefully taken, are the best sample and that mud cuttings are the least desirable types of samples for oil yield research.

4.1.2. Surface Sampling

For both oil yield and fracture/joint patterns, surface sampling can be used in lieu of a drilling/coring program. These methods provide only preliminary determinations of where and how deep to sample using drilling methods. Two types of surface sampling were used in the early days of oil shale strata exploration and occasionally these techniques are still used to explore remote areas suspected to contain oil shale outcrops. The first and least desirable is surface skim where only the outer layer or the weathered and broken rocks are harvested, usually without the aid of tools. These rocks have been weathered and such samples can be used only for determining the type of rock in place and its relative oil yield to similar samples. However, these surface samples are valuable from the standpoint of indicating both large and small fracture and joint patterns and zones of weakness at the surface and probably at some depth in the subsurface. The second method is surface crop where hand or mechanical tools are used to obtain samples at some horizontal distance into an outcrop of oil shale, depending on the amount of weathering and the kind of oil shale affected. Massive vertical outcrops of comparatively non-laminated oil shale will produce good samples within one to a few feet of the outcrop surface whereas laminated and non-massive oil shales will be weathered to distances of up to a dozen or so feet into the outcrop. Where no to little overburden is found on oil shale outcrops, weathering effects may exist as deep as $15 \cdot 3$ m (50 ft). Samples from weathered surface outcrops will yield from 5–20% less in Fischer assay for western oil shales and up to 50–70% less in eastern Devonian oil shales.

If one's sampling program is initially limited to outcrop sampling, penetration of the rock to where little or no visible weathering effects are observed is the preferred method of obtaining usable samples. Trenching is a method of sampling a vertical section of oil shale which involves digging back from the outcrop face and sampling continuously every one or more feet of comparatively unaltered rock section. However, if significant lithologic changes occur in the section, sampling on a less than 1 ft length may be desirable over short distances.

Sampling, logging and in situ *stresses of oil shale* 129

4.1.3. Subsurface Sampling
Three methods for sampling subsurface oil shale beds in any formation are, in order of desirability: coring, air/foam drilling and mud drilling. Each of these is discussed.

4.1.3.1. Core
Coreholes to sample oil shale are drilled by a rotary rig using a diamond or tungsten carbide bit on a core barrel. These bits are diamond-studded or tungsten carbide-tipped tooth circles which cut rock from an annular ring, leaving an untouched cylinder of rock in the center. The bit is cooled and drilling debris is removed by the flow of drilling mud, foam or compressed air out through the bit and up the hole. As drilling progresses, the core—the central rock cylinder— passes up into the core barrel (essentially a metal tube 3–27·5 m (10 to 90 ft) long) where it is retained and lifted intact from the hole. At the surface the core is removed from the core barrel, laid out, reassembled and carefully measured. Any lost core is logged at the bottom of the interval drilled, unless it is obvious where the section was lost. In hard, non-fractured formations like the Green River Formation, complete core recovery is usually obtained (Gatlin, 1960). Oil shale cores which are obtained range from 3·8 to 15·2 cm ($1\frac{1}{2}$ to 6 in) in diameter, with the most common size being NX size, 5·4 cm ($2\frac{1}{8}$ in) in diameter.

Core taken perfectly is a continuous column of rock, an actual section of the formation. Because of the lateral uniformity of the oil shale formations, cores represent closely the rock in the same stratigraphic position in the surrounding area. Thus, core samples provide oil-yield data representative of the formation in their vicinity. Donnell (1961) used oil-yield data from core samples as the basis for determining indicated oil shale reserves within a 3·2 km (2 mile) radius.

4.1.3.2. Air/Foam Cuttings
Pneumatic rotary drilling procedures are quite similar to those used with hydraulic drilling, except that the mud stream is replaced by a flow of compressed gas—usually air—along with small amounts of injected water and detergent creating foam. For fire protection and cleanliness, the gas stream carrying drill cuttings from the borehole is discharged at a safe distance from the derrick area. Samples of cuttings are collected in a cloth bag attached to a small diameter pipe branching at a small angle off the discharge pipe. At regular intervals determined from kelly progress, the bag is changed and labeled.

Pneumatic drilling and this method of sample collection eliminates most of the factors affecting formation representation by hydraulically drilled cutting samples. No significant sorting of cuttings occurs because of high gas velocity and rapid removal of cuttings from the borehole. The sample diversion and collection method makes representative sample selection probable and fine particles are retained by the cloth bag. Cuttings from pneumatic drilling do not require washing and are not contaminated by mud components as occurs with hydraulic drilling; borehole wall scalings are the only contaminant. Because of the rapid rate of cuttings removal, the depth measured from kelly progress accurately reflects the sample depth. Even human difficulties are minimized by the simplicity of the sampling method. Water flowing into the borehole from a formation aquifer may require a change in the sample collection method. Even with this complication, however, collecting representative samples is simpler than with hydraulic drilling.

Substantially better representation of the formation oil yield should be afforded by air/foam cutting samples than by mud-cutting samples.

4.1.3.3. Mud Cuttings

Mud cuttings are the least desirable and least reliable method for obtaining oil shale samples for scientific evaluation. Gatlin's discussion (Gatlin, 1960) of hydraulic rotary drilling suggests characteristics of drilling and sampling procedures which may affect the representative quality of the samples. During drilling, mud is pumped down the drill stem and out through the bit to carry the broken rock fragments up and out of the hole. A typical procedure for collecting drill-cutting samples involves diverting part of the mud flow into a catch box drained by overflow. As the mud velocity is reduced, the entrained cuttings settle and are trapped in the catch box. At regular intervals determined by the distance drilled (usually $3 \cdot 1 - 9 \cdot 2$ m (10–30 ft)), the debris collected in the catch box is removed, and a sample up to 1 lb is cut out, washed, bagged and labeled. This is a general description of the process which has many variations. Sample catching equipment is frequently varied because catch boxes are of diverse designs made to fit particular installations. If the drill rig is equipped with a shale shaker (a screening device inserted in the mud stream), cutting samples may be taken directly from the debris accumulated by the shaker.

Ideally, a cutting sample from hydraulic drilling should provide a representative composite sample of the formation rock penetrated. The quality of representation actually obtained, however, may be affected by many factors in the drilling and sampling procedures.

Groups of factors appearing to present the most significant effects are: sorting of the cuttings during entrainment in the mud stream and during sample collection; sample contamination; mechanical, procedural and human errors.

Sorting of cuttings in the mud stream results from selective displacement of rock fragments from the sequence in which they were cut. As fragments are carried up the borehole, they settle in the mud; thus their rate of travel up the hole is less than the mud velocity. This difference in rate of travel, the fall rate of the particles, is affected by the specific gravity, shape and size of the rock particles, by the physical properties of the fluid, and by the rate and nature of the fluid flow. Because the specific gravity of Green River oil shale ranges from 2·7 for barren shale to about 1·6 for very rich shale (Smith, 1958), and the specific gravity of commonly used drilling muds ranges from 1 to 1·2, lean particles will fall somewhat faster than rich particles. Consequently, there is a continuously operating selective force at work on the cuttings during their travel, tending to smooth out the effect of natural stratigraphic variations in oil yields of the shale beds. Lean oil shale appears to break into larger particles than rich oil shale during drilling, tending to slow the fall of leaner particles in opposition to the effect of specific gravity. The high viscosity of drilling fluid also tends to minimize sample sorting as the higher the viscosity the slower the rate of fall of the cuttings. The overall effect of factors influencing the sorting of cuttings in the mud stream cannot be evaluated directly. During drilling in a continuous oil shale sequence, however, a collected sample representing a lean 3·1 m (10 ft) interval would probably contain fragments of richer shale from sections below and a sample representing a rich 3·1 m (10 ft) section would contain lean shale from sections above. As a result, in continuous oil shale, the oil yields of mud cuttings from rich and lean strata will be altered toward a median value and extreme deviations from this median will not appear. In effect this would rotate a relationship between oil yields from mud cuttings and core samples around this median so that above the median the oil yields indicated by cuttings would be lower than formation yield and below the median they would be higher.

Cuttings sorting during collection in a catch box or on a shale shaker is produced largely by differences in the size of particles. In a catch box with normal flow rates, large particles (+10 mesh) make up most of the collected debris, because fine material is washed from the box before it can settle. On a shale shaker, large particles are retained while smaller particles are carried through the screen with the mud flow. Since rich shales break into smaller particles than lean shales, or are more readily pulverized under action of the rotary bit, collection of leaner material is favored.

Sample contamination may arise from borehole wall scalings and from components of the drilling mud. Action of the drill stem may break out rock particles from friable rock beds above the drilling point, inserting scalings of inert material into the mud streak. Oil shale itself is a tough rock which does not scale readily. From caliper logs, boreholes in oil shale are bit size with little variation. Consequently, any scaling material would dilute the collected sample. In most locations in the Green River Formation, sandstones and siltstones occur above the oil shale beds so some scaling dilution probably occurs. Samples may be contaminated by drilling mud components added to control lost circulation, by incomplete washing of the cuttings and possibly by chemical effects of muds. Drilling muds are of variable composition and most contain materials added to obtain desirable viscosity, density and pH characteristics. Various fibrous or shredded materials—cellophane, tree bark, cane, wood, walnut shells—used to control lost circulation may appear in collected cuttings. These act as diluents since they yield little oil. Incomplete washing of cutting samples can leave bentonite from the drilling mud behind, also diluting the sample. Chemical action on the organic matter in the oil shale chips may result from the basicity of the mud (usually pH of about 10). The effect probably would be to lessen the apparent oil yield of the sampled interval, because treatment of oil shale with basic solutions (sodium carbonate) has been found to reduce oil production. The contaminants and the possible chemical action tend to decrease the oil yield of a mud cutting sample.

Mechanical, procedural and human errors affecting oil yield representation by cutting samples include difficulties encountered in correlating samples with depths, in obtaining representative samples of the cuttings and in maintaining a comprehensive sampling procedure. Depth measurements controlling the taking of samples are usually made from progress of the kelly during drilling. When the kelly has gone down $3 \cdot 1$ m (10 ft), a sample is taken and labeled as representing this $3 \cdot 1$ m (10 ft) of drilling progress. The cuttings actually collected in this sample, however, have traveled up the borehole; consequently, the footage level on the sample may be deeper than the depth at which the samples were cut. No exact correction for this error is minimized.

Representative samples of rock penetrated by the borehole would best be obtained by carefully sampling all of the material removed by the bit during drilling of 10 ft of hole. However, rock fragments from a 15·2 cm (6 in) hole drilled in ordinary rock would weigh about 136 kg (300 lb), requiring a reduction in the amount of sample caught. In the catch box system, diversion of part of the mud stream provides adequate reduction so that debris collected amounts to about 4·5 kg (10 lb); from the shale shaker system, the material collected is usually 13·6–22·7 kg (30–50 lb). Because finer particles of drill cuttings are not collected by either of these systems, neither can provide a truly representative fraction of the debris. Additional errors may be introduced during subsequent washing and sampling of this debris to obtain 0·5 kg (1 lb) samples.

In spite of all the possible variables, however, some degree of representation of oil yields of the formation rock is obtained from mud cuttings. The thickness and richness of oil-yielding sections of drill cutting samples from wells in the same area correspond.

4.2. DISCUSSION

Since coring is the best method of sampling oil shale formations for total scientific analyses, further discussion is warranted.

4.2.1. Fracture Record

If detailed fracture/joint information is of value in the subsurface, several factors should be considered in obtaining cores. Penetration rates should be slower than in other drilling to preserve the natural fractures and to help eliminate mechanically induced fractures due to high rotation speeds, vibration of the drill string, wobbling of the drill bit and weight on the drill bit and string. Cores taken with an excess of any of the above will tend to break up easier and occasionally to have a corkscrew surface.

Another method of aiding in the fracture of joint preservation is the use of a split inner core barrel. A standard solid barrel, where a core must be extracted from one end, may become clogged and the core inside broken during attempts to beat or pound the core out of the barrel. A split barrel is disengaged wherein one half is removed in a core tray and the core inside lies as it was cut and can be examined in its undisturbed state. The joints and fractures can be then recorded,

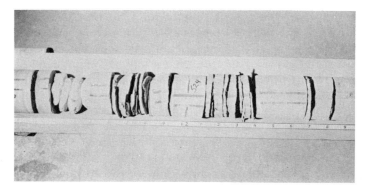

Fig. 4.1. 3 in (7·6 cm) core of Green River Basin oil shale from the Green River Formation. The rounded or thinly-spaced fractures are *usually* natural fractures. The sharply defined horizontal fractures are *usually* mechanically induced.

and the well-site geologist has a much better opportunity to determine whether the fractures are natural or are mechanically induced (see Fig. 4.1).

4.2.2. Core Handling

To assure the greatest possible detailing of the core for various purposes, treatment of the core after it is cut is a very important step. Upon removal of the upper half of the core barrel, the core should be examined to determine the frequency and type of fractures, the amount of rounding or turning of the core in the core barrel during the coring operation and the possible ways of fitting the core together for proper measurement and marking. If only a few fractures are evident, the core may be assembled prior to the recording of the fractures. If a large number of fractures exist in the column, a preliminary record should be made prior to reassembling or fitting the core so that if spilled, dropped, etc., some initial knowledge would be retained.

In either case, after the core is assembled by fitting the end pieces together it is measured and the length compared to the interval cut according to the driller's measurements or the Geolograph recording instrument present on most drilling rigs. If the rock column in the tray is shorter or longer than the cored interval, adjustments are made based on core recovery for the previous coring run or for sections of core which may be missing due to (1) grinding of core, (2) solution cavities or (3) core left as a stub near the bottom of the hole.

Sampling, logging and in situ *stresses of oil shale* 135

After that judgment is made, the core is marked with twin, closely-spaced lines from top to bottom. Wide felt-tipped magic markers with indelible ink are taped together for uniform line marking; one is red (always on the right side) and the other is any darker color. The second step in marking is drawing a line perpendicular to and through both vertical lines at each 1 ft mark and recording that drilling depth on either or both sides of the vertical lines. If the footage is marked twice, it will appear on each half after it is sawed or split. These types of markings serve the following purposes: (1) if a core or box of core is dropped or otherwise jumbled, it is much easier to reassemble it in proper sequence; (2) when the core is split, it is done so between the colored lines so that the same half of the core can be kept intact for lithologic and stratigraphic examination and description and the other half can be used for other analyses such as Fischer assay, x-ray diffraction for mineral content, etc., both destructive and non-destructive testing.

Unless field measurement and marking of the core is done with care, accuracy and final results suffer. It is very difficult to reconstruct the core after shipment. Careful and precise field marking is essential.

If mechanical and physical property characteristics are desired, the core is not split until those tests are complete because such testing requires (1) a full-diameter core or (2) a smaller diameter plug taken from a larger diameter core, depending on size. On very large diameter cores ($8 \cdot 9$–$15 \cdot 2$ cm, $3\frac{1}{2}$–6 in) the core is normally sawed first then half is split to reveal two fresh faces of rock, one of which is used for detailed lithologic description.

During sawing or splitting, new fractures which are created should be recorded to distinguish them from those noted earlier immediately following the emergence of the core from the core barrel.

After measurement and marking of the core in the field situation, a brief field description should be made of the lithology and its vertical changes so that some record is available (1) for determining coring depths and marker beds in the subsurface; (2) for comparison with known stratigraphy in the nearest available well; and (3) in case of the loss of core for any reason.

If the gas or moisture content of the core is an objective, the core is then dipped in wax or wrapped in cellophane or aluminium foil and/or placed in plastic bags, and on rare occasions the core is frozen prior to shipment. It is then placed in properly marked cardboard or wooden boxes for shipment and/or storage. Rich oil shale stored in boxes with no other cover will retain its structure and oil content for years if placed in a non-weathering atmosphere. Lean shale or clayey shales will desiccate, fracture and fall apart along planes of weakness (bedding planes) under the same storage conditions. Crushed samples, if not protected, will weather and within a year or two will lose some hydrocarbon yield value up to 30%.

If auxiliary minerals are present in the shale in the form of carbonates or evaporites, these will slowly effervesce to the surface of a core, usually in the form of white powdery crystals. In unsealed cores, these minerals will begin to show up within hours of exposure to air. Tuffs in a core usually contain such minerals in greater concentration than are found in oil shale.

4.2.3. Kinds of Fractures, Joints, and Faults

The beginning of a study of these features consists of examination of both large and small scale aerial photographs of areas underlain by oil shale deposits. Fault traces and joint patterns can be distinguished in various parts of the basins and the density and frequency of joints and fractures on a large scale can be mapped and located for field checks. Contact prints, either color or black and white, aid in following fault and fracture traces due to their three-dimensional aspect. Areas of stress and strain as indicated by these features can be determined for future detailed investigation.

Following aerial investigations, field work will permit measurements of dip, strike, direction, numbers of sets of fractures and joints, and occasionally even the throw of a fault with surface expression.

In cores both macro and micro joints or fractures are observed. Three principal joint and fracture types are found. The first is diagonal, many of which are induced in the coring process. These diagonal fractures are either planar or have a conchoidal surface. The richer the shale, the more likely a conchoidal breakage surface will be found or created. A fairly common occurrence is a series of small *en echelon* fractures with common angles and directions of fracture surfaces. The second joint type is horizontal along bedding planes. These most commonly appear in lighter colored varves or on the top and bottom of tuff layers. Because these are not so strongly bound by matrix content, they separate first and most frequently. It is most difficult to determine with any certainty that these fractures are natural or if they are induced. On rare occasions, secondary crystallization along a bedding

Sampling, logging and in situ *stresses of oil shale*   137

plane will indicate that the fracture is natural and has existed for a long enough time for percolating water to deposit and grow crystals such as calcite, shortite, carbonates and even rare quartz. The third joint type is vertical, perpendicular to bedding planes. These are infrequently found in cores because of the horizontal extent of area covered by a small corehole. However, in an area with frequent vertical fractures, usually found at shallow depths, they can be detected in cores drilled at closely spaced locations. This phenomenon was observed at US Department of Energy, Laramie Energy Technology Center experimental sites 4 and 12 near Rock Springs, Wyoming. Both surface and subsurface studies were made of the fracture and joint systems of the area. Effects of weathering created fractures of both vertical and horizontal extent as detected at depths up to about 16·8 m (55 ft).

In a number of outcrops, vertical fractures and joints can be examined. Often they may be measured throughout their vertical length because they terminate at various levels producing a continuous joint through a group of oil shale beds but failing to penetrate the strata above and below. Most often, the lower limits of these types of fractures cannot be pinpointed because of alluvium, talus or the fact that they have considerable extent in depth below the outcrop. Also found are joints/fractures which are obvious on outcrops at or near ground level but do not penetrate very great depths and die out on the visible outcrop face. However, in large scale aerial photographs where areas of fracture are large, obvious and traceable, the subsurface has been found to contain fracture and joints and small macrofaults. In the late 1970s, the US Bureau of Mines drilled and cored observation well no. 1 in Colorado's Piceance Creek Basin in an area of evident surface fractures. Their purpose in drilling the hole was selecting a site for building a mine shaft to and through the zone of sodium minerals. Excess water flows below the bottom of the surface casing created a problem in the hole, and a second hole was drilled 1·9 km (1·2 miles) to the northeast in a relatively undisrupted area to obtain the results they sought.

When larger fractures or faults are expressed at the surface and volumes of water pass through them, travertine and other calcium carbonate deposits are found. An example of such a fracture and deposit is described by Dana (1970) of an occurrence in the southeastern Green River Basin, Wyoming (see Figs 4.2 and 4.3). In subsurface fractures secondary minerals may restrict or even seal them to water flows.

Fig. 4.2. Exterior view of large vertical fracture in Laney shale member of the Green River Formation, west of Rock Springs, Wyoming. The exposed horizontal distance of the fracture is about 20 ft (6·1 m).

4.2.3.1. Micro

Both outcrop and core samples, when cut and polished or split to reveal a fresh face, exhibit very small faulting, sharp features, vertical cracks filled with tuff or ash flows and other visible but very small structural features. These occurred as adjustments in the strata due to differential loading of overburden, to horizontal and vertical stresses and strains and to different rates of compaction. They occurred during consolidation when the planktonic ooze was solidified sufficiently to maintain varves and bedding configuration yet mobile enough to react to the above pressures. Displacement in the small faults can be measured in millimeters and these also die out in short vertical distances.

4.2.3.2. Macro

The larger, more evident fractures and joints occurred after consolidation and at a younger geologic age than most of the micro fractures. Some, such as tension fractures, are the result of erosion and removal of overburden. Others are created from tectonic movement, small as it may be. When the oil shale beds were deposited in their lacustrine environments, they were laid flat on a level lake bed. It is only the

Fig. 4.3. Interior of large vertical fracture which was mined for a distance of 70 ft (21·4 m) into the outcrop. Secondary minerals recovered were principally calcium carbonate and pyrite. Abandoned tools and equipment include this vintage ore cart.

tectonism which occurred in the post-consolidation phase which created fractures of that origin. Dips in the centers of most of the oil-shale bearing basins range up to 5° but average closer to 2°. Around the edges of the basins, dips of up to 20° have been measured but the average again is very low, probably averaging not more than 5–6°.

Studies of joint and fracture patterns have been conducted by Sandia and Los Alamos National Laboratories in association with their 1982 oil shale rock fragmentation tests in the US Department of Energy Anvil Points, Colorado mine. The work, done in cooperation with the Science Applications Inc. consortium of interested oil companies, was conducted to research the potential for creating rock fragmentation

Fig. 4.4. Explosive preparation of a near surface oil shale zone for horizontal *in situ* retorting at the Geokinetics Inc. site, Uinta Basin, Utah. The uplifting of the surface creates fracture porosity in the horizontal oil shale strata.

methods which would produce uniform sizes and as much uniform permeability as possible within potential subsurface oil shale retorts. A secondary and longer range benefit was to create models and optimum methods for open pit mining and blasting of large amounts of oil shale (see Fig. 4.4). Since rich oil shale is much more resilient than coal beds or sandstones containing uranium, the methods used in mining those commodities cannot be fully extrapolated or applied to oil shale.

Figure 4.5 is a plan view of the floor of the Colony Mine in the southern Piceance Creek Basin. These joints exist in the Mahogany Zone as well as in both underlying and overlying strata. The circular hole in the top center of the picture is a small diameter (4·8 cm, $1\frac{7}{8}$ in) corehole, taken to provide pre-shot information on the strata to be experimentally fractured.

Figure 4.6 is an oblique photograph showing a mine level surface (upper third of picture) and a vertical wall where the face has been removed. The joints can be traced from the surface exposures downward along the vertical wall (lower two-thirds of the picture). The horizontal lines are holes drilled to pre-split surface.

Figure 4.7 shows a wall in the Colony Mine exhibiting natural joints and fractures. These are generally vertical and diagonal, some with curbed traces. Horizontal fractures also occur in the mine walls but those are not so obvious. The dark hole in the left center of the picture is a small washed-out vug about 20–30 cm across. Considerable natural permeability is observed in this photograph.

Sampling, logging and in situ stresses of oil shale 141

Fig. 4.5. Plan view of joints in floor of Colony Mine.

Fig. 4.6. Joints in wall of Colony Mine. Face has been partially removed showing joints below surface (bottom half of figure). Horizontal lines are drilled holes to presplit surface.

Fig. 4.7. Natural joints and vug at Colony Mine (mine wall).

4.3. LOGGING

Several types of log have proven beneficial in formation evaluation and should be made during the exploration phase of oil shale evaluation.

4.3.1. Drilling Log

This log is kept of the time, footage, penetration rate and any predicted or unpredicted occurrences during the drilling and coring of oil shale. It consists of two types of logs:

(a) A mechanical logging instrument which, when used properly, indicates the drilling depth at all times, each foot cored and the amount of time used to core each foot. Under uniform conditions, the hardness or softness of the subsurface formations can be determined by the speed with which drilling or coring is accomplished. A number of other factors, such as weight of the drilling string, speed of rotation and kinds of drilling fluid used, must be taken into account but they can be corrected for in most cases. When saline deposits are drilled in the subsurface, the penetration rate increases because the formation is softer and the drilling fluids usually wet the soluble minerals in the strata, placing them in solution. Differences in penetration rates can also indicate fractured and broken layers such as the Leached Zone in the subsurface of the central and northern Piceance Creek Basin.

(b) The second log is the geologist's field lithologic log, made on a full diameter core just after it is removed from the core barrel. Photography of cores, preferably in color, at this time is an available and desirable supplement to the description. This field log may prove to be valuable because it provides information on the core in its closest to natural state. In extreme cases, the core may be lost, dropped or otherwise changed from its fresh condition, so first impressions are very important. Another valuable aspect of the core which should be recorded is the number, location and type of fractures in the strata and the geologist's best determination of whether the fracture(s) are natural or induced. Some types of oil shale cores, usually those with high clay content, will dry quickly, desiccate to a certain extent and fracture from exposure to these drier conditions. Even when placed in plastic, or otherwise wrapped, free water in cores is exuded due to rock or internal or release pressures.

Since the core's outer surface is a scarred and more or less polished face, the core should be split to expose a fresh, non-disturbed rock surface from which to make the laboratory lithologic log. The fresh face will reveal many features which are obscure to indistinguishable on the core surface providing the geologist with details on many kinds of features, both lithologic and structural. These interiors of core are almost always unaffected by drilling fluids because of the extremely low permeabilities of non-fractured oil shale. Where fractures are found, drilling fluid effects can be detected and taken into account in the geologist's description.

4.3.2. Mineralogical Log

This log too should be made in two parts; that in field upon initial recovery, and the laboratory log made from the x-ray diffraction analyses of representative core samples. In addition to the stratigraphic information, the analyses of both matrix minerals and secondary minerals is of two-fold importance, to determine: (1) the potential commercial value of any accessory minerals such as dawsonite and nahcolite as found principally in the Piceance Creek Basin and trona in the Green River Basin; and (2) any minerals which may either enhance or be detrimental to any sort of technological process used to produce shale oil. These determinations are of environmental concern to the entire concept of fossil fuel production. Since heat in oil shale retorting changes many of the original minerals in the matrix, spent shale should be analyzed to identify newly formed minerals or altered forms of original minerals, both qualitatively and quantitatively. It has been through this type of analysis of the Green River Formation that a number of new minerals and mineral compounds have been identified and named in the literature. A surprising number of these are unique to the Green River Formation or at least were discovered in these lacustrine environments.

4.3.3. Geophysical Logs

The word 'geophysical' as used in this heading encompasses a number of borehole tools used to determine various properties of subsurface rocks by making measurements while the tool is traversing the borehole. Results of these measurements are relayed to the surface and recorded during the logging operation. Although much has been written about the application of geophysical logging methods to standard or conventional exploration for oil and gas, little is found in the literature regarding specific or special application of the suite of geophysical logs for oil shale exploration or research. The most common geophysical logs used in oil shale drill holes and coreholes are discussed.

(*i*) *Caliper:* The caliper log measures hole diameter, an essential piece of data for interpreting readings from other logs. It is especially useful in the following two instances: (1) to locate the Leached Zone in the Piceance Creek Basin where nahcolite has been dissolved from portions of the bedrock, either from vugs or layers of nahcolite (Thomas and Smith, 1970); and (2) to determine the extent of void space in post-burn oil shale experiment situations. The tool has a limited diameter-determining radius beyond which it is ineffective.

(*ii*) *Resistivity:* The resistivity log is principally used to detect zones containing water saturation in some existing porosity. Not much of the bedrock in the oil shale bearing portions of the Green River Formation contain the large volumes of water detectable by the use of this instrument. However, much of the Green River Formation shows high resistivity.

Logs (iii)–(v) are particularly useful for correlating the formation stratigraphy:

(*iii*) *Sonic:* The sonic log is created by sending sound pulses into the formation and recording the time required for their return. The time varies with the continuous density of the formation since the sound travels only through solid rock and ignores voids. Sonic log signals correlate well with the density, with the richer grades showing longer transit times. Because the sonic log must be run in a fluid-filled hole in order to obtain coupling with the formation, its application is somewhat limited.

(*iv*) *Neutron:* The neutron log provides a rough measurement of total hydrogen in a formation because hydrogen is the primary neutron absorber in most formations. While the organic matter in oil shale reveals concentrations of hydrogen, the log is not sensitive enough to provide oil yield estimates in oil shale.

(*v*) *Physical property:* This three-dimensional (multi-acoustical) log has been adapted to estimating physical properties of oil shale in place. By correlating the return time on several sound pulses, estimates of properties such as Young's modulus and Poisson's ratio can be obtained by applying computer methods.

Sampling, logging and in situ *stresses of oil shale* 147

(*vi*) *Gamma ray:* The application of gamma ray logging to oil shale exploration is limited to its use as a lithologic and stratigraphic tool. Radioactive peaks delineate zones containing uranium or potassium-40 in clay in both eastern and western oil shales. Differences in types of rock can be determined by the gamma ray log and correlations over comparatively great distances can be achieved by its use.

(*vii*) *Density:* The gamma-gamma density log measures the gamma rays reradiated from a formation after irradiation. The rate of reradiation depends on formation density. These logs make excellent estimates of the potential oil content of oil shales in gallons per ton. Densities from the log can be used to estimate yield from density–oil yield relationships (Smith, 1969, 1976). The density log can be calibrated to produce these oil yield estimates directly.

(*viii*) *Hot-film flow:* In 1970 and 1971 H. E. Thomas, of the US Bureau of Mines in Laramie, Wyoming, developed a method for evaluating shallow *in situ* fracturing and retorting experiments in Green River oil shale (Thomas and Smith, 1970; Thomas, 1971). The instrument measures gas flow rates.

(*ix*) *Radioactive pulse:* A radioactive pulse method has been used to evaluate the degrees of fracturing in an underground formation (Lorenz, 1973). A chemically inert tracer is injected into fluids flowing through a fractured retort and recorded by detectors in boreholes. The system was not satisfactory because of channeling, dispersion of tracer, inaccuracy of measuring devices and methods of analysing results.

4.4. ANALYSES

4.4.1. Fischer Assay
The most widely used and accepted (ASTM Designation D-3904-80) method of analyzing oil shale for oil yield is the destruction distillation method known as Fischer assay. This laboratory technique heats a sample of known size, usually 100 g, at a standardized rate to convert kerogen to a liquid hydrocarbon (Fig. 4.8). Water, oil and spent shale are measured, and the gas evolved is estimated by difference. The initial sample is crushed to 8 mesh for uniformity. Oil yield by Fischer assay is an essential part of evaluating an oil shale resource (Smith, 1981). If the conversion factors are known, this oil yield can also be

Fig. 4.8. View of a typical Fischer assay laboratory showing the following components: (a) temperature controller; (b) condensers; (c) 2000-W ovens; (d) assembled glass adaptors and 100 ml centrifuge tube; (e) cooling bath; (f) cast aluminum alloy retorts.

used to estimate organic content. The determination of organic content is an important part of evaluating oil shale for process technologies for both surface and subsurface commercial oil production. The organic content also has its influence on stress, strain and tensile measurements as discussed elsewhere in this book.

4.4.2. X-ray Diffraction
The most common laboratory technique for determining the kinds of minerals that make up the mineral fraction of oil shale is x-ray diffraction. The process involves irradiation of a powdered oil shale sample and interpreting the resulting diffraction pattern. Considerable experience is desirable for interpretation of peaks, especially in the Green River Formation where a number of exotic and unique minerals are found. Even greater experience is necessary to provide some

Sampling, logging and in situ *stresses of oil shale* 149

qualitative values for the total analysis. The minerals may occur as primary or secondary, but primary minerals are a very high percentage of the matrix total. The minerals occur as (1) part of the oil shale matrix; (2) in layers between oil shale strata or varves; (3) as crystal growths in both vugs and bedding; and (4) as secondary minerals in veinlets, fractures and joints, with the minerals precipitated and deposited by percolating waters. Normally these joints and fractures containing minerals are zones of weakness and are least resistant to efforts in artificial fracturing.

4.5. STRESS RELATIONSHIPS

Information about the *in situ* stress state is essential to the development of suitable mine designs for oil shale mines. This *in situ* stress information is essential because the stability of such structural members as the mine roof, floor and pillars will be affected by the magnitudes and directions of these *in situ* stresses. In general, the magnitudes of both the vertical and horizontal *in situ* stresses increase with depth and, although the vertical stress magnitudes are directly governed by the density and thickness of the overburden, the magnitudes and directions of the horizontal *in situ* stresses will be strongly dependent on the tectonic history of the oil shale in a particular location. Clues about the horizontal *in situ* stress state can be provided, however, by a knowledge of the oil shale failure properties and the geologic structure. For example, knowledge of the type of fault and the strike and dip of the fault plane can be used to make direct inferences about the *in situ* stress state at the fault location. In the Piceance Creek Basin, the tectonic history is sufficiently complex that it is somewhat hazardous to make inferences about the *in situ* stress state, at a particular location, based on measurements made at other locations. As a result, it is desirable to make direct measurements of the *in situ* stress state at the particular locations where underground mines are planned. Prior to discussing particular *in situ* stress measurement techniques, however, it is convenient to make some general observations about *in situ* stresses.

4.5.1. General Discussion of *in situ* Stresses

To describe the state of stress, for the most general situation, it is necessary to identify the reference coordinate system being used (usually three orthogonal axes) and all six distinctive components of the symmetric stress tensor: three normal and three shear stress components. For at least one particular orientation of the reference coordinate system, for which there are no shear stresses, the stress state is completely specified by the three normal stresses. The direction of the axes of this particular coordinate system are termed the principal stress directions. The normal stresses for these directions are termed the principal stresses. Since only three principal stress magnitudes and directions are required to completely specify the state of stress, when the principal stresses are used, it is convenient to discuss the *in situ* stress state in terms of these principal stresses. In rock it is customary to adopt the convention that compressive stresses are positive and the principal stress magnitudes are specified by: σ_1, σ_2 and σ_3 (where $\sigma_1 > \sigma_2 > \sigma_3$). Further, it is often assumed, perhaps without justification in particular cases, that one of the principal stress directions is vertical and the other two are horizontal. This assumption occurs, in part, as a result of some of the conceptual models used to describe the *in situ* stress situation.

Consider, for example, the situation where the surface of the overburden is a flat plane. Near the surface, one of the principal stress directions will be vertical and the other two principal stress directions will be horizontal because shear stresses are not generally transmitted across the free surface. It is then assumed that, in the absence of tectonic forces, this is the *in situ* stress situation at depth. The next item to address is the estimation of the magnitudes of these *in situ* stresses. The vertical stress, averaged over a sufficiently large area, must necessarily be equal to the stress represented by the load of the overburden. That is:

$$\sigma_v = \gamma h \tag{4.1}$$

where σ_v is the vertical stress, γ is the average density of the overburden rock and h is the overburden thickness. Typically the magnitude of the vertical stress is of the order of 1 psi per foot of overburden thickness.

Regarding the horizontal stresses, one of the simplest assumptions that can be made is that no horizontal displacements have been allowed to occur, i.e. a plane strain condition. Under these conditions the horizontal stress will be:

$$\sigma_h = \sigma_v(\nu/(1-\nu)) \tag{4.2}$$

where σ_h is the horizontal stress and ν is Poisson's ratio. Poisson's ratio is between 0·20 and 0·33 for most rock. This formula predicts that the horizontal stress is between 0·25 and 0·50 of the vertical stress. Generally the horizontal stress to vertical stress ratio is greater than the ratio predicted by eqn (4.2), thus this formula represents a lower bound for the horizontal stresses. Another assumption that is commonly made is that the rock creeps sufficiently, or is sufficiently viscoelastic, that shear stresses cannot be supported for an extended period of geologic time (assuming that little tectonic activity occurs during this time period). Under these conditions, the *in situ* state of stress, at depth, will be hydrostatic, i.e. $\sigma_h = \sigma_v = \sigma_1 = \sigma_2 = \sigma_3$. In general, some tectonic forces are active and, as a result, it is difficult to predict the magnitudes and directions of the horizontal *in situ* stresses.

It is possible to obtain some clues about the magnitude and the direction of the horizontal *in situ* stresses by examining pertinent geologic structures in the region of interest. This is particularly true if there are some active faults in the area. For the idealized situation where the principal stress directions are vertical and horizontal, faults can be divided into three categories: normal faults, thrust faults and strike slip faults. For normal faults, σ_1 is vertical and the σ_2 direction is parallel to the strike, so σ_3 is perpendicular to the strike direction. For thrust faults, σ_3 is vertical, and again σ_2 is parallel to the strike, so the σ_1 direction is perpendicular to the strike direction. For strike slip faults, σ_2 acts in the vertical direction so both σ_1 and σ_3 act in the horizontal direction. In addition to predicting the direction of action for the *in situ* stresses, the fault observations can also be used to estimate the magnitudes of these *in situ* stresses. Assuming that the rock mass failure properties have been determined, or can be inferred from laboratory tests, the magnitudes of the *in situ* stresses are estimated to be the stresses required to induce the failure in the rock mass that is represented by the observed fault.

From an *in situ* stress prediction standpoint, another useful structural feature is a fold. It can be inferred that the maximum principal stress direction is the direction perpendicular to the fold axis.

Clearly it is possible to make inferences about the *in situ* stress state from study of geologic structural features. These inferences should be treated with caution, however, because the present tectonic situation may be significantly different from the tectonic situation at the time the structural feature of interest was formed. It is very desirable to make actual measurements of the *in situ* stress state, when the detailed mine design stage is reached, so that accurate *in situ* stress information is available for mine design purposes. Accordingly, the next topic will be a discussion of selected *in situ* measurement techniques.

4.5.2. *In situ* Stress Measurement Techniques

Some general items are herein considered for discussion of *in situ* stress measurement techniques. It is generally extremely difficult to make direct measurements of stress. For most engineering applications, and rock mechanics applications in particular, it is common practice to infer the state of stress from strain and/or displacement measurements using the elastic constants of the material to relate stress to strain. To use strain measurements, it is necessary to make measurements on the structural member in both the stressed and completely unstressed condition so that all the strain that is induced by the application of the stress can be measured. On common engineering structural members, the initial stress state is the unstressed condition. A second measurement, made on the structural member in the stressed condition, is compared to the initial measurement to determine the strain induced by the applied load and provides an estimate of the applied stresses. In contrast, for the *in situ* stress measurement, the initial, or given, stress condition is the *in situ* stress state. As a result it is necessary to detach a portion, or block, of the rock from its surroundings to obtain rock in an unstressed condition. Again, a second measurement, made on the stress free block, is compared to an initial measurement made on the block subjected to the *in situ* stresses to determine the strain that was produced by relieving the *in situ* stresses. It is then generally assumed that the *in situ* stress state is that stress state required to reverse the process, i.e. transform the strain state of the unstressed block to the strain state of the block subjected to the *in situ* stresses.

In situ stress measurement techniques can be divided into three categories according to the maximum distance from the observer that the stress measurement can be made. The three somewhat arbitrary categories are: surface (0 m, 0 ft), moderate depth (30·5–61 m, 100–200 ft) and great depth (the maximum depth to which holes can be drilled).

The first category considered here will be the surface stress measurement techniques. At least two such techniques are commonly used: the stress relief and flat jack techniques.

The flat jack stress measurement technique will be considered first.

Sampling, logging and in situ *stresses of oil shale* 153

The first step in the flat jack stress measurement process is to precisely measure the distance between two points on the surface of the mine opening. Then a slot perpendicular to both the surface of the mine opening and the line connecting the two measurement points is cut in the rock, midway between the two measurement points. Then the flat jack is placed in the slot, usually with grout to insure good contact with the rock material. Finally, the flat jack is inflated with sufficient pressure so the original distance between the measurement points is re-established. It is then assumed that the pressure in the flat jack is equal to the normal stress perpendicular to the plane of the flat jack. Normal stresses in any given direction can be measured by using a surface (roof, floor or wall) that is perpendicular to the direction of interest and orienting the flat jack perpendicular to the direction for which the normal stress measurement is desired.

The other surface *in situ* stress measurement technique to be considered is a stress or strain relief technique which utilizes a strain rosette. As commonly practiced, the first step in utilizing this technique is to make precise measurements of the length of three legs of an equilateral triangle. The stress is then relieved by cutting slots on two or more sides of the block of material that contains the three measurement points. The difference between the lengths of the legs of the triangle, in the stressed and the unstressed block respectively, are used to determine the elongation strain in three directions oriented 60° from one another. These elongation strains are used to determine the directions and magnitudes of the two principal surface strains using well-known equations for a strain rosette. The elastic constants of the rock material are then used to estimate the *in situ* stress state that gave rise to these strains. The roof and floor of the mine openings are available, in addition to the wall, so that the normal stresses in various directions can be determined.

Surface *in situ* stress measurement techniques have several advantages. By definition, the site of a stress measurement should be readily accessible so it is convenient to inspect the site and to avoid some of the problems associated with obtaining measurements from locations that are not representative of the rock mass (near faults, intrusions, etc.). Easy access also makes it possible to perform a series of tests in a relatively short period of time. It is also the case that the equipment used to perform the surface stress measurement tests is relatively rugged and inexpensive. Skills required to operate the equipment and perform the tests competently are readily acquired by semi-skilled personnel. These are significant advantages when operating a mine environment.

Unfortunately, some very significant disadvantages to using surface *in situ* stress measurement techniques exist. One of the most significant disadvantages to using these surface techniques is that the measurement devices are responding to stresses that are only indirectly related to the *in situ* state of stress. In general, the mine opening required to gain access to the measurement site will induce a stress concentration in the local region immediately surrounding the mine opening. Thus, the surface stresses may be significantly different from the *in situ* stresses simply because of the stress concentration effect. Another possibility is that the properties of the material immediately surrounding the mine opening have been greatly altered as a result of disturbance by the mining required to create the mine opening or the local mine opening induced stress concentration. For example, it is quite possible that failure could be induced in a local region immediately surrounding the mine opening. It is expected that the properties of this failed material will be significantly different than the properties of the rock mass. As a result, the stresses measured in this failed zone will not be very closely related to the *in situ* stress state. The net result of the local stress concentration and the disturbance in the immediate vicinity of the mine opening is that it is extremely difficult to relate the measured surface stresses to the *in situ* stress state. Another disadvantage of the surface stress measurement technique is that mine access is required to the site where stress measurement is required. Unless mine development occurs in place, obtaining this mine access may be very costly. In summary, it is difficult to relate the stresses measured at a surface to the *in situ* stress state. As a result, surface stress measurement techniques are only rarely used to estimate *in situ* stresses.

The next *in situ* stress measurement category considered is the moderate depth category. The moderate depth *in situ* stress measurement technique most commonly used in this country is the US Bureau of Mines overcoring technique. This technique is actually a strain or stress relief technique using a borehole deformation gauge to monitor the deformation of a borehole as the strain in the rock immediately surrounding the borehole is relieved by detaching this rock from its surroundings by an overcoring bit. As usually practiced, the overcoring technique involves drilling a hole 15·2 cm (6 in) in diameter into the rock to be tested, drilling an EX (3·8 cm, 1·5 in) diameter pilot hole in the center of the end of the original 15·2 cm (6 in) access hole, placing

Sampling, logging and in situ *stresses of oil shale*   155

the borehole deformation gauge at the location where the *in situ* stress measurement is to be made and monitoring the response of the borehole deformation gauge while overcoring the gauge with a 15·2 cm (6 in) thin wall overcoring bit. The 15·2 cm (6 in) diameter core, with the 3·8 cm (1·5 in) diameter center hole, is then removed from the hole and placed in a biaxial pressure cell capable of applying a uniform radial pressure to the outside of the core. The relationship between the response of the borehole deformation gauge and the magnitude of the pressure applied to the outside of the core is then established and used to determine the elastic properties of the rock. These field determined elastic properties are used to convert the borehole deformation measurements into estimates of the *in situ* stress state.

The borehole deformation gauge is basically a strain rosette device in that measurements of the deformation across the diameter of the borehole are made in three different directions that are 60° apart. The borehole deformation gauge sensing elements are strain gauges mounted on small cantilever beams, the free ends of which are connected to the borehole wall by small pistons. Thus, movement of the piston induced by deformation of the borehole wall is translated into bending of the cantilever beam which in turn produces a response in the strain gauge. The strain gauge response is typically monitored using an off-the-shelf strain indicator measuring device. As is the case with the surface strain relief technique, a second set of borehole deformation gauge measurements, made on the detached stress free core, are compared to the initial set of measurements made in the borehole in the stress rock to determine the effect of relieving the *in situ* stresses. It is then assumed that the *in situ* stress state is that stress state which is required to reverse the process, i.e. transform the borehole deformations of the unstressed core to the borehole deformations of the core subjected to the *in situ* stresses. Like any other strain rosette, the borehole deformation gauge can only be used to estimate the *in situ* stress state in the plane of measurement which is the plane perpendicular to the axis of the borehole. To obtain the complete stress ellipsoid it is necessary to make borehole deformation gauge measurements in boreholes that are drilled in three different directions, preferably orthogonal to one another.

There are, of course, advantages and disadvantages to using the moderate depth overcoring technique. One significant advantage of this technique is that by simply conducting tests at sufficient distances from the mine opening (about one mine opening diameter), it is possible to avoid both the mine opening induced stress concentration and the local disturbed zone, and to obtain measurements of actual *in situ* stresses. Another advantage is that it is possible to obtain a direct measurement of the complete three-dimensional stress ellipsoid. Thus, it is not necessary to make assumptions about the directions of the principal stresses to utilize this technique.

One of the major disadvantages of the overcoring technique is that the borehole deformation gauge is not an off-the-shelf item produced in large quantities. The manufacturing process involves a great deal of precision machine work, increasing the cost of the gauge. Another problem involves some non-standard drilling techniques requiring skilled personnel. As was the case with the surface *in situ* stress measurement techniques, mine access is generally required to conduct the tests. This may be costly to obtain.

To summarize, the overcoring technique can be used to obtain measurements of actual *in situ* stresses. It is one of the few techniques capable of producing an estimate of the complete three-dimensional stress ellipsoid. The borehole deformation gauge is expensive to produce, skilled test personnel are required to insure that the test results are meaningful, and costly mine access is required. Thus, assuming that mine access and test equipment is readily available, the overcoring *in situ* stress measurement technique is usually the technique of choice for obtaining estimates of *in situ* stresses.

The final *in situ* stress measurement category discussed is the great depth category. The hydraulic fracturing technique is the sole method of measuring *in situ* stresses at great depth. This technique is primarily used to stimulate gas or oil production wells by creating a fracture path for the flow of hydrocarbons from the formation into the wellbore. To use this technique to estimate *in situ* stresses, it is first necessary to isolate a portion ($\cong 3$ m, 10 ft) of wellbore, usually with a pair of inflatable wellbore packers. Fluid pressure is then applied to this isolated region and increased until breakdown or rupture of the rock around the wellbore occurs. Finally fluid is pumped into the fracture until the desired fracture size is attained. Measurement of the instantaneous shut in pressure (ISIP), which is the fluid pressure in the fracture and fluid delivery system immediately after the pumps have been shut down, provides a measure of the pressure required to keep the fracture open.

Three important parameters for the purpose of obtaining hydraulic fracturing *in situ* stress measurements are: the ISIP, the fracture orientation and the breakdown or rupture pressure. It is generally assumed that the hydraulic fracture propagation is in the direction perpendicular to the least principal stress (σ_3). The ISIP is the pressure required to just keep the hydraulic fracture open, i.e. counteract the normal *in situ* stress perpendicular to the crack. Thus, the ISIP is a direct measure of the least principal stress, and the least principal stress direction is perpendicular to the fracture. It then follows that data on the hydraulic fracture orientation provides useful information about the principal directions for the *in situ* stresses.

It is generally assumed that the hydraulic fracture orientation will be either horizontal or vertical. This assumption is a necessary companion to the assumption that one of the principal *in situ* stress directions is vertical. This vertical principal stress direction assumption is made to facilitate the determination of the theoretical elastic stress field around the well bore. It follows that the final orientation of the hydraulic fracture will depend on whether the least principal stress is vertical or horizontal. If the least principal stress is vertical (and the resulting hydraulic fracture is horizontal) the hydraulic fracturing test only provides a measure of the vertical *in situ* stress state. Fortunately, in most cases, the least principal stress direction is horizontal and a vertical hydraulic fracture is formed.

When vertical fractures are formed it is possible to specify the *in situ* stress state completely by describing the magnitudes and directions of the principal stresses. For directions, one of the principal stress directions is assumed to be vertical, the minimum horizontal principal stress (the least principal stress) direction is normal to the hydraulic fracture and the direction of the maximum horizontal stress is orthogonal to the minimum horizontal stress, i.e. parallel to the fracture. To determine the orientation of the maximum and minimum horizontal *in situ* stresses, it is necessary to determine the orientation of the hydraulic fracture. This orientation is determined by measuring the orientation of the fracture in the wellbore, by use of an oriented impression packer. The magnitude of the vertical stress is the stress required to support the overburden, the minimum horizontal stress is given by the ISIP, and the minimum horizontal stress is determined using the breakdown pressure and the mathematical solution for the elastic stresses around a wellbore.

The formula for determining the minimum horizontal principal *in situ* stresses is as follows:

$$\sigma_{Hmax} = T_0 + 3ISIP - P_b - P_0 \tag{4.3}$$

where σ_{Hmax} is the maximum horizontal stress, T_0 is the tensile strength of the rock material, ISIP is the instantaneous shut in pressure = σ_{Hmin} = σ_3, P_b is the breakdown or rupture pressure, and P_0 is the formation pore pressure. Thus, to determine the maximum horizontal *in situ* stress, it is necessary to obtain estimates of the tensile strength of the rock and the formation pore pressure, in addition to determining the ISIP and the breakdown pressure.

A significant advantage of the hydraulic fracturing *in situ* stress measurement technique is that a mine opening is not required at the measurement site. Thus hydraulic fracturing can be used to measure *in situ* stresses at any location at which a wellbore can be drilled. Another advantage is that the ISIP provides a direct measure of the least principal stress and it is not necessary to make inferences from strain measurements using elastic rock properties. It is also worth noting that exotic equipment is not required for the application of the hydraulic fracturing *in situ* stress measurement technique. The required equipment is off-the-shelf or is readily constructed from commercially available hardware.

One disadvantage of the hydraulic fracturing technique is that the axis of the wellbore must be drilled in the direction of one of the principal stresses for the theory to be strictly applicable. As the tests are usually conducted, it must be assumed that one of the principal stress directions is vertical. The test results are only as good as the field conditions fitting this assumption. Another disadvantage is that no new information is produced if the least principal stress direction is vertical. In this instance a horizontal fracture is produced and the ISIP is a measure of the vertical normal stress which can be accurately estimated by calculating the load applied by the overburden.

It is now possible to make some generalizations about the *in situ* stress measurement techniques. The technique chosen depends on various factors including the location of the *in situ* stress measurement site with respect to the nearest access point. Obviously, if the measurement site is a great distance from the access location, only the hydraulic fracturing *in situ* stress measurement technique can be used. If, however, access exists to a location near the desired stress measurement point, any of the three categories of stress measurement techniques can be used and other factors govern the choice of measurement method. For example, a verified accurate model has been produced for relating the local stresses in the vicinity of the access opening to the *in situ* stress state and a surface *in situ* stress measurement technique would be the most appropriate method. Conversely, if such a model were not available, the overcoring *in situ* stress measurement technique would be preferred particularly if actual measurement of the complete three-dimensional ellipsoid is desired. Due to difficulties involved in drilling a wellbore precisely in the direction of either the σ_1 or σ_2 principal stress axis, the hydraulic fracturing *in situ* stress measurement technique is usually only used when the measurement site is remote from the measurement location.

The *in situ* stress measurement technique most appropriate for a given location depends on various factors including measurement versus access location, the existence of a model for relating local stress to *in situ* stresses and the need for a complete three-dimensional ellipsoid.

4.5.3. *In situ* Stresses and Mine Stability

One of the main objectives of the mine design process is to develop mine designs that allow the structural stability of the various mine structures to be maintained. This structural stability must be maintained to insure the safety of mine personnel and to allow the maximum recovery of the in place resource. The various mine structures for which stability must be maintained include the mine roof, floor and pillars or walls. The stability of these structures is often interdependent so it is important to prevent failure. Failure of one could trigger failure of the others resulting in a serious mine stability problem. Many factors affect the stability of these mine structures; the main effort herein will be discussion of the effects of the *in situ* stresses.

One of the more significant factors that affect the stability of mine structures is the vertical *in situ* stress which is directly proportional to the overburden thickness. The structural features most strongly affected by the vertical *in situ* stress are the pillars and/or walls of the mine openings. For an isolated mine opening, failure of the opening wall can be a hazard to personnel and an impediment to travel and removal of the resource. It is possible to avoid some of the *in situ* stress induced failure problems by selecting the proper shape for the mine opening. For extensive mine workings it is necessary to maintain the stability of the supporting mine pillars. Proper pillar sizes are selected to prevent catastrophic pillar failure as induced by the vertical *in situ* stress. To this end it is important to recognize that the compressive strength of rock materials depends on the specimen size: the larger the rock sample being tested, the less the compressive strength. Laboratory measures of compressive strength for small rock samples will generally greatly overestimate the strength for full scale pillars. It is obviously desirable to determine the actual compressive strength for full scale pillars, preferable to large scale pillar tests, to insure long term mine stability.

The horizontal *in situ* stresses will be a major factor in the stability of the mine roof and, to some extent, the mine floor. The preferred orientation of the mine opening with respect to the maximum horizontal *in situ* stress direction depends on the relative strength of the roof (floor) rock compared to the *in situ* stress magnitude. If the roof and floor rock is significantly stronger than the horizontal *in situ* stresses, the proper orientation is to align the long axis of the mine opening perpendicular to the maximum horizontal *in situ* stress direction. One conceptual model for the mine roof is a beam uniformly loaded by its own weight. Using this conceptual model and the empirical observation that rock materials are much weaker in tension than in compression, roof failure is often found to be the result of roof sag induced tensile stresses. By aligning the long axis of the mine opening perpendicular to the maximum horizontal *in situ* stress direction, it is possible for the *in situ* stresses to counteract the roof sag induced tensile stresses and aid in maintaining the stability of the mine roof. It is also possible that the magnitude of the maximum horizontal *in situ* stress is sufficiently great that this stress actually induces failure in the mine floor or roof. Under these conditions, it is more appropriate to orient the axis of the mine opening parallel to the direction of the maximum principal stress. Thus the effects of the maximum horizontal *in situ* stress on the roof and floor rock can be minimized.

Several general observations can be made about the relationship between mine stability and *in situ* stresses. For example, the appropriate pillar size for a particular mine location will be directly related to the overburden thickness because the vertical *in situ* stress is directly related to the overburden thickness. The proper orientation for the axis of the mine opening with respect to the direction of the maximum horizontal *in situ* stress will depend on the relative strength of the rock compared to the magnitude of this *in situ* stress. The orientation of the mine opening axis should be perpendicular to the maximum *in situ* stress direction if the *in situ* stress magnitude is significantly less than the rock strength. The axis should be oriented parallel to the maximum stress direction if the rock strength is significantly less than the stress magnitude. To take advantage of these general principles, it is

Sampling, logging and in situ *stresses of oil shale* 161

necessary to determine the effect of pillar size on pillar strength. It is also obviously necessary to determine the directions and the magnitudes of the horizontal *in situ* stresses.

4.5.4. *In situ* Stresses in the Piceance Creek Basin

Several general features of *in situ* stress fields will be examined to provide a basis for discussion of the *in situ* stress field in the Piceance Creek Basin. The first such feature emphasizes that the *in situ* stress field is often oriented so one of the principal stress directions is vertical while the other two principal stress directions are horizontal. The *in situ* stress state can then be described in terms of vertical and horizontal principal stress magnitudes and the horizontal principal stress directions. Another general feature of *in situ* stress fields is the magnitudes of both the vertical and horizontal *in situ* stresses increase with depth. One pertinent characteristic of *in situ* stress fields is the nature of this *in situ* stress versus depth relationship. The other pertinent feature of *in situ* stress fields is that near the surface the *in situ* stress field is strongly influenced by the topography. While the topographically influenced *in situ* stress field may be important for the purpose of developing mine designs for mines located near the surface, great care must be used when extrapolating from near surface *in situ* stress measurements to any regional trends in the *in situ* stress field.

The main emphasis of the Piceance Creek Basin *in situ* stress field is regional trends that have been observed. Accordingly the *in situ* stress measurements that have been made in existing oil shale mines near outcrops will not receive strong emphasis. The most useful source of information about these regional trends in the *in situ* stress field is a paper reporting hydraulic fracturing tests in the Piceance Creek Basin.

The first item to be discussed is the relationship between the magnitudes of the *in situ* stresses and the overburden thickness or depth. It appears that for depths less than 122 m (400 ft) the topography is exerting an influence on the *in situ* stress field in the Piceance Creek Basin. At these shallow depths, the vertical stress is the least principal stress and the greatest and intermediate principal stresses are horizontal. As a result, shallow hydraulic fracturing tests only provide information about the magnitude of the vertical *in situ* stress and no information about either the magnitude or the direction of the horizontal principal *in situ* stresses. These shallow tests indicate that the magnitude of the vertical *in situ* stress is something like $1 \cdot 1 \, h$ psi (where h is the depth in feet and $1 \, \text{psi} = 6 \cdot 9 \, \text{kN m}^{-2}$). At this shallow depth the magnitudes of both horizontal principal *in situ* stresses are greater than $1 \cdot 1 \, h$ psi, at the relatively few test sites where shallow tests were performed. However, below this 122 m (400 ft) depth, regional trends in the *in situ* stress field were observed and the least principal stress is horizontal. Below the 122 m (400 ft) depth hydraulic fracturing tests provide information about the regional trends in the magnitude and directions of the horizontal *in situ* stresses. Data from several wells throughout the basin indicate that below 122 m (400 ft) the magnitude of the maximum and the minimum horizontal *in situ* stresses are linearly related to the depth. The magnitude of the minimum principal stress is $0 \cdot 7 \, h$ psi and the magnitude of the maximum horizontal stress is $1 \cdot 0 \, h$ psi (where again h is the depth in feet).

The next item to consider is the direction of the principal horizontal *in situ* stresses. The orientations of both hydraulic fractures and some major normal faults in the basin provide information about these principal stress directions. The direction of the least principal stress will be perpendicular and the direction of the intermediate principal stress (the maximum horizontal *in situ* stress) will be parallel to the strikes of both the hydraulic fractures and the normal faults. The strikes of both features are slightly north of west to slightly south of east (almost east–west). Thus the maximum horizontal *in situ* stress is oriented slightly north of west to slightly south of east (approximately east–west). The minimum horizontal *in situ* stress is oriented slightly east of north to just slightly west of south (approximately north–south).

4.6. CONCLUSION

To conclude the following observations should be noted. Assuming that the previously described relationship between the magnitudes of the principal *in situ* stresses and the depth is valid through the basin and the *in situ* stress state does not differ sufficiently from the hydrostatic state of stress, orienting the axis of the mine openings is expected to have a significant effect on roof stability for the deeper mines. However, assuming that orienting the mine openings will slightly affect the stability of mine openings, it would be advantageous to orient the axis of the mine openings north–south. In any case, due to uncertainties in regional trends for the *in situ* stresses and the generally unknown influence of topography on the *in situ* stress field near the surface, it is definitely desirable to obtain *in situ* stress measurements from the site where mine development is contemplated for the purpose of providing input for the mine design.

ACKNOWLEDGMENT

Acknowledgment is herein given to John N. Edl, Jr. of Laramie Energy Technology Center for contributions to the stress–strain relations.

REFERENCES

Dana, G. F. (1970). Mineral deposit in a 'cave' in the Laney Member of the Green River Formation. Unpublished report.

Donnell, J. R. (1961). Tertiary geology and oil-shale resources of the Piceance Creek Basin between the Colorado and White Rivers, northern Colorado. *US Geological Survey Bulletin 1082-L*, pp. 835–91.

Gatlin, C. (1960). *Petroleum Engineering: Drilling and Well Completions*, Prentice-Hall Inc., Englewood Cliffs, New Jersey, pp. 52–93.

Lorenz, P. B. (1973). Radioactive tracer pulse method of evaluating fracturing of underground oil shale formations. *US Bureau of Mines Report of Investigations 7791*.

Smith, J. W. (1958). Applicability of a specific gravity–oil yield relationship to Green River oil shale. *Chem. and Eng. Data Series*, **3** (2), pp. 306–10.

Smith, J. W. (1969). Theoretical relationships between density and oil yield from oil shales. *US Bureau of Mines Report of Investigations 7248*.

Smith, J. W. (1976). Relationship between rock density and volume of organic matter in oil shales. *Laramie Energy Research Center Report of Investigations 76/6*.

Smith, J. W. (1981). Oil shale resource evaluation. In: *Synthetic Fuels from Oil Shale II*, ed. R. D. Matthews, IGT, Chicago, pp. 123–42.

Smith, J. W., Trudell, I. G. and Stanfield, K. E. (1963). Comparison of oil yields from core and drill-cutting sampling of Green River oil shales. *US Bureau of Mines Report of Investigations 6299*.

Thomas, H. E. (1971). The hot film flow log—a formation evaluation tool designed for shallow *in situ* oil shale experiments. *46th Ann. Meeting SPE, Am. Inst. of Min. Met. and Ret. Eng.*, New Orleans, Preprint SPE 3501.

Thomas, H. E. and Smith, J. W. (1970). Caliper location of leached zones in Colorado oil shale. *Log Analyst*, **11** (4), 12–16.

Chapter 5

Mechanical Characterization of Oil Shale

KEN P. CHONG

Professor of Civil Engineering, University of Wyoming, Laramie, Wyoming, USA

and

JOHN WARD SMITH

Consultant, Laramie, Wyoming, USA

SUMMARY

Mechanical characterization of oil shale, including testing techniques, methodology, analyses and concepts are presented in this chapter. Properties investigated include non-linear three-dimensional stiffnesses, tensile strengths, strain-rate effects, creep behavior and compressive strengths. Samples representing the carbonate and the clay-rich oil shales were tested. The carbonate shales tested came from the Parachute Creek Member in Utah's Cowboy Canyon and the Mahogany Zone near Anvil Points, Colorado. The clay-rich shales came from the Tipton Member west of Rock Springs, Wyoming and Silver Point Quad in DeKalb County, Tennessee. The mineralogy which influenced mechanical properties is discussed.

5.1. GENERAL INTRODUCTION

This chapter discusses the mechanical characterization of oil shales from testing concepts through testing techniques, methodology, analysis and evaluation. Since any energy development from oil shale requires modification of the rock, testing of oil shale's mechanical behavior is needed to provide the predictive capabilities necessary for designing and costing oil shale treatments.

Properties investigated include compressive strength, non-linear three-dimensional stiffness, tensile strength along bedding planes, creep phenomena, triaxial strength and shear modulus. Equations capable of predicting these properties from the volume of the organic matter in the rocks as the primary independent variable were developed and evaluated from test results. All of the mechanical properties evaluated are linearly related to organic volume which is also linearly and directly related to absolute rock density. A non-linear relationship to stress level was also demonstrated for several properties. The influence of strain rate through the range 10^{-4}–$10^2\,\mathrm{s}^{-1}$ was investigated and shown to have a linear influence on ultimate strains and on Young's modulus. The influence of major variations in the types of minerals in oil shale is also evaluated.

Mathematical models and test methods for mechanical properties specifically adapted to meet oil shale problems are discussed. Three-dimensional stiffness coefficients can be derived from test results on comparable representative samples cut perpendicular and parallel to the bedding planes. The theory behind this and the preparation of the comparable samples are presented. A new method to determine shear modulus for oil shales using the anticlastic bending of anisotropic plates is described. Application of the split cylinder test to the determination of average bedding plane tensile strengths for oil shale specimens is presented. A failure mechanism dependent on strain rate is discussed to explain the correlation between ultimate compressive stresses, strain rate and oil shale composition. Linear extrapolatability of strain rate effects on oil shale into the range found in explosions is demonstrated.

Oil shale sample sets to which the testing methods were applied include two dolomite cemented Green River Formation cores from the Mahogany Zone in Colorado and Utah, a clay-rich Green River Formation core from the Tipton Member in Wyoming, and a core whose minerals are primarily quartz and illite from the Devonian black shale in Tennessee and Ohio. These shale types bracket most of the world's oil shales.

5.2. NON-LINEAR THREE-DIMENSIONAL STIFFNESSES AND ULTIMATE STRESSES

5.2.1. Introduction

Constitutive laws governing three-dimensional stress/strain relationships for oil shale are fundamental to predicting the mechanical behavior of the rock when subjected to modifications incurred during mining, fragmentation, creation of in-place permeability, etc. Mechanical testing of oil shales from Wyoming's Tipton Member and the Mahogany Zone near Anvil Points, Colorado and the Parachute Creek Member in Utah's Cowboy Canyon provided the data to evaluate the three-dimensional non-linear coefficients of the stiffness matrix for transversely isotropic materials (Chong *et al.*, 1979*b*, 1980*c*, 1982*b*). These twelve non-zero coefficients are expressed in terms of the elastic constants consisting of Young's modulus, Poisson's ratio and shear modulus, related by regression analysis to the independent variables consisting of organic volume and stress level. Sample sets consisted of square prisms taken from the same horizon but cut both perpendicular and parallel to bedding. Elastic constants and, also, compressive strengths perpendicular and parallel to bedding are determined from uniaxial compression tests, yielding all information necessary to characterize the mechanical behavior of oil shale in three-dimensions. A comparison is made against the behavior of oil shales from the western United States.

Comprehensive testing conducted on core samples of oil shale exploits the lateral homogeneity of oil shale to determine the ultimate strengths and elastic constants of Young's moduli and Poisson's ratios under loading perpendicular (σ_{uz}, E_z, ν_{zx}) and parallel (σ_{ux}, E_x, ν_{xy}) to the rock's bedding planes and shear modulus G_{xz}. Regression analysis is used to produce prediction equations which estimate the elastic coefficients as well as ultimate strength of oil shale from organic volume (O_c) and stress level (S). These prediction equations are presented, along with their statistical evaluations. Young's modulus and Poisson's ratio are found to be linear functions of organic volume (Smith, 1969*a*, 1976; Stanfield and Frost, 1949), and more complex functions of stress level, expressed as a percentage of ultimate stress (Chong *et al.*, 1980*c*, 1982*b*). Ultimate strengths are found to be linear functions of organic volume only. It is mathematically demonstrated that the 12 non-zero coefficients in the stiffness matrix for oil shale can be calculated from the predicted elastic constants (Barden, 1963; Lekhnitskii, 1963). Equations yielding these coefficients are presented. The results of this study provide a complete characterization of the three-dimensional mechanical properties of Western oil shale.

5.2.2. Theory

A significant physical characteristic of Green River Formation oil shale is that bedding is laterally constant. The smallest of the regular layers

Fig. 5.1. Oil shale profile.

($\sim 30\ \mu$m) are called varves. Varves are pairs of fine laminations alternatively higher and lower in content or organic material. Larger layer variations called laminations also exist. The properties of individual varves and laminations such as mineralogy and organic content may be variable in the direction normal to bedding (Fig. 5.1). In the lateral direction, however, the properties of individual layers are remarkably uniform (Trudel *et al.*, 1970; Smith and Lee, 1982). Thus, in the layered oil shale there exists an axis of symmetry (z-axis) which is perpendicular to the bedding planes. The horizontal axes (x and y) are equivalent (or interchangeable) due to lateral homogeneity, forming planes of isotropy. Therefore, oil shale can be adequately characterized as a transversely isotropic material.

Conventionally, transversely isotropic materials are completely described by the following five constants as indicated by Lekhnitskii (1963) and Chen and Saleeb (1982): E_x, E_z, ν_{xy}, ν_{zx} and G_{xz}, where E_x, E_z are the Young's moduli along the x- and z-axes, respectively, ν_{ij} is Poisson's ratio, denoting contraction/expansion in the j-direction due to loading along the i-direction, and G_{xz} is the shear modulus in the xz plane. Using the symmetrical properties of the stiffness matrix (setting coefficients C_{13} equal to C_{31}) the Poisson's ratio ν_{yz} can be expressed as a function of E_x, E_z and ν_{zx}. Thus, ν_{yz} is a dependent constant as follows:

$$\nu_{yz} = \frac{E_x}{E_z}\,\nu_{zx} = n\nu_{zx} \tag{5.1}$$

where n = degree of anisotropy = E_x/E_z.

A recent report by Chong *et al.* (1980a) presents a new method to determine the shear modulus of transversely isotropic materials. Based on twisting tests of rectangular plates (sample C, Fig. 5.2(a)), it is demonstrated that the independent shear modulus G_{xz} can be computed by the following expression:

$$G_{xz} = \frac{3Pab}{h^3 w_p} \tag{5.2}$$

where P is a load applied to the corner of an anticlastically supported plate of length a, width b and thickness h, and w_p is the deflection at load point. Correlation of the above formula for materials with known shear moduli indicates the method is highly duplicable and reliable. Using results from the testing of oil shale plates, an attempt was made to find an expression for shear modulus as a function of organic

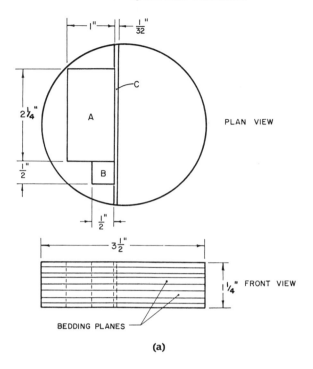
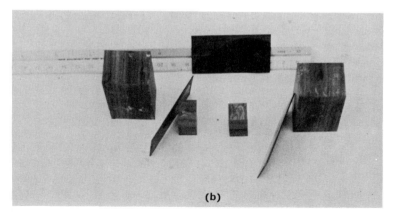

Fig. 5.2.(a) Sampling technique; (b) finished samples.

content by volume (O_c in %). Regression analysis yields the following typical relationship for shear modulus (G_{xz} in GPa) of Utah oil shale:

$$G_{xz} = 19 \cdot 43 - 0 \cdot 31 O_c$$

$$r^2 = 0 \cdot 84 \qquad (5.3)$$

$$N = 5$$

where r^2 = coefficient of determination, N = number of samples. Notice for the particular stiffness of G_{xz}, no dependency of the stress level is observed. After G_{zx} is found, it is necessary to determine the remaining four constants. The generalized Hooke's law is:

$$\{\sigma\} = [C]\{\varepsilon\} \qquad (5.4)$$

$$\{\sigma\} = \{\sigma_{xx}, \sigma_{yy}, \sigma_{zz}, \sigma_{xz}, \sigma_{yz}, \sigma_{xy}\} \qquad (5.5)$$

$$\{\varepsilon\} = \{\varepsilon_{xx}, \varepsilon_{yy}, \varepsilon_{zz}, \varepsilon_{xz}, \varepsilon_{yz}, \varepsilon_{xy}\} \qquad (5.6)$$

and the stiffness matrix takes the following form:

$$[C] = \begin{bmatrix} C_{1111} & C_{1122} & C_{1133} & 0 & 0 & 0 \\ C_{1122} & C_{1111} & C_{1133} & 0 & 0 & 0 \\ C_{1133} & C_{1133} & C_{3333} & 0 & 0 & 0 \\ 0 & 0 & 0 & C_{2323} & 0 & 0 \\ 0 & 0 & 0 & 0 & C_{2323} & 0 \\ 0 & 0 & 0 & 0 & 0 & \frac{1}{2}(C_{1111} - C_{1122}) \end{bmatrix} \qquad (5.7)$$

Using engineering constants (Lekhnitskii, 1963) and eqns 5.1 and 5.3, the stiffness coefficients are:

$$C_{1111} = \frac{E_x(1 - n\nu_{zx}^2)}{(1 + \nu_{xy})(1 - \nu_{xy} - 2n\nu_{zx}^2)} \qquad (5.8)$$

$$C_{1122} = \frac{E_x(\nu_{xy} + n\nu_{zx}^2)}{(1 + \nu_{xy})(1 - \nu_{xy} - 2n\nu_{zx}^2)} \qquad (5.9)$$

$$C_{1133} = \frac{E_x \nu_{zx}}{(1 - \nu_{xy} - 2n\nu_{zx}^2)} \qquad (5.10)$$

$$C_{3333} = \frac{E_x(1 - \nu_{xy})}{(1 - \nu_{xy} - 2n\nu_{zx}^2)} \qquad (5.11)$$

$$C_{2323} = G_{xz} = G_{yz} = 19 \cdot 43 - 0 \cdot 31 O_c \qquad (5.12a)$$

$$\text{(for Utah oil shale)}$$

172   Ken P. Chong and John Ward Smith

For other oil shales where the shear modulus (G_{xz} or G_{yz}) is not found by the plate testing as shown in eqn (5.2), the following approximate formula can be used (Chong *et al.*, 1980c):

$$G_{xz} = G_{yz} = \frac{E_z n}{(1 + n + 2n\nu_{zx})} \tag{5.12b}$$

Based on the positive definiteness of the strain energy function, the following inequality (Barden, 1963; Dooley, 1964) must be satisfied:

$$1 - \nu_{xy} > 2\nu_{xz}\nu_{zx} \tag{5.13}$$

or using eqn 5.1:

$$1 - \nu_{xy} > 2n\nu_{zx}^2 \tag{5.14}$$

Note that all elastic 'constants', except the shear modulus, used to compute the coefficients in $[C]$ depend on the stress levels applied and on organic volume in the test specimens. Thus the stiffnesses are also non-linear due to stress dependency.

Using the linear generalized Hooke's law (eqn 5.4), the strain energy function, U, can be obtained by numerical integration as follows:

$$U = \int \sigma_{ij} \, \mathrm{d}\varepsilon_{ij} \tag{5.15}$$

But, from eqn 5.4:

$$\sigma_{ij} = C_{ijkl}\varepsilon_{kl} \tag{5.16}$$

and

$$C_{ijkl} = C_{ijkl}(O_c, S) \tag{5.17}$$

where S = stress level (equal to actual longitudinal stress, σ, divided by ultimate longitudinal stress, σ_u) expressed as a percentage. That is,

$$S = \sigma/\sigma_u \tag{5.18}$$

Also,

$$S \cong \varepsilon/\varepsilon_u \tag{5.19}$$

where ε = actual longitudinal strain and ε_u = ultimate longitudinal strain. Using eqns 5.16–5.19, eqn 5.15 becomes

$$U = \int [C_{ijkl}(O_c, \varepsilon, \varepsilon_u)]\varepsilon_{kl}[\mathrm{d}\varepsilon_{ij}] \tag{5.20}$$

5.2.3. Experiments

Samples for the experiments were obtained from cores drilled in Wyoming's Tipton Member, the Mahogany Zone near Anvil Points,

Colorado and Utah's Cowboy Canyon. The stratigraphic characteristics of oil shale allow a single bed to be identified over an area of many square miles. Sampling is the key on the transverse isotropy of oil shale, allowing a selection of similar (oil yield) samples from a given horizon. Organic content is the prime independent variable affecting the mechanical behavior of oil shale, varying with depth from layer to layer. Therefore, layers were selected to obtain a broad range of organic content (about 15–60% for Colorado oil shale) as determined by modified Fischer assay (Stanfield and Frost, 1949; Smith, 1969a,b, 1976). For Green River Formation oil shale, the organic volume (as a percent of the total volume) is (Smith, 1969a):

$$O_c = \frac{164\cdot9M}{M+111\cdot8} \qquad (5.21)$$

where M is the oil yield in gallons per ton (1 gallon ton^{-1} = 4·17 liters tonne^{-1}).

Conventionally, oil shale samples have been cut mechanically by diamond studded saw blades. The procedure is known to induce disturbances to the shale, quite often resulting in hairline cracks. To avoid these unwanted effects, the specimens were cut using a specially designed wire saw (Smith *et al.*, 1977; Chong *et al.*, 1980b, 1980c). The model 2008 wire saw (Fig. 5.3), designed by Laser Technology Inc., uses a diamond impregnated wire (0·25 mm (0·01 in) diameter in this case) of extremely high strength (over 3450 MN m^{-2}) as the cutting tool. The wire cuts smoothly and accurately to 0·025 mm (0·001 in). A significant feature of the wire saw is the very low force exerted on the specimen from the wire. This force usually falls within the range 0·2–2 N. A sprinkling system developed and attached to the wire saw kept the samples wet during cutting. This acted as a lubricant, coolant and carrier of shale dust and debris as well. An automatic control panel for easy operation made the wire saw a fast and efficient tool for sample preparation.

Each sample 'set' consisted of four prisms from the same horizon, two cut with the long axis normal to bedding, measuring 12·7 mm × 12·7 mm × 31·75 mm ($\frac{1}{2}$ in × $\frac{1}{2}$ in × $1\frac{1}{4}$ in) (set 'B') and two cut with the long axis parallel to bedding, measuring approximately 25 mm × 32 mm × 58 mm (1 in × $1\frac{1}{4}$ in × $2\frac{1}{4}$ in) (set 'A'). Figure 5.2(a) indicates half of the set of samples taken and Fig. 5.2(b) shows the cut samples. Although specimens A and B were of different sizes, they had roughly the same slenderness ratio. As observed by Dhir and Sangha (1973),

Fig. 5.3. Wire saw for sample preparation.

the axial stress/strains are independent of specimen size as long as the slenderness ratios are about equal.

Uniaxial compression tests were performed using an Instron Universal 1251 static/dynamic testing machine which features an electrohydraulic closed loop servo system composed of a loading frame, an electric console, and a hydraulic power pack. A servo-controlled hydraulic ram supplies the axial load and to ensure uniform contact pressure, a hemispheric steel ball resting on a lubricated socket was fabricated and positioned on the ram. The specimen was placed on the flat surface of the hemispheric ball which was free to rotate to compensate for any lack of complete parallelism between the two

Fig. 5.4. General test set-up.

loading surfaces. A Nicolet Explorer digital oscilloscope with four storage registers was connected externally to the Instron to display and store the output consisting of load, longitudinal displacement and their relationship with time. A Hewlett Packard plotter was later connected to the oscilloscope to trace curves of load versus time and longitudinal displacement versus time. To measure lateral displacement, a full bridge extensometer was constructed, calibrated and connected to the plotter, allowing a lateral displacement versus time curve to be traced during the test. Figure 5.4 shows the general test set-up.

5.2.4. Data Evaluation
Raw data obtained from testing consisted of curves showing load versus time and lateral displacement versus time. For a given value of time, instantaneous values of load and displacements could be read directly from the graphs for each tested specimen, enabling multilinear modeling of the tangent moduli and Poisson's ratios (Desai and Christian, 1977). To employ such a piecewise linear method requires the material to be quasi-linear (Hahn and Tsai, 1973), meaning that no coupling occurs between stress components in the non-linear range, permitting superposition of stresses. Stress level here is a normalized quantity

expressed as a percentage of applied stress of the ultimate stress. Therefore, each incremental tangent modulus and Poisson's ratio is associated with a unique stress level. The objective of these experiments is to relate these elastic constants to the independent variables stress level and organic content. A multivariable stepwise regression computer program (Dixon, 1976) yielded equations expressing the desired relationships. The flexibility of the program allowed investigation of the relative significance of the many possible mathematical combinations involving S and O_c.

The non-linear behavior of the stress–strain curves can be modeled using higher-order hyperelastic constitutive laws (Desai and Christian, 1977) instead of the generalized (linear) Hooke's law as shown in eqn 5.4. The main disadvantage of the higher-order constitutive laws is that the number of independent elastic constants increases drastically. For example, using cubic order in the stress tensor (Cleary, 1978), 35 independent constants (compared with five that are used in this chapter) have to be determined. Besides, due to the usual fitting, the solution for the cubic stress tensor still remains non-unique (Desai and Christian, 1977). It is the experience of the authors that to have good controllable determination of five independent constants is most difficult; to do the same for 35 independent constants is untenable indeed. Thus, throughout this investigation, the piecewise linear generalized Hooke's law is used, and the strain energy function can be obtained by numerical integration (Chong *et al.*, 1980c).

The results of testing and analysis show that the elastic coefficients E_x, E_z, ν_{xy} and ν_{zx} for oil shale are highly stress sensitive, that is, they can change drastically depending on the stress level being applied.

It was observed from the raw data (Fig. 5.5) that the Poisson's ratios vary linearly with respect to O_c and are directly proportional to S. Physically, this indicates that oil shale dilates laterally faster than it contracts longitudinally for higher stress levels and for richer oil shale.

Maximum values for Young's moduli occur at various stress levels. A study of the raw data indicated for lean oil shale the maximum moduli are higher than those of richer oil shale, occurring at higher stress levels (Fig. 5.6).

Physically, oil shale is slightly viscoelastic (Chong *et al.*, 1982a,b,c). The richer oil shale is more 'spongy'—hence, it is less stiff, and it absorbs more energy than the lean oil shale. Even though the maximum Young's moduli occur at lower stress levels for rich oil shale, it actually takes longer to achieve those levels than for the lean ones.

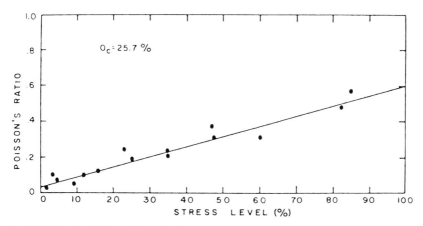

Fig. 5.5. Typical Poisson's ratio in x–y plane versus stress level for carbonate shales as demonstrated by the Utah oil shale.

Numerical analysis showed that the stress levels S_m, corresponding to the maximum moduli, are roughly proportional to the cube root of O_c:

$$S_m \propto O_c^{1/3}$$

Thus, let

$$S_m = [A + BO_c]^{1/3} \tag{5.22}$$

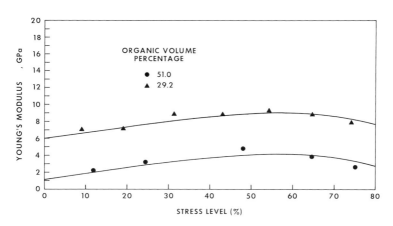

Fig. 5.6. Typical Young's modulus in z-direction versus stress level for carbonate shales as demonstrated by the Colorado oil shale.

Equation 5.22 is equivalent to:

$$\frac{\partial E}{\partial S} = A + BO_c - S^3 = 0 \qquad (5.23)$$

where

$$E = E_x \quad \text{or} \quad E_z$$

and

$$S = S_m \quad \text{when} \quad \frac{\partial E}{\partial S} = 0 \qquad (5.24)$$

Integrating eqn 5.24 with respect to S:

$$E = AS + BO_c S - CS^4 + D \qquad (5.25)$$

where A, B, C and D are integration constants. However, it was also observed that E varies linearly with O_c. Thus, the final form of E is selected as:

$$E = D + HO_c + AS + BO_c S - CS^4 \qquad (5.26)$$

where A, B, C, D and H are coefficients to be determined by regression. The equations and coefficients relating Young's moduli (E_x, E_z) and Poisson's ratios (ν_{zx}, ν_{xy}) to organic volume (O_c) and stress level (S) are developed by regressional analysis of the data, as indicated in Table 5.1. In addition, the uniaxial ultimate compressive stress in the x-direction (σ_{ux} in MPa) and z-direction (σ_{uz} in MPa) are also listed in Table 5.1. Using Utah oil shale for illustration, typical prediction equations for Poisson's ratios compared to the raw data are shown in Fig. 5.5. Curves showing a comparison between prediction equations and actual experimental results for Young's moduli are shown in Fig. 5.6 for Colorado oil shale. Using eqns 5.8–5.12, the five coefficients completely defining the stiffness matrix can be calculated from values for E_z, E_x, ν_{zx}, ν_{xy} and G_{xz}.

Experimental determinations of ultimate stress are compared with the plot of predicted ultimate stress in Fig. 5.7, for carbonate shales typified by the Colorado oil shale.

Examples of calculated stiffness matrix coefficients for Colorado oil shale are given in Tables 5.2a and 5.2b, in comparison with existing literature, using ultrasonic methods which are not stress dependent. It is observed that the shear moduli (C_{2323}) as calculated by eqn 5.12b are on the low side. Plate testing (Chong et al., 1980a) yielding equations such as eqn 5.12a should remedy this discrepancy.

TABLE 5.1
Three-dimensional Stiffness and Compressive Strengths for Different Oil Shales

	Colorado	Utah	Wyoming
(a) Young's modulus in x-direction, $E_x = D + HO_c + AS + BO_cS - CS^4$:			
Coefficient D	10·45	7·51	6·069
Coefficient H	−0·174	−0·124	−0·178
Coefficient A	0·384	0·315	0·163
Coefficient B	−0·005 19	−0·003 79	−0·001 50
Coefficient C	$1·883 \times 10^{-7}$	$2·418 \times 10^{-7}$	$1·000 \times 10^{-7}$
Coefficient of determination, r^2	0·826 5	0·702 2	0·852 8
Number of samples, N	709	99	127
Mean value, $\bar{E}_x$ (GPa)	10·072	9·097	5·407
Mean value, $\bar{O}_c$ (%)	33·48	27·79	25·90
Mean value, $\bar{S}$ (%)	51·31	41·77	51·90
Standard error of estimate, s (GPa)	2·868	2·794	1·249
(b) Young's modulus in z-direction, $E_z = D + HO_c + AS + BO_cS - CS^4$:			
Coefficient D	12·34	6·97	4·276
Coefficient H	−0·219 6	−0·142 6	−0·093 79
Coefficient A	0·746 1	0·172 7	0·075 17
Coefficient B	$-6·82 \times 10^{-5}$	$-1·899 \times 10^{-3}$	$-9·448 \times 10^{-4}$
Coefficient C	$9·869 \times 10^{-8}$	$1·126 \times 10^{-7}$	$5·241 \times 10^{-8}$
Coefficient of determination, r^2	0·712 7	0·701 6	0·753 5
Number of samples, N	662	94	134
Mean value, $\bar{E}_z$ (GPa)	4·882	5·683	2·841
Mean value, $\bar{O}_c$ (%)	37·18	28·23	27·30
Mean value, $\bar{S}$ (%)	65·22	54·52	57·70
Standard error of estimate, s (GPa)	1·861	2·334	0·894

TABLE 5.1—contd.

(c) *Poisson's ratio in zx-plane,* $v_{zx} = A + BO_c + CS$:

	Colorado	Utah	Wyoming
Coefficient A	−0.044 19	−0.2747	0.196 6
Coefficient B	0.003 85	0.011 62	0.000 875
Coefficient C	0.006 45	0.006 42	0.002 54
Coefficient of determination, r^2	0.777 4	0.704 4	0.781 7
Number of samples, N	633	87	130
Mean value, $\bar{v}_{zx}$	0.383	0.358	0.323
Mean value, $\bar{O}_c$ (%)	30.13	27.89	23.44
Mean value, $\bar{S}$ (%)	48.33	48.03	48.94
Standard error of estimate, s	0.112	0.177	0.037 8

(d) *Poisson's ratio in xy-plane,* $v_{xy} = A + BO_c + CS$:

	Colorado	Utah	Wyoming
Coefficient A	−0.033 07	−0.020 9	0.082 2
Coefficient B	0.003 33	0.002 23	−0.001 14
Coefficient C	0.004 80	0.005 55	0.002 58
Coefficient of determination, r^2	0.787 3	0.744 4	0.800 6
Number of samples, N	700	102	189
Mean value, $\bar{v}_{xy}$	0.261	0.270	0.225
Mean value, $\bar{O}_c$ (%)	28.44	27.25	24.62
Mean value, $\bar{S}$ (%)	41.62	41.48	52.71
Standard error of estimate, s	0.084 1	0.101	0.041 2

(*e*) *Uniaxial ultimate compressive stress in x-direction*, $\sigma_{ux} = A + BO_c$:

Coefficient A	127·73	152·14	103·70
Coefficient B	−1·121 5	−1·672 5	−2·041 0
Coefficient of determination, r^2	0·651 0	0·848 0	0·873 8
Number of samples, N	41	11	11
Mean value, $\bar{\sigma}_{ux}$ (MPa)	88·0	105·1	51·79
Mean value, $\bar{O}_c$ (%)	35·3	28·2	25·5
Standard error of estimate, s (MPa)	10·30	7·69	7·31

(*f*) *Uniaxial ultimate compressive stress in z-direction*, $\sigma_{uz} = A + BO_c$:

Coefficient A	161·60	160·39	103·20
Coefficient B	−1·541 5	−2·042 0	−1·474 0
Coefficient of determination, r^2	0·743 0	0·874 0	0·746 6
Number of samples, N	55	8	13
Mean value, $\bar{\sigma}_{uz}$ (MPa)	104·0	104·3	68·1
Mean value, $\bar{O}_c$ (%)	37·3	27·5	23·8
Standard error of estimate, s (MPa)	10·6	10·5	7·4

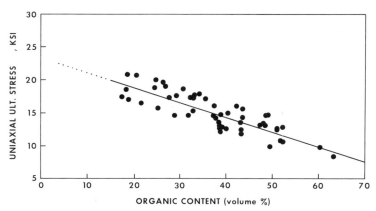

Fig. 5.7. Ultimate compressive stress in z-direction for carbonate shales as demonstrated by the Colorado oil shale. (Note: 1 ksi = 6·9 MPa.)

TABLE 5.2(a)
Comparison of Elastic Coefficients with Morris (1978) for 2 g cm^{-3} Oil Shale (Equivalent to 44% O_c)

		Authors' results		Morris, stress level unknown
		33% stress level	41% stress level	
C_{1111}	(GPa)	13·5	25·4	24·5
C_{1122}	(GPa)	7·4	18·8	8·5
C_{1133}	(GPa)	7·1	17·2	6·2
C_{3333}	(GPa)	9·7	18·8	15·1
C_{2323}	(GPa)	2·1	2·2	15·1

TABLE 5.2(b)
Comparison of Elastic Coefficients with Podio et al. (1968) for 2·23 g cm^{-3} Oil Shale (Equivalent to 30% O_c)

		Authors' results		Podio et al., stress level unknown
		37% stress level	47% stress level	
C_{1111}	(GPa)	20·4	39·9	43·4
C_{1122}	(GPa)	9·6	28·2	12·4
C_{1133}	(GPa)	9·3	25·5	9·7
C_{3333}	(GPa)	14·0	27·8	41·4
C_{2323}	(GPa)	3·7	3·7	15·2

In comparison with the ultrasonic methods the present results agree reasonably well at about 33–47% stress levels (see Tables 5.2a and 5.2b). However, ultrasonic methods usually give higher values than static methods (Schreiber et al., 1973; Jaeger and Cook, 1979). Thus much lower stress levels are actually encountered in ultrasonic testing. In Tables 5.2a and 5.2b, last column, C_{2323} (which is equal to G_{xz}) seems too high in comparison with the Young's modulus (E_x or E_z). Strong dependency on stress levels makes it difficult to determine the mechanical properties by acoustical methods.

5.2.5. Mineralogical Considerations

Comparison of results obtained for Utah's Parachute Creek Member, Green River Formation, oil shale from Wyoming's Tipton Member and from the Mahogany Zone in Colorado's Piceance Creek Basin shows basic differences in the mechanical behavior of rocks, all of which are classified as oil shale. In general, oil shales tested from Utah and Colorado are stronger and stiffer than those from Wyoming. Figure 5.8 shows curves for Young's moduli at 35% organic volume for the carbonate and the clay-rich oil shales identified specifically by the Green River Formation oil shales. The carbonate shales are represented by Utah and Colorado oil shales. The clay-rich oil shale

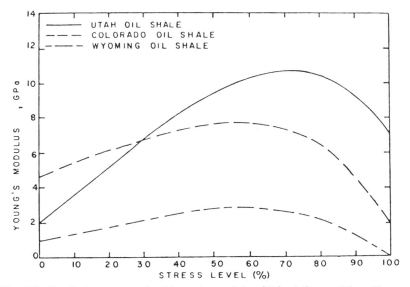

Fig. 5.8. Prediction curves for Young's modulus (E_z) of Green River Formation oil shale at 35% organic volume.

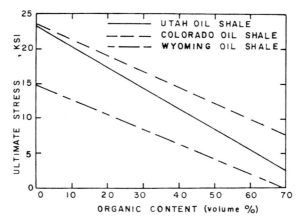

Fig. 5.9. Prediction curves for ultimate stress (σ_{uz}) of Green River Formation oil shale. (Note: 1 ksi = 6·9 MPa.)

typified by the Wyoming oil shale exhibits lower moduli than the carbonate shales at all stress levels. Figure 5.8 demonstrates emphatically the differences between carbonate and clay-rich oil shales. Similarly, Poisson's ratio shows higher values for clay-rich oil shale at any given stress level. The higher Poisson's ratios mean that physically the clay-rich Wyoming oil shale is more 'spongy' and less resistant to lateral deformation when subjected to a longitudinal load. Poisson's ratios for carbonate shales (Utah and Colorado oil shales) are roughly the same. Finally, Fig. 5.9 shows that all oil shales decrease in ultimate strength as organic content increases. The carbonate shales as demonstrated by Utah and Colorado oil shales show higher strengths than the clay-rich (Wyoming) oil shales throughout the range of organic contents. The dolomite forms a strength-producing structure in carbonate-rich shales.

5.2.6. Conclusions

A method is presented for the complete three-dimensional mechanical characterization of oil shale under static loading. To relate these properties to dynamic loading, strain rate effects have been previously investigated (Chong *et al.*, 1980b, 1981). Prediction equations for E_x, E_z, ν_{xy}, ν_{zx}, G_{xz}, σ_{ux} and σ_{uz} are developed, and all are shown to vary linearly with respect to organic volume, the prime independent variable; while stress level, the other major variable, enters as linear and

Mechanical characterization of oil shale 185

higher order terms, depending on the physical behavior of oil shale. Constitutive laws are examined, and it is demonstrated that non-zero coefficients in the stiffness matrix can be calculated from the predicted elastic constants. Comparison is made against the mechanical behavior of the carbonate and clay-rich oil shales of Wyoming and Colorado.

5.3. AVERAGE TENSILE STRENGTHS

5.3.1. Introduction

The ultimate average tensile strength of carbonate and clay-rich oil shales typified by the Green River Formation and central Tennessee (Chong *et al.*, 1979*a*, 1982*a*) oil shale along the bedding planes is a mechanical property parameter important to predicting how oil shale will break. This is particularly important to *in situ* fragmentation. The split cylinder test was reviewed and critically evaluated in detail to assure its applicability to this study. Test specimens representing the carbonate oil shale of the Mahogany Zone, sections of cores taken from the Naval Oil Shale Reserve No. 1 in Colorado and from the Bonanza area in Utah, the clay-rich oil shale of the Tipton Member of Wyoming and of Silver Point Quad in Central Tennessee, were subjected to the split cylinder test. Linear regression equations relating ultimate tensile strength along the bedding planes to volume percent of organic matter in the rock were developed from the test data. The Utah and Colorado equations are statistically similar, corresponding to their comparable mineralogy. Similar equations representing the clay-rich, dolomite-poor Tipton Member of Wyoming's Green River Formation and the Tennessee oil shale differed sharply from the carbonate oil shale, demonstrating an emphatic mineralogical influence with major variations.

In situ recovery of oil from oil shale minimizes materials handling, a major cost in surface processing. *In situ* retorting requires fracturing the oil shale to create permeability. Fracturing shales in place entails breaking the bedding planes. Since oil shale and most rocks are weaker in tension than in compression, practically all fractures across the bedding planes are caused by tension. Hence ultimate strengths across the bedding planes are important in fracture studies. The mineralogy of the two Mahogany Zone cores is quite similar, while the Tipton Member oil shale of Wyoming and the Tennessee oil shale contain substantially more clay than the Colorado/Utah oil shales. The effect of this mineralogic difference on ultimate tensile strength is examined.

Split cylinder testing, which measures average tensile strength across many bedding planes, yields more representative information than direct tension testing (Chong et al., 1979a, 1982a). Consequently, split cylinder testing was used to obtain indirect ultimate tensile strengths along (not across) bedding planes. Test cylinders were cut perpendicular to bedding planes from Mahogany Zone sections of core taken from US Naval Oil Shale Reserve, Core Hole No. 21 near Rifle, Colorado (Smith et al., 1979), from Cowboy Canyon (sec. 33, T9S, R25E, SLM) in Uinta County, Utah, from upper parts of the Tipton Member and the Tennessee oil shale in Silver Point Quad. Samples were cored at many different depths in order to obtain comprehensive representation and a broad range of organic contents. The oil yield of the samples tested ranged from about 10 to 73 gallons ton^{-1} (41·7–304·4 liters tonne^{-1}) for Colorado/Utah oil shale; leaner for Wyoming oil shale (to about 55 gallons ton^{-1}, 229·4 liters tonne^{-1}).

5.3.2. Theory

A critical review and analysis of the split cylinder test will now be presented. The split cylinder test, or the 'Brazilian' test, is an established method used primarily to determine the tensile strength of concrete (ASTM, 1973) and other brittle materials having much higher compressive strengths than tensile strengths. Basically a circular cylinder or disk is compressed with concentrated line loads, P, across a diameter (Fig. 5.10). Ideally the tensile failure will occur along the loaded diameter, splitting the cylinder into two halves.

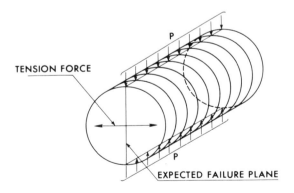

Fig. 5.10. Loading with bedding planes vertical.

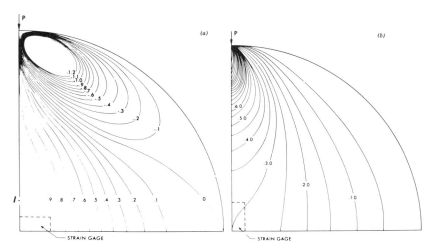

Fig. 5.11. Contours of normal stresses: (a) x-direction; (b) y-direction. $[\times 2p/(\pi d)]$.

Since the split disk has very simple boundary conditions and geometry, it has been a classic problem since 1883, attracting the attention of many mathematicians and elasticians alike (Love, 1944; Timoshenko and Goodier, 1951). Nevertheless, the failure behavior is still a hot topic requiring evaluation to determine if the test really measures tensile strength along the bedding planes. The classical theory assumes that the line load is applied over an infinitesimally small width. This cannot be realized in practice. Another discrepancy of the classical theory is that it implies a constant tensile stress acting across the loaded diameter (Fig. 5.11). If half of the disk is taken as a free body along the vertical diameter, this tensile stress cannot be balanced since no external horizontal force is applied. Thus the equilibrium condition is violated (Fairhurst, 1964). To overcome these difficulties, Hondros (1959) developed a modified theory assuming negligible body forces and a finite width of loading applied radially. The theoretical predictions agreed closely with the experimental results monitored by strain gages (Hondros, 1959). Using Hondros' theory and Griffith's fracture criterion, Fairhurst (1964) investigated the validity of the split cylinder test on brittle materials and concluded the finite width of the applied load should be about a quarter of the disk radius.

Bouwkamp (1963) used the moiré method to analyze the split

cylinder. His results compared favorably with the classical theory in regions a quarter of the disk radius away from the load points. Recently Pindera *et al.* (1978) conducted extensive experiments using isodynes, isochromatics and strain gages. Their findings compared favorably with those developed by finite element analysis (Chong *et al.*, 1982*a*).

For long cylinders (plane strain case) and thin disks (plane stress case), the stress expressions given by the classical theories of Timoshenko and Goodier (1951) and Hondros (1959) remain unchanged. However, the stress–strain relationships are different. The stress contours according to the classical theories are shown in Fig. 5.11.

The finite element method can model the split cylinder test quite simply. Due to symmetry only a quarter of the disk needs to be considered. The finite elements and boundary conditions are shown in Fig. 5.12. A total of 250 two-dimensional elements with 146 nodes were used. Each node had two degrees of freedom. The nodal stresses were computed using consistent stress distributions (Oden and Brauchli, 1971). The load of $P/2$ was assumed to act at the apex node alone, since the classical theory is based on concentrated line loading. Practically, a finite width of contact designated a, will exist between the loading plate and the disk. The contact width is a function of the Young's moduli and Poisson's ratios of the loading plate (E_1, ν_1,

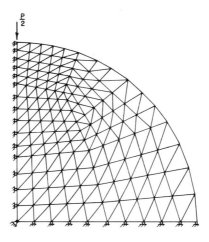

Fig. 5.12. Finite element model.

respectively) and the disk (E_2, ν_2, respectively). It is given by the following formula (Roark, 1965):

$$a = 1 \cdot 6 \left[\frac{Pd}{d} \left(\frac{1-\nu_1^2}{E_1} + \frac{1-\nu_2^2}{E_2} \right) \right]^{1/2} \quad (5.27)$$

Since the apex of the disk is subjected to maximum deformation, the maximum contact pressure will result. The results of the finite element analysis, presented in the next section, compare favorably with experimental observations (Bouwkamp, 1963; Pindera et al., 1978).

The stress distributions from the above theories, experiments and finite element method, along the vertical diameter ($\bar{\sigma}_x, \bar{\sigma}_y$) and the horizontal diameter ($\bar{\bar{\sigma}}_x, \bar{\bar{\sigma}}_y$) are presented in Fig. 5.13. These stresses have been normalized (divided) by the quantity, $\sigma_0 = P/(bd)$, for comparison with other references. Four different methods are compared in Fig. 5.13: (a) classical theory according to Love (1944), Timoshenko and Goodier (1951), Muskhelishvili (1975); (b) finite element analysis; (c) isodynes method (Pindera et al., 1978); and (d) Hondros' theory (Hondros, 1959) with bearing width, a, equal to a sixth of the disk

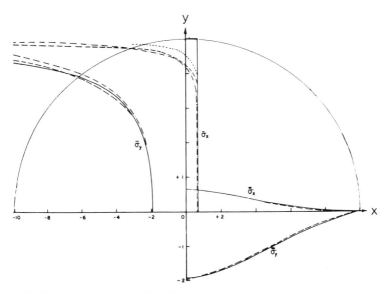

Fig. 5.13. Normalized stress distributions along the vertical and horizontal diameters. (———) Theoretical (classical); (– –) finite element method; (- - - -) isodynes method, Pindera et al. (1978); (– · – ·) theoretical, Hondros (1959).

radius. Methods (a) and (b) are plotted for all four curves. For simplicity, methods (c) and (d) are shown only if they deviate from the classical theory.

It can be seen that the classical theory agrees well with all methods except for the tensile stress across the loaded diameter, $\bar{\sigma}_x$. For $\bar{\sigma}_x$, methods (b), (c) and (d) show good agreement, indicating a huge compressive stress close to the load. This represents the reversal of stresses necessary for equilibrium and balance of internal horizontal forces. Both methods (b) and (d) indicate zero stress at 0·85 of the disk radius measuring from the center, whereas method (c) measures zero stress at 0·90 of the disk radius.

Physically the region under the load experiences huge uniform compressive pressure in $\bar{\sigma}_x$ and $\bar{\sigma}_y$ (as can be seen from the finite element analysis). Apparently this region tries to wedge its way into the disk, causing an ultimate tensile failure in brittle materials. This wedging action can be seen in the displacement contours based on finite element analysis (Fig. 5.14).

In summary, the classical theory is valid for most areas of the disk except in the proximity of load points. Along the loaded (vertical) diameter the classical theory fails to predict the reversal of tensile

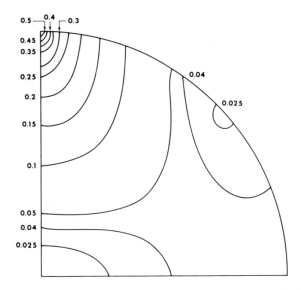

Fig. 5.14. Deflection contours by finite element method.

stresses required for the equilibrium condition. Finite element analysis agrees well with Hondros' modified theory and the experimental data by Pindera *et al.* By calculating the contact width from eqn 5.27, Hondros' theory can be used with confidence even though uniform radial loads are assumed for non-uniform vertical contact loads.

Since the classical theory is valid for most areas of the disk it is used to compute the average ultimate tensile strength of the rock. Using the classical theory, along the loading plane under ultimate load, the ultimate tensile stress in the x-direction is

$$\sigma_u = \frac{2P_u}{\pi d} \tag{5.28}$$

where $P_u =$ ultimate (maximum) load per unit thickness of the disk. Subsequently eqn 5.28 is used in this chapter. Wang and Chong (1980) showed that the above formula also holds for transversely isotropic materials if plane elasticity is assumed.

5.3.3. Experimental

5.3.3.1. Sample Preparation
Field cores measuring $3\frac{1}{2}$ in (8·9 cm) diameter were drilled from the geologically similar Colorado and Utah Mahogany Zone oil shale deposits by the Laramie Energy Technology Center. Field cores from Wyoming and Tennessee are 6 in (15·2 cm) in diameter. Using a diamond-tipped, water-cooled core drill, test cores were drilled from the field cores perpendicular to bedding planes. Nominal diameters of the test cores were $1\frac{3}{4}$ in (4·5 cm) and lengths were 1 in (2·5 cm). These specimens provided a range of organic contents. Layers with obviously varying composition, with major elastic inclusions, or with existing faults were avoided (Chong *et al.*, 1979*a*). The over-cores were saved to maintain a record of each sample's stratigraphy and structure.

5.3.3.2. Instrumentation
A Tinius Olsen Super 'L' Universal testing machine with ranges up to 400 000 lb (1780 kN) was used for loading. The loading head consisted of a steel plate connected to a ball bearing so that the load was applied uniformly along the side of a sample. Load was applied to the cylindrical specimens diametrically along the side. Two narrow bearing strips of plywood were placed between the specimen and the upper and lower bearing blocks of the testing machine. Load was applied

192   *Ken P. Chong and John Ward Smith*

with a strain rate of approximately 0.5% min^{-1} until tensile cracks appeared and failure was complete. Test data with the failure planes deviated from the loading planes were neglected, since the coincidence of failure and loading planes indicated competent rocks whose strength governs in modifying the rocks in place (Colback, 1966; Chong *et al.*, 1979*a*).

5.3.4. Relationship of Test Results to Oil Shale Parameters

Three oil shale parameters which may be associated with variations in the mechanical properties of oil shale are: (1) organic matter content; (2) mineralogy; and (3) anisotropy perpendicular to the bedding planes (Smith, 1976). The testing method described here eliminates anisotropy as a significant parameter because the principal planes coincide with the bedding planes (Wang and Chong, 1980). Statistically, the effects of mineralogic variations are insignificant among geologically similar samples (Chong *et al.*, 1979*a*). Consequently relationships of test results with organic matter volume (as determined by eqn 5.21) in the tested samples were examined.

The coincidence of failure and loading planes (Chong *et al.*, 1979*a*) indicates competent rocks whose strength governs in modifying the rocks in place. Thus only the test data with such failure modes were retained. Applying regressional analysis on these data, linear relationships between ultimate tensile stress (σ_u) in MPa and the organic matter volume (O_c) in percentages were obtained and listed in Table 5.3 for different oil shales, taken from Colorado, Utah, Tennessee and Wyoming, representing carbonate shales and clay-rich oil shales. Statistically there is no significant difference between the relationships for the ultimate tensile strength of the carbonate shales (the Colorado and Utah oil shales). This is expected from the geological similarity of these oil shales. As for the clay-rich oil shale of the Green River Formation's Tipton Member in Wyoming the slope of 0.3517 is more than three times those for the carbonate oil shales. This difference is definitely statistically significant. The organic matter in the Colorado–Utah samples is virtually identical with that in the Wyoming Tipton Member. Consequently the primary difference must be in mineralogy. Illite and mixed-layer clay are the primary components ($>50\%$) of the mineral material in the Tipton Member while dolomite is only about 10% of the mineral matter (Robb and Smith, 1976). In contrast, dolomite is the primary component ($>40\%$) of the mineral matter in the Mahogany Zone cores and illite is less than 20%. Of the Green

TABLE 5.3
Average Tensile Strengths for Different Oil Shales

	Colorado[a]	Utah[a]	Tennessee[b]	Wyoming[c]
Form of regressional equation, $\sigma_u = A - BO_c$:				
Coefficient A	14·78	13·64	15·27	23·17
Coefficient B	0·092 83	0·121 2	0·078 36	0·351 7
Coefficient of determination, r^2	0·575 5	0·566 4	0·606 5	0·811 0
Number of samples, N	56	28	15	40
Mean value, $\bar{\sigma}_u$ (MPa)	11·5	10·1	13·2	13·5
Mean value, $\bar{O}_c$ (%)	35·4	29·2	25·9	27·6
Standard error of estimate, s (MPa)	1·0	1·1	0·33	1·9

[a] Chong et al., (1982a).
[b] Chong et al., unpublished data.
[c] Chong et al. (1979a).

River Formation minerals only dolomite formed in such a way that it forms a structural constituent in the oil shale (Smith and Robb, 1966). An increasing volume of organic matter apparently dilutes and weakens the structure of the dolomite matrix. All other mineral particles are extremely fine and discrete (not interconnected). Apparently dolomite provides a structure matrix to the Mahogany Zone rock capable of providing tensile strength. The Tipton Member's dolomite amounts are apparently too low to exert a similar influence at a significant level. This demonstrates that major differences in the ultimate tensile strength of oil shale may be produced by major variations in mineral composition.

The Tennessee oil shale, a Devonian black shale, is clay-rich as is the Wyoming oil shale. However only about one third of the organic carbon is converted to oil (Smith and Young, 1967) instead of about two-thirds of the Green River oil shale.

Based on computer testing on the convergence of organic volume from oil yield and from densities (details of which are described in Chapter 1) the following preliminary formula is derived:

$$O_c = \frac{188 \cdot 4M}{M + 73 \cdot 6} \tag{5.29}$$

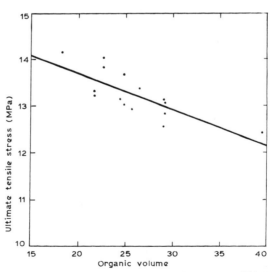

Fig. 5.15. Typical tensile strength versus organic volume (%) for clay-rich oil shales exemplified by the Tennessee oil shale.

Mechanical characterization of oil shale 195

A typical plot of the tensile strength versus organic volume is shown in Fig. 5.15 for the clay-rich oil shale (Tennessee oil shale).

5.3.5. Conclusions

Theoretical and numerical bases for the validity of the split cylinder test are presented. Ultimate tensile strengths for carbonate and clay-rich oil shales, typified by Colorado, Utah, Tennessee and Wyoming oil shales, are tested and correlated. Volume of organic matter is the primary oil shale variable, bearing a linear relationship with the tensile strength. The two linear relationships developed for predicting ultimate tensile strength from organic volume for carbonate shales are statistically identical, a result expected from their mineralogical similarity. The mineralogically different clay-rich oil shales show a much steeper slope in this relationship with organic volume, indicating the weakening effect contributed by the lesser amount of mineral dolomite.

5.4. STRAIN-RATE EFFECTS

5.4.1. Introduction

One scheme for extracting oil from shale in a manner which is both economical and environmentally acceptable is to heat it in place. Efficient *in situ* retorting of oil shale requires that the shale be broken and void volume increased and distributed. The purpose is to provide permeability so combustion gases and hydrocarbon products can flow. To optimize the fragmentation process, knowledge of the strain-rate effects on mechanical properties and fracture characteristics of the oil shale is needed. The principal goal of the investigation reported here was to determine the influence of strain-rate on the ultimate compressive stress, the ultimate strain and the Young's modulus for the clay-rich and carbonate oil shales from Tipton Member in Wyoming and Cowboy Canyon in Utah, respectively (Chong *et al.*, 1980*b*, 1981). An attempt was made to provide accurate data in the badly needed conventional explosive range. The test was performed at strain-rates ($\dot{\varepsilon}$) between 1×10^{-4} and $1 \times 10^{1} \, \text{s}^{-1}$ in a total of 500 specimens. The volume fraction of organic matter in the test specimens ranged from about 15% by volume (11·6 gallons ton^{-1}, 48·4 liters tonne^{-1}) through 47% by volume (44·6 gallons ton^{-1}, 186 liters tonne^{-1}). Depths varied from 454 ft (138·5 m) through 512 ft (156·2 m) for the Utah oil shale

196   *Ken P. Chong and John Ward Smith*

and from the surface to 40 ft (12·2 m) for the Wyoming oil shale. A multi-parameter statistical analysis demonstrated that the ultimate stresses were strongly influenced by the organic volume as well as the strain-rate, while the ultimate strain was influenced by the organic volume only. Young's modulus showed strong dependency on both organic volume and strain-rate. Failure modes ranged from brittle for lean samples at fast strain-rates to ductile for rich samples at slow strain-rates. A strain-rate dependency failure mechanism is proposed to explain the observed behavior. A linear relationship was observed between the ultimate stress, organic volume and logarithmic strain-rate. The linear relationship enables the ultimate stress to be extrapolated to the higher explosive strain-rates with confidence.

5.4.2. Review of Literature

In the past, there have been various experimental investigations into the physical and mechanical properties of oil shale. However, Lankford (1976) is the only reference the authors found that dealt with the strain rate dependence of mechanical properties of oil shale.

The strength of most rocks is known to depend strongly on strain-rate, increasing with increased rate of loading (Green and Perkins, 1968). Little was known of the significance of this effect on oil shale until recently. In 1976 Lankford investigated the strain rate dependence of strength and ductility for oil shale from the US Bureau of Mines test mine at Anvil Points, Colorado. A total of 98 specimens were tested at various confining pressures (0, 5, 10 and 20 ksi; 0, 35, 69 and 138 MPa) and various strain-rates (from $1·73 \times 10^{-4}$ to $1·87 \times 10^{3} \text{ s}^{-1}$). Typical lean, medium and rich samples (with oil yields of 10·7, 32 and 45·7 gallons ton^{-1}; 44·6, 133·4 and 190·6 liters tonne^{-1}, respectively) were tested. The slow and intermediate strain-rates were performed with a servo-controlled hydraulic ram supplying the axial load, while the high strain-rate was done with a split Hopkinson pressure bar apparatus. Lankford developed a failure criterion in accordance with the general first-order stresses from the data. The experimental results agreed reasonably well, over the range of strain-rates and confining pressures, with the failure criterion. Of the three parameters varied (static confining pressure, strain-rate and organic content) in Lankford's experiment, strain-rate dominated consideration of fracture strength. From 18 unconfined tests Lankford concluded that failure strength increased with strain-rate, while the inelastic strain to failure was essentially invariant with respect to strain-rate.

The actual strength of the shale nearly trebled, increasing from about 62·1 to 165·6 MPa (9–24 ksi) as the strain-rate varied from $1·7 \times 10^{-4}$ to $2 \times 10^3 \, s^{-1}$. Strength and ductility increase with static confining pressure, but since static confining pressures in place (overburden pressures) are not likely to exceed 13·8–20·7 MPa (2–3 ksi), confining pressure is not included in this section.

5.4.3. Experimentation
Samples for the experiment were obtained from cores drilled in Wyoming's Tipton Member and Utah's Cowboy Canyon. From the original 6 in (15·24 cm) diameter core, small samples were cut perpendicular to the bedding planes at selected depths throughout the core. To obtain a number of specimens with the same composition, 32–40 individual samples were cut from the identical stratigraphic horizon. The specimens were cut using a specially designed wire saw (Fig. 5.3). Test specimens measuring $1·27 \times 1·27 \times 2·54$ cm ($\frac{1}{2} \times \frac{1}{2} \times 1$ in) were cut with the long axes perpendicular to the oil shale's bedding plane. Special effort was made to have smooth cuts so that the friction between the loading surface and the specimen was minimized. Since oil shale is stratified rock, care was taken to get as representative a sample as possible as described in Sections 5.2.3 and 5.3.3. After cutting, the samples were marked and their dimensions measured.

Strain-rate tests were performed with a servo-controlled hydraulic ram supplying the axial load. The testing instrument was an Instron Universal 1251 static/dynamic testing machine. The specimen was placed on the flat surface of a hemispheric ball which is free to rotate to compensate for any lack of complete parallelism between the two loading surfaces.

A Nicolet Explorer digital oscilloscope with four storage registers was connected externally to the Instron to display and store the output. This allowed for a close investigation of the load versus time, the displacement versus time and the load versus displacement curves. A Hewlett Packard plotter was connected to the oscilloscope to trace the load versus displacement curves. The test set up is similar to the one shown in Fig. 5.4.

5.4.4. Data Evaluation
The fracturing behavior of oil shale with different compositions and loaded at varying strain rates, has been documented (Chong *et al.*, 1980*b*). For the lean samples (about 15% in organic volume), all the

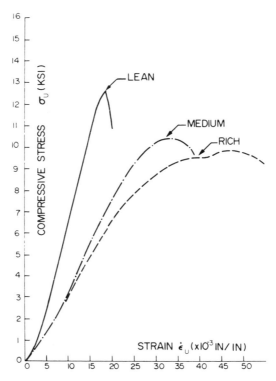

Fig. 5.16. Stress versus strain curves.

failures were sudden (Fig. 5.16), without any indication of yield prior to failure. The ultimate strain ranged typically from about 1·5 to 2·0%. Most of the specimens showed rather explosive failures at the higher strain rates. It is well known that during the unloading process (after the specimen yields) inadequate stiffness of the testing machine creates an instability which releases huge energy from the testing machine, resulting in violent specimen failure. However, the testing machine used (Instron 1251) has an overall frame and load cell stiffness of 6·75 MN cm^{-1}, compared with the stiffnesses of the lean oil shale, ranging from 0·44 to 0·61 MN cm^{-1} (for low to high strain rates), based on the median Young's modulus which is on the high side compared with the unloading Young's modulus. Thus the energy stored in the testing machine at any load after yielding is always less than the energy required for further compression of the specimen and

Mechanical characterization of oil shale

the situation is stable. It can be concluded that the fragmentation of lean samples is an intrinsic mode of failure not affected by this test machine. Most of the specimens in this category failed in shear.

For the medium samples (about 27% in organic volume), the failures were less sudden with some indication of yield prior to failure. The ultimate strain typically ranged from about 2·0 to 4·0%. The modes of failure seen were the same as for the lean samples. The brittleness of the failures increased as the strain-rate increased.

The rich samples (about 46% in organic volume) showed a clearly more ductile failure than the two previous categories. A distinct yield plateau followed by a strain softening zone was observed for most of the specimens. The ultimate strain typically ranged from about 3 to 7%.

For this investigation samples perpendicular to the bedding planes were selected since parallel samples tend to fail by peeling, lowering yield strength (Chong *et al.*, 1976). This phenomenon can best be explained by modeling with a series of Euler columns connected along each layer. The surface layers (columns) buckle first because they are only braced on the inside. Thus column buckling followed by tension failure between the layers occurs. Without free surfaces, such as is the case for rocks under overburden, the observed 'peeling' failure will not occur. This agrees with previous investigations (Chong *et al.*, 1980*b*), which concluded that subject to confining pressure, the ultimate strength of oil shale was not dependent upon orientation.

Young's modulus at one half the ultimate stress was determined from the raw test data. As noted by earlier investigators (Sellers *et al.*, 1972; Chong *et al.*, 1976, 1979*a*), Young's modulus was observed to be strongly dependent upon the organic content, with the leanest samples showing the highest Young's modulus. It was also observed that Young's modulus was dependent upon the strain-rate, with the higher strain-rate showing the higher Young's modulus.

5.4.5. Results and Discussion

The basic purpose of this investigation was to determine the relation of strain rate and oil shale composition to the ultimate compressive strength, the ultimate strain and the Young's modulus for oil shale from the carbonate and the clay-rich oil shale. Therefore, the following variables were used in this study: σ_u = ultimate compressive strength (MN m^{-2}, MPa); ε_u = ultimate strain (cm cm^{-1}); $E_{1/2}$ = Young's modulus at one half σ_u (MN m^{-2}); O_c = organic content (vol%); and

$\dot{\varepsilon}$ = net strain-rate (cm cm^{-1} s^{-1}, or s^{-1}). In order to avoid an excessive range of numbers, the common logarithm of the strain rate was used in the regression program.

Regression analysis reveals that for oil shale, a linear relation between the variables O_c and $\log \dot{\varepsilon}$ yields the best correlations. A stepwise regression program was used for each of the three dependent variables, σ_u, ε_u, and $E_{1/2}$. The results are summarized in Table 5.4. Figure 5.17 is a plot of the prediction equation in Table 5.4 versus the raw data from different grades of carbonate oil shale from Utah. The

TABLE 5.4
Strain Rate Effects for Different Oil Shales

	Utah	Wyoming
(a) Ultimate compressive strength, $\sigma_u = A + BO_c + \log_{10} \dot{\varepsilon}$:		
Coefficient A	195·87	123·59
Coefficient B	−1·78	−1·74
Coefficient C	13·35	4·02
Coefficient of determination, r^2	0·75	0·78
Number of samples, N	147	221
Mean value, $\bar{\sigma}_u$ (MPa)	129·08	464·10
Mean value, $\bar{O}_c$ (%)	28·88	29·38
Mean value, $\log \bar{\dot{\varepsilon}}$	−1·410 1	−1·289 7

	Utah oil shales
(b) Ultimate strain, $\varepsilon_u = A + BO_c$:	
Coefficient A	−0·005 7
Coefficient B	0001 0
Coefficient of determination, r^2	0·64
Number of samples, N	146
Mean value, $\bar{\varepsilon}_u$ (cm/cm)	0·021 9
Mean value, $\bar{O}_c$ (%)	26·62
Mean value, $\log \bar{\dot{\varepsilon}}$	−1·432 0
(c) Young's modulus at one-half ultimate compressive strength, $E_{1/2} = A + BO_c + C \log_{10} \dot{\varepsilon}$:	
Coefficient A	11125·82
Coefficient B	−153·77
Coefficient C	330·62
Coefficient of determination, r^2	0·62
Number of samples, N	145
Mean value, $\bar{E}_{1/2}$ (MPa)	6475·77
Mean value, $\bar{O}_c$ (%)	27·17
Mean value, $\log \bar{\dot{\varepsilon}}$	−1·4267

Mechanical characterization of oil shale 201

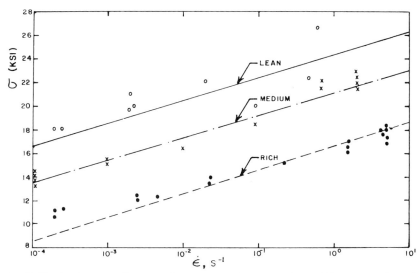

Fig. 5.17. Strain rate effects on the compressive strengths of different grades of carbonate shales exemplified by the Utah oil shale. (Note: 1 ksi = 6·9 MPa.)

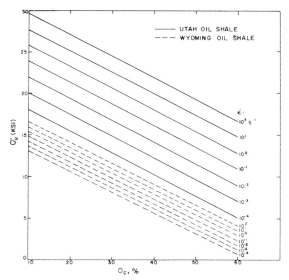

Fig. 5.18. Ultimate compressive strengths of a typical clay-rich oil shale (Wyoming's Tipton Member) and a typical carbonate shale (Utah's Parachute Creek Member). (Note: 1 ksi = 6·9 MPa.)

straight lines are for organic volumes of 15·4% (lean), 27·6% (medium) and 46·7% (rich).

Figure 5.18 shows a graphical representation of prediction equations for ultimate compressive strength in Table 5.4, for clay-rich oil shales (Wyoming's Tipton Member) and carbonate shales (Utah's Parachute Creek Member in Cowboy Canyon). The carbonate oil shale shows higher ultimate strength for any given strain rate, while the slopes are about the same for both groups, indicating that the effect of organic content is roughly the same. The effects of strain-rate are much more important for carbonate shales than clay-rich ones.

The ultimate strain to failure was found to depend only upon organic volume and to be independent of the strain-rate. For the median Young's moduli ($E_{1/2}$) the leanest samples showed the highest values, while the higher strain-rates produced the highest Young's moduli for the same organic content. The influence of mineralogy is discussed in Section 5.2.5.

5.4.6. Strain Rate Dependency Failure Mechanism

The failure mechanism has been investigated by Chong et al. (1980b). The unloading model suggested by Janach (1976) does not apply for the following reasons:

1. The strain-rates tested were between $10^{-4}\,\text{s}^{-1}$ and $10^{1}\,\text{s}^{-1}$, considerably lower than Janach's critical strain-rate. Thus the lateral unloading velocity will be high enough to unload the specimen as soon as the failure stress is reached, resulting in no confining pressure due to bulking.
2. The failure patterns were not entirely brittle.
3. As indicated by Achenbach (1973), the inertia effects (including lateral inertia) are only important if the duration of load application, t_a, is of the same order as the time taken for a disturbance to travel at propagation velocity, c, over the largest distance, r, within the sample (measured from the load point). In the present study:

$$t_a \approx 2 \times 10^{-3}\,\text{s (for fastest } \dot{\varepsilon})$$
$$c \approx 3\cdot 2\,\text{km s}^{-1}$$
$$r/c \approx 0\cdot 4 \times 10^{-5}\,\text{s}$$

Thus $t_a > r/c$ and the inertia effects can be neglected.

The model proposed by Kumar (1968) is primarily for brittle fracture. It is given by:

$$\dot{\varepsilon} = KNCV_c \quad (5.30)$$

where K = orientation constant; $N = N(\sigma)$ = number of microcracks; σ = stress; C = average length of microcracks; $V_c = V_c(\sigma)$ = crack propagation velocity. If $\dot{\varepsilon}$ increases, then $N(\sigma)$ and/or $V_c(\sigma)$ has to increase, either of which require higher σ_u. For oil shale, N and V_c are also functions of O_c; and the failure pattern is not just brittle.

To explain the strain-rate dependency oil shale is modeled from the macroscopical and microscopical points of view. Oil shale, like most solids, is slightly viscoelastic and can be characterized as a standard non-linear solid (Chong et al., 1982c). Referring to Fig. 5.19, the viscosity, η, is furnished primarily by the intermolecular forces before nucleation of microcracks and by frictional forces during the crack propagation stage. Under the low strain-rates the dashpot will relax gradually, leaving the spring E_2 to carry the total load F. The springs store the strain energy which finally reaches the ultimate shear stress and the rock fails. For high strain-rates, the dashpot stiffens and springs E_1 and E_2 act in unison. In addition part of the input energy will be dissipated through the dashpot. Thus a much higher force (or stress) is required to fail the sample.

Microscopically let each particle be represented by a minute standard non-linear solid (as in Fig. 5.19). Let: η_p = viscosity of the particle; e_1 = spring constant in series with η_p; e_2 = spring constant parallel to η_p and e_1; V = volume of test sample; t = time variable; s = space variable; and x = longitudinal distance (along load direction).

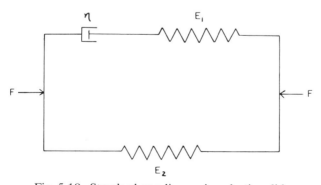

Fig. 5.19. Standard non-linear viscoelastic solid.

Then the input energy (W_{in}) is given by

$$W_{in} = F \iint \dot{\varepsilon} \, dx \, dt \qquad (5.31)$$

The strain energy (W_s) is given by

$$W_s = \frac{1}{2} \int k(e_1 + e_2) \left(\iint \dot{\varepsilon} \, ds \, dt \right)^2 dV \qquad (5.32)$$

where $K = $ orientation constant.

The dissipative energy (W_d) is given by

$$W_d = \iiint K \eta_p \dot{\varepsilon}^2 \, ds \, dt \, dV \qquad (5.33)$$

From the energy balance

$$W_{in} = W_s + W_d \qquad (5.34)$$

For lowest strain-rates

$$\dot{\varepsilon}^2 \to 0$$

Thus, $W_d \to 0$ and due to relaxation of the dashpot, spring e_1 is not acting. Thus

$$F \iint \dot{\varepsilon} \, dx \, dt = \frac{1}{2} \int_V Ke_2 \left(\iint \dot{\varepsilon} \, ds \, dt \right)^2 dV \qquad (5.35)$$

At ultimate load and strain

$$F = F_u \qquad (5.36)$$

$$e_2 = e_{2u} \qquad (5.37)$$

and

$$\sigma_u = F_u/A \qquad (5.38)$$

where $A = $ area of cross-section of sample.

For higher strain rates, W_d does not approach zero and contributes to eqn 5.33. Since springs e_1, e_2 and the dashpot act together, and the observed ultimate strains remain relatively constant, it requires much higher ultimate load F_u to fracture the rock. Thus σ_u is also much higher. Also the Young's modulus is higher since it is a measure of the stiffness of the springs which are now acting together.

For richer samples e_{1u} and e_{2u} are weaker and η_p increases. The net effect is that σ_u varies linearly with respect to the organic volume O_c (Figs 5.17 and 5.18). In short, the non-linear solid model (non-linearity

Mechanical characterization of oil shale 205

due to stress level dependency) presented explains the observed data on σ_u, ε_u and $E_{1/2}$ quite well.

5.4.7. Conclusions

A non-linear viscoelastic model is presented which explains the strain-rate dependency failure mechanism for different grades of oil shale subject to strain-rates of 10^{-4}–$10 \, \text{s}^{-1}$. It also predicts the increase in Young's moduli at higher strain-rates.

The ultimate fracture stress of oil shale under uniaxial compressive conditions was observed to be influenced linearly by the volumetric organic content and the logarithmic strain rate. The richest shale consistently showed the lowest strength, while the leanest shale clearly was the strongest. This monotonic increase in strength is consistent with earlier investigations (Sellers *et al.*, 1972; Chong *et al.*, 1976) but different from Lankford's (Lankford, 1976). For the same richness of shale the higher strain-rates of oil shale consistently showed higher strength than that of the lower strain-rates.

The prediction equation for σ_u in Table 5.4 agreed well with the data observed over the range of strain-rates and organic contents used. The linear relationship found between the ultimate stress, the organic content and the logarithmic strain-rate enables the ultimate stress to be extrapolated into the explosive strain-rates (10^1–$10^3 \, \text{s}^{-1}$) for different oil yields. Compared to a similar expression for oil shale from Wyoming's Tipton Member, the Utah Parachute Creek Member shows much higher dependency on strain-rate, while the influence of organic content is the same.

The ultimate strain to failure was found to be independent of strain-rate, but clearly dependent upon the organic volume. The relationship was observed to be linear with the richer specimens showing the larger strains.

Young's modulus at one-half the ultimate stress was influenced by volumetric organic content and strain-rate. The richest shale yielded the lowest Young's modulus, while the leanest shale showed the highest modulus. If the richness was kept constant, the higher strain-rates showed higher Young's moduli than that of the lower strain-rates.

5.5. CREEP BEHAVIOR

5.5.1. Introduction

As discussed in previous sections, knowledge of the mechanical properties of oil shale is needed for any form of development, including room-and-pillar mine design, in which the pillars are subjected to constant overburden over an extended period of time. Under these conditions, the oil shale tends to creep. This article presents results of a study on the creep behavior of Utah oil shale (Chong *et al.*, 1982*c*) and Ohio oil shale (Foust *et al.*, 1984). A Conbel model 355 pneumatic driven testing machine was used. The set of duplicate test specimens required for creep testing were cut perpendicular to bedding planes of the same horizon using a wire saw. A rheological model was developed for creep behavior of oil shale as a function of stress level and organic content. Data from creep testing and Fischer assay analyses were used to demonstrate correlation between various stress levels and organic contents for carbonate-rich samples taken from the Mahogany Zone of the Parachute Creek Member in Utah's Cowboy Canyon.

5.5.2. Theoretical Considerations

This section provides the theoretical background and the type of rheological model used (Bland, 1960; Fung, 1965; Courant and Hilbert, 1966; Christensen, 1971; Findley *et al.*, 1976). Oil shale is nearly linearly elastic under small strains as measurements of the load–displacement relationship reveal (Chong *et al.*, 1976, 1979*a*, 1982*b*). However, under constant loading it creeps slightly. Thus the model chosen is for materials which retain initial linearity between load and deflection and a non-linear relationship depending on time. Figure 5.20 illustrates typical curves of longitudinal strain versus time. The initial elastic portions are instantaneous, usually occurring within the first few minutes. The richer the oil shale, the higher are the initial strains.

Based on the observation of the raw data on ε versus time plots (Fig. 5.20) the following non-linear model (Findley *et al.*, 1976) is used:

$$\varepsilon(O_c, S, t) = \varepsilon^\circ(O_c, S) + \varepsilon^+(O_c, S)t^n \qquad (5.39)$$

where ε = total strain perpendicular to bedding planes = ε_{33}; ε° = initial strain; ε^+ = creep related strain; t = non-dimensionalized time; O_c = organic content expressed as a volume percentage; S = stress level expressed as a percentage of ultimate stress; and n = coefficient depending on O_c and S.

To determine the functions ε°, ε^+ and n, the following procedure is used:

Taking the logarithms of eqn 5.39 yields

$$\log (\varepsilon - \varepsilon^\circ) = \log \varepsilon^+ + n \log t \qquad (5.40)$$

Mechanical characterization of oil shale

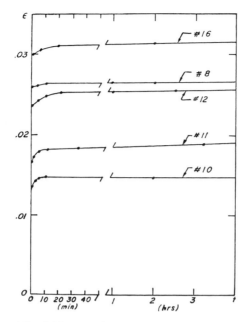

Fig. 5.20. Longitudinal strain versus time.

Figure 5.21 shows a typical plot of $\log(\varepsilon - \varepsilon^\circ)$ versus $\log t$ resulting in a straight line with slope n and the intercept at unit time ε^+. On the other hand, ε° is the intercept at 0 time taken from Fig. 5.20.

Based on the present uniaxial testing, a non-linear pseudo three-dimensional creep model can be constructed, relating the strains to stresses.

For pure axial compression, the stress tensor is

$$[\sigma] = \begin{bmatrix} 0 & 0 & 0 \\ 0 & 0 & 0 \\ 0 & 0 & \sigma_3 \end{bmatrix} \quad (5.41)$$

For simplicity, let

$$\sigma_3 = \sigma \quad (5.42)$$

where axes 1 and 2 indicate directions along the bedding planes and axis 3 is perpendicular to the bedding planes, forming a right-handed Cartesian coordinate system. The non-linear strains are (Findley *et al.*,

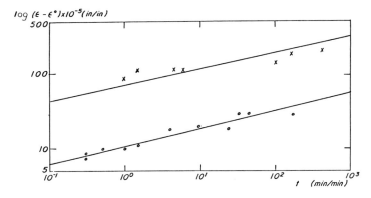

Fig. 5.21. Log $(\varepsilon - \varepsilon^\circ)$ versus non-dimensionalized time. ($\times$) $O_c = 27\cdot 8\%$, $S = 25\%$; ($\bigcirc$) $O_c = 18\cdot 2\%$, $S = 75\%$.

1976)

$$\varepsilon_{33}(O_c, S, t) = F_1\sigma + F_2\sigma^2 + F_3\sigma^3$$
$$\varepsilon_{22}(O_c, S, t) = \varepsilon_{11}(O_c, S, t) = \nu_{31}\varepsilon_{33}(O_c, S, t)$$
$$= (F_1 - G_1)\sigma + (F_2 - G_3)\sigma^2 + (F_3 - G_4)\sigma^3$$
$$\varepsilon_{32} = \varepsilon_{31} = \varepsilon_{21} = 0 \tag{5.43}$$

From previous investigations (Chong et al., 1982b)

$$\nu_{31} = -0\cdot 2747 + 0\cdot 01162 O_c + 0\cdot 00642 S \tag{5.44}$$

where $F_1, F_2, F_3, G_1, G_3, G_4$ are kernel functions involving the various O_c and t values. These six functions can be solved from the following six simultaneous equations:

$$\varepsilon_{33}(O_c, 25\%, t) = F_1\sigma_a + F_2\sigma_a^2 + F_3\sigma_a^3 \tag{5.45}$$
$$\varepsilon_{33}(O_c, 50\%, t) = F_1\sigma_b + F_2\sigma_b^2 + F_3\sigma_b^3 \tag{5.46}$$
$$\varepsilon_{33}(O_c, 75\%, t) = F_1\sigma_c + F_2\sigma_c^2 + F_3\sigma_c^3 \tag{5.47}$$
$$\nu_{31}(O_c, 25\%)\varepsilon_{33}(O_c, 25\%, t) = (F_1 - G_1)\sigma_a$$
$$+ (F_2 - G_3)\sigma_a^2 + (F_3 - G_4)\sigma_a^3 \tag{5.48}$$
$$\nu_{31}(O_c, 50\%)\varepsilon_{33}(O_c, 50\%, t) = (F_1 - G_1)\sigma_b$$
$$+ (F_2 - G_3)\sigma_b^2 + (F_3 - G_4)\sigma_b^3 \tag{5.49}$$
$$\nu_{31}(O_c, 75\%)\varepsilon_{33}(O_c, 75\%, t) = (F_1 - G_1)\sigma_c$$
$$+ (F_2 - G_3)\sigma_c^2 + (F_3 - G_4)\sigma_c^3 \tag{5.50}$$

F_1, F_2 and F_3 can be found by Cramer's rule or other methods from eqns 5.45–5.47, whereas G_1, G_3 and G_4 can be solved from eqns 5.48–5.50. Subscripts a, b and c correspond to stress levels of 25, 50 and 75%, respectively. Once these six kernels are found, the complete time history of the strains are (Findley *et al.*, 1976)

$$\varepsilon_{33}(O_c, S, t) = \int_0^t F_1 \dot{\sigma}(\zeta_1)\, d\zeta_1 + \int_0^t \int_0^t F_2 \dot{\sigma}(\zeta_1)\dot{\sigma}(\zeta_2)\, d\zeta_1\, d\zeta_2$$

$$+ \int_0^t \int_0^t \int_0^t F_3 \dot{\sigma}(\zeta_1)\dot{\sigma}(\zeta_2)\dot{\sigma}(\zeta_3)\, d\zeta_1\, d\zeta_2\, d\zeta_3 \quad (5.51)$$

$$\varepsilon_{22}(O_c, S, t) = \varepsilon_{11}(O_c, S, t) = \int_0^t (F_1 - G_1)\dot{\sigma}(\zeta_1)\, d\zeta_1$$

$$+ \int_0^t \int_0^t [(F_2 - G_3)\dot{\sigma}(\zeta_1)\dot{\sigma}(\zeta_2)]\, d\zeta_1\, d\zeta_2$$

$$+ \int_0^t \int_0^t \int_0^t [(F_3 - G_4)\dot{\sigma}(\zeta_1)\dot{\sigma}(\zeta_2)\dot{\sigma}(\zeta_3)]$$

$$\times d\zeta_1\, d\zeta_2\, d\zeta_3 \quad (5.52)$$

$$\varepsilon_{32} = \varepsilon_{31} = \varepsilon_{21} = 0 \quad (5.53)$$

and ζ_1, ζ_2, ζ_3 are dummy variables.

5.5.3. Experimental Procedure

5.5.3.1. Sample Preparation
The set of three to four duplicate carbonate oil shale samples required for creep testing, obtained by exploiting the lateral uniformity of oil shale, were cut from horizontal layers (3·175 cm thick) from a core of the upper Green River Formation. The test sets were selected with different oil yields. The nominal size of each specimen used for creep testing was 1·27 cm square in cross-sectional area and 3·175 cm long.

5.5.3.2. Instrumentation
Prior to creep testing, a Tinius Olsen universal testing machine was used to determine the ultimate compressive strength of material from the same horizon as the materials to be tested for creep. For creep tests, specimens were placed between the spherical seat and the fixed plate of the creep frame. The creep machine consists of a Conbel model 355 testing machine, pneumatically driven. Dial gages readable to 0·002 54 mm (0·000 1 in) measured the displacement of the specimens.

Three to four duplicate samples were tested with various loads, ranging from 25% through 75% of the ultimate compressive strength of the material in the layers. In this manner, the stress effect can be determined on the specimens of the same oil and mineral content. Specimens were placed between the spherical seat and the fixed plate of the creep frame. The spherical seat resting on a lubricated socket was used to insure a uniform contact pressure during testing. An initial load of approximately 890 N (200 lbs) was applied to check for eccentricity and for specimen alignment. The exact load applied to the specimen was read from the dial gage of the creep machine which was calibrated before it was used as a loading device. Also dial gages were placed between a fixed bar and the spherical seat, and the air was introduced slightly, just enough to set the dial gage into motion. At this instant, the exact dial reading was taken. The load was then increased to full load and maintained throughout the testing by an air pump.

The specimens were surface-dried, weighed, measured and then placed in the creep frame for testing. The vertical displacement of the specimen was measured after the initial load was applied and also when the full load was applied. The time interval for measuring the displacement depended on the displacement rate of the specimen tested.

Due to the small displacements, great care was taken in measuring them. Typical data of creep strain versus time are shown in Fig. 5.20. It shows the creep behavior with an initial non-linear range and gradually assumes a constant strain rate.

5.5.4. Test Results and Analysis

5.5.4.1. Parameters

The following variables were found significant in the statistical analysis: O_c = organic volume and S = stress level.

From the oil yield values of M in gallons per ton (1 gallon ton^{-1} = 4·17 liters tonne^{-1}) determined by Fischer assay, the organic content in percentage by volume, O_c, was calculated using the relation derived by Smith (1969a,b) for Green River Formation oil shale as shown in eqn 5.21.

5.5.4.2. Statistical Analysis

From the graphs of the raw data of Utah oil shale it was found that the exponent n varied linearly with respect to O_c and S; that the parameter ε^+ varied parabolically with S but linearly with O_c; and that the initial strain ε° varied parabolically with O_c but linearly with S. Thus the form of the regressional equations was determined.

Based on regressional analysis, the following equations for Utah oil shale were obtained which relate the parameters n, ε^+ and ε° from eqn 5.39 with the organic contents and stress levels:

$$n = 0.2801 - 0.008\,245O_c - 0.001\,750S + 0.000\,158\,27O_cS \tag{5.54}$$

Coefficient of determination: $r^2 = 0.7440$

Number of samples: $N = 13$

Standard error of estimate: $s = 0.0286$

Mean values: $\bar{n} = 0.1736$

$\bar{O}_c = 18.70\%$

$\bar{S} = 48.08\%$

$$\varepsilon^+ = -0.0005 + 0.000\,027S + 0.000\,004O_c - 0.000\,000\,26S^2 + 0.000\,000\,34O_cS \tag{5.55}$$

Coefficient of determination: $r^2 = 0.8383$

Number of samples: $N = 13$

Standard error of estimate: $s = 0.0001$

Mean values: $\bar{\varepsilon}^+ = 0.0004$

$\bar{O}_c = 16.12\%$

$\bar{S} = 48.08\%$

$$\varepsilon^\circ = 0.0141 + 0.000\,885O_c - 0.000\,041\,3O_c^2 + 0.000\,010\,85O_cS \tag{5.56}$$

Coefficient of determination: $r^2 = 0.8462$

Number of samples: $N = 12$

Standard error of estimate: $s = 0.0023$

Mean values: $\bar{\varepsilon}^\circ = 0.0225$

$\bar{O}_c = 16.27\%$

$\bar{S} = 48.08\%$

Behavior of Ohio oil shale (Foust *et al.*, 1984) is depicted in Figs 5.22–5.27. Figures 5.22–5.24 illustrate the influence of organic volume (O_c) and stress level (S) on the initial strain ε° and parameters ε^+ and

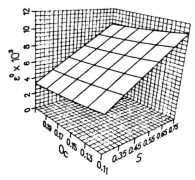

Fig. 5.22. Initial strain versus O_c and S ($O_c \times 10^2$; $S \times 10^2$).

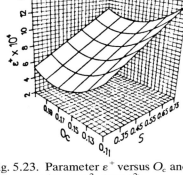

Fig. 5.23. Parameter ε^+ versus O_c and S ($O_c \times 10^2$; $S \times 10^2$).

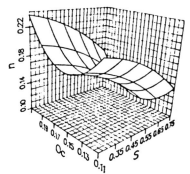

Fig. 5.24. Parameter n versus O_c and S ($O_c \times 10^2$; $S \times 10^2$).

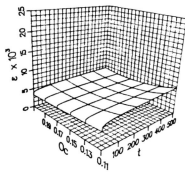

Fig. 5.25. Total strain versus O_c and t for $S = 25\%$ ($O_c \times 10^2$).

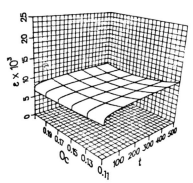

Fig. 5.26. Total strain versus O_c and t for $S = 50\%$ ($O_c \times 10^2$).

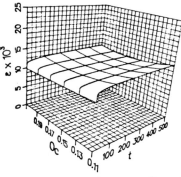

Fig. 5.27. Total strain versus O_c and t for $S = 75\%$ ($O_c \times 10^2$).

Mechanical characterization of oil shale 213

n, respectively. The role of stress levels on the total strain is shown in Figs 5.25–5.27, with the non-dimensional time t normalized with respect to 1 min.

5.5.4. Conclusions

The mechanical properties of oil shale under creep tests were influenced by the volumetric organic content and by stress levels. The organic content is an important parameter. The cross product of organic content and stress level also improved the multiple regression fit. The creep can be represented by a non-linear strain–time model as indicated by eqn 5.39, yielding the parameters ε°, ε^{+} and n. From these parameters, a non-linear pseudo three-dimensional creep model can be built, relating the strains to the non-linear stresses.

5.6. TRIAXIAL TESTING

5.6.1. Introduction

Triaxial testing is one of the most realistic laboratory testing methods for simulating the mechanical behavior of rocks (in this case, oil shale) under *in situ* conditions, with the overburden providing the confining pressure in an actual situation. During triaxial testing, confining pressure is provided hydraulically to a prescribed magnitude, and the specimen is compressed axially to failure (Jaeger and Cook, 1979).

Triaxial tests with confining pressures of up to 14 MPa (2000 psi) to simulate the overburden pressure, etc., are performed on duplicate sets of clay-rich oil shale cores taken from Silver Point Quad in DeKalb County, Tennessee. Oil yields in the specimens ranged from 35 liters tonne^{-1} (8·4 gallons ton^{-1}) to 65 liters tonne^{-1} (15·6 gallons ton^{-1}). A Mohr–Coulomb failure model was developed making use of p–q plots and statistical analysis. A significant finding of this study is that the failure of Tennessee oil shale under confining pressures is independent of the richness of the oil shale. Cohesion and the angle of internal friction of oil are obtained with high accuracy.

5.6.2. Theory

Triaxial testing is an established method (Lambe and Whitman, 1969; Jaeger and Cook, 1979) to investigate the behavior of rocks, soils, or other material under confined pressure. Generally, the axial stress (σ_1) and lateral confining stress (σ_3) are applied in the principal planes of the specimen. Thus, σ_1 and σ_3 are principal stresses. For oil shale specimens, the bedding planes are some of the principal planes since there exists an axis of symmetry perpendicular to the bedding planes. Cylindrical oil shale specimens are cored perpendicular to the bedding planes. Therefore, σ_1 is applied perpendicular to bedding planes and σ_3 along bedding planes.

The stress σ_1 represents the ultimate axial stress applied by the Tinius Olsen testing machine, whereas σ_3 is the confining pressure applied. Both σ_1 and σ_3 are generally plotted as Mohr circles for different confining pressures and different oil yields. The envelope which touches the Mohr circles represents the Mohr–Coulomb failure criterion (Jaeger and Cook, 1979). The equation of the envelope, which is the Mohr–Coulomb failure law, can be expressed as:

$$\tau = c + \sigma \tan \phi \tag{5.57}$$

where τ = shear stress; σ = normal stress; c = cohesion (intercept at τ-axis); and ϕ = angle of internal friction.

Alternatively, instead of plotting the envelope, it is simpler to plot the locus of the maximum τ-coordinates. These coordinates are $(\sigma_1 + \sigma_3)/2$ and $(\sigma_1 - \sigma_3)/2$, respectively, for horizontal and vertical axes, at the top point of each circle. Let

$$p = (\sigma_1 + \sigma_3)/2 \tag{5.58}$$

$$q = (\sigma_1 - \sigma_3)/2 \tag{5.59}$$

The p–q diagram (Lambe and Whitman, 1969), with slope 'α' and intercept (with the vertical axis) equal to 'a' is related to the Mohr–Coulomb envelope as follows:

$$\sin \phi = \tan \alpha \tag{5.60}$$

$$c = a/\cos \phi \tag{5.61}$$

The p–q diagram is used in subsequent analysis and statistical correlation.

5.6.3. Sample Preparation

Sampling is the key in the transverse isotropy of oil shale. The samples tested in this experiment were taken from the Gassaway and Dowelltown Members at DeKalb County, Tennessee. From the original $15 \cdot 2$ cm (6 in) diameter core, three duplicate samples of 5 cm (2 in) diameter and approximately 10 cm (4 in) in length were cored perpen-

Mechanical characterization of oil shale 215

dicular to the bedding planes at selected depths. Special effort was made to have flat and smooth cuts of both ends of the cylinders of the samples so that the load could be uniformly distributed between the loading surfaces and the friction was minimized.

5.6.4. Results and Discussion

Since most oil shale beds have relatively shallow overburden, a maximum confining pressure of 14 MPa (2000 psi) was used, representing an overburden of 610 m (2000 ft). Oil yields of the specimens ranged from 35 liters tonne^{-1} (8·4 gallons ton^{-1}) to 65 liters tonne^{-1} (15·6 gallons ton^{-1}). The test data and results are summarized in Table 5.5. Under confined conditions, the failure is quite ductile.

Based on test data and statistical analysis (Chong *et al.*, 1983), the following Mohr–Coulomb failure law is obtained:

$$q = 1330·24 + 0·6684p \tag{5.62}$$

and coefficient of determination, $r^2 = 0·9131$

Number of samples: $N = 12$

Mean values: $\bar{q} = 6026·73$ psi (41·58 MPa)

     $\bar{p} = 7026·57$ psi (48·48 MPa)

Standard error of estimate: $s = 472$ psi (3·3 MPa)

Using eqns 5.60 and 5.61

Cohesion: $c = 1788·43$ psi (12·34 MPa)   (5.63)

Angle of internal friction: $\phi = 41·94°$ (5.64)

Chang (1979) had done testing on *one grade* of Colorado oil shale. Details are described in Chapter 6. His results were:

$$c \approx 4000 \text{ psi } (27·6 \text{ MPa}) \tag{5.65}$$

$$\phi \approx 28° \tag{5.66}$$

Comparing the clay-rich Tennessee oil shales with the carbonate oil shales of Colorado, the differences are quite dramatic. However, for the carbonate oil shales, more research is needed to establish the dependence or non-dependence of the oil yields. In eqn 5.62 the coefficient of determination of 0·9131 indicated strong correlation between p and q, independent of oil yields for the clay-rich Tennessee oil shales.

TABLE 5.5
Triaxial Testing Data

Number	Depth (ft)	M (gallons ton^{-1})	Confining pressure σ_3 (psi)	Failure stress σ_1 (psi)	$p = \left(\dfrac{\sigma_1 + \sigma_3}{2}\right)$, psi	$q = \left(\dfrac{\sigma_1 - \sigma_3}{2}\right)$, psi
3	40·9	8·4	0	9 526·33	4 763·2	4 763·2
1	40·9	8·4	1 000	12 524·05	6 762·0	5 762·0
2	40·9	8·4	2 000	16 502·93	9 251·5	7 251·5
6	36·8	15·6	0	10 623·34	5·311·7	5 311·7
4	36·8	15·6	1 000	12 437·20	6 718·6	5 718·6
5	36·8	15·6	2 000	15 344·16	8 672·1	6 672·1
11-1	21·1	9·1	0	5 853	2 926·5	2 926·5
11-2	21·1	9·1	1 000	11 801	6 400·5	5 400·5
11-3	21·1	9·1	2 000	13 605	7 802·5	5 802·5
9-1	16·65	10·4	0	11 790·7	5 895·4	5 895·4
9-3	16·65	10·4	1 000	17 494·6	9 247·3	8 247·3
9-2	16·65	10·4	2 000	19 138·9	10 569·5	8 569·5

Note: 1 ft = 0·305 m; 1 gallon ton^{-1} = 4·17 liters $tonne^{-1}$; 1 psi = 6·9 kN m^{-2}.

5.7. FRACTURE MECHANICS

5.7.1. Introduction

The fracture mechanics of oil shale (Schmidt, 1977) has been based on conventional fracture mechanics using notched samples and assuming the material to be isotropic and elastic (Kobayashi, 1975; Caddell, 1980). Oil shale is a layered material more accurately characterized as transversely isotropic (Chong *et al.*, 1980c, 1982b). In addition, it behaves non-linearly due to stress dependency. Furthermore, the notches upon which cracks initiate are extremely sensitive to the location of the bedding planes (layers). Consequently, a more promising fracture mechanism should involve the average behavior of a section of oil shale (for example, a 50 mm thick specimen sampled by oil yield) instead of the breaking strength of a particular layer.

5.7.2. Approaches

For anisotropic rocks, such as oil shale, the fracture toughness is not a uniform material property; rather, it depends on the crack orientation with respect to the layers or bedding planes. However, there is a definite need for the establishment of a standard test procedure for the determination of the fracture toughness of layered materials. Such an endeavor must be backed by developments of appropriate theoretical and numerical methods. For the analysis of fracture properties of layered rocks, it may be necessary to determine the fracture toughness for complex crack geometries as well.

Analytical solutions related to the crack border stress intensity factors are available for transversely isotropic materials where the crack lies in a plane of elastic symmetry. However, types of cracks encountered in practical applications of layered materials are general in nature; therefore, the use of numerical methods is more appropriate. Among the various approaches reported to determine the stress intensity factors, use of special finite elements to represent the stress singularities at the crack tip and strain energy methods are more prominent. Furthermore, orthotropic or transversely isotropic material properties can be handled. The quarter point singular crack tip elements are easily incorporated in ordinary finite element programs employing isoparametric elements and have been shown to yield satisfactory results (Kuruppu and Woo, 1982; Woo and Kuruppu, 1982). The stress intensity factor associated with a crack has a strong relationship with the crack opening profile as it is the governing parameter of the local environment of the crack tip. The opening profile of a sharp crack embedded in an elastic solid may be assumed to satisfy an elliptic displacement function and a solution for fracture toughness is derived (Woo and Kuruppu, 1982).

Input data for an analytical model and a finite element method would be derived by testing oil shale specimens to fracture. The experimental methods suggested by previous investigators for the determination of the fracture toughness of rock materials make use of test specimens designed for testing metallic materials (Schmidt, 1977; Ouchterlony, 1980). However, there are several problems associated with oil shale which do not qualify it for customary fracture mechanics test methods such as the ASTM standard test method (ASTM, 1981). First, the specimens that are used for standardized fracture mechanics test procedures require a great deal of machining which is expensive and difficult to accomplish in rock. Second, oil shale, being a rock material, is difficult to load in tension, and tests should be done with compressive loadings where tensile fractures are induced. Third, the fracture toughness needs to be related to the orientation of bedding planes as the fracture properties of a layered material are highly sensitive to the relative directions of cracks with respect to the bedding planes. In order to overcome these problems, Chong proposes to use a circular core specimen with a center crack subjected to a diametrical compressive loading system similar to the split cylinder tests (Chong *et al.*, 1982*a*). The disk which is drilled and precracked at the center (Fig. 5.28) is investigated (Chong, 1984; Kuruppu and Chong, 1984). The uncracked specimen, commonly used in the split cylinder test, is subjected to relatively uniform horizontal tensile stresses along the loaded vertical diameter. The horizontal tensile stresses will open up the precracked cylinders in Fig. 5.28. The load–displacement behavior up to fracture can be determined; using the knowledge of crack geometry, fracture load and material properties, fracture toughness can be evaluated by means of analytical and finite element methods (Kuruppu and Chong, 1984). Alternatively, fracture toughness may be determined by other methods or by evaluating the J-integral along an appropriate path of the finite element idealization (Woo and Kuruppu, 1982). These precracked cylinders can be easily fabricated from oil shale cores, requiring a minimum amount of machining. Proposed by Chong (Chong, 1984; Kuruppu and Chong, 1984), a refinement of the precracked cylinder in Fig. 5.28 is to use half of the specimen subject to bending in a three-point loading test (with the crack at the center as

Mechanical characterization of oil shale

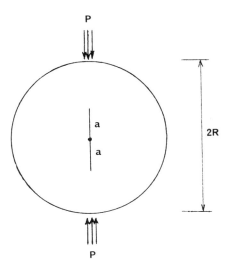

Fig. 5.28. Diametrically loaded disk, drilled at the center and precracked.

shown in Fig. 5.29). This Chong–Kuruppu cracked specimen satisfies the above three criteria for rocks.

All test specimens must satisfy certain minimum dimensional requirements for valid plane strain fracture toughness results (ASTM, 1981). However, these requirements have not been well established for most rock materials, particularly for layered materials. The effects of

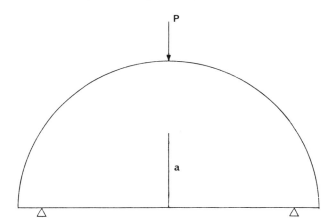

Fig. 5.29. The Chong–Kuruppu cracked specimen.

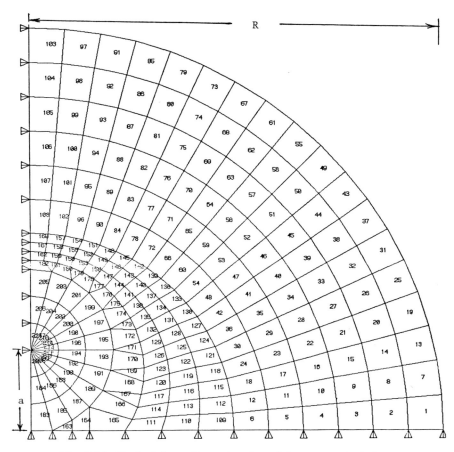

Fig. 5.30. Typical three-dimensional mesh for precracked disks.

the variation of such major dimensions as crack length, uncracked ligament and thickness, on the fracture toughness, are investigated.

5.7.3. Results

One-quarter of the disk was analyzed because of symmetry. A typical finite element idealization is shown in Fig. 5.30. The material was assumed to possess elastic transversely isotropic properties with the z-direction being the axis of symmetry. The precracked disk is modeled under displacement loading. The effect of crack length on the non-dimensionalized stress intensity factor is investigated by (a) the strain energy method; (b) the ellipse method; and (c) the stress method (using Westergaard's solution for stress at the crack tip element). Very close agreements are found, except for $a/R < 0.3$ as shown in Fig. 5.31. Figure 5.32 illustrates the influence of center hole radius on the stress intensity factor. Again, three independent methods are used, and the results agree with each other. The horizontal normalized stress distribution across the loaded (vertical) diameter is shown in Fig. 5.33 with and without the crack. It is apparent that the presence of a center hole does not affect the stress distribution away from the crack tip. Furthermore, high tensile stresses (several times higher than the solid

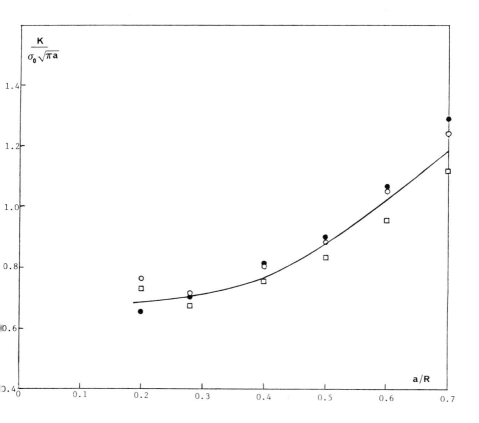

Fig. 5.31. Stress intensity factor versus crack length. $R_i = 0$; displacement loading. (○) Strain energy method; (●) ellipse method; (□) stress method.

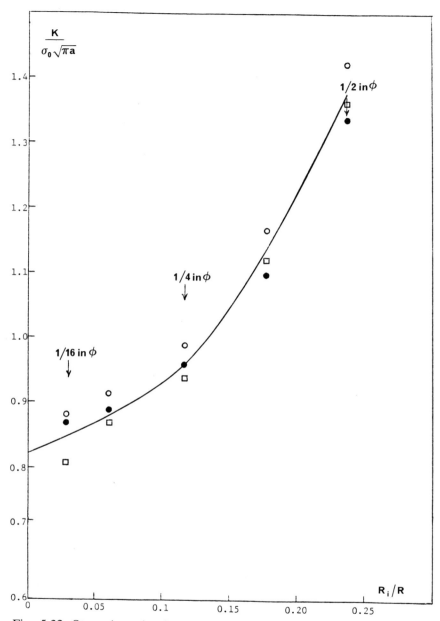

Fig. 5.32. Stress intensity factor versus center hole radius. Displacement loading (0·3 mm); $a/R = 0·5$. (○) Strain energy method; (●) ellipse method; (□) stress method. (Note: 1 in = 25·4 mm.)

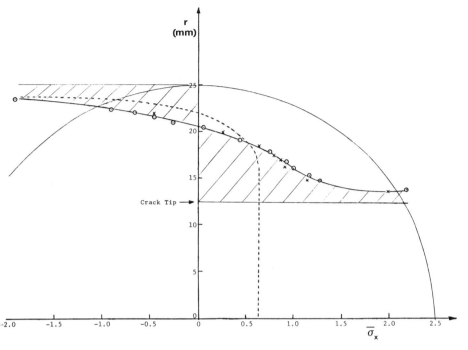

Fig. 5.33. Normalized horizontal stress distributions. $a/R = 0.5$ (◯) $R_i/R = 0.118$; (×) $R_i = 0$; (----) solid disk.

disk) are found close to the crack tip as expected. These high tensile stresses will open the crack when the load is increased. The fracture toughness is obtained by compressing the specimen in Fig. 5.28 or Fig. 5.29 to failure.

5.8. APPLICATIONS OF MECHANICAL PROPERTIES

Developing the fracturing plans necessary for production of oil from oil shale in place, planning and interpreting oil shale mine blasting and modeling these plans to establish prediction capabilities for evaluating variables in these oil shale studies all require complete and realistic values for the mechanical properties of oil shales. This apparently indicates that a formidable amount of testing is required. However, it has been demonstrated that the volume of organic matter in oil shale is the primary variable influencing mechanical properties (Chong *et al.*, 1976, 1979*b*, 1979*c*, 1980*c*, 1982*b*). For the Green River Formation this organic volume can be estimated from oil yield (Smith, 1969*a,b*). Prediction equations based on organic volume are available for oil shale elastic coefficients, Young's modulus and Poisson's ratio, both perpendicular and parallel to the bedding planes (Chong *et al.*, 1979*c*). Ultimate compressive strength both parallel and perpendicular to the bedding planes can also be estimated (Chong *et al.*, 1979*b*, 1980*c*, 1982*b*). Because much of the fracturing behavior must involve tensile failure along the bedding plane, this information is also required. The testing method outlined here and the prediction equations developed can provide this information. An excellent illustration of application of these techniques to modeling oil shale fracturing is presented by Trent *et al.* (1981). They incorporated the appropriate tensile strength prediction equation developed in this chapter together with elastic properties, densities and apparent yield surfaces predicted from organic volume to develop an explicit fracture model for fragmentation of Green River Formation oil shale. All of the mechanical property values were estimated as accurate as those obtainable by direct measurement. All (except the triaxial behavior of the clay-rich Tennessee oil shale) were developed from Fischer assay oil yield data.

ACKNOWLEDGMENTS

The authors would like to thank the following graduate research assistants who have been involved in various aspects of research in this chapter: Messrs K. O. Engen, S. Roine, K. Uenishi, A. B. Khaliki, B. Paulsen, K. L. Costello, J. P. Turner, J. L. Chen, J. Fillerup, S. H. Foust, postdoctor M. D. Kuruppu and others. Undergraduate assistance was given by Messrs E. S. Borgman and J. Weber. Expert technical support was provided by Engineer D. Logan. Technical assistance was given by Messrs G. F. Dana, W. Robb, L. Trudell and G. Mason and Miss D. Sinks of the Laramie Energy Technology Center. In addition, acknowledgment is due to Professor A. P. Boresi, Head of the Civil Engineering Department at the University of Wyoming, for providing the assistance and atmosphere conducive to scholarly endeavor. Last but not the least, the authors are indebted to Mrs M. E. Ayres for the expert typing of the manuscript using a word processor.

REFERENCES

Achenbach, J. D. (1973). *Wave Propagation in Elastic Solids*, North-Holland, Amsterdam.

ASTM (1973). Designation C 496–71: Splitting tensile strength of cylindrical concrete specimens.

ASTM (1981). Designation E 399–81: Standard method of test for plane strain fracture toughness in metallic materials.

Barden, L. (1963). Stresses and displacements in a cross-anisotropic soil, *Geotechnique, London*, **13**(3), 198–210.

Bland, D. R. (1960). *The Theory of Linear Viscoelasticity*, Pergamon Press, Oxford.

Bouwkamp, J. G. (1963). Analysis of two-dimensional stress problems by the moiré method, In: *Experimental Mechanics*, ed. B. E. Rossi, Pergamon Press, Oxford, pp. 195–219.

Caddell, R. M. (1980). *Deformation and Fracture of Solids*, Prentice-Hall, Englewood Cliffs.

Chang, N. Y. (1979). Mechanical properties of oil shale, *Geotechnical Publication 102*, University of Colorado, Denver.

Chen, W. F. and Saleeb, A. F. (1982). *Constitutive Equations for Engineering Materials*, Vol. 1, Wiley-Interscience, New York.

Chong, K. P. (1984). Fracture mechanics of layered rocks, *Research Report*, University of Wyoming, Laramie.

Chong, K. P., Smith, J. W., Chang, B., Hoyt, P. M. and Carpenter, H. C. (1976). Characterization of oil shale under uniaxial compression, *Proc. 17th US Symp. on Rock Mech.*, Univ. of Utah, Salt Lake City, pp. 5C5-1–5C5-8.

Chong, K. P., Smith, J. W., Chang, B. and Roine, S. (1979a). Oil shale properties by split cylinder method, *J. Geotechnical Eng. Div.*, ASCE, **105**(GT5), proc. paper 14567, pp. 595–611.

Chong, K. P., Costello, K. L. and Smith. J. W. (1979b). Three-dimensional characterization of the mechanical properties of Green River Formation oil shale of Wyoming's Tipton Member, *Proc. 1979 SESA Spring Meeting*, SESA, Brookfield Center, Connecticut, paper no. R79-118.

Chong, K. P., Uenishi, K. and Smith, J. W. (1979c). Complete elastic constants and stiffness coefficients for oil shale, *DOE/LETC/RI-79/8*, USDOE Technical Information Center, Washington DC.

Chong, K. P., Chen, J. L., Uenishi, K. and Smith, J. W. (1980a). A new method to determine the independent shear moduli of transversely isotropic materials, *Proc. 4th Int. Congress on Experimental Mech.*, SESA, Boston, paper no. R121.

Chong, K. P., Hoyt, P. M., Smith, J. W. and Paulsen, B. Y. (1980b). Effects of strain rate on oil shale fracturing, *Int. J. Rock Mech. & Geomech. Abstr.*, **17**, 35–43.

Chong, K. P., Uenishi, K., Smith, J. W. and Munari, A. C. (1980c). Non-linear three dimensional mechanical characterization of Colorado oil shale, *Int. J. Rock Sci. & Geomech. Abstr.*, **17**, 339–47.

Chong, K. P., Turner, J. P. and Dana, G. F. (1981). Strain rate effects on the mechanical properties of Utah oil shale, *Mechanics of Structured Media*, Part A, Elsevier, Amsterdam, pp. 431–45.

Chong, K. P., Smith, J. W. and Borgman, E. S. (1982a). Tensile strengths of Colorado and Utah oil shales, *J. Energy, AIAA*, **6**(2), 81–5.

Chong, K. P., Turner, J. P. and Dana, G. F. (1982b). 3-D non-linear characterization of Utah oil shale, *Proc. 1982 Joint Conf. on Expt. Mech.*, Vol. 1, SESA, Brookfield Center, Connecticut, pp. 454–9.

Chong, K. P., Dana, G. F. and Chen, J. L. (1982c). Creep behavior of Utah oil shale subject to uniaxial loading, *Report of Investigations*, Department of Energy, Washington DC.

Chong, K. P., Chen, J. L. and Dana, G. F. (1983). Triaxial testing of oil shale from DeKalb County, Tennessee, *DOE/LETC/10877-4*, USDOE Technical Information Center, Washington DC.

Christensen, R. M. (1971). *Theory of Viscoelasticity*, Academic Press, New York.

Cleary, M. P. (1978). Some deformation and fracture characteristics of oil shale, *Proc. 19th US Symp. on Rock Mech.*, Vol. 2, University of Nevada, Reno, pp. 72–82.

Colback, P. S. B. (1966). An analysis of brittle fracture initiation and propagation in the Brazilian test, *Proc. 1st Int. Congress of the Soc. of Rock Mech.*, Vol. 2, Lisbon, Portugal, pp. 343–9.

Courant, R. and Hilbert, D. (1966). *Methods of Mathematical Physics*, Vol. I, Interscience Publishers, New York.

Dana, G. F. and Smith, J. W. (1972). Oil yields and stratigraphy of the Green River Formation's Tipton Member at Bureau of Mines Sites near Green River, WY, *US Bureau of Mines Report of Investigations 7681*.

Desai, C. S. and Christian, J. T. (eds) (1977). *Numerical Methods in Geotechnical Engineering*, Article 2-3, McGraw-Hill, New York, pp. 79–92.

Dhir, R. K. and Sangha, C. M. (1973). Relationships between size deformation and strength for cylindrical specimens loaded in uniaxial compression, *Int. J. Rock Mech. & Geomech. Abstr.* **10**, 669–712.

Dixon, W. F. (1976). *BMD, Biomedical Computer Programs*, University of California Press, Los Angeles, pp. 305–52.

Dooley, J. C. (1964). Discussion of stresses and displacements in a cross-anisotropic soil, *Geotechnique, London*, **14**, 278–9.

Fairhurst, C. (1964). On the validity of the Brazilian test for brittle materials, *Int. J. Rock Mechanics and Mining Sciences*, **1**, 535–46.

Findley, W. N., Lai, J. W. and Onaran, K. (1976). *Creep and Relaxation of Nonlinear Viscoelastic Materials*, North-Holland, Amsterdam.

Foust, S. H., Chong, K. P. and Dana, G. F. (1984). Non-linear creep characteristics of Devonian oil shale, *Proc. 9th Int. Congress on Rheology*, Acapulco, Mexico.

Fung, Y. C. (1965). *Foundations of Solid Mechanics*, Prentice-Hall, Englewood Cliffs, New Jersey.

Green, S. J. and Perkins, R. D. (1968). Uniaxial compression tests at strain-rates from 10^{-4} to $10^{4} \mathrm{s}^{-1}$ on three geological materials, *Proc. 10th Symposium on Rock Mechanics*.

Hahn, H. T. and Tsai, S. W. (1973). Non-linear elastic behavior of unidirectional composite laminae, *J. Composite Materials*, **7**, 102–18.

Mechanical characterization of oil shale 227

Hondros, G. (1959). The evaluation of Poisson's ratio and the modulus of materials of a low tensile resistance by the Brazilian test, *Australian J. Applied Science* **10**(3), 243–68.

Jaeger, J. C. and Cook, N. G. W. (1979). *Fundamentals of Rock Mechanics*, 3rd edn, Chapman and Hall, London, Chapter 6.

Janach, W. (1976). The role of bulking in brittle failure of rocks under rapid compression, *Int. J. Rock Mech. Min. Sci. & Geomech. Abstr.*, **13**, 177–86.

Kobayashi, A. S. (ed.) (1975). *Experimental Techniques in Fracture Mechanics*, SESA, Stamford, Connecticut.

Kumar, A. (1968). The effect of stress rate and temperature on the strength of basalt and granite, *Geophysics*, **33**, 501–10.

Kuruppu, M. D. and Chong, K. P. (1984). Fracture mechanics of layered rocks, *Proc. 5th ASCE-EMD Specialty Conference*, Laramie, Wyoming.

Kuruppu, M. D. and Woo, C. W. (1982). Use of isoparametric crack tip elements for the analysis of linear elastic and elastic-plastic fracture problems, *Proc. Int. Conf. on Finite Element Methods*, Shanghai, China.

Lambe, T. W. and Whitman, R. V. (1969). *Soil Mechanics*, John Wiley, New York.

Lankford, J., Jr. (1976). Dynamic strength of oil shale, *Soc. Petrol. Engrs. J.*, 17–22.

Lekhnitskii, S. G. (1963). *Theory of Elasticity of an Anisotropic Elastic Body*, trans. P. Fern, ed. J. J. Brandstatter, Holden-Day, San Francisco.

Love, A. E. H. (1944). *Mathematical Theory of Elasticity*, 4th edn, Dover Publications, New York.

Morris, C. E. (1978). Elastic constants of oil shale, *Explosively Produced Fracture of Oil Shale, LA-7164-PR*, compiled by W. J. Carter, Los Alamos Scientific Lab., New Mexico.

Muskhelishvili, N. I. (1975). *Some Basic Problems of the Mathematical Theory of Elasticity*, Article 80a, 4th edn, trans. J. R. M. Radok, Noordhoff International, Leyden, The Netherlands.

Oden, J. T. and Brauchli, H. J. (1971). On the calculation of consistent stress distributions in finite element approximations, *Int. J. Numerical Methods in Engineering*, **3**, 317–25.

Ouchterlony, F. (1980). Review of fracture toughness testing of rock, *Swedish Detonic Research Foundation Report DS 1980:15*.

Pindera, J. T., Mazurkiewicz, S. B. and Khattab, M. A. (1978). Stress field in circular disk loaded along diameter: discrepancies between analytical and experimental results, *SESA Spring Meeting*, Wichita, Kansas, paper no. Cr. 10.

Podio, A. L., Gregory, A. R. and Gray, K. E. (1968). Dynamic properties of dry and water-saturated Green River shale under stress, *Soc. Petrol. Engrs. J.*, 389–404.

Roark, R. J. (1965). *Formulas for Stress and Strain*, 4th edn, McGraw-Hill, New York, p. 320.

Robb, W. A. and Smith, J. W. (1976). Mineral profile of Wyoming's Green River Formation, *Earth Science Bull.*, **9**(1), 1–8.

Schmidt, R. A. (1977). Fracture mechanics of oil shale—unconfined fracture toughness, stress corrosion cracking and tension test results, *Proc. 18th US Symp. on Rock Mech.*, Keystone, Colorado, pp. 2A2-1 to 2A2-6.

Schreiber, E., Orson, L. A. and Soga, N. (1973). *Elastic Constants and their Measurement*, McGraw-Hill, New York.

Sellers, J. B., Haworth, G. R. and Zamas, P. G. (1972). Rock mechanics research on oil shale mining, *Society of Mining Engineers Trans. AIME*, **252**, 222–32.

Smith, J. W. (1969*a*). Theoretical relationship between density and oil yield for oil shales, *US Bureau of Mines Report of Investigations 7248.*

Smith, J. W. (1969*b*). Geochemistry of oil shale genesis, Green River Formation, Wyoming, *21st Annual Field Conference, Wyoming Geological Association Guidebook*, pp. 185–90.

Smith, J. W. (1976). Relationship between rock density and volume of organic matter in oil shales, *Laramie Energy Research Center Report of Investigations 76/6.*

Smith, J. W. and Lee, K. K. (1982). Geochemistry and physical paleolimnology of Piceance Creek Basin oil shales, *15th Oil Shale Symposium Proc.*, Colorado School of Mines, pp. 101–14.

Smith, J. W. and Robb, W. A. (1966). Ankerite in the Green River Formation's Mahogany Zone, *J. Sedimentary Petrology*, **36**(2), 486–90.

Smith, J. W. and Young, N. B. (1967). Organic composition of Kentucky's New Albany shale: determination and uses, *Chem. Geol.*, **2**, 157–70.

Smith, J. W., Chong, K. P. and Khaliki, B. A. (1977). Preparation of oil shale samples for mechanical testing, *1st Annual Oil Shale Conversion Symposium*, Laramie, Wyoming.

Smith, J. W., Beard, T. N. and Trudell, L. G. (1979). Oil shale resources of the naval oil shale reserve no. 1, Colorado, *DOE/LETC/RI-79/2, USDOE* Technical Information Center, Washington DC.

Stanfield, K. E. and Frost, I. C. (1949). Method of assaying oil shale by a modified Fischer retort, *US Bureau of Mines Report of Investigations 4477.*

Timoshenko, S. and Goodier, J. N. (1951). *Theory of Elasticity*, Article 37, 2nd edn, McGraw-Hill, New York.

Trent, B. C., Young, C., Barbour, T. G. and Smith, J. W. (1981). A coupled gas pressurization explicit fracture model for oil shale fragmentation, *Proc. 22nd Symposium on Rock Mechanics*, Cambridge, Massachussetts, pp. 198–204.

Trudel, L. G., Beard, T. N. and Smith, J. W. (1970). Green River Formation lithology and oil shale correlations in the Piceance Creek Basin, CO. *US Bureau of Mines Report of Investigations 7357.*

Wang, K. A. and Chong, K. P. (1980). Diametrical compression of orthotropic or transversely isotropic cylinders, *Proc. of the 21st US Symposium on Rock Mechanics*, Rolla, Missouri, pp. 243–8.

Woo, C. W. and Kuruppu, M. D. (1982). Use of finite element method for determining stress intensity factors with a conic-section simulation model of crack surface, *Int. J. Fracture*, **20**, 163–78.

Chapter 6

Statistical Analysis and Modeling of Physical, Mechanical and Strength Properties of Oil Shale

NIEN-YIN CHANG

Associate Professor, Department of Civil and Urban Engineering, University of Colorado at Denver, USA

and

ELLEN J. BONDURANT

Engineer, Structure Stability Branch, US Bureau of Reclamation, Boise, Idaho, USA

SUMMARY

The oil shale in the tristate area of Colorado, Utah and Wyoming covers a large geographical area and contains varying amounts of kerogen and different minerals, such as nahcolite, dawsonite, dolomite, etc. The physical, mechanical and strength properties of oil shale may be greatly affected by its kerogen and mineral contents and its spatial location.

Results of a comprehensive statistical analysis of some laboratory-determined engineering properties of oil shale are presented. Data for over 1800 tested specimens from seven locations in Colorado—Corehole CE-706, US Bureau of Mines Tract, C-a Tract, Colony Property, Union Property, Anvil Points Mine and Occidental Property as shown in Fig. 6.1—was used in the analysis. 29 distinct engineering properties were studied, including common physical and mechanical properties obtained in uniaxial, triaxial, Brazilian tensile and modulus of rupture tests. Also included were Mohr–Coulomb strength parameters obtained for individual specimens by using multiple-stage triaxial testing techniques. Presented here are: (1) general statistical and distributional characteristics of some engineering properties; (2) the correlations between various properties and several easily obtainable or readily available properties; and (3) several significant bivariate and multiple regression equations which may be used for the prediction of various properties from the easily obtainable or readily available properties.

Also presented are the multiple-stage triaxial testing technique used in determining the Mohr–Coulomb strength parameters and the limited information on the uniaxial thermal creep compliance and elasticity moduli obtained from cubic triaxial tests.

6.1. INTRODUCTION

The recovery of shale oil in the Piceance Creek Basin can be accomplished through four general schemes: conventional mining with surface retorting, open-pit mine with surface retorting, true *in situ* (TIS) retorting and modified *in situ* (MIS) retorting. A competent engineering design using rock mechanics principles for any of these schemes requires many laboratory-determined engineering properties of oil shale. These laboratory properties can be obtained by one of three general methods. First, the properties can be obtained in a detailed laboratory testing program on samples from the site. Second, average values for the properties can be obtained by compiling the data generated by previous researchers. The probable range of values can then be used in a sensitivity analysis. Third, the properties can be predicted from some easily obtainable properties determined for samples from the site, by using regression equations (models) formulated by previous researchers. These three methods of determining the engineering properties will, of course, give results of varying reliability. The properties so obtained may be used in various stages of design. Sometimes approximate values may be adequate, particularly for a preliminary feasibility study.

An extensive amount of prior research has been done to characterize some engineering properties of oil shale in the Piceance Creek Basin. Several properties of oil shale have previously been found to have strong correlation with other properties and many regression equations have been developed by previous researchers. However, most of these previous studies dealt only with oil shale from the Anvil Points Mine and the Colony Pilot Mine, which are located in the Mahogany Zone near the southern edge of the Basin. Thus, it was apparent that additional research was needed to investigate properties of oil shale from zones other than the Mahogany, and from the deeper, thicker deposits near the center of the Basin (Hoskins *et al.*, 1976). Furthermore, strength parameters which characterize the strength of oil shale under various *in situ* stress states are needed for various design purposes (Abel and Hoskins, 1976). The Mohr–Coulomb strength parameters, cohesion and angle of internal friction, had not been adequately investigated for oil shale from any location in the Basin. They were difficult to obtain by conventional testing techniques in that several identical specimens are required. Furthermore, because specimens used are normally not identical, the strength parameters so obtained cannot be readily included in regression analyses.

In this chapter, 29 properties of oil shale have been included in various statistical analyses. The specimens are from various zones within the Parachute Creek Member of the Green River Formation as shown in Fig. 6.1. The multiple-stage triaxial testing technique (Chang and Jumper, 1978; Chang and Bondurant, 1979; Chang, 1980) has been used to determine the Mohr–Coulomb strength parameters (cohesion and angle of internal friction) of oil shale. This technique provides a simple method for determination of the Mohr–Coulomb strength parameters for both peak and residual strengths for each individual specimen. This technique thus enables the inclusion of these strength parameters in various statistical analyses to study their distributional characteristics, functional relationships with physical properties and spatial variations. The multiple-stage triaxial testing technique is further discussed in Section 6.3.

Because of the obvious varved stratification of oil shale, some degree of anisotropy in mechanical properties is expected. This anisotropy obtained from a cubic triaxial testing apparatus (Hawk, 1979) is also presented. Because of insufficient data, no statistics are provided.

All elements of rock mass are subjected to some initial *in situ* stresses caused by overburden and tectonic stresses. As a mining activity proceeds, these stresses are progressively redistributed and altered. In MIS retorting, stresses in mine structures are further changed due to a sustained heating process. Most importantly, under elevated temperatures, oil shale mine structures such as retort pillars and roofs undergo a progressive weakening process. Thus, in the MIS retort process, mine structures which are safe initially may become

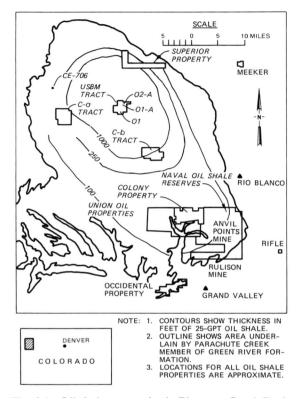

Fig. 6.1. Oil shale properties in Piceance Creek Basin.

unstable and eventually rupture under sustained high temperature and stress. To assess the effect of temperature on the performance of mine structures and potential ground subsidence over a sustained period of time requires the thermal mechanical properties (or thermal creep compliance) of oil shale. Results of some previous studies (Chu and Chang, 1980; Chang, 1982) are presented.

Also presented are the general statistical and distributional characteristics of the engineering properties of oil shale, the coefficient of correlation between various pairs of properties and the functional relationship for those properties with strong correlation. These statistics and functional relationships can be very useful in the feasibility study of an oil shale operation.

Statistical analysis of properties of oil shale   233

6.2. PREVIOUS LITERATURE ON PHYSICAL, MECHANICAL AND STRENGTH PROPERTIES OF OIL SHALE

A substantial amount of research has been conducted to characterize the engineering properties of oil shale and to develop functional relationships between these properties. This literature review, which includes the publications up to 1982, deals only with those properties presented in this chapter.

6.2.1. Assay, Organic Content and Density

The specific gravity of the organic material in oil shale is much less than that of its mineral constituents. Thus the bulk density of oil shale decreases with an increase in organic content (or assay grade). The above relationships have been observed and quantified by several researchers (Stanfield *et al.*, 1951; Smith, 1956, 1966; Smith *et al.*, 1968). Smith (1969, 1976) derived some theoretical functional relationships between density, oil yield, organic content by weight and by volume. These theoretical relationships were found to compare well with previously obtained empirical relationships.

6.2.2. Minerals in Oil Shale

Moore (1950) found that the amount of authigenic albite increased with organic matter. Hite and Dyni (1967) found that albite varied inversely with the quartz and dawsonite content of oil shale and the dawsonite content was directly proportional to the quartz content. Dyni (1974) presented a regression equation relating weight percent dawsonite to weight percent quartz for oil shale in the nahcolite-bearing sequence in the northern part of the Piceance Creek Basin.

Robb and Smith (1974, 1977) investigated relationships between volume of organic matter and eight major minerals in oil shale from the northern part of the Piceance Creek Basin and reported strong relationships between the volume of organic matter and the amount of various silicate minerals.

Chong *et al.* (1979a) stated that the authigenic carbonate mineral dolomite is the cementing agent holding oil shale together, while all other minerals occur as individual microcrystals. Thus they theorized that dolomite may be the mineral of primary significance to the mechanical properties of oil shale.

6.2.3. Engineering Properties of Oil Shale

Various physical and mechanical properties of oil shale have been studied in detail by various researchers. Organic and mineral contents and assay grade were found to have a marked effect on physical and mechanical properties of oil shale.

Wright and Bucky (1949) found that rich oil shale exhibited plastic deformation at stress levels as low as 21 MPa (3000 psi) while lean oil shale behaved elastically to stresses beyond 34·5 MPa (5000 psi). They noted also that the modulus of elasticity decreased, and Poisson's ratio increased, with an increase in the assay value. Results of their centrifuge modeling indicated that a pillar stressed to 75% its ultimate compressive strength would stand for many years.

Studying the physical and mechanical property data for specimens from the Bureau of Mines Anvil Points Mine, Windes (1950) found that as kerogen content increased, specific gravity, compressive strength, velocity of sound, Young's modulus and porosity decreased and impact toughness increased and Raju (1961) found: (1) the velocity of the longitudinal wave in oil shale decreased with increasing kerogen content; (2) the compressive strength was lower for rich shale than for lean shale; and (3) lean shale behaved as a brittle material while rich shale behaved plastically under load.

Sellers *et al.* (1972) measured elastic properties, compressive strength and creep of various grades of oil shale from the vicinity of the Bureau of Mines Anvil Points mine. Regression equations were presented for the prediction of Young's modulus and Poisson's ratio from oil content at various stress levels and for the prediction of unconfined compressive strength from assay and the height-to-width ratio of a sample. Young's modulus was found to decrease, and Poisson's ratio was found to increase with an increase in oil content. Unconfined compressive strength decreased with an increase in assay or an increase in the height-to-width ratio. Results of creep tests indicated that creep generally did not occur at uniaxial compressive stresses below 14 MPa (2000 psi). At stress levels from 14 MPa (2000 psi) to 41·4 MPa (6000 psi), creep deformation exhibited a logarithmic relationship with time. At stress levels of 55 MPa (8000 psi), specimens with grades of 125 liters tonne^{-1} (30 gallons ton^{-1}) higher underwent continuing deformation to failure.

Schmidt and Schuler (1974) presented results of unconfined compression tests on oil shale from the Bureau of Mines Anvil Points Mine. Their results showed that kerogen-rich oil shale had a lower Young's modulus, lower fracture strength, greater fracture strain and

Statistical analysis of properties of oil shale 235

ductility, greater dilatation and required greater energy to fracture. Further, they found that the stress at maximum compressive volume strain, denoted as the 'yield stress' decreased with increasing kerogen content between 12.5 and 62.6 liters tonne^{-1} (3 and 15 gallons ton^{-1}) but differed little for richer oil shale. The yield stress was equal to approximately 75% of the fracture stress for 12.5 liters tonne^{-1} (3 gallons ton^{-1}) oil shale and 50% the fracture stress for both 62.6 and 166.8 liters tonne^{-1} (15 and 40 gallons ton^{-1}) oil shale.

Agapito (1972) presented results of uniaxial compression, direct tension and modulus of rupture tests on oil shale from the Colony Mine. Results showed that a higher oil content corresponded to a lower modulus of elasticity and a higher Poisson's ratio. The uniaxial compressive strength was found to decrease with an increase in kerogen content up to 125 liters tonne^{-1} (30 gallons ton^{-1}). Above this grade the strength remained constant at approximately 90 MPa (13 000 psi). Regression equations relating assay grade to compressive strength and elasticity modulus were also presented.

The Gulf Oil Corporation and Standard Oil Company of Indiana (1976a, 1976b, 1977) reported the results of research on the engineering properties of the oil shale including Mahogany Zone oil shale from C-a Tract and their relationships with oil content. Similar relationships between oil content and Young's modulus of elasticity and unconfined compressive strength were concluded. The Brazilian tensile strength increased with an increase in oil content between 0 and 117 lites tonne^{-1} (28 gallons ton^{-1}). Tensile strength was almost constant for grades above 117 liters tonne^{-1} (28 gallons ton^{-1}). The modulus of rupture was three–four times as large as the Brazilian tensile strength. Poisson's ratio was independent of oil content and highly dependent on stress levels. The angle of internal friction for intact rock decreased with an increase in oil content.

Horino and Hooker (1976a) studied various engineering properties and developed regression equations for oil shale from the Mahogany Zone on Union Oil property near Parachute Creek and for oil shale from Federal Lease Tracts U-a and U-b (Horino and Hooker, 1976b). Their results showed that an increase in yield corresponded to a decrease in specific gravity, compressive strength and Young's modulus and an increase in Poisson's ratio. No correlations were found, however, between specific gravity and Brazilian tensile strength.

Hoskins *et al.* (1975) reviewed published data on the engineering properties of oil shale, and compiled some published and unpublished data regarding the relationships between assay and various engineering properties of oil shale. Equations were presented for the prediction of compressive strength, tensile strength determined from flexural, direct or indirect tension tests, Young's modulus, angle of internal friction for intact rock, cohesion for intact rock and p-wave velocity both parallel and perpendicular to the bedding plane. Results showed that an increase in assay corresponded to a decrease in compressive strength, tensile strength, deformation modulus, angle of internal friction, and p-wave velocity both parallel and perpendicular to the bedding plane and an increase in cohesion and Poisson's ratio. The correlation coefficient between assay and angle of internal friction showed that a very strong negatively-sloped linear relationship existed; however, only five data points were available to define the relationship. The correlation coefficient between assay and cohesion was fairly low.

Horino and Hooker (1978) compiled data on the properties of oil shale from the unleached saline zone in the Piceance Creek Basin and performed several statistical analyses. The result showed no strong correlation between kerogen yield (or specific gravity) and the uniaxial compressive strength, Young's modulus and Poisson's ratio. However, the mechanical properties were somewhat dependent upon the percentage of nahcolite and dawsonite.

Smith (1976) suggested that engineering properties be related to volumetric organic content rather than oil yield. Further, Smith formulated a linear equation between the volumetric organic content and the density of Green River Formation oil shales.

Chong *et al.* (1976) related the volumetric organic content to the ultimate uniaxial strength, initial tangent Young's modulus, the strain at ultimate load and the offset strain which was used to measure non-linear ductility. Further, X-ray diffraction analysis was used to detect mineral variations in samples. Mineral content and organic content were found to have an important influence on the mechanical properties of oil shale.

Chong *et al.* (1977, 1979a) performed modified split cylinder tests and found that Poisson's ratio had a positively-sloped linear relationship and tensile strength and Young's modulus had negatively-sloped linear relationships with organic content. The inclusion of dolomite as an independent variable led to a small but statistically significant increase in the quality of the regression equation for the prediction of Poisson's ratio.

Because of its varved stratification, oil shale is expected to be anisotropic in its mechanical properties and strengths. Chenevert and

Gatlin (1965) studied the anisotropic behavior of oil shale. Triaxial tests were performed. Results showed that Young's modulus was higher along the bedding plane than perpendicular to it, the angle of internal friction varied with orientation and the cohesion was constant at a value of 51·8 MPa (7500 psi). McLamore and Gray (1967) reported the variation of maximum strength, cohesion and angle of internal friction with changes in the angle between the bedding plane and the axial load for confining pressures from 6·9 to 172·5 MPa (1000 to 25 000 psi). Further, they observed a brittle–ductile transition at a confining pressure of 34·5 MPa (5000 psi) for 162·6 liters tonne^{-1} (39 gallons ton^{-1}) oil shale.

Agapito (1972, 1974) found that uniaxial compressive strengths of Colony oil shale were approximately the same for samples loaded parallel and perpendicular to the bedding plane and were 25 times as large as tensile strengths measured in direct tension perpendicular to the bedding plane. Young's modulus of elasticity was approximately 26% higher measured along the bedding plane than that measured perpendicular to it. Further, values for the *in situ* modulus of elasticity in both pre-failure and post-failure states and *in situ* pillar strengths were calculated from data obtained during the pillar loading and failure at the Colony Mine.

Raju (1961) found, from testing oil shale from the Bureau of Mines Anvil Points Mine, that the velocity of a longitudinal wave was higher along the bedding plane than across layers; and the compressive strength along the bedding plane was greater than that normal to it. Results of direct tensile and triaxial tests on oil shale from the same mine by Schmidt (1977) and Schuler and Schmidt (1977) showed anisotropy of oil shale in strength, as well as in Young's modulus. The tensile strength of one grade of oil shale stressed parallel to the bedding plane was found to be four times as large as that stressed perpendicular to the bedding plane. The Young's modulus measured along the bedding plane was higher than that measured perpendicular to it. Further it was noted that the observed tensile strength was remarkably high compared with the compressive strength (approximately 20% compressive strength).

Chong *et al.* (1979*b*) conducted uniaxial tests on cubic oil shale specimens from the Naval Oil Shale Reserve near Rifle, Colorado to determine the three-dimensional compliance of oil shale and found that the compliance coefficients are stress sensitive. Hawk (1979) conducted true triaxial tests on Mahogany Zone oil shale from the

Colony Mine to determine its three-dimensional compliance. Under confining pressures less than 6·9 MPa (1000 psi), Hawk reported that oil shale is transverse isotropic or nearly isotropic within linear elastic range.

Horino and Hooker (1971) presented results of their study on the strength of Mahogany Zone oil shale from the Colony Mine. Both solid core specimens and specimens with a plane of weakness were tested under uniaxial and triaxial conditions. They found that the unconfined compressive strength of the specimens containing planes of weakness was as low as 27% of that of solid specimens. The results were later compared with the *in situ* behavior of pillars in the Colony Mine by Brady *et al.* (1975), Hardy and Agapito (1975) and Agapito and Page (1976). The Gulf Oil Corporation and Standard Oil Company of Indiana (1976a, 1976b, 1977) also reported values of angle of internal friction and cohesion of samples containing planes of weakness.

Abel and Hoskins (1976) calculated values for the cohesion of the oil shale rock mass in the Colony Mine from measured pillar stresses during various failure conditions. They found the cohesion of the rock mass to be less than the cohesion obtained in laboratory tests on oil shale specimens.

Chang and Jumper (1978) and Chang and Bondurant (1979) investigated and validated the multiple-stage triaxial (MST) test for evaluating the Mohr–Coulomb strength parameters of oil shale. The strength parameters obtained from MST tests were found to be in excellent agreement with those obtained from conventional single-stage triaxial tests. Further, both cohesion and angle of internal friction were found to be normally distributed.

Chong *et al.* (1978) used a rheological model to simulate creep and relaxation behavior of oil shale. Chu and Chang (1980) and Chang (1982) investigated the effect of elevated temperatures on the creep behavior of oil shale. They found that the creep behavior is very sensitive to temperature increase. A sample undergoing steady state creep could enter tertiary creep and eventually fail in creep rupture when temperature was increased.

6.3. MULTIPLE-STAGE TRIAXIAL TESTING TECHNIQUE

6.3.1. Mohr–Coulomb Failure Criterion

The strength of rock is often characterized by a Mohr envelope, which is obtained experimentally as the envelope drawn tangent to the Mohr circles corresponding to failure under a variety of confining pressures (Jaeger and Cook, 1976).

Frequently it is assumed that a linear Mohr envelope adequately fits the Mohr circles. This linear Mohr envelope is equivalent to the Coulomb criterion of failure. Thus this failure criterion may be termed the Mohr–Coulomb failure criterion. The equation for the linear Mohr envelope is

$$\tau = c + \sigma \tan \phi \qquad (6.1)$$

where τ = shear stress on plane of failure; σ = normal stress on plane of failure; c = cohesion of rock material; ϕ = angle of internal friction of rock material.

The Mohr–Coulomb strength parameters cohesion and angle of internal friction for peak and residual (post-peak) strengths (c_p, ϕ_p, c_r and ϕ_r) thus define the Mohr–Coulomb failure criterion for a rock at failure and after failure. It may be noted that the residual cohesion is more appropriately referred to as the apparent residual cohesion, as the post-failure residual strength envelope goes through the origin (Hendron, 1968). Typical Mohr–Coulomb failure envelopes for peak and residual strengths are shown in Fig. 6.2(a).

Test data which normally would be presented in the form of a linear Mohr envelope drawn tangent to several Mohr circles can instead be presented in the form of a q versus p line (Lambe and Whitman, 1969). Typical q versus p lines for peak and residual strength are shown in Fig. 6.2(b), in which p and q represent the center and radius, respectively, of a Mohr circle: $p = (\sigma_1 + \sigma_3)/2$; $q = (\sigma_1 - \sigma_3)/2$; $\alpha = \tan^{-1} \sin \phi$; and $A = c \cos \phi$. The Mohr–Coulomb strength parameters, c and ϕ, can be obtained from the above relationship for the known A, and α from experimental data. The use of q–p lines better enables one to consider the scatter of experimental values.

6.3.2. Single-Stage Triaxial Testing

The Mohr–Coulomb strength parameters are conventionally obtained from several single-stage triaxial (SST) tests on several specimens. In an SST test a rock specimen is loaded to failure under one confining pressure (σ_3), and the peak strength at that confining pressure is determined. Further, complete stress–strain curves of the rock, including post-peak behavior, can be determined if a stiff testing machine is used (Vutukuri *et al.*, 1974). Additionally, it has been shown that after failure the specimen can be subjected to several different confining pressures and the residual strength at each can be determined (Hobbs,

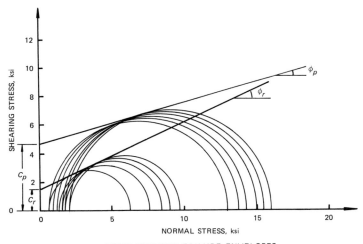

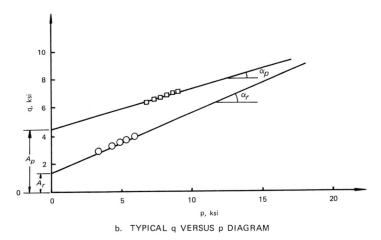

Fig. 6.2. Typical multiple-stage triaxial test results.

1966). A typical deviator stress versus longitudinal strain curve, and the associated q versus p stress path are shown in Figs 6.3(a) and (b), respectively. Thus a Mohr envelope may be constructed for the residual strength from each test. In order to obtain a Mohr envelope for peak strength, one must conduct several tests on several identical specimens at different confining pressures. However, specimens are

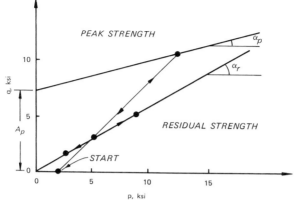

a. STRESS PATH IN p-q STRESS SPACE

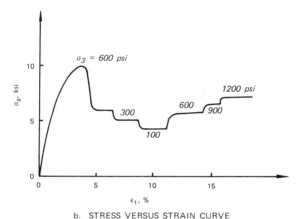

b. STRESS VERSUS STRAIN CURVE

Fig. 6.3. Single-stage triaxial test.

hardly identical, and results would be so varied that it would be difficult to draw a representative Mohr envelope and its associated strength parameters, c_p and ϕ_p. To obtain average strength parameters for peak strength, thus, would require many samples. This deficiency of SST tests can be alleviated by conducting multiple-stage triaxial (MST) tests.

Specimens cored from the same horizon of a large oil shale block sample from the Colony Mine were subjected to SST tests. The

TABLE 6.1
Statistics of Peak Strengths from SST Tests

Confining pressure (ksi)	Sample size (statistical)	Statistics of peak strengths, q_{peak} (ksi)			
		Range	Mean	Standard deviation	Coefficient of variation
0·30	4	6·38–6·90	6·62	0·22	0·03
0·60	16	6·73–7·72	7·04	0·28	0·04
0·90	5	7·02–7·30	7·11	0·12	0·02
1·20	3	7·09–7·54	7·27	0·24	0·03
1·50	9	7·48–8·24	7·72	0·26	0·03

Note: 1 ksi = 6·9 MPa.

statistics of the peak strengths of these samples are summarized in Table 6.1. In Section 6.3.5 these statistics will be compared with those from MST tests on the samples from the same block to investigate the validity of MST tests in determining the peak Mohr–Coulomb strength parameters.

6.3.3. Multiple-Stage Triaxial Test

Multiple-stage triaxial (MST) testing techniques have been developed as a valuable extension of the conventional SST test. When using these testing procedures, a specimen is stressed to near-failure under several confining pressures and the peak strength of the specimen at each confining pressure is determined. After failure, the residual strength at several confining pressures can be determined. Thus, several Mohr circles for peak strength as well as several Mohr circles for residual strength are obtained, and a Mohr envelope for both peak and residual strength can be drawn. The four strength parameters, c_p, ϕ_p, c_r, ϕ_r, can thus be obtained for each individual specimen. These procedures reduce the number of specimens required to determine the Mohr–Coulomb strength parameters. Thus the procedures may be used to advantage when only a limited number of specimens are available or where the specimens are very variable. The use of the procedures also enables one to study the variation of the strength parameters throughout a corehole and/or a deposit, and to include the strength parameters in analyses of the relationships with other more easily obtainable properties.

Heuzé (1967) performed several MST tests on Chino limestone.

Kovari and Tisa (1975) discussed the use of two types of triaxial tests to determine the strength parameters of individual rock specimens: the Multiple Failure State Triaxial Test and the Strain-Controlled Triaxial Test (or Continuous Failure State Triaxial Test). In the first test, the confining pressure was increased each time a peak load was obtained. Thus the rock approached a state of failure many successive times under different confining pressures. In the second type of test, the rock was loaded and the confining pressure adjusted so that the specimen was continuously kept in a near-failure state. The strength of the specimen over a continuous range of confining pressures was therefore determined. Both procedures were shown to be valid for Buchberg sandstone and Carrara marble and results were also presented for Gotthard granite and concrete. The technique was seen as a means of obtaining strength parameters from fewer specimens. Standardized test procedures for the Continuous Failure State Triaxial Test have been prepared by the International Society for Rock Mechanics (1978). Duvall (1976) indicated the need for additional research to validate, for many rock types, the methods developed by Kovari and Tisa.

Kim (1977) and Kim and Ko (1979) presented results of Multiple Failure State Triaxial Tests on Pierre shale, Raton shale and Lyons sandstone. In their tests, a rock specimen was loaded until a zero slope on the stress–strain curve was reached. The confining pressure was then increased and subsequent peaks were obtained. The test procedure was found to be valid for Pierre shale and Raton shale.

Atkinson (1977) followed the procedures of Kim and Ko in testing coal measure rocks. The resulting data were significantly different than those of conventional single-stage tests. He attributed the problem to the substantial deterioration and alteration of specimens by micro-cracking which occurs when specimens are loaded to high stress levels, and referred to the work of Peng (1970) in this regard.

Chang and Jumper (1978) and Chang and Bondurant (1979) reported the results of their studies on the validity of several MST testing techniques for oil shale. Three distinct test procedures were developed which were designated as MST-I, MST-II and MST-III tests. Results from the three types of MST tests were compared with results obtained from SST tests on identical specimens. It was found that the MST-II test produced results which were closest to those obtained from SST tests. Further, it was found that approaching the peak strength of oil shale at several confining pressures did not significantly alter the measured strength.

6.3.4. Test Procedures

Three types of MST test have been tried by Chang and Jumper (1978) and Chang and Bondurant (1979) at the University of Colorado at Denver for testing oil shale specimens. These have been designated as MST-I, MST-II and MST-III.

In the MST-I test a specimen was subjected to a confining pressure σ_3 and was then loaded axially. When the axial load neared the peak strength of the specimen at that confining pressure, as determined by changes in the slope of the load–deformation curve, the confining pressure was immediately increased, while vertical loading was continued. The process was repeated through several loading segments until the strength at each of several confining pressures was determined and the specimen was then allowed to fail. After failure the residual strength at each of several confining pressures could be obtained.

In the MST-II test a specimen was subjected to a confining pressure σ_3 and was then loaded axially. When the axial load neared the peak strength of the specimen at that confining pressure, initially the specimen was quickly unloaded to about 70% of the peak strength and then slowly unloaded until the axial stress was equal to the confining pressure to be used in the next cycle. The confining pressure was then increased and the specimen was again loaded axially. The process was repeated through several loading–unloading cycles until the strength at each of several confining pressures was determined. After failure, the residual strength at each of several confining pressures could be obtained.

The MST-III test was quite similar to the MST-II test. The confining pressure was decreased rather than increased between the loading–unloading cycles.

The typical deviator stress versus longitudinal strain curves and the associated stress paths in the q versus p stress space for these three types of MST tests are shown in Figs 6.4, 6.5 and 6.6, respectively. The Mohr–Coulomb failure envelopes and the associated q versus p diagrams for both peak and residual strengths can be prepared for each individual specimen as shown in Fig. 6.2.

The decisive factor which governs the success of any of the three types of MST test is the slope of the axial load versus longitudinal deformation curve at which the specimen is to be immediately unloaded during each loading–unloading cycle (MST-II and MST-III tests), or at which the confining pressure is to be increased immediately during each loading segment (MST-I tests). In the studies by Chang

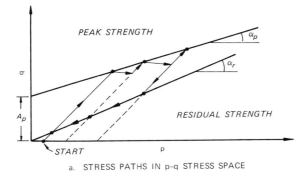

a. STRESS PATHS IN p-q STRESS SPACE

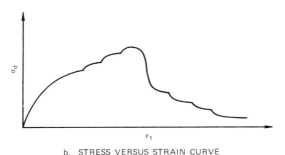

b. STRESS VERSUS STRAIN CURVE

Fig. 6.4. MST-I triaxial test.

and Jumper (1978) and Chang and Bondurant (1979), the critical slope was determined in advance, and was based on the slopes occurring near failure in previous SST tests on similar material using the same plotting scales. Upon the occurrence of this slope during an MST test the operator presumes that the maximum load-carrying capacity of the specimen at a particular confining pressure has very nearly been reached and that failure is imminent.

Experience in performing SST tests on various grades of oil shale can greatly assist an operator in determining the minimum slope obtainable before a specimen will fail. Brittle specimens (lean oil shale) have to be unloaded at larger slopes than ductile specimens (rich oil shale) or failure will occur. It should be noted here that since the cross-sectional area of a rock specimen expands as it is loaded axially to failure, a zero slope on a load–deformation curve indicates a decreasing stress sustained by the specimen. Therefore, care must be

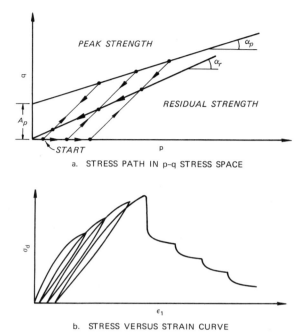

Fig. 6.5. MST-II triaxial test.

taken so that each loading cycle or stage is not continued to the zero slope on the load–deformation curve as this would mean that the specimen has already reached the post-peak portion of the stress–strain curve.

The action of unloading in MST-II and MST-III tests and increase in confining pressure in MST-I tests are to be immediate when the vertical stress of a specimen approximates its peak strength. It appears that at a stress near the peak strength, microcracking occurred at an increased rate and at a higher rate in brittle rocks than in ductile ones. Without the immediate and quick action of unloading in MST-II and MST-III tests or increasing confining pressure in MST-I tests, a rock specimen can deteriorate to an extent that the peak strengths so determined are no longer valid estimates of its true peak strengths at various confining pressures. Thus the resulting Mohr–Coulomb strength parameters for peak strengths are incorrect. Sometimes, a specimen may even fail before a sufficient number of peak strengths is

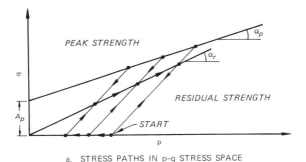

a. STRESS PATHS IN p-q STRESS SPACE

b. STRESS VERSUS STRAIN CURVE
Fig. 6.6. MST-III triaxial test.

obtained. The findings suggest that the non-success of the MST tests reported by Atkinson (1977) and Kim (1977) and Kim and Ko (1979) could have been avoided if a correct slope on the load–deformation curve had been chosen and immediate and quick action had been taken.

6.3.5. Discussion of Test Results

In the study by Chang and Bondurant (1979), modified Hoek cells were used and the vertical stress, axial deformation, lateral deformation and confining pressures were continuously monitored. The specimens were cored from the same horizon of a large oil shale block sample from the Colony Mine. The data were then processed to produce the initial tangent Young's modulus (E_i); tangent Young's modulus at 50% peak strength (E_{t50}); secant Young's modulus at peak strength (E_{sp}); and the corresponding Poisson's ratios (μ_i, μ_{t50}, μ_{sp}) and deviatoric stress at the initiation of dilation [$(\sigma_d)_{dlt}$] as shown in Fig. 6.7.

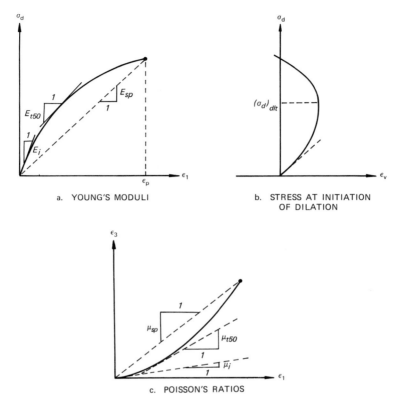

Fig. 6.7. Some engineering properties of oil shale.

The statistics of peak strengths from MST-I, MST-II and MST-III tests are summarized in Tables 6.2, 6.3 and 6.4, respectively. In Tables 6.5 and 6.6, the mean peak strengths and Mohr–Coulomb strength parameters obtained from MST tests were compared with those from SST tests. From Table 6.5 one may conclude that MST-II and MST-III tests closely estimate the true peak strength of oil shale while MST-I underestimates the strength at high confining pressures. Figure 6.8 and Table 6.6 also show that the MST-II test produces the best estimators of the true Mohr–Coulomb peak strength parameters of oil shale, with the MST-III closely behind. However, the MST-I test underestimated the angle of internal friction, ϕ_p, by 13%.

From the above comparison, it is concluded that the MST-II test is most feasible for the characterization of the peak strength of oil shale

TABLE 6.2
Statistics of Peak Strengths from MST-I Tests

	Statistics of peak strengths, q_{peak} (ksi)				
Confining pressure (ksi)	95% Confidence interval	Range	Mean	Standard deviation	Coefficient of variation
0·30	6·47–6·81	6·44–6·79	6·64	0·14	0·02
0·60	6·63–7·05	6·59–7·04	6·84	0·17	0·03
0·90	6·80–7·30	6·78–7·29	7·05	0·20	0·03
1·20	6·99–7·53	6·98–7·55	7·26	0·22	0·03
1·50	7·16–7·78	7·18–7·80	7·47	0·25	0·03
1·80	7·32–8·02	7·37–8·05	7·67	0·28	0·04
2·10	7·50–8·26	7·57–8·31	7·88	0·31	0·04

Statistical sample size = 5.

from the Piceance Creek Basin. The MST-II test was used to determine the Mohr–Coulomb strength parameters of oil shale from the corehole CE 706 (Chang and Bondurant, 1979; Bondurant and Chang, 1980; Chang, 1980; Chang and Bondurant, 1983; Bondurant, 1984), the Naval Oil Shale Reserve at Anvil Points, the Occidental property at Logan Wash and the Union Oil property in Garfield County, Colorado (Dolinar et al., 1979; Horino et al., 1981; Horino et al., 1982). The statistics of the strength parameters from these studies are presented in Section 6.6.

TABLE 6.3
Statistics of Peak Strengths from MST-II Tests

	Statistics of peak strengths, q_{peak} (ksi)				
Confining pressure (ksi)	95% Confidence interval	Range	Mean	Standard deviation	Coefficient of variation
0·30	6·35–6·97	5·80–7·24	6·66	0·40	0·06
0·60	6·67–7·17	6·26–7·32	6·92	0·32	0·05
0·90	7·00–7·36	6·73–7·46	7·18	0·24	0·03
1·20	7·29–7·55	7·20–7·67	7·42	0·17	0·02
1·50	7·56–7·82	7·46–7·95	7·69	0·17	0·02
1·80	7·78–8·12	7·64–8·22	7·95	0·22	0·03
2·10	7·98–8·44	7·72–8·59	8·21	0·30	0·04

Statistical sample size = 9.

TABLE 6.4
Statistics of Peak Strengths from MST-III Tests

Confining pressure (ksi)	Statistics of peak strengths, q_{peak} (ksi)				
	95% Confidence interval	Range	Mean	Standard deviation	Coefficient of variation
0·30	6·37–7·07	5·49–7·09	6·72	0·33	0·05
0·60	6·67–7·29	5·92–7·32	6·98	0·30	0·04
0·90	6·96–7·54	6·35–7·56	7·25	0·28	0·04
1·20	7·26–7·78	6·78–7·79	7·52	0·25	0·03
1·50	7·54–8·04	7·21–8·04	7·79	0·24	0·03
1·80	7·82–8·30	7·64–8·31	8·06	0·23	0·03
2·10	8·09–8·55	8·07–8·57	8·32	0·22	0·03

Statistical sample size = 6.

TABLE 6.5
Mean Peak Strengths

Confining pressure (ksi)	Mean peak strengths (ksi)				Mean peak strength ratios[a]		
	MST-I	MST-II	MST-III	SST	MST-I	MST-II	MST-III
0·30	6·64	6·66	6·72	6·62	1·00	1·01	1·02
0·60	6·84	6·92	6·98	7·04	0·97	0·98	0·99
0·90	7·05	7·18	7·25	7·11	0·99	1·01	1·02
1·20	7·26	7·42	7·52	7·27	1·00	1·02	1·03
1·50	7·47	7·69	7·79	7·72	0·98	1·00	1·01
1·80	7·67	7·95	8·06	8·04	0·95	0·99	1·00
2·10	7·88	8·21	8·32	8·30	0·95	0·99	1·00
			Average ratios:		0·98	1·00	1·01

[a] Mean peak strength from MST tests/mean peak strength from SST tests under the same confining pressure.

TABLE 6.6
Mohr–Coulomb Strength Parameters

Type of test	Cohesion c_p (ksi)	Angle of internal friction ϕ_p (degrees)	Strength parameter ratios	
			c ratio[a]	ϕ ratio[b]
SST	3·90	27·80	1·00	1·00
MST-I	4·16	24·10	1·07	0·87
MST-II	4·00	26·62	1·03	0·96
MST-III	3·66	29·42	0·94	1·06

[a] $= c_p/c_{SST}$.
[b] $= \phi_p/\phi_{SST}$.

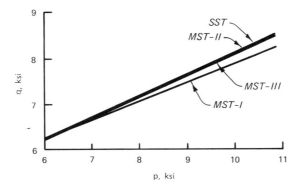

Fig. 6.8. q versus p diagram for SST and MST tests.

6.4. ANISOTROPIC MECHANICAL PROPERTIES OF OIL SHALE

Oil shale is not actually a true shale, but rather a marlstone, a fine-grained sedimentary rock containing a solid carbonaceous material known as kerogen. As are most sedimentary rocks, oil shale is stratified and varved. Because of this structural anisotropy, its mechanical properties are dependent on the direction in which the rock is stressed. Few previous studies have dealt with this subject of the anisotropic mechanical properties of oil shale. Morris (1978) presented his study on the elastic constants of oil shale. Because of its containing varves of lateral homogeneity, the transversely isotropic stiffness matrix was used by Chong et al. (1979b) to characterize its three-dimensional mechanical properties. Uniaxial tests on cubic specimens of oil shale were conducted to determine all associated elastic constants.

The anisotropic mechanical property was also investigated by Hawk (1979) at the University of Colorado at Boulder. Figure 6.9 shows the cubical test cell designed by Atkinson (1972) and later improved by Sture (1973). This test cell was used to test oil shale specimens. Three 10·2-cm (4-in) cubic specimens were cut from an oil shale block from the Colony Mine near Rifle, Colorado. To determine all elements of a three-dimensional stiffness matrix, C_{ij}, requires four specimens, one normal and three rotated. A normal specimen is required to determine the normal and Poisson's elastic constants of the 6×6 matrix. Because

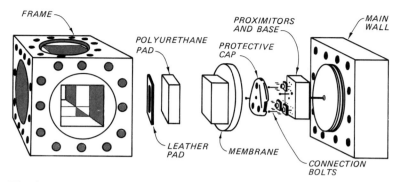

Fig. 6.9. Exploded view of a true triaxial test cell. (After Atkinson, 1972.)

of the inability of the cubical test cell in applying shear stresses, three rotated specimens, one rotated about each of the three axes, are needed to indirectly determine the shear elastic constants of oil shale. In this study, however, one normal cube along with one x-rotated specimen and one y-rotated specimen were used. The shear modulus C_{66} was calculated from C_{11} and C_{12} by the following relationship:

$$C_{66} = 1/2(C_{11} - C_{12}) \qquad (6.2)$$

Three confining pressures of 1·7 MPa, 3·5 MPa and 6·9 MPa (250 psi, 500 psi and 1000 psi, respectively) and the maximum principal stress difference of 13·8 MPa (2000 psi) were used. Thus the oil shale specimens were tested at stresses well below their proportional limits and they behaved as a linear elastic material. Test results indicated that the confining pressure had little or no effect on the elastic constants of oil shale.

Average elastic constants from these three tests were summarized in the following stiffness matrix:

$$(C_{ij})_{av} = \begin{bmatrix} 1·30 & 0·40 & 0·45 & & & \\ 0·30 & 1·22 & 0·42 & & & \\ 0·37 & 0·38 & 0·96 & & & \\ & & & 0·417 & & \\ & & & & 0·306 & \\ & & & & & 0·455 \end{bmatrix} \times 6900 \text{ MPa}$$

$$(6.3)$$

Statistical analysis of properties of oil shale 253

TABLE 6.7
Comparison of Elastic Constants for Oil Shale

Items	Source				
	Chong et al. (1979b)			Morris (1978)	Hawk (1979)
Stress level (%)	48	51	54	unknown	15
c_{11} (10^6 psi)	2·00	2·52	3·60	3·55	1·26
c_{12} (10^6 psi)	1·22	1·73	2·78	1·23	0·35
c_{13} (10^6 psi)	1·28	1·78	2·86	0·90	0·41
c_{33} (10^6 psi)	1·75	2·26	3·32	2·19	0·96
c_{44} (10^6 psi)	0·29	0·29	0·29	2·19	0·36

Hawk (1979) concluded that Mahogany Zone oil shale can be represented remarkably well by a transversely isotropic model.

The results of these three studies are summarized in Table 6.7. There are significant differences between the results presented by the different authors. These can be attributed to the difference in rock specimens tested, stress levels and testing techniques and equipment. Further investigation is needed to identify the cause of these differences. The data base of the three-dimensional stiffness matrix of oil shale is obviously very limited. A more accurate method to determine the shear modulus (C_{44}) has been presented by Chong (1983).

6.5. THERMAL MECHANICAL PROPERTIES OF OIL SHALE

To achieve maximum shale oil recovery while maintaining mine safety requires careful planning and engineering. The recurrence of the past ups and downs of the oil shale industry can be attributed, at least partly, to poor planning and insufficient research to improve our understanding of the mechanical properties of oil shale and their effects on mine structure stability. Thus, before the initiation of any mining activity, if it is to be successful, all potential mining schemes should be carefully analyzed with the input of proper mechanical properties of oil shale and its underlying and overlying rocks. An optimum mining scheme should emerge from this investigation.

Modified *in situ* (MIS) retorting scheme is one of the currently available methods of processing oil shale. It involves heating approximately 70% of the crushed oil shale in place and the other 30% above ground. The MIS process requires disposal of a smaller amount of spent shale (residue of a retort process) than a surface retorting process and is considered to be environmentally more acceptable. However, because little experience is available, the MIS process requires a great deal more research. Experiments on the MIS process were performed within a pilot project in Tract C-a and on a commercial scale in Tract C-b in the Piceance Creek Basin.

Oil shale and its surrounding rocks creep under prevailing *in situ* stresses caused by overburden. Mining and its subsequent underground retort in an MIS process can cause the stress to redistribute and the deformation rate (or creep rate) to increase. Under certain stress and temperature conditions, creep may eventually lead to mine structure failure. The creep of mine structures of oil shale and their underlying and overlying rocks will reflect, cumulatively, on the ground surface, as time-dependent ground subsidence. Mine structure failure can lead to catastrophic ground subsidence which can further cause safety and environmental concerns and hamper an energy production process.

An example of such creep in mine structures is illustrated in Fig. 6.10. Under a sustained stress, a mine structure deforms at a decreasing rate and the deformation rate gradually approaches a constant value as shown in curve A. These two portions of the creep are named

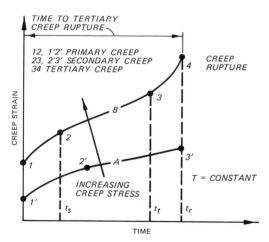

Fig. 6.10. Constant temperature creep.

primary and secondary creep, respectively. However, the creep process can change from secondary creep to tertiary creep if the sustained creep stress is greater than a critical fraction of the ultimate strength of a mine structure as shown in curve B in Fig. 6.10. Once a mine structure enters a tertiary creep state, its eventual rupture is imminent.

The modified *in situ* retorting process causes the temperature in oil shale mine structures to increase. This temperature increase may accelerate the creep process and lead to eventual creep rupture of mine structures because of the strength reduction of oil shale at an elevated temperature. Oil shale contains kerogen which can be substantially softened at high temperatures. As a result, the strength of oil shale is greatly reduced as the temperature is elevated. Therefore, a mine structure which is safe at an ambient temperature may become unstable at an elevated temperature. As shown in Fig. 6.11, a rock structure which is safe and undergoes secondary creep can reach the tertiary creep and eventually fail by creep rupture under an elevated temperature. Figure 6.12 shows the constant stress creep behavior of a metal under different temperatures. Figure 6.13 demonstrates the effect of stress and temperature on rupture time, t_r, i.e. time required to reach creep rupture. As the sustained creep stress and/or temperature increase, the rupture time decreases and the life expectancy of mine structures is shortened accordingly.

Because of the advancement of computer hardware, finite element analysis has emerged as a powerful numerical analysis technique. The versatility of finite element analysis is attributed to its ability to solve

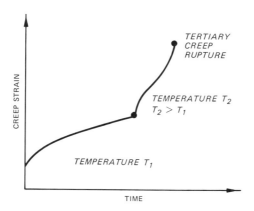

Fig. 6.11. Creep under two different temperatures.

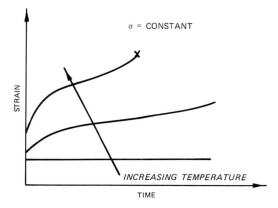

Fig. 6.12. Effect of temperature on the creep at constant stress.

engineering problems of complex boundary conditions and material compliance. However, to conduct such an analysis on selected mine plans in a high temperature environment requires a comprehensive understanding of the thermal mechanical properties of oil shale and its underlying and overlying rocks. Data on the thermal mechanical properties of oil shale are still very limited.

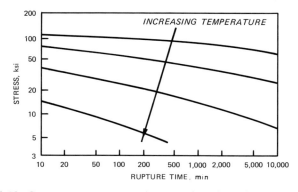

Fig. 6.13. Stress versus rupture time as a function of temperature.

6.5.1. Effect of Temperature on the Strength and Elasticity Modulus of Oil Shale

Existing literature on the effect of temperatures on the strength and elasticity modulus of oil shale is very scarce. Miller *et al.* (1979), Chu and Chang (1980), Zeuch (1982) and Chang (1982) all reported that both strength and the elasticity modulus are extremely sensitive to any change in temperature. As the temperature increases, the strength and Young's modulus of oil shale decrease drastically. Zeuch (1982) subjected oil shale specimens to undrained triaxial tests at an elevated temperature and found that the ductility of oil shale is greatly enhanced by the increase in confining pressure.

6.5.2. Creep Function

Sellers *et al.* (1972) investigated the time-dependent behavior of oil shale. Agapito and Page (1976) reported that the creep of oil shale pillars in the Colony Mine is a linear function of vertical pillar stress. Chong *et al.* (1978), Olsson (1980) and Horino *et al.* (1981) reported that the creep behavior of oil shale can be represented by a linear viscoelastic constitutive law. However, the above experimental studies were all conducted at room temperature.

Horino *et al.* (1981) concluded that creep-induced mine closure should not be a problem. However, the situation may be quite different in high-temperature environments. Chu and Chang (1980), Chang (1982) and Sinha *et al.* (1982) concluded from their studies that the creep behavior of oil shale is extremely temperature sensitive. As the temperature is increased, the creep strain rate also increases. Thus, closure may become a problem in high temperature environments.

The creep deformation of rocks can be described by the following general function of stress, time and temperature

$$\varepsilon_c = f(t, \sigma, T) \tag{6.4a}$$

where ε_c = creep strain, t = time, σ = applied creep stress, and T = temperature. The above creep function is frequently expressed as the product of separate stress, time and temperature functions:

$$\varepsilon_c = f_1(t) \cdot f_2(\sigma) \cdot f_3(T) \tag{6.4b}$$

For oil shale, the effect of oil content can be accounted for either through a separate independent function $f_4(O_c)$ or by expressing the material constants of time, stress and temperature function as functions of oil content.

Various empirical and rheological models are available which simulate the creep behavior of different materials. Many forms of stress function have been used in the past. Penny and Marriott (1971) have listed the following forms of stress function:

$$\text{Norton (1929):} \quad f(\sigma) = k\sigma^m \tag{6.5}$$

$$\text{McVetty (1943):} \quad f(\sigma) = A \sinh (\sigma/\sigma_0) \tag{6.6}$$

$$\text{Dorn (1955):} \quad f(\sigma) = C \exp (\sigma/\sigma_0) \tag{6.7}$$

where k, A, C and m are material constants and σ and σ_0 are the applied and reference stresses, respectively.

As for the time function, the most widely used expressions for characterizing primary creep are of exponential and power form (Cruden, 1971a, 1971b). Andrade (1910) used a combination of power and exponential laws. Griggs (1936, 1939, 1940) used a combination of logarithmic and power functions to simulate the creep behavior of limestone, talc, shale and other mineral crystals in compression.

Upon an increase in temperature, the creep of a rock may change from primary to secondary creep. This is the result of two temperature effects:

(i) it changes the material constants in a selected creep function; and

(ii) it encourages the change in material structure.

Dorn (1955) suggested the following temperature function:

$$\varepsilon_c \propto \exp \left(-\frac{U}{KT} \right) \tag{6.8}$$

where $U =$ activation energy, $K =$ Boltzmann's constant and $T =$ absolute temperature. Misra and Murrell (1965) conducted creep tests on different rock types at temperatures up to 700°C and suggested:

$$\varepsilon \propto T, \text{ at } T < 0 \cdot 2 T_m$$

and

$$\varepsilon \propto \exp \left(-\frac{\sigma^m}{T} \right), \text{ at } T > 0 \cdot 2 T_m$$

where $T_m =$ absolute melting temperature of rocks, $\sigma =$ stress and $m =$ constant. Durham et al. (1977) also used the above exponential temperature function to describe the steady state creep of olivine in the temperature range 1250–1600°C.

Rheological models of discrete elastic spring and viscous dashpot elements are used to represent material behavior. Jaeger (1972) summarized the models used by various authors, as shown in Table 6.8. In

Statistical analysis of properties of oil shale

TABLE 6.8
Rheological Models for Different Rock Types (after Lama and Vutukuri, 1978)

Rock type	Rheological model	Behavior	Source
Solid hard rock	Hookean	Elastic	Obert and Duvall (1967)
Rock in general	Kelvin	Visco-elastic	Salustowicz (1958)
Rock at greater depths	Maxwell	Visco-elastic	Salustowicz (1958)
Rock loaded for short interval	Generalized Kelvin or Nakamura	Visco-elastic	Nakamura (1949)
Sandstone, limestone and other rocks	Hooke's model parallel with Maxwell	Visco-elastic	Ruppeneit and Libermann (1960)
Coal	Modified Burger	Visco-elastic	Hardy (1959); Bobrov (1970)
Dolomite, claystone and anhydrite	Hooke's model + number of Kelvin models in series	Visco-elastic	Langer (1966, 1969)
Carboniferous rocks	Kelvin	Visco-elastic	Kidybinski (1966)
Carboniferous rocks	St. Venant parallel with Newtonian	Elastic, visco-plastic	Loonen and Hofer (1964)
Rock salt		Elastic, plastic	

their study on oil shale Chu and Chang (1980) found that a Kelvin–Voigt element in series with an elastic element (three-parameter model) satisfactorily fitted their data:

$$J(t) = \frac{1}{E_0} + \frac{1}{E_1} (1 - \exp(-t/\tau_1)) \qquad (6.9)$$

where $J(t)$ = creep function; E_0, E_1 = elastic spring constants; $\tau_1 = F_1/E_1$ = retardation time; and F_1 = coefficient of viscosity. However, this was based on creep test data with a test duration of not more than 30 days. As the duration increases, the model characterizing the creep behavior can be a Maxwell element in series with a generalized

Kelvin–Voigt model. The creep compliance for this model is:

$$J(t) = \frac{1}{E_0} + \frac{t}{F_0} + \sum_{n=1}^{N} \frac{1}{E_n}(1 - \exp(-t/\tau_n)) \qquad (6.10)$$

where τ_n is the retardation time with $\tau_n = F_n/E_n$ for $n = 1, 2, 3, \ldots, N$ and E_n and F_n are the elastic modulus and viscous damping coefficient, respectively, of the nth chain.

In the tertiary creep range, the formulation of creep compliance becomes more difficult, since the constitutive relationship becomes highly non-linear. Volterra (1959) represented the non-linear function as a series of multiple integrals in order to construct a non-linear theory of hereditary elasticity. If a constant load is suddenly applied, the Volterra–Frichet relation can be used to represent the non-linearity in tertiary creep.

6.5.3. Uniaxial Creep of Oil Shale at Elevated Temperatures

Very few studies involving high-temperature creep tests have been conducted. Under the sponsorship of the US Department of Energy, Chu and Chang (1980) and Chang (1982) conducted some high temperature uniaxial creep tests to investigate the effect of high temperature on the uniaxial creep behavior of oil shale. Oil shale specimens cored from large blocks of oil shale from the Colony Mine were tested. Results of this investigation are summarized in Table 6.9. The following conclusions were drawn from these results.

As the stress ratio and/or temperature increase (see Fig. 6.14) the secondary creep strain rate of oil shale also greatly increases, where the stress ratio is the ratio between the applied creep stress and the peak uniaxial strength at the corresponding temperature. This implies that the stress increase during a mining process and the temperature increase in an MIS process can accelerate pillar failure and room closure in an oil shale mine. It was also observed that, at the same temperature and stress ratio, the secondary (or steady state) creep strain rate is higher for rich oil shale than for lean.

Substantial temperature and/or stress increases can greatly reduce the creep rupture time and, thereby, force the early closure of an oil shale mine. At a stress ratio of 0·7, such a reduction in rupture time is of the order of several hundred times, when the temperature is increased from 24 to 204°C (75 to 400°F). The rupture time also decreases substantially with a stress ratio increase. At 121°C (250°F), as the stress ratio increases from 0·7 to 0·9, the rupture time is

Results of Creep Tests and Associated Material Parameters

Sample number	Unit weight (g cm^{-3})	Temperature (°F)	Stress ratio	Creep stress (ksi)	Secondary creep strain rate (10^{-6} in/in day)	Tertiary creep rupture time (min)	Elasticity modulus, E_1 (ksi)	Elasticity modulus, E_2 (ksi)	Relaxation time, τ (min)
3-12	2·204	75	0·33	3·52	72·0	—	2 800	7 000	15·9
3-7	2·174	75	0·50	5·28	237·0	—	4 400	3 800	18·6
3-8	—	75	0·67	7·04	240·5	—	5 900	2 800	21·8
3-9	2·211	75	0·83	8·80	432·0	6 237·00	2 900	1 800	17·0
4-9	2·145	75	0·31	3·14	47·5	—	3 900	2 100	34·8
4-10	2·142	75	0·47	4·71	96·5	—	2 800	6 300	32·0
4-8	—	75	0·78	7·90	—	—	5 600	1 960	32·0
4-11	—	75	0·77	7·84	840·0	1 714·00	—	—	—
4-16	—	75	0·31	3·15	168·5	—	5 600	2 000	32·0
3-6	2·174	75	0·50	5·30	47·5	—	—	—	—
2-1	2·081	100	0·26	3·10	240·5	—	610	280	28·8
2-2	2·109	100	0·40	4·70	—	—	345	207	45·0
2-3	2·081	100	0·27	3·04	299·5	—	610	280	28·8
2-5	2·109	100	0·41	4·57	479·5	—	760	240	37·1
2-6	2·098	100	0·54	6·09	960·5	—	310	180	50·0
2-9	2·119	100	0·68	7·61	—	1 320·00	—	—	—
2-10	2·109	200	0·27	3·04	144·0	—	350	210	45·0
4-15	2·178	200	0·34	2·71	504·0	—	1 700	680	7·6
4-6	—	200	0·52	4·07	432·0	—	1 200	620	6·6
4-13	2·141	200	0·63	4·97	497·5	—	200	100	12·0
4-17	—	200	0·68	5·40	—	—	—	—	—
4-1	2·141	200	0·34	2·71	240·5	—	680	1 600	7·0
4-2	2·178	250	0·63	5·00	—	—	455	865	7·0
3-5	2·209	250	0·38	3·00	119·5	—	2 000	500	13·5
3-10	—	250	0·76	6·00	2 405·0	450·00	—	—	—
3-11	1·902	300	0·66	4·46	299·5	5 790·00	1 400	520	36·1
3-19	—	400	0·38	2·80	911·5	—	—	—	—
3-22	2·227	400	0·71	3·87	—	4·02	—	—	—
3-21	2·317	400	0·94	5·15	—	0·24	—	—	—
3-20	2·239	400	0·47	2·58	1 200·0	—	820	500	27·7

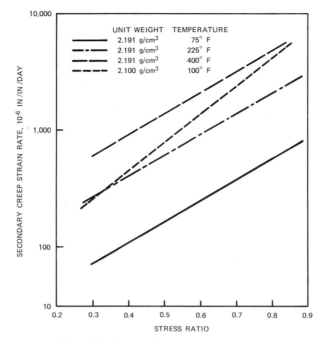

Fig. 6.14. Secondary creep strain rates.

reduced by around 150 times. At 204°C (400°F), however, the decrease in rupture time is about 15 times given the same stress ratio increase. Thus, the effects of temperature and stress on the life expectancy of an oil shale mine must be comprehensively studied before constructing an MIS retorting facility.

A three-parameter rheological model, an elastic spring in series with a Kelvin–Voigt element, was found to satisfactorily simulate the creep behavior of oil shale at elevated temperatures. However, it must be noted that the uniaxial creep tests conducted lasted no longer than 30 days. These three parameters are the retardation time, τ, and elasticity moduli, E_0 and E_1. Retardation time was observed to decrease with temperature increase. However, its dependence on stress ratio is still not clear. The retardation time for a rich oil shale was found to be longer than that for lean. The elasticity moduli, E_0 and E_1, decrease substantially with increases in the temperature and the stress ratio and are larger for lean oil shale than for rich.

Statistical analysis of properties of oil shale 263

Thus, in summary, the creep behavior of oil shale is extremely sensitive to any changes in temperature, stress level or oil content, and deserves extensive and systematic study.

6.6. STATISTICAL ANALYSIS AND MODELING

The engineering properties of oil shale vary throughout the deposits. These properties can be loosely divided into two groups: physical/index properties and mechanical/strength properties. Some of these properties are strongly correlated. A comprehensive study of the statistical analysis and modeling of the engineering properties of oil shale was conducted at the University of Colorado at Denver (UCD) (Bondurant and Chang, 1980; Bondurant, 1983; Chang and Bondurant, 1983) under the sponsorship of the National Science Foundation. Presented in this section are the results of this study including: (1) statistical and distributional characteristics; (2) statistical correlation between various properties; (3) some significant bivariate and multivariate regression equations. In most of the equations presented herein, the mechanical/strength properties are expressed as functions of easily obtainable or readily available physical and index properties, such as Fischer assay, organic content, hardness and density.

Subsequent to the UCD studies, the US Bureau of Mines also conducted statistical research on the mechanical properties of oil shale from the Anvil Points Mine (Dolinar *et al.*, 1979), the Union's Oil Shale Mine (Horino *et al.*, 1982) and the Logan Wash Oil Shale Site of Occidental Oil Shale Inc. (Horino *et al.*, 1981). The bivariate regression equations from these studies are also presented in this section.

During the preliminary feasibility study and design phases of a full-scale oil shale project, these regression equations can be used to estimate the mechanical and strength properties of oil shale in a mine. The precision of such predicted properties depends on the reliability of the regression equation and the precision of the independent variable(s) entering the prediction. The higher the correlation coefficient, the more reliable is the predicted value of the dependent variable.

6.6.1. Data Sources
Oil shale data used in the statistical studies at UCD were obtained from two testing programs, one at UCD and the other at the US Bureau of Mines. 52 EX-size specimens from corehole CE706 were

tested at UCD. The location of the corehole is shown in Fig. 6.1. Specimens were obtained at various depths from 55 to 397 m (180 to 1300 ft) throughout the corehole. 35 MST and 17 conventional SST tests were performed. In addition to the parameters obtained in the triaxial tests, other properties such as density were determined. Specimens were cored perpendicular to the bedding plane.

Test data for 1060 specimens from coreholes 01-A, 02-A, and 01 were provided by the Bureau of Mines. Most of the tests were conducted by its personnel at the Twin Cities Research Center. The locations of these coreholes are shown in Fig. 6.1. The specimens were from various zones throughout the Parachute Creek Member of the Green River Formation. Many were from an area in the unleached saline zone at a depth of approximately 610 m (2000 ft). Included were test data from 209 unconfined compression tests, 147 SST tests, 381 Brazilian tensile tests parallel to bedding plane, 280 Brazilian tensile tests perpendicular to bedding plane and 43 flexural (modulus of rupture) tests. Several other properties such as density, Fischer assay and properties determined from acoustic methods, were determined for some of the specimens. Test procedures generally followed those specified by Lewis and Tandanand (1974). Specimens used for the uniaxial, triaxial and dynamic testing were cored perpendicular to the bedding plane. The specimens used for uniaxial and triaxial testing generally ranged in diameter from 4·3 to 5·1 cm (1·7 to 2·0 in) and had aspect ratios of approximately 2·0. The reported values for unconfined compressive strength were normalized to a common aspect ratio of 1·0 by using the following equation:

$$C_0 = \frac{C_p}{0·778 + 0·222\left(\dfrac{D}{L}\right)} \tag{6.11}$$

where C_0 = compressive strength normalized to an aspect ratio of 1·0, C_p = measured compressive strength of specimen with an aspect ratio between 0·333 and 2·0, D = diameter of specimen, and L = length of specimen.

The confining pressures used in the triaxial tests ranged from 1·6 to 18·3 MPa (225 to 2650 psi). The flexural strength was determined during three-point loading of the cylindrical specimens. Specimens used for modulus of rupture tests were cored parallel to the bedding plane and generally had diameters of approximately 1·9 cm (0·75 in) and lengths of approximately 8·9 cm (3·5 in).

As mentioned previously, the Bureau of Mines has conducted additional studies on the mechanical properties of oil shale from three oil shale mine sites: the Anvil Points Mine, the Union Mine and the Logan Wash Oil Shale Site. The locations of these mine sites are also shown in Fig. 6.1. The studies involved 758 tests including 214 specific gravity tests, 214 Fischer assay tests, 208 uniaxial compression tests, and 123 MST tests. Data for these tests were not included in the University of Colorado study. Instead, bivariate regression equations were formulated independently. These equations are also reported in this section.

6.6.2. Properties Analyzed
The properties determined in the testing program at the University of Colorado were generally not the same as those determined in the Bureau of Mines testing program.

The following 13 parameters were available for most of the specimens tested at the University of Colorado:

(1) depth;
(2) density;
(3) Young's modulus at three points along the stress–strain curve;
(4) Poisson's ratio at three points along the stress–strain curve;
(5) longitudinal (axial) strain at peak stress for first loading cycle,
(6) deviator stress at which first dilation of the sample occurred in the first loading cycle;
(7) deviator stress at dilation divided by peak deviator stress of the first loading cycle;
(8) deviator stress at dilation divided by the confining pressure of the first loading cycle;
(9) maximum shear stress for the first loading cycle (peak deviator stress divided by two, i.e. the radius of the Mohr circle);
(10) peak cohesion;
(11) peak angle of internal friction;
(12) apparent residual cohesion; and
(13) residual angle of internal friction.

The three values of Young's modulus and Poisson's ratio were determined in order to characterize, to some extent, the non-linear behavior of oil shale. The stress at which initial sample dilation occurred, which may indicate the onset of unstable fracture propagation, was determined by following the procedures described by Bieniawski (1967).

The parameters available for each of the specimens tested by the Bureau of Mines included two or more of the following:

(1) depth;
(2) Fischer assay;
(3) density;
(4) dry density;
(5) Shore hardness;
(6) porosity;
(7) dynamic Young's modulus, from acoustic methods;
(8) dynamic Poisson's ratio, acoustic methods;
(9) dynamic shear modulus, acoustic method;
(10) Brazilian indirect tensile strength, parallel or perpendicular to bedding;
(11) flexural strength (modulus of rupture);
(12) unconfined compressive strength;
(13) Young's modulus determined in uniaxial testing;
(14) Poisson's ratio determined in uniaxial testing;
(15) Young's modulus determined in triaxial testing; and
(16) maximum shear stress (peak deviator stress divided by two, i.e., the radius of the Mohr circle).

From the two data sources, there were 29 properties available for statistical analysis and statistical modeling. These properties, their symbols, and units of measure are shown in Table 6.10. In this table, the variables have been divided according to six property classifications:

(1) physical and index properties;
(2) mechanical properties obtained from acoustic methods;
(3) mechanical properties obtained from tensile and flexural tests;
(4) mechanical properties obtained from uniaxial tests;
(5) mechanical properties obtained from SST and MST tests; and
(6) strength parameters obtained from MST tests.

The results of all tests on specimens from the four coreholes and both testing programs were combined into one data set. The resulting data matrix consisted of data for 1112 specimens, and two or more of the 29 engineering properties were available for each specimen tested.

6.6.3. Analytical Methods and Procedures

Several statistical methods were used to analyze the data. First, the general statistical and distributional characteristics of the properties

Statistical analysis of properties of oil shale 267

TABLE 6.10
Variables and Symbols Used in Statistical Analysis

Symbol	Description
	Physical and index properties
A	Modified Fischer assay (gallons ton^{-1})
D	Density (g cm^{-3})
D_d	Dry density (g cm^{-3})
S	Shore hardness
P	Porosity (%)
G_s	Specific gravity
	Mechanical properties obtained from acoustic methods
E_{ac}	Dynamic Young's modulus (10^6 psi)
μ_{ac}	Dynamic Poisson's ratio
G_{ac}	Dynamic shear modulus (10^6 psi)
	Mechanical properties obtained from tensile and flexural tests
$T_{\parallel}$	Brazilian tensile strength corresponding to a directly applied tensile stress parallel to bedding plane (psi)
$T_{\perp}$	Brazilian tensile strength corresponding to a directly applied tensile stress perpendicular to bedding plane (psi)
F	Flexural strength (psi)
	Mechanical properties obtained from uniaxial tests
C_0	Unconfined compressive strength perpendicular to bedding plane (psi)
E_{uni}	Young's modulus, maximum tangent value (10^6 psi)
μ_{uni}	Poisson's ratio, maximum tangent value
	Mechanical properties obtained from single-stage and multiple-stage triaxial tests
E_i	Young's modulus, initial tangent value (10^6 psi)
E_{t50}	Young's modulus, tangent at 50% of peak load (10^6 psi)
E_{sp}	Young's modulus, secant at peak load (10^6 psi)
μ_i	Poisson's ratio, initial tangent value
μ_{t50}	Poisson's ratio, tangent at 50% of peak load
μ_{sp}	Poisson's ratio, secant at peak load
ε_p	Longitudinal strain to peak stress for loading cycle
$(\sigma_d)_{dlt}$	Deviator stress at which first dilation of sample occurs in first loading cycle (psi)
$\dfrac{(\sigma_d)_{dlt}}{\sigma_d}$	$(\sigma_d)_{dlt}$ divided by peak deviator stress, first loading cycle
$\dfrac{(\sigma_d)_{dlt}}{\sigma_3}$	$(\sigma_d)_{dlt}$ divided by confining pressure, first loading cycle
q	Maximum shear stress for first loading cycle (peak deviator stress divided by 2) (psi)
	Strength parameters obtained from multiple-stage triaxial tests
C_p	Peak cohesion (ksi)
ϕ_p	Peak angle of internal friction (degrees)
C_r	Residual cohesion (ksi)
ϕ_r	Residual angle of internal friction (degrees)

were determined. Second, some properties were chosen as dependent and some as independent variables, and the correlations between the two groups of variables were analyzed. Third, bivariate regression equations were formulated, in which a dependent variable was expressed as a function of an independent variable. Fourth, multiple regression equations, for the prediction of a dependent variable from more than one independent variable, were developed.

For the analysis of correlations and for the subsequent regression analyses, many non-linear relationships between rock properties were investigated. In order to investigate these relationships, many new parameters were formed by transforming and multiplying some of the 29 original properties, as shown in Section 6.6.5.

Most of the statistical analyses at UCD were performed using the Statistical Package for the Social Sciences (SPSS) system of computer programs. These programs are described by Nie *et al.* (1975).

6.6.4. General Statistical and Distributional Characteristics

General statistical and distributional characteristics were determined for the 29 properties listed in Table 6.10. These characteristics included the mean, standard deviation, coefficient of variation, range of values, 95% confidence interval for the true mean, mode, median, skewness and kurtosis. The general statistical and distributional characteristics of the 29 properties are shown in Table 6.11. For each variable, the number of cases (sample size) available for the computations is also listed.

The coefficient of variation is defined as the standard deviation (s) divided by the mean ($\bar{x}$). The 95% confidence interval (CI) for the true mean, for a normal population with unknown population standard deviation, σ, is computed by using the following equation (Miller and Freund, 1965):

$$0 \cdot 95 CI = \bar{x} + t\left(\frac{s}{\sqrt{n}}\right) \tag{6.12}$$

where $t = 0 \cdot 025$ t statistic for $(n-1)$ degrees of freedom and $n =$ sample size (i.e. number of observations, or cases).

The mode is the value of a variable which occurs most frequently in an observed set of data. The median is the value at which the number of cases with larger values is equal to the number with smaller values.

Skewness is a measure of the deviations of an observed frequency distribution from symmetry. For this study, a coefficient of skewness g_1

was expressed by the following equation:

$$g_1 = \frac{n \sum_{i=1}^{n} (x_i - \bar{x})^3}{(n-1)(n-2)s^3} \tag{6.13}$$

For a symmetric distribution, this coefficient of skewness is equal to zero. A positive coefficient indicates that the extreme values (the long tail) of an observed distribution are greater than the mean, while a negative value indicates that they are less than the mean.

Kurtosis is a measure of the peakedness of an observed frequency distribution. For this study, a coefficient of kurtosis g_2 was calculated with the following equation:

$$g_2 = \frac{n(n+1) \sum_{i=1}^{n} (x_i - \bar{x})^4 - 3 \sum_{i=1}^{n} (x_1 - \bar{x})^2 \sum_{i=1}^{n} (x_i - \bar{x})^2 (n-1)}{(n-1)(n-2)(n-3)s^4} \tag{6.14}$$

For a normal distribution, this coefficient of kurtosis is equal to zero. A positive value indicates that the observed distribution is more peaked, and narrower than a normal distribution, while a negative value indicates that it is flatter.

6.6.5. Correlation Coefficients

The bivariate correlation coefficients between the dependent variables and 11 selected independent variables were computed by using standard equations (Walpole and Myers, 1972). The 11 selected independent variables were:

(1) assay (A);
(2) assay squared (A^2);
(3) assay cubed (A^3);
(4) density (D);
(5) density squared (D^2);
(6) dry density (D_d);
(7) Shore hardness (S);
(8) Shore hardness squared (S^2);
(9) the product of Shore hardness and dry density (SD_d);
(10) the square of the product of Shore hardness and dry density $((SD_d)^2)$; and
(11) the confining pressure of the first loading cycle or segment in triaxial tests (σ_3).

TABLE 6.11
General Statistics and Distributional Characteristics of Engineering Properties of Oil Shale

Variable	Data set	Number of cases	Mean	Standard deviation	Coefficient of variation	Range of values	95% confidence interval	Mode	Median	Skewness	Kurtosis
Physical and index properties											
A	BM3	763	26·50	14·30	0·54	0·00–84·60	25·50–27·50	0·00	27·10	0·075	−0·070
D	BM3	365	2·03	0·18	0·09	1·49–2·85	2·01–2·05				
	CE706	49	2·10	0·19	0·09	1·67–2·44	2·04–2·15				
	4 holes	414	2·04	0·18	0·09	1·49–2·85	2·02–2·05	2·05	2·03	0·328	1·083
D_d	BM3	186	2·00	0·17	0·08	1·66–2·58	1·97–2·02	1·91	1·97	1·109	2·141
S	BM3	244	33·50	7·60	0·23	8·20–55·20	32·50–34·40	36·90	34·50	−0·379	0·373
P	01A	44	0·80	0·50	0·66	0·00– 2·10	0·70– 1·00	0·60	0·90	0·094	−0·687
Mechanical properties from acoustic methods											
E_{ac}	01A	58	1·61	1·00	0·62	0·49–5·12	1·34–1·87	1·36	1·36	1·927	4·054
μ_{ac}	01A	58	0·35	0·04	0·11	0·26–0·43	0·34–0·36	0·33	0·35	−0·156	0·110
G_{ac}	01A	58	0·67	0·40	0·60	0·25–1·96	0·56–0·77	0·39	0·54	2·013	3·997
Mechanical properties from tensile and flexural tests											
$T_{\parallel}$	BM3	381	838·00	346·00	0·41	56–2039	803–873	712·00	825·00	0·639	0·899
$T_{\perp}$	BM3	280	577·00	287·00	0·50	128–1844	544–611	464·00	506·00	1·265	2·132
F	BM3	43	3377·00	1388·00	0·41	527–7700	2950–3804	3378·00	3246·00	1·156	2·314
Mechanical properties from uniaxial tests											
C_0	BM3	208	8999·00	4451·00	0·49	1633–36537	8391–9608	—	8262·00	2·813	13·197
E_{uni}	BM3	159	0·91	0·74	0·81	0·19–5·16	0·80–1·03	0·49	0·65	2·603	8·495
μ_{uni}	BM3	147	0·27	0·12	0·46	0·03–0·92	0·25–0·29	0·24	0·25	1·634	6·099

Mechanical properties from single-stage and multiple-stage triaxial tests

E_i	CE706	52	0·87	0·45	0·51	0·09–2·01	0·74–0·99	0·65	0·76	0·687	0·174
E_{t50}	BM3	31	1·41	1·74	1·24	0·20–7·60	0·77–2·05				
	CE706	52	0·99	0·48	0·48	0·34–2·50	0·86–1·13				
	4 holes	83	1·15	1·14	0·99	0·20–7·60	0·90–1·40	0·86	0·76	3·226	13·405
E_{sp}	CE706	52	0·70	0·37	0·54	0·26–1·82	0·60–0·80	0·42	0·51	1·112	0·547
μ_i	CE706	49	0·35	0·20	0·57	0·00–0·93	0·29–0·40	—	0·33	0·516	0·532
μ_{t50}	CE706	49	0·64	0·20	0·31	0·00–1·25	0·58–0·69	—	0·62	−0·228	2·730
μ_{sp}	CE706	49	0·85	0·23	0·28	0·48–1·86	0·78–0·91	—	0·81	1·828	6·441
ε_p	CE706	52	0·02	0·01	0·34	0·01–0·05	0·02–0·03	0·03	0·02	0·779	1·073
$(\sigma_d)_{dlt}$	CE706	49	4400·00	4056·00	0·92	100–21400	3235–5566	100·00	3370·00	1·777	5·073
$\dfrac{(\sigma_d)_{dlt}}{\sigma_d}$	CE706	49	0·29	0·23	0·79	0·01–0·86	0·23–0·36	—	0·25	0·808	−0·039
$\dfrac{(\sigma_d)_{dlt}}{\sigma_3}$	CE706	49	10·47	11·21	1·07	0·13–46·65	7·25–13·69	0·33	6·60	1·808	3·083
q	BM3	146	6241·00	2110·80	0·34	2297–14631	5896–6586				
	CE706	52	7095·50	1644·00	0·23	4172–12399	6638–7553				
	4 holes	198	6465·40	2030·10	0·31	2297–14631	6181–6750	5315·50	5918·80	1·471	3·099

Strength parameters from multiple-stage triaxial tests

C_p	CE706	35	4·05	1·01	0·25	1·67– 6·97	3·70–4·40	4·12	4·03	0·662	1·768
ϕ_p	CE706	35	24·80	9·10	0·37	8·40–51·40	21·60–27·90	21·30	22·60	0·899	1·247
C_r	CE706	35	0·78	0·47	0·60	0·00– 1·87	0·62–0·94	0·00	0·75	0·492	0·044
ϕ_r	CE706	35	37·10	7·40	0·20	23·50–52·20	34·60–39·70	42·40	37·30	0·102	−0·598

The first 10 independent variables were selected either because they are easily obtainable from simple tests, such as density and Shore hardness, or because they are ordinarily obtained during evaluation of an oil shale deposit and thus are readily available, such as assay. Thus, these variables would be very useful as 'index' properties for the prediction of other properties which are more difficult to determine, such as mechanical properties and strength parameters. The confining pressure was selected as an independent variable in order to ascertain its effect on various mechanical properties.

The correlation coefficient (R) is used to measure the strength of the linear relationships between two variables. When a straight line is a perfect fit to the data, the correlation coefficient is equal to $\pm 1 \cdot 0$. A negative value of R implies that the regression line has a negative slope. It is generally considered that if the absolute value of the correlation coefficient is equal to or greater than $0 \cdot 7$, the correlation is considered significant and one variable can be predicted from the other through a linear bivariate regression equation.

The correlation coefficients between various properties and the 11 selected independent variables are presented in Table 6.12. Because of the sparseness of the data matrix, coefficients could not be determined between many parameters of interest. For each pair of variables, the correlation coefficient is listed and the number of data pairs available for the computation is shown in parentheses.

6.6.6. Bivariate Regression Analysis

Bivariate regression analysis was performed by using a standard least-square regression procedure. The equations formulated express the relationships between various properties and the 10 selected easily obtainable or readily available properties used as independent variables.

Some of the bivariate regression equations developed are shown in Table 6.13. Equations have been reported for most of those pairs of variables for which the absolute value of the correlation coefficient was $\geqslant 0 \cdot 65$. A few equations, which are considered important, are also listed even if their correlation coefficients are $<0 \cdot 65$. The number of data pairs (cases) used in the development of the equation and the value of the correlation coefficient R also are listed. Also included in Table 6.10 are the bivariate regression equations recently formulated by colleagues with the Bureau of Mines from their studies on the mechanical properties of oil shale from three oil shale development

TABLE 6.12
Correlation Coefficients from BME and CE706

Dependent variable	Independent variables										
	A	A^2	A^3	D	D^2	D_d	S	S^2	(SD_d)	$(SD_d)^2$	σ_3
A				-0.748	-0.735	-0.600	0.373	0.346	0.102	0.014	
				(187)	(187)	(185)	(182)	(182)	(180)	(180)	
A^2				-0.746	-0.720	-0.605	0.405	0.380	0.124	0.037	
				(187)	(187)	(185)	(182)	(182)	(180)	(180)	
A^3				-0.706	-0.673	-0.585	0.394	0.378	0.120	0.043	
				(187)	(187)	(185)	(182)	(182)	(180)	(180)	
D	-0.748	-0.746	-0.706			0.958	-0.127	-0.086	0.247	0.335	
	(187)	(187)	(187)			(186)	(223)	(223)	(180)	(180)	
D^2	-0.735	-0.720	-0.673			0.961	-0.109	-0.067	0.267	0.354	
	(187)	(187)	(187)			(186)	(223)	(223)	(180)	(180)	
D_d	-0.600	-0.605	-0.585	0.958	0.961		0.036	0.079	0.392	0.463	
	(185)	(185)	(185)	(186)	(186)		(180)	(180)	(180)	(180)	
P	-0.203	-0.203	-0.199	0.055	0.052	0.185	-0.047	-0.050	0.010	0.009	
	(26)	(26)	(26)	(42)	(42)	(24)	(42)	(42)	(24)	(24)	
E_{ac}	-0.800	-0.701	-0.600	0.680	0.684	0.436	-0.512	-0.472	-0.499	-0.440	
	(38)	(38)	(38)	(58)	(58)	(39)	(57)	(57)	(38)	(38)	
μ_{ac}	0.193	0.163	0.124	-0.246	-0.250	-0.133	0.161	0.163	0.210	0.200	
	(38)	(38)	(38)	(58)	(58)	(39)	(57)	(57)	(38)	(38)	
G_{ac}	-0.763	-0.677	-0.581	0.656	0.659	0.403	-0.482	-0.441	-0.468	-0.407	
	(38)	(38)	(38)	(58)	(58)	(39)	(57)	(57)	(38)	(38)	
$T_{\parallel}$	0.266	0.225	0.178	0.069	0.100	—	—	—	—	—	
	(313)	(313)	(313)	(25)	(25)						
$T_{\perp}$	-0.104	-0.112	-0.103	—	—	—	—	—	—	—	
	(258)	(258)	(258)								

TABLE 6.12—contd.

Dependent variable	Independent variables										
	A	A^2	A^3	D	D^2	D_d	S	S^2	(SD_d)	$(SD_d)^2$	σ_3
F	—	—	—	0·439 (14)	0·447 (14)	—	—	—	—	—	
C_0	−0·218 (142)	−0·164 (142)	−0·126 (142)	0·367 (189)	0·397 (189)	0·574 (137)	0·452 (146)	0·532 (146)	0·644 (132)	0·742 (132)	
$\log C_0$	−0·131 (142)	−0·075 (142)	−0·049 (142)	0·259 (189)	0·284 (189)	0·501 (137)	0·422 (46)	0·472 (146)	0·597 (132)	0·644 (132)	
E_{uni}	−0·588 (143)	−0·502 (143)	−0·421 (143)	0·637 (140)	0·646 (140)	0·524 (137)	0·033 (147)	0·080 (147)	0·213 (133)	0·299 (133)	
$\log E_{uni}$	−0·648 (143)	−0·601 (143)	−0·541 (143)	0·692 (140)	0·691 (140)	0·590 (137)	0·017 (147)	0·054 (147)	0·227 (133)	0·297 (133)	
μ_{uni}	0·334 (135)	0·309 (135)	0·275 (135)	−0·271 (136)	−0·267 (136)	−0·268 (133)	0·311 (143)	0·280 (143)	0·166 (129)	0·112 (129)	
E_i	—	—	—	0·477 (49)	0·478 (49)	—	—	—	—	—	−0·017 (52)
$\log E_i$	—	—	—	0·403 (49)	0·404 (49)	—	—	—	—	—	−0·159 (52)
E_{t50}	—	—	—	0·623 (80)	0·635 (80)	—	—	—	—	—	−0·192 (83)
E_{sp}	—	—	—	0·837 (49)	0·852 (49)	—	—	—	—	—	−0·178 (52)
$\log E_{sp}$	—	—	—	0·855 (49)	0·863 (49)	—	—	—	—	—	−0·193 (52)
μ_i	—	—	—	−0·169 (48)	−0·164 (48)	—	—	—	—	—	−0·103 (49)
μ_{t50}	—	—	—	0·068 (48)	0·066 (48)	—	—	—	—	—	−0·197 (49)

μ_{sp}	—	—	—	0·211 (48)	0·203 (48)	—	—	—	—	—	−0·050 (49)
ε_p	—	—	—	−0·605 (49)	−0·614 (49)	—	—	—	—	—	0·445 (52)
$(\sigma_d)_{dlt}$	—	—	—	0·286 (48)	0·285 (48)	—	—	—	—	—	0·203 (49)
$(\sigma_d)_{dlt}/\sigma_d$	—	—	—	0·100 (48)	0·095 (48)	—	—	—	—	—	0·165 (49)
$(\sigma_d)_{dlt}/\sigma_3$	—	—	—	0·234 (48)	0·231 (48)	—	—	—	—	—	−0·279 (49)
q	−0·547 (49)	−0·426 (49)	−0·342 (49)	0·589 (184)	0·596 (184)	0·773 (48)	0·259 (96)	0·283 (96)	0·699 (47)	0·757 (47)	−0·110 (198)
C_p	—	—	—	0·426 (35)	0·430 (35)	—	—	—	—	—	−0·579 (35)
ϕ_p	—	—	—	0·334 (35)	0·334 (35)	—	—	—	—	—	0·486 (35)
$\sin \phi_p$	—	—	—	0·330 (35)	0·330 (35)	—	—	—	—	—	0·474 (35)
$\cos \phi_p$	—	—	—	−0·351 (35)	−0·350 (35)	—	—	—	—	—	−0·503 (35)
$\tan \phi_p$	—	—	—	0·331 (35)	0·330 (35)	—	—	—	—	—	0·508 (35)
C_r	—	—	—	−0·109 (33)	−0·107 (33)	—	—	—	—	—	−0·312 (35)
ϕ_r	—	—	—	0·007 (33)	0·003 (33)	—	—	—	—	—	0·370 (35)
$\sin \phi_r$	—	—	—	0·015 (33)	0·011 (33)	—	—	—	—	—	0·359 (35)
$\cos \phi_r$	—	—	—	0·007 (33)	0·011 (33)	—	—	—	—	—	−0·384 (35)
$\tan \phi_r$	—	—	—	−0·010 (33)	−0·014 (33)	—	—	—	—	—	0·389 (35)

TABLE 6.13
Bivariate Regression Equations

Dependent variable	Independent variable	Data set	Equation	Number of cases	R
Physical and index properties					
A	D	BM3	$A = -65 \cdot 8D + 162$	187	$-0 \cdot 748$
	D_d	BM3	$A = -52 \cdot 0D_d + 131$	185	$-0 \cdot 600$
	S	01A	$A = 1 \cdot 14S - 10 \cdot 2$	66	$0 \cdot 630$
D	A	BM3	$D = -8 \cdot 50 \times 10^{-3} A + 2 \cdot 28$	187	$-0 \cdot 748$
D_d	A	BM3	$D_d = -6 \cdot 91 \times 10^{-3} A + 2 \cdot 18$	185	$-0 \cdot 600$
G_s	A	Anvil 1	$G_s = 1/(0 \cdot 379 + 0 \cdot 003 A)$	20	$0 \cdot 980$
		Anvil 2	$G_s = 1/(0 \cdot 378\ 6 + 0 \cdot 003\ 3A)$	39	$0 \cdot 990$
		Union	$G_s = 1/(0 \cdot 356\ 5 + 0 \cdot 003\ 7A)$	91	$0 \cdot 973$
		Logan	$G_s = 1/(0 \cdot 392\ 4 + 0 \cdot 002\ 9A)$	64	$0 \cdot 940$
Mechanical properties from acoustic methods					
E_{ac}	A	01A	$E_{ac} = -0 \cdot 066\ 2A + 3 \cdot 77$	38	$-0 \cdot 800$
	D	01A	$E_{ac} = 5 \cdot 38D - 9 \cdot 26$	58	$0 \cdot 680$
G_{ac}	A	01A	$G_{ac} = -0 \cdot 025\ 7A + 1 \cdot 50$	38	$-0 \cdot 763$
	D	01A	$G_{ac} = 2 \cdot 08D - 3 \cdot 53$	58	$0 \cdot 656$

Mechanical properties from uniaxial tests

C_0	$(SD_d)^2$	BM3	$C_0 = 1 \cdot 40 (SD_d)^2 + 2020$	132	0·742
	A	Anvil 1	$C_0 = 1\,038 + 332\,774/A$	20	0·841
		Anvil 2	$C_0 = 6\,735 + 200\,781/A$	38	0·827
		Union	$C_0 = 9\,205 + 172\,663/A$	86	0·795
		Logan	$C_0 = 6\,650 + 165\,948/A$	63	0·860
E_{uni}	S	01	$E_{uni} = -0 \cdot 009\,66 S + 0 \cdot 796$	11	−0·679
	A	Anvil 1	$E_{uni} = -0 \cdot 75 + 60 \cdot 08/A$	20	0·880
		Anvil 2	$E_{uni} = -0 \cdot 13 + 40 \cdot 61/A$	38	0·926
		Union	$E_{uni} = 0 \cdot 21 + 28 \cdot 95/A$	86	0·877
		Logan	$E_{uni} = -0 \cdot 000\,7 + 28 \cdot 80/A$	64	0·940
$\log E_{uni}$	A	BM3	$\log E_{uni} = -0 \cdot 011\,5 A + 0 \cdot 173$	143	−0·648
	D	BM3	$\log E_{uni} = 1 \cdot 23 D - 2 \cdot 65$	140	0·692
μ_{uni}	S	01	$\mu_{uni} = 0 \cdot 005\,64 S + 0 \cdot 093\,4$	11	0·610
	A	Union	$\mu_{uni} = 0 \cdot 215\,3 + 0 \cdot 002\,9 A$	84	0·531
		Logan	$\mu_{uni} = 0 \cdot 195\,4 + 0 \cdot 003\,9 A$	62	0·580

Mechanical properties from single-stage and multiple-stage triaxial tests

$\log E_{t50}$	D	BM3	$\log E_{t50} = 2 \cdot 57 D - 5 \cdot 38$	31	0·891
		CE706	$\log E_{t50} = 0 \cdot 908 D - 1 \cdot 95$	49	0·847
		4 holes	$\log E_{t50} = 1 \cdot 40 D - 2 \cdot 98$	80	0·776
$\log E_{sp}$	D	CE706	$\log E_{sp} = 1 \cdot 01 D - 2 \cdot 31$	49	0·855
ε_p	D	CE706	$\varepsilon_p = -0 \cdot 026\,6 D + 0 \cdot 079\,1$	49	−0·605
q	D	CE706	$q = 7320 D - 8220$	49	0·810
	D_d	BM3	$q = 8820 D_d - 11\,300$	48	0·773
C_p	A	Anvil 2	$C_p = 4 \cdot 871 - 0 \cdot 298 A$	38	0·497
		Union	$C_p = 6 \cdot 441 - 0 \cdot 056 A$	85	0·430
ϕ_p	A	Anvil 2	$\phi_p = 21 + 291/A$	28	0·794
		Union	$\phi_p = 24 + 181/A$	84	0·545

TABLE 6.14
Multiple Regression Equations

Dependent variable	Data set	Variables available and regression equation	Number of cases	Multiple R
Physical and index properties				
D	BM3	Variables: $A\,A^2\,A^3$	187	
		Equation: $D = -0.006\,23A - 0.875\times10^{-6}A^3 + 2.25$	187	0.756
A	BM3	Variables: $D\,D^2\,D_d S\,S^2\,(SD_d)\,(SD_d)^2$	180	
		Equation: $A = -141D + 121D_d + 0.037\,7S^2$		
		$\qquad -0.004\,48(SD_d)^2 - 0.432(SD_d) + 81.2$	180	0.873
Mechanical properties from acoustic methods				
E_{ac}	01A	Variables: $A\,A^2\,A^3 D\,D^2 D_d\,S\,S^2\,(SD_d)\,(SD_d)^2$	38	
		Equation: $E_{ac} = -0.153A + 0.001\,43A^2 + 4.81$	38	0.847
G_{ac}	01A	Variables: $A\,A^2\,A^3\,D\,D^2\,D_d\,S\,S^2\,(SD_d)\,(SD_d)^2$	38	
		Equation: $G_{ac} = -0.042\,2A + 0.522\times10^{-5}\,A^3 + 1.77$	38	0.800
Mechanical properties from uniaxial tests				
C_0	BM3	Variables: $A\,A^2\,A^3\,D\,D^2\,D_d\,S\,S^2\,(SD_d)\,(SD_d)^2$	132	
		Equation: $C_0 = 2.70(SD_d)^2 - 480S = 12\,000$	132	0.832
E_{uni}	BM3	Variables: $A\,A^2\,A^3\,D\,D^2\,D_d\,S\,S^2\,(SD_d)\,(SD_d)^2$	133	
		Equation: $E_{uni} = -2.06D^2 - 6.66D_d + 0.000\,110(SD_d)^2$		
		$\qquad -0.140A + 0.004\,45A^2 - 0.380\times10^{-4}A^3$		
		$\qquad + 17.9D - 13.3$	133	0.855

| μ_{uni} | 01 | Variables: $S \, S^2$ | 11 | |
| | | Equation: $\mu_{uni} = 0.031\,6S - 0.578 \times 10^{-3}S^2 - 0.147$ | 11 | 0.822 |

Mechanical properties from single-stage and multiple-stage triaxial tests

E_i	CE706	Variables: $D \, D^2 \, \sigma_3$	49	
		Equation: $E_i = 0.282D^2 - 0.381$	49	0.478
$\log E_{t50}$	BM3	Variables: $D \, D^2 \, \sigma_3$	31	
		Equation: $\log E_{t50} = 2.57D - 5.38$	31	0.891
	CE706	Variables: $D \, D^2 \, \sigma_3$	49	
		Equation: $\log E_{t50} = 0.219D^2 - 1.02$	49	0.854
	4 holes	Variables: $D \, D^2 \, \sigma_3$	80	
		Equation: $\log E_{t50} = 0.337D^2 - 1.53$	80	0.780
E_{sp}	CE706	Variables: $D \, D^2 \, \sigma_3$	49	
		Equation: $E_{sp} = 3.65D^2 - 13.5D - 0.199 \times 10^{-3}\sigma_3 + 12.9$	49	0.927
ε_p	CE706	Variables: $D \, D^2 \, \sigma_3$	49	
		Equation: $\varepsilon_p = -0.006\,63D^2 + 0.104 \times 10^{-4}\sigma_3 + 0.0473$	49	0.782
q	BM3	Variables: $A \, A^2 \, A^3 \, D \, D^2 \, D_d \, S \, S^2 \, (SD_d) \, (SD_d)^2 \sigma_3$	47	
		Equation: $q = 11\,000D^2 + 0.554(SD_d)^2 - 30\,800D$		
		$\qquad -10\,500D_d + 41\,000$	47	0.932
c_p	CE706	Variables: $D \, D^2$	35	
		Equation: $C_p = 0.547D^2 + 1.69$	35	0.430
ϕ_p	CE706	Variables: $D \, D^2$	35	
		Equation: $\phi_p = 3.84D^2 + 8.25$	35	0.334

sites: Anvil Points Mine in the Naval Oil Shale Reserve (Dolinar *et al.*, 1979), the Union Oil Shale Mine (Horino *et al.*, 1982) and Logan Wash Mine of Occidental Oil Shale Inc. (Horino *et al.*, 1981).

6.6.7. Multiple Regression Analysis

A stepwise multiple regression analysis (Draper and Smith, 1966) was performed. The equations formulated expressed the relationships between various properties and the 11 selected independent variables. It is generally considered that a multiple regression equation can be used for engineering predictions if the multiple correlation coefficient $R \geqslant 0.7$.

In the stepwise multiple regression, at each step an additional independent variable was brought into the equation. The variable selected for inclusion at any step was the one which would account for the greatest amount of variance unaccounted for by those variables already in the equation. As many of the 11 selected independent variables as reasonably possible were considered for inclusion in the equation. The reported equation is the one for which the inclusion of each independent variable had been significant at the 0.05 level. Furthermore, the equation was based on data from as many specimens as possible.

The multiple regression equations are shown in Table 6.14. Most equations reported have a multiple correlation coefficient >0.65. In the first line of the table for any dependent variable, all independent variables considered for inclusion as predictors are listed. The developed equation, the number of data sets used and the multiple correlation coefficient are listed in the second line. The order in which each independent variable appears in any reported equation is the same as the order in which they were brought into the equation. In this way, the relative order of importance of each independent variable as a predictor is indicated.

6.6.8. Discussion of Results

The results shown in Table 6.11 indicate that many properties of oil shale varied over wide ranges and many exhibited much dispersion, as evidenced by large coefficients of variation.

The results of Brazilian tension tests showed that the average tensile strength parallel to the bedding plane ($T_{\parallel} = 5.8$ MPa, 838 psi) was 1.5 times as large as that perpendicular to the bedding plane ($T_{\perp} = 4$ MPa,

577 psi). The average value of the modulus of rupture ($F = 23 \cdot 3$ MPa, 3377 psi) was four times as large as the average value of $T_{\parallel}$.

The average value for the unconfined compressive strength perpendicular to the bedding plane ($C_0 = 62$ MPa, 8999 psi) was $10 \cdot 7$ times as large as the average value of $T_{\parallel}$ and $15 \cdot 6$ times as large as the average value of $T_{\perp}$.

The average value of peak cohesion ($c_p = 27 \cdot 9$ MPa, 4050 psi) obtained from corehole CE706 was 45% of the average value of C_0 obtained from the three Bureau of Mines coreholes. The average residual angle of internal friction was larger than the average peak angle of internal friction.

Almost all the variables had positive coefficients of skewness. Thus, the extreme values of the observed frequency distributions were greater than the mean. Furthermore, most of the variables had positive coefficients of kurtosis. Thus, most of the observed frequency distributions were more peaked than a normal distribution. In this regard, investigations were made concerning the observed distributions of the four strength parameters determined from MST tests, and results have shown that these parameters were normally distributed (Chang and Bondurant, 1979; Bondurant, 1983; Chang and Bondurant, 1983).

The results shown in Table 6.12 indicated the existence of strong correlations between many properties and the selected independent variables.

The bivariate and multiple regression equations presented in Tables 6.13 and 6.14 can be used for the prediction of various properties. Some multiple regression equations have much improved correlation coefficients over the corresponding bivariate regression equations and are, therefore, more reliable predictive equations.

6.7. CONCLUSIONS

In this chapter the results of several studies on the engineering properties of oil shale have been presented. The primary study involved the statistical analysis of data on oil shale from zones other than the Mahogany Zone and from various locations, including the deep, thick deposits at the depositional center of the Piceance Creek Basin. Included in this statistical analysis were data on the Mohr–Coulomb strength parameters of oil shale obtained from multiple-stage triaxial tests. The general statistical and distributional data presented may be useful as a valuable 'handbook' to engineers involved in various early stages of planning and feasibility studies for all four stages of oil shale recovery. The correlation coefficients and the regression equations presented herein may be useful for the prediction of various engineering properties from some easily obtainable or readily available properties. Correlation coefficients between density and the Mohr–Coulomb strength parameters and density and mechanical properties are very low. Further research is needed to identify other significant predictors. Whenever substantial additional data become available, the statistical analysis and modeling should be performed to upgrade the regression equations.

Results of a study to investigate the validity of the MST test in determining the Mohr–Coulomb strength parameters of oil shale have also been presented. The MST testing technique, especially MST-II, has been identified as being extremely reliable for evaluating Mohr–Coulomb strength parameters of oil shale. Further research should be conducted to: (1) standardize testing procedures; (2) use the MST-II testing technique to evaluate the Mohr–Coulomb strength parameters of oil shale of various grades and mineral contents; (3) formulate better predictive equations for the strength parameters; and (4) extend the testing technique to other rock types.

Results of studies on the anisotropic mechanical properties and thermal mechanical properties of oil shale have also been presented. Substantial research efforts are still required in these areas.

In summary, a successful oil shale project requires a sound data base on the engineering properties of oil shale. Much work has been done, yet the necessary data on mechanical properties are still scarce. Before our understanding of the mechanical properties of oil shale and their effects on the performance of various recovery schemes is substantially improved, any full-scale oil shale venture will be exceedingly costly.

ACKNOWLEDGMENTS

The research performed at the UCD was sponsored by the US Department of Energy and the National Science Foundation under grant nos DE-FG01-78ET-12226 and ENG77-06438, respectively. The Colony Mine and the Marathon Oil Co., Denver Research Center provided oil shale specimens for the above research projects. The Denver Mining Research Center of the US Bureau of Mines provided a large amount of the data included in the statistical analysis and modeling of the engineering properties of oil shale performed at UCD. Colleagues from the Bureau of Mines also conducted statistical research and provided additional information for inclusion in this chapter.

REFERENCES

Abel, J. F., Jr. and Hoskins, W. N. (1976). Confined core pillar design for Colorado oil shale. *Colorado School of Mines Quart.*, **71**(4) (October), 287–308. (*Proc. 9th Oil Shale Symposium.*)

Agapito, J. F. T. (1972). Pillar design in competent bedded formations. PhD Thesis, Colorado School of Mines, Golden.

Agapito, J. F. T. (1974). Rock mechanics applications to the design of oil shale pillars. *Mining Engineering*, **26**(5), 20–5.

Agapito, J. F. T. and Page, J. B. (1976). A case study of long-term stability in the Colony Oil Shale Mine, Piceance Creek Basin, Colorado. *Proc. 17th US Symposium on Rock Mechanics*, University of Utah, Salt Lake City, pp. 3A4.1–3A4.6.

Andrade, E. N. da C. (1910). Viscous flow in metals. *Proc., Roy Soc.*, **A-84**, 1–12.

Atkinson, R. H. (1972) A cubical test cell for multiaxial testing of materials,. PhD Thesis, University of Colorado at Boulder.

Atkinson, R. H. (1977). Results of the multi-stage triaxial technique applied to coal measure rock. Internal report, University of Colorado at Boulder.

Bieniawski, Z. T. (1967). Mechanism of brittle fracture of rock. *CSIR Report MEG 580*, National Mechanical Engineering Research Institute, Pretoria, South Africa.

Bobrov, G. F. (1970). Anisotropy and rheological properties of Kuzbass coal. *Sov. Min. Sci.*, **2**, 159–63.

Bondurant, E. J. (1983). Physical, mechanical, and strength properties of oil shale. MSc Thesis, University of Colorado.

Bondurant, E. J. and Chang, N.-Y. (1980). Statistical analysis and modeling of the physical, mechanical, and strength properties of oil shale. *Proc. 21st US Symposium on Rock Mechanics*, University of Missouri at Rolla.

Brady, B. T., Hooker, V. E. and Agapito, J. F. T. (1975). Laboratory and in situ mechanical behavior studies of fractured oil shale pillars. *Rock Mechanics*, **7**(2) (June), 101–20.

Chang, N.-Y. (1980). Validity of multiple-state triaxial tests for oil shale strength characterization. Final report to National Science Foundation for grant no. ENG 7706438.

Chang, N.-Y. (1982). Effect of elevated temperatures on uniaxial creep of oil shale. Final report to Department of Energy for grant no. ET-78-G-01-3033, University of Colorado at Denver.

Chang, N.-Y. and Bondurant, E. J. (1979). Oil shale strength characterization through multiple stage triaxial tests. *Proc. 20th US Symposium on Rock Mechanics*, Austin, Texas, pp. 393–401.

Chang, N.-Y. and Bondurant, E. J. (1983). Statistical modeling of physical and engineering properties of oil shale. Final report to National Science Foundation for grant no. ENG 7706438, University of Colorado at Denver.

Chang, N.-Y. and Jumper, A. L. (1978). Multiple-stage triaxial test on oil shale. *Proc. 19th US Symposium on Rock Mechanics*, pp. 520–2.

Chenevert, M. E. and Gatlin, C. (1965). Mechanical anisotropies of laminated sedimentary rocks. *J. Soc. Petrol. Engineers*, **5**(1) (March), 67–77.

Chong, K. P. (1983). Shear modulus of transversely isotropic materials. *Applied Math. & Mechanics*, **4**.

Chong, K. P., Smith, J. W., Chang, B., Hoyt, P. M. and Carpenter, H. C. (1976). Characterization of oil shale under uniaxial compression. *Proc. 18th US Symposium on Rock Mechanics*, University of Utah, Salt Lake City, pp. 5C5.1–5C5.8.

Chong, K. P., Smith, J. W., Roine, S. and Chang, B. (1977). Mechanical properties of oil shale by modified split cylinder testing. *Proc. 18th US Symposium on Rock Mechanics*, Colorado School of Mines, Golden, pp. 5B3.1–5B3.5.

Chong, K. P., Smith, J. W. and Khaliki, B. (1978). Creep and relaxation of oil shale. *Proc. 19th US Symposium on Rock Mechanics*, pp. 414–22.

Chong, K. P., Smith, J. W., Chang, B. and Roine, S. (1979*a*). Oil shale properties by split cylinder method. *ASCE, GT Journal*, **105**(GT5), 595–611.

Chong, K. P., Uenishi, K., Munari, A. C. and Smith, J. W. (1979*b*). Three dimensional characterization of the mechanical properties of Colorado oil shale. *Proc. 20th US Symposium on Rock Mechanics*, Austin, Texas, pp. 369–79.

Chu, M.-S. and Chang, N.-Y. (1980). Uniaxial creep of oil shale under elevated temperatures. *Proc. 21st US Symposium on Rock Mechanics*, University of Missouri at Rolla.

Cruden, D. M. (1971*a*). The form of the creep law for rock under uniaxial compression. *Int. J. Rock Mech. Min. Sci.*, **8**(2) (March), pp. 105–26.

Cruden, D. M. (1971*b*). Single-increment creep experiments on rock under uniaxial compression. *Int. J. Rock Mech. Min. Sci.*, **8**, 127–42.

Dolinar, D. R., Horino, F. G. and Hooker, V. E. (1979). Mechanical properties of oil shale and overlying strata, Naval Oil Shale Reserve, Anvil Points, Colorado, *US Bureau of Mines Progress Report 10024*, USBM.

Dorn, J. E. (1955). Some fundamental experiments on high temperature creep. *J. Mech. Phys. Solids*, **3**.

Draper, N. R. and Smith, H. (1966). *Applied Regression Analysis*, John Wiley and Sons, New York.

DuBow, J., Nottenburg, R., Rajeshwar, K. and Wang, Y. (1978). The effects of moisture and organic content on the thermophysical properties of Green River oil shale. *Proc. 11th Oil Shale Symposium*, ed. J. H. Gary, Colorado School of Mines, pp. 350–63.

Durham, W. B., Goetze, C. and Blake, B. (1977). Plastic flow of oriented single crystals of olivine, 2, observations and interpretations of the dislocation structures. *J. G. Res.*, **82**(36), 5755–70.

Duvall, W. I. (1976). General principles of underground opening design in competent rock. *Proc. 17th US Symposium on Rock Mechanics*, University of Utah, Utah Engineering Experiment Station, pp. 3A1.1–3A1.11.

Dyni, John R. (1974). Stratigraphy and nahcolite resources of the saline facies of the Green River Formation in Northwest Colorado. *Energy Resources of the Piceance Creek Basin, Colorado: Guidebook for the 25th Field Conference Rocky Mountain Assoc. Geologists*, ed. D. K. Murray, RMAG, Denver, pp. 111–12.

Griggs, D. T. (1936). Deformation of rocks under high confining pressures—1: Experiments at room temperature. *J. Geol.*, **44**(5) (July–August), 541–77.

Griggs, D. T. (1939). Creep of rocks. *J. Geol.*, **47**(3) (April–May), 225–51.

Griggs, D. T. (1940). Experimental flow of rocks under conditions favouring recrystallisation. *Geol. Soc. Am. Bull.*, **51**, 1001–22.

Gulf Oil Corp. and Standard Oil Co. of Indiana (1976a). *Rio Blanco Oil Shale Project, Progress Report No. 5—Summary, Sept., Oct, Nov. 1975, Tract C-a Oil Shale Development*. Submitted to area oil shale supervisor, USGS.

Gulf Oil Corp. and Standard Oil Co. of Indiana (1976b). *Rio Blanco Oil Shale Project, Detailed Development Plan, Tract C-a*. Submitted to area oil shale supervisor, USGS.

Gulf Oil Corp. and Standard Oil Co. of Indiana (1977). *Rio Blanco Oil Shale Project, Revised Detailed Development Plan, Tract C-a*. Submitted to the area oil shale supervisor, USGS.

Hardy, H. R. (1959). Time-dependent deformation and failure of geologic materials. *Proc. 3rd Symposium on Rock Mechanics*, Golden, Colorado, pp. 135–75.

Hardy, M. P. and Agapito, J. F. T. (1975). Pillar design in underground oil shale mines. *Proc. 16th US Symposium on Rock Mechanics*, University of Minnesota at Minneapolis, pp. 257–66.

Hawk, D. J. (1979). A study of orthotropy of coal and other rock materials. MS Thesis, University of Colorado at Boulder.

Hendron, A. J., Jr. (1968). Mechanical properties of rock, *Rock Mechanics in Engineering Practice*, eds K. G. Stagg and O. C. Zienkiewicz, John Wiley and Sons, London, pp. 21–53.

Heuze, F. E. (1967). Mechanical properties and *in situ* behavior of the Chino Limestone. MS Thesis, University of California at Berkeley, 165 pp.

Hite, R. J. and Dyni, J. R. (1967). Potential resources of dawsonite and nahcolite in the Piceance Creek Basin, Northwest Colorado. *Colorado School of Mines Quart.*, **62**(3) (July), 25–38.

Hobbs, D. W. (1966). A study of the behaviour of a broken rock under triaxial compression and its application to mine roadways. *Int. J. Rock Mech. Min. Sci.*, **3**(1), 11–43.

Horino, F. G. and Hooker, V. E. (1971). The mechanical properties of oil shale and *in situ* stress determinations, Colony Mine. *US Bureau of Mines Progress Report DMRC 10001.*

Horino, F. G. and Hooker, V. E. (1976a). The mechanical properties of oil shale, Union Oil Company of California, Garfield County, Colorado. *US Bureau of Mines Progress Report 10015.*

Horino, F. G. and Hooker, V. E. (1976b). The mechanical properties of oil shale, White River Oil Shale Corp., Uinta County, Utah. *US Bureau of Mines Progress Report 10014.*

Horino, F. G. and Hooker, V. E. (1978). Mechanical properties of cores obtained from the unleached saline zone, Piceance Creek Basin, Rio Blanco County, Colorado. *US Bureau of Mines Report RI-8297.*

Horino, F. G., Dolinar, D. R. and Bickel, D. L. (1981). Mechanical properties, *in situ* stress and temperature measurements, Logan Wash Oil Shale Site. *US Bureau of Mines Progress Report 10025.*

Horino, F. G., Dolinar, D. R. and Tadolini, S. C. (1982). Mechanical properties of the Mahogany Zone at the Union Oil Shale Mine, Garfield County, Colorado. *US Bureau of Mines Progress Report 10031.*

Hoskins, W. N., Wright, F. D., Tobie, R. L., Bills, J. B., Upadhyay, R. P. and Sandberg, C. B. (1975). A technical and economic study of candidate underground mining systems for deep, thick oil shale deposits, Phase 1 report, *US Bureau of Mines Contract No. S0241074* by Cameron Engineers Inc., OFR 23-76.

Hoskins, W. N., Upadhyay, R. P., Bills, J. B., Sandberg III, C. R., Wright, F. D. and Tobie, R. L. (1976). A technical and economic study of candidate underground mining systems for deep, thick oil shale deposits, final report. *US Bureau of Mines Contract No. S0241074* by Cameron Engineers Inc., OFR 9-77.

International Society for Rock Mechanics (1978). (Commission on Standardization of Laboratory and Field Tests.) Suggested method for determination of the triaxial compressive strength of rock materials using continuous failure state triaxial tests, final draft.

Jaeger, C. (1972). *Rock Mechanics and Engineering*, Cambridge University Press, UK.

Jaeger, J. C. and Cook, N. G. W. (1976). *Fundamentals of Rock Mechanics*, 2nd Edn, Chapman and Hall, London, 585 pp.

Kidybinski, A. (1966). Rheological models of upper Silesian carboniferous rocks. *Int. J. Rock Mech. Min. Sci.*, **3**(4), 279–306.

Kim, M. M. (1977). Multistage triaxial testing. MS Thesis, University of Colorado at Boulder.

Kim, M. M. and Ko, H.-Y. (1979). Multistage triaxial testing of rocks. *Geotechnical Testing J.*, *ASTM*, **2**, 98–105.

Kovari, K. and Tisa, A. (1975). Multiple failure state and strain controlled triaxial tests. *Rock Mechanics*, **7**(1) (March), 17–33.

Lama, R. D. and Vutukuri, V. S. (1978). *Handbook on Mechanical Properties of Rocks*, Vol. III, Trans Tech Publications, Clausthal-Zellerfeld.

Lambe, T. W. and Whitman, R. V. (1969). *Soil Mechanics*, John Wiley and Sons, New York.

Langer, M. (1966). Grundlagen einer theoretischen Gebirgskorpermechanik. *Proc. 1st Cong. Int. Soc. Rock Mech.*, Vol. 1, Lisbon, pp. 277–82.

Langer, M. (1969). Rheologie der Gesteine. *Duet. Geol. Ges. Z.*, **119** (August), 313–425.

Lewis, W. E. and Tandanand, S. (1974). Bureau of Mines test procedures for rocks. *US Bureau of Mines IC 8628.*

Loonen, H. E. and Hofer, K. H. (1964). *Proc. 6th Meeting Int. Bur. Rock Mech.*, Leipzig.

McLamore, R. and Gray, K. E. (1967). The mechanical behavior of anisotropic sedimentary rocks. *Trans. ASME (J. Eng. Ind.)*, **89**(1) (February), 62–76.

McVetty, P. G. (1943). Creep of metals at elevated temperatures—the hyperbolic sine relation between stress and creep rate. *Trans. ASME*, **65**.

Miller, I. and Freund, J. E. (1965). *Probability and Statistics for Engineers*, Prentice-Hall, Englewood Cliffs, New Jersey.

Miller, R. J., Wang, F.-D. and DuBow, J. (1978). Mechanical and thermal properties of oil shale at elevated temperatures. *Proc. 11th Oil Shale Symposium*, ed. J. H. Gary, Colorado School of Mines Press, Golden, pp. 135–46.

Miller, R. J., Wang, F.-D., Sladek, T. and Young, C. (1979). The effect of *in situ* retorting on oil shale pillars. *Colorado School of Mines Contract No. H0262031*, interim report.

Misra, A. K. and Murrell, S. A. F. (1965). An experimental study of the effect of temperature and stress on the creep of rocks. *Geophys. J. Roy. Astr. Soc.*, **9**(5) (July), 509–35.

Moore, Fred E. (1950). Authigenic albite in the Green River oil shales. *J. Sedimentary Petrology*, **20**(4) (December), 227–30.

Morris, C. E. (1978). Elastic constants of oil shale. *Explosively Produced Fracture of Oil Shale*, compiled by W. J. Carter, Los Alamos Scientific Lab., New Mexico (LA-7164-PR).

Nakamura, S. T. (1949). On visco-elastic medium. *Sci. Rep. Tokyo Univ., 5th Series: Geophysics*, **1**(2), 91–5.

Nie, N. H., Hull, C. H., Jenkins, J. G., Steinbrenner, K. and Bent, D. H. (1975). *SPSS Statistical Package for the Social Sciences*, 2nd Edn, McGraw-Hill, New York.

Norton, F. H. (1929). *The Creep of Steel at High Temperatures*, McGraw-Hill, New York.

Obert, L. and Duvall, W. I. (1967). *Rock Mechanics and the Design of Structures in Rock*, Wiley, New York.

Olsson, W. A. (1980). Stress-relaxation in oil shale. *Proc. 21st US Symposium on Rock Mechanics*, University of Missouri at Rolla.

Peng, S.-D. (1970). Fracture and failure of Chelmsford granite. PhD Thesis, Stanford University.

Penny, R. K. and Marriott, D. L. (1971). *Design for Creep*, McGraw-Hill (UK) Ltd, London.

Raju, N. Murty (1961). Elastic, static, and dynamic behaviour of layered Rifle oil shale and coal. MSc Thesis No. 941, Colorado School of Mines.

Robb, W. A. and Smith, J. W. (1974). Mineral profile of oil shales in Colorado Core Hole No. 1, Piceance Creek Basin, Colorado. *Energy Resources of the Piceance Creek Basin, Colorado: Guidebook for 25th Field Conference Rocky Mountain Assoc. Geologists*, ed. D. K. Murray, RMAG, Denver, pp. 91–100.

Robb, W. A. and Smith, J. W. (1977). Mineral and organic relationships through Colorado's Green River Formation across its saline depositional center. *Proc. 10th Oil Shale Symposium*, ed. J. B. Reubens, Colorado School of Mines Press, Golden, pp. 136–147.

Ruppeneit, K. V. and Libermann, Y. M. (1960). *Einfuhrung in die Gebirgsmechanik.*

Salustowicz, A. (1958). Rock as an elastic viscous medium. *Archiwum Gornictwa*, **3**, 141–72. (In Polish.)

Schmidt, R. A. (1977). Fracture mechanics of oil shale—unconfined fracture toughness, stress corrosion cracking, and tension test results. *Proc. 18th US Symposium on Rock Mechanics*, Colorado School of Mines Press, Golden, pp. 2A2.1–2A2.6.

Schmidt, R. A. and Schuler, K. W. (1974). Mechanical properties of oil shale from Anvil Points under conditions of uniaxial compression. *Sandia Report, No. SAND-74-0035.*

Schuler, K. W. and Schmidt, R. A. (1977). Mechanical properties of oil shale of importance to *in situ* rubblization. *Energy and Mineral Resource Recovery, ANS Topical Meeting, CONF-770440*, pp. 381–91.

Sellers, J. B., Haworth, G. R. and Zambas, P. G. (1972). Rock mechanics research on oil shale mining. *Trans. AIME*, **252** (June), 222–32.

Sinha, K. P., Borschel, T. F., Schatz, J. F. and Demu, S. (1982). Triaxial creep of oil shale and deformation of pillars in the *in situ* retorting environment. *Proc. 23rd US Symposium on Rock Mechanics*, University of California at Berkeley.

Smith, J. W. (1956). Specific gravity–oil yield relationships of two Colorado oil shale cores. *Ind. Eng. Chem.*, **48**, 441–4.

Smith, J. W. (1966). Conversion constants for Mahogany Zone oil shale. *Bull. Am. Assoc. Petrol. Geol.*, **50**(1) (January), 167–70.

Smith, J. W. (1969). Theoretical relationship between density and oil yield for oil shales. *US Bureau of Mines RT-7248.*

Smith, J. W. (1976). Relationship between rock density and volume of organic matter in oil shales. *Laramie Energy Research Center RI-76/6.*

Smith, J. W., Trudell, L. G. and Stanfield, K. E. (1968). Characteristics of Green River Formation oil shales at Bureau of Mines Wyoming corehole no. 1. *US Bureau of Mines RI-7172.*

Stanfield, K. E., Frost, I. C., McAuley, W. S. and Smith, H. N. (1951). Properties of Colorado oil shale. *US Bureau of Mines RI-4825.*

Sture, S. (1973). An improved multiaxial cubical cell and its application to the testing of anisotropic materials, MS Thesis, University of Colorado at Boulder.

Tisot, P. R. and Sohns, H. W. (1970). Structural response of rich Green River oil shales to heat and stress and its relationship to induced permeability. *Chem. Eng. Data*, **15**(3), 425–34.

Tisot, P. R. and Sohns, H. W. (1971). Structural deformation of Green River oil shale as it relates to *in situ* retorting. *US Bureau of Mines RI-7576.*

Volterra, V. (1959). *Theory of Functionals and of Integral and Integro-Differential Equations*, Dover Publications, New York.

Vutukuri, V. S., Lama, R. D. and Saluja, S. S. (1974). *Handbook on Mechanical Properties of Rocks*, Vol. I: Testing Techniques and Results, Trans Tech Publications, Clausthal-Zellerfeld.

Walpole, R. E. and Myers, R. H. (1972). *Probability and Statistics for Engineers and Scientists*, Macmillan, New York.

Windes, S. L. (1950). Physical properties of mine rock, Part II. *US Bureau of Mines RI-4727*.

Wright, F. D. and Bucky, P. B. (1949). Determination of room and pillar dimensions for the oil-shale mine at Rifle, Colorado. *Trans. AIME*, **181,** 352–9. (Also *AIME Tech. Publ. No. 2489.*)

Zeuch, D. H. (1982). The mechanical behavior of Anvil Points oil shale at elevated temperatures and confining pressures. *Sandia National Laboratories SAND-82-0164.*

Chapter 7

Stratigraphic Variations in Fracture Properties

CHAPMAN YOUNG

Sunburst Recovery Inc., Steamboat Springs, Colorado, USA

BRUCE C. TRENT

Department of Civil and Mineral Engineering, University of Minnesota, Minneapolis, USA

and

NANCY C. PATTI

Science Applications Inc., Steamboat Springs, Colorado, USA

SUMMARY

The prediction, control and evaluation of explosive oil shale fracturing require a detailed knowledge of tensile strength behavior as a function of shale grade and stratigraphic position. Direct-pull tensile tests, point-load pinch tests and four-point-bend fracture toughness tests have been utilized to develop detailed logs of the relevant fracture properties for the 37 m thick Mahogany Zone section of the Green River Formation near Anvil Points, Colorado and for the rich, upper 13 m of the Tipton Member near Rock Springs, Wyoming. Detailed statistical analyses were performed on these data and on Fischer assay oil yield data to establish the correlations between them. Data from both tensile strength and fracture energy tests correlate well with lithologic and oil yield charac-teristics of the Mahogany Zone shale while poor correlations are found for the Tipton shale.

The results of the fracture property testing on oil shale were incorpo-rated into an explicit finite-difference wave propagation code to study explosive-induced fracture characteristics for the purpose of void volume generation prior to retorting. The fracture model provided for crack pressurization by the explosive gases and concordant crack opening and growth. A special time-step scaling technique was utilized so that calculations were carried out in a quasi-static mode for up to 1 s. The presence of gas pressure and relatively weak bedding plane partings influenced the ultimate three-dimensional fracture distribution.

7.1. INTRODUCTION

With few exceptions the methods being presently considered for the exploitation of oil from oil shale require extensive fracturing and fragmentation of the rock. Where the shale is to be processed by surface retorting, the rock must be mined and crushed. Although relatively conventional mining and crushing operations can be employed, some consideration must be given to the specific fracturing properties of oil shale for process optimization. Many oil shale processing technologies are designed to fracture and retort the rock *in situ*. The *in situ* processing methods are considered as being either true *in situ*, wherein the rock is adequately fractured by the placement of explosives in wellbores or hydraulic type fractures, or modified *in situ*, wherein mining operations are employed to provide an initial void volume into which the rock may be explosively fractured. Several research programs have been conducted to evaluate various true *in situ* approaches with the results being generally disappointing (Stevens *et al.*, 1975; Beasley and Boade, 1980; Parrish *et al.*, 1980). At present, serious consideration is only being given to modified *in situ* technologies with the void volume required for adequate fragmentation and permeability development being provided either by surface heave or partial mining (Ridley, 1978). Successful *in situ* retorting of oil shale requires that the fractured and fragmented rock has adequately uniform and homogeneous permeability which in turn requires that the explosive fragmentation process is well controlled. The large stratigraphic variations in oil shale, especially with respect to the organic content of the rock, could play an important role in controlling the *in situ* fragmentation process. If *in situ* fragmentation is to be controlled and optimized for the recovery of shale oil, the stratigraphic variations in oil shale fracture properties must be described and understood.

Although significant oil shale deposits exist in the eastern United States, most consideration has been given to the shales of the Green River Formation in Colorado, Wyoming and Utah. The Green River shales are found in three separate basins representing three distinct inland lakes or seas in which the rocks were deposited during the

Eocene Epoch of the Tertiary Period. Most considerations of true *in situ* processing techniques have been done for the Tipton Member of the Green River Formation in the Green River Basin of Wyoming. These research efforts were conducted by the Department of Energy at the Rock Springs field site (Stevens *et al.*, 1975; Parrish *et al.*, 1980) and by Talley Energy Systems Inc. at a site near Green River, Wyoming (Beasley and Boade, 1980). Most studies of modified *in situ* methods have been carried out in the Mahogany Zone portion of the Parachute Creek Member of the Green River Formation in the Piceance Creek Basin, Colorado (Ridley, 1978). A modified *in situ* technology involving near surface deposits and surface heave during the explosive fragmentation process has been extensively studied by Geokinetics Inc. in shales stratigraphically equivalent to the Mahogany Zone at a site 30 miles south of Vernal, Utah in the Uinta Basin (Lekas, 1979). Because of the interest in oil shale processing in these three basins, it is desirable to have sufficient data on the fracture properties of rock from each of these three areas. Core material was obtained by the Laramie Energy Technology Center of the US Department of Energy for the specific purpose of obtaining data on the mechanical properties of oil shale and their stratigraphic variations at each of these areas.

An extensive data base exists on the compressive strength and elastic properties of most oil shales of interest (Johnson and Simonson, 1977; Olinger, 1977; Chong *et al.*, 1980). In addition, a high degree of correlation has been found between these properties and the organic content or oil yield of the rock. A complete data base does not exist for the tensile or fracture properties of oil shale. Since the tensile fracture of the rock would be expected to be controlled by subtle variations in mineralogy (especially the clay mineralogy) it would not be expected that these properties would be as readily correlated to organic content. To the extent that tensile fracture properties cannot be correlated to other rock properties such as organic content these properties must be described on a site specific and stratigraphically controlled basis.

Many standard techniques exist for determining the tensile and fracture properties of a variety of materials including rocks. Most of these techniques, however, are designed for obtaining high precision data on a relatively small number of samples and require significant sample preparation. As a proper description of the stratigraphic distribution of oil shale fracture properties would require an unusually large number of tests, it would be desirable if not necessary to develop modified test procedures that could be performed on a production basis with a minimal amount of sample preparation effort. Due to the inherent variability of the mechanical properties of rocks, and especially the fracture properties, it is not necessary for such modified techniques to provide the degree of precision often required in mechanical property testing. For the fracture properties of oil shale it is more meaningful to have a large, stratigraphically distributed data base than a few unnecessarily precise data points.

Laboratory data on rock properties are generally utilized in predicting some future response. Such predictions can take the form of empirical data related to specific engineering problems. For example, by knowing the length and width of an underground opening along with the depth and thickness of a particular seam, reasonable estimates of surface subsidence can be made based upon the observed results of numerous case studies. Predictions are also made by analytical methods, which utilize closed-form solutions to arrive at specific practical predictions. Simple heat transfer, fluid flow and linear-elastic deformation responses may often be addressed adequately with these types of techniques. Very often the complexity of the engineering problem under consideration or the local and global *in situ* environments make the empirical and analytical approaches inadequate or of limited value. The rapid advancements realized in the last two decades in large digital computers have provided new cost-effective tools which can be used to solve problems which previously would have been impossible. Large computer programs based upon explicit finite-difference techniques can now treat in detail the time-dependent solution of complicated and non-linear rock mechanics problems. An excellent example of the application of these programs is the simulation of the explosive fragmentation of rock masses by the detonation of explosive charges placed in shot holes. The proper execution of these programs requires that suitably detailed and site specific data on the rock mechanical properties be available. The results of explicit finite-difference calculations can demonstrate the relative importance or parameter sensitivity of various rock properties and the potentially important influences of stratigraphically distributed properties.

7.2. MODELING METHODS

Digital computers have been developed to the point where complicated problems may now be solved with the use of a variety of

Stratigraphic variations in fracture properties 295

numerical techniques. The finite-element method (FEM) is an approximation method for studying continuous physical systems. The system is modeled by being broken down into discrete elements that are interconnected at distinct node points (Trent and Langland, 1983). Mathematically, the solutions are obtained by solving a number of system variables at the node points that represent a minimization of an integral function, such as potential energy, summed over all the discrete elements. The explicit finite-difference method (EFD) is an approximation method in which the difference between the values of a function at two discrete points is used to approximate a system. Mathematically, the system variables at the discrete points are numerical solutions to the differential equations governing the system performance. These differential equations are represented by difference equations that use the system variables at the node points.

There is a mathematical equivalence between the minimization of an integral functional and the associated governing differential equation. However, the numerical solution techniques are significantly different. It is usually easy to formulate coupled processes (such as thermal, mechanical and chemical) with differential equations and, therefore, with finite-difference techniques. Obtaining solutions to the resulting equations can be costly and tricky. An equivalent integral formulation of the same processes is more difficult to obtain, but solving the resulting set of implicit equations can often be easy and more efficient.

The selection of one solution method over another often depends on the specific process being represented.

7.2.1. Input Data Requirements for EFD and FEM
The basic input data required for each modeling technique are the same: modulus of elasticity, Poisson's ratio, density, mechanical strength and boundary conditions. Thermal modeling requires conductivity, emissivity and thermal boundary conditions. Solutions using the FEM can be either implicit or explicit. The same is true of finite-difference techniques, but the EFD techniques reported here are explicit and therefore require special solution techniques such as density scaling that are discussed in the sections that follow. The major advantage of the FEM for quasistatic and static problems is that the solutions are implicit, and a straightforward, once-through Gaussian elimination solution scheme is usually required. Special techniques such as density scaling and special 'artificial' damping are not required.

The STEALTH (Hofmann, 1976) system of computer codes is EFD-based and uses a Lagrangian formulation. The spatial frame of reference at any time is constantly moving, reflecting updated zonal or grid point values. The reader is referred to any basic text on Lagrangian methods for further information. The motion in an EFD system is determined by solution of equations for the conservation of momentum, mass and energy in three principal directions.

Initial conditions give rise to forces exerted on grid points. An average mass (considering the four quadrilateral zones adjacent to any grid point) may be determined and a finite increment of acceleration of the grid point in question may be readily calculated through conservation of momentum. With the current time step (explained in greater detail below), an explicit time integration may be performed twice to obtain the new grid-point velocity and position. Since the old positions are known, a constitutive law may be applied to calculate increments to the stress tensor. This process leads to new forces and the cycle begins again.

The EFD calculational sequence is illustrated in Fig. 7.1. Note the dotted box around the strain and stress tensor blocks. It is here that

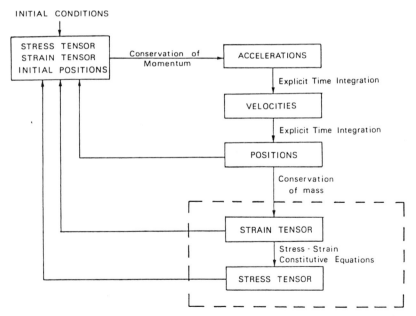

Fig. 7.1. Generalized calculational cycle for the STEALTH explicit finite difference programs.

any special constitutive laws (such as tensile failure, compaction, etc.) are inserted into the code. The special constitutive laws developed for the tensile fracture of oil shale consist of several thousand card images and are described in greater detail in Sections 7.5.1 and 7.5.2. Special consideration must be given to the order that physical phenomena are represented in time (tensile failure, joint slip, plasticity, creep, etc.).

The STEALTH solutions are explicit in both space and time. At any grid point, center spatial differencing allows only the four adjacent zones to influence its motion in any cycle; thus solutions are spatially explicit. Only adjacent zones influence the momentum calculations for a given point. A backward temporal differencing technique only allows previously determined values to govern motion in the current cycle; thus, solutions are explicit in time.

The time step for a purely dynamic problem is governed by the minimum transit time across any calculational zone. This transit is defined by:

$$dt = L/C_l \tag{7.1}$$

where C_l is the longitudinal wave speed of the material and L is a characteristic length of that zone. Clearly, each different zone will have a specific length dimension, and in general the sound speed will vary with the material properties.

7.3. EXPERIMENTAL METHODS

Numerous experimental methods have been developed and evaluated for determining the tensile strength and fracture toughness properties of rocks. Although some of these methods have become standardized, it has been necessary to develop methods especially applicable to the testing of oil shale. Three basic types of tests have been developed for determining the stratigraphic variations in oil-shale fracture properties. As the tensile strength is the most basic description of a rock's fracture behavior, a standardized direct-pull tensile strength method was modified for application to the oil shales of interest. Direct-pull tensile strength tests provide data on the weakest zones in a rock. Such tests cannot provide data on the stronger rock between weaker zones. An indirect tensile test method could provide data on the stronger portions of the rock and could be applied more readily to rock samples of limited dimensions. Due to the difficulty in preparing specimens for the more traditional Brazilian indirect tensile strength test, a point-load test method was adopted and developed for application to oil shale. While tensile strength values can provide a good measure of the rock's ability to support certain applied stresses, tensile strength data alone are not suitable for fracture mechanics calculations of fracture growth and propagation. Consequently, a simple fracture energy test was modified so that fracture energy values could be readily obtained on a production basis and on relatively small oil shale samples.

7.3.1. Direct-Pull Tensile Tests

In determining the stratigraphic variations in tensile strength of an oil-shale section it is necessary to have data both on the stratigraphic occurrence of bedding planes, parting planes and other weak zones and on the tensile strength of the more competent rock lying between the planes of easy separation. Once the naturally occurring parting planes have been logged then the location and strength of weak bedding planes can be best determined by direct-pull tensile tests. Although a standard tensile testing procedure (ASTM, 1972) has been established for rocks, it was necessary to modify this procedure for application to oil shale. The most significant modification involved the use of square prism specimens required to minimize desiccation and cracking of the samples, as discussed in Section 7.3.4. For some of the Tipton tests and all of the Colorado and Utah Mahogany Zone tests, it was possible to utilize a 3·65 cm diameter core. Flexural loading of the specimens was minimized by using a modified universal joint arrangement, whereby two orthogonal axes of rotation served to keep the tensile loads oriented along the axis of the sample. Details of the universal-joint gimbals and the sample loading testing machine are given by Young *et al.* (1982).

Development of a system for gripping the sample so that multiple tensile breaks could be obtained was the most difficult task. This task was complicated for the Tipton shale by the need to maintain sample moisture during preparation and testing. Several Hysol (Hysol Division, Dexter Corp., Olean, NY, USA) epoxies and a cyanoacrylate cement (Super-Glu-3) were evaluated. Although glue joints of moderate strength were obtained with all of the glues tested, it was found that consistently strong joints with moist samples were only obtained with the epoxy Hysol 608. As this epoxy was clear, it was possible to observe that the epoxy had an excellent wetting effect on the clay-rich

Tipton shale. It is believed that this wetting effect, coupled with the very short 3-min curing time for this epoxy, was responsible for the excellent quality of the glue bonds. Multiple breaks, often with markedly increasing tensile strength values, were obtained by recementing each tensile break with the Hysol 608 epoxy. The fracture surfaces were cleaned of any loose chips and fragments by scratching the surfaces on an orthogonal, 1-cm grid spacing pattern with a sharp knife point.

For some of the testing on the Tipton shale and all of the testing on the Mahogany Zone shale from Colorado and Utah, test specimens were made by coring 3·65 cm diameter core specimens out of the 8·9 and 15·2 cm diameter core pieces obtained for this purpose. For performing tensile tests on the 3.65 cm diameter material the aluminum gripping blocks were remachined so that a 60° included-angle cone served as a transition between the 3·65 cm diameter gripping surface and the 5·1 cm square bearing end of the block. The higher strength values noted for the Geokinetics and especially the NOS 26 shales as compared to the Tipton shale also required that higher strength epoxies be employed in the direct-pull tensile strength tests. Two other Hysol epoxies were found to be quite suitable for attaching dry oil shale samples to the aluminum grip blocks and to recementing fractures so that multiple fracture tensile strength data could be obtained. Hysol 615 epoxy was found to be quite satisfactory for tensile strengths up to 10 MPa (1450 psi). For unusually strong oil shale test specimens (such as regularly encountered in the NOS 26 core material) Hysol 907 was found to be the most satisfactory. This epoxy provided adequate adhesion of the oil shale to the aluminum grip blocks for tensile strength measurements up to 16 MPa (2320 psi). Both the Hysol 907 and 615 epoxies required that the samples be dry for adequate adhesion to the rock to be achieved. In addition, the Hysol 907 epoxy required that the prepared sample be allowed to cure overnight at room temperature or that the sample and aluminum blocks be heated to 60°C for approximately 2 h.

By regluing and retesting the specimens, efforts were made to obtain at least one direct-pull tensile strength break per 3 cm of the Tipton shale. For the Colorado Mahogany Zone shale obtained in NOS corehole 26, approximately one tensile break per 5 cm of specimen was obtained and for the Geokinetics Mahogany Zone shale only one tensile break per 10 cm of specimen was obtained. For the Tipton, this usually required that 5–7 breaks be made for each 10 cm long test specimen and that unusually weak and friable zones be cut out of the test specimens once tensile data were obtained on them.

7.3.2. Indirect Tensile Strength Tests

Because of the difficulty in preparing and loading direct-pull tensile specimens, indirect testing methods are often used for evaluating the tensile strength properties of rocks. The Brazilian test, wherein a cylindrical disk of rock is loaded diametrically, is the most commonly utilized method (Jaeger and Cook, 1969). The Brazilian test is attractive because sample preparation is simple and because linear elastic analyses indicate that a large portion of the sample is subjected to a relatively uniform tensile stress. The Brazilian test is limited in that the inferred tensile strengths are usually calculated for an isotropic linear elastic material and any deviation of the rock from such ideal response (including localized plastic deformation where the load is applied to the sample circumference) will result in unknown deviations in the calculated stress. The stresses and strains in a transversely isotropic Brazilian test specimen have been calculated for the case where the axis of the disk is parallel to the axis of symmetry in the sample (Wang and Chong, 1980).

A second indirect tensile strength method, often referred to as a 'pinch' test, involves the loading to failure of a cylindrical rock specimen by two hardened-steel rollers oriented perpendicular to the cylindrical axis of the specimen (Reichmuth, 1963). As for the Brazilian test, the tensile stresses acting within the rock specimen are calculated utilizing isotropic linear elastic theory. The indirect tensile strength value, obtained with the pinch test is given by:

$$\sigma_t = 0 \cdot 96 \frac{P_{max}}{d^2} \qquad (7.1a)$$

where P_{max} is the load at failure and d is the sample diameter (Reichmuth, 1963). Again any non-linear non-isotropic response of the sample and especially any plastic deformation of the sample in contact with the hardened-steel rollers, will result in an unknown variation in the actual stress causing failure. In an effort to evaluate the importance of local plastic yielding where the rollers contact the sample, a series of experiments were performed on $1 \cdot 27$ cm diameter core material using hardened-steel rollers $0 \cdot 64$ and $0 \cdot 95$ cm in diameter. By using two parallel (stratigraphically identical) $1 \cdot 27$ cm diameter cores it was observed that variations in tensile strength along the

Stratigraphic variations in fracture properties 301

sample were comparable for both diameters of hardened-steel rollers. The larger, 0·95 cm rollers, however, did yield higher tensile strength values, indicating less plastic deformation of the sample during loading. All of the indirect tensile strength data reported upon here were obtained using 0·95 cm diameter hardened-steel rollers on the 1·27 cm diameter core material of the rock. In general, the indirect tensile strength tests were performed on ≈2·5 cm long segments of the 1·27 cm diameter core remaining after the fracture energy tests discussed below were performed.

7.3.3. Fracture Energy Tests

Numerous experimental techniques have been developed and evaluated for determining the fracture toughness of brittle materials such as rock. Fracture toughness values, whether expressed in units of energy per square surface area ($J\,m^{-2}$) or in terms of the critical stress intensity factor, K_{IC} ($N\,m^{-3/2}$) are a measure of the work required to propagate a given fracture. Small scale borehole pressurization experiments (Clifton *et al.*, 1976), pressure wedge splitting of 1·27 cm diameter V-notched specimens (Barker, 1977) and fatigue-crack initiation in three-point-bend specimens (Schmidt, 1976; Schmidt, 1977) have all been used successfully to measure fracture toughness values in a variety of rock types. Unfortunately, all of these methods require large volumes of sample material and/or extensive sample preparation procedures. For the large number of tests required to develop a stratigraphic log of fracture toughness values, a simple and readily interpreted test is required. Three-point-bend tests, wherein the work or energy to fracture a sample is of known dimensions, have been developed and extensively utilized on a variety of rock types (Friedman *et al.*, 1972; Cooper, 1977). The only requirement for performing such energy dissipation fracture experiments is that fracture propagation be stable throughout fracture growth so that the work represented by the load–displacement curve of the testing machine can be directly correlated to the energy of fracture. Cooper (1977) analyzed in detail the energy requirements for stable crack propagation. By carefully matching sample geometry to loading machine characteristics, Cooper was able to obtain stable crack propagation in brittle rocks such as Solenhofen limestone.

In order to expedite sample preparation and permit the use of small pieces of sample material, a jig for holding and loading a standard 1·27 cm diameter core piece was designed and built. Details of this test jig and its placement in the testing machine are given by Young *et al.* (1982). The 1·27 cm diameter pieces to be tested were notched so that the V-shaped segment of material remaining would provide both a sharp stress concentration for easy fracture initiation and an increasing fracture length with propagation for increased stability. As noted by Cooper (1977) this V-shaped notch is beneficial, although not always essential, for obtaining stable breaks. The fracture toughness tests were performed on 1·27 cm diameter core pieces up to 10 cm long. These cores were then V-notched at approximately 2·54 cm intervals so that, on average, three fracture toughness values were obtained for each 10 cm of core. The residual, approximately 2·54 cm long, core pieces recovered from the fracture toughness tests were then used for the indirect tensile strength tests discussed above. Sample drying and desiccation cracking of the water sensitive shales were minimized during both the fracture toughness and indirect tensile strength tests by carefully keeping the sample material moist at all times.

7.3.4. Sample Preparation and Handling

In obtaining the core material to be utilized for the fracture property testing, special precautions were taken to preserve the natural *in situ* quality of the rock. Most notably, the Tipton Member shale near Rock Springs, Wyoming had been observed in earlier cores to undergo significant desiccation shrinkage and cracking. So as to minimize the effects of desiccation cracking upon test results, a large diameter (15·2 cm) core was taken and was protected from desiccation by wrapping the core in polyethylene bags with a few deciliters of drilling mud. Although the Mahogany Zone shale from the Geokinetics site was not as moisture sensitive, comparable precautions were taken to protect this core which was also 15·2 cm in diameter. As the Mahogany Zone shale of the Piceance Basin has not been noted to display any moisture sensitive behavior, a smaller 8·9 cm diameter core was taken of this rock and techniques to prevent the core from drying were not required.

In order to maximize the amount of material available for testing and minimize the possibilities for desiccation and other mechanical and chemical degradation of the Tipton shale, most sample preparation was done with a diamond wire saw. The basic 15·2 cm core was first sectioned into pieces 10·2 cm long with the wire saw and these pieces were in turn cut so as to yield four 5·1 cm square prisms and four crescent shaped segments. When natural core breaks or tuff beds precluded cutting the 15 cm core into prisms then the required core segment was delivered to Colorado State University and test specimens were cored directly from the 15 cm core.

As the Geokinetics and Colorado NOS 26 Mahogany Zone shales did not contain extensive mixed-layer clays and were relatively insensitive to desiccation cracking, the extensive sample preparation procedures outlined above were not employed for these two rocks. Efforts were made to keep the original 15·2 cm diameter core obtained from the Geokinetics site moist during all phases of sample preparation. 3·65 cm diameter cores of this rock were prepared for the direct-pull tensile tests and a few smaller 1·27 cm diameter samples were prepared for fracture toughness and indirect tensile strength tests. All of these test specimens were kept moist up to the time of actual test execution. As the NOS 26 Mahogany Zone shale contained essentially no mixed-layer clays nor propensity for desiccation cracking, this core material was neither protected as it was recovered in the field nor were special procedures utilized in test specimen preparation to protect sample moisture content.

7.4. EXPERIMENTAL RESULTS

The fracture properties of the NOS 26 Mahogany Zone shale, the rich upper Tipton shale and the Geokinetics shale have been obtained utilizing the experimental methods discussed in the preceding section. These properties were obtained as continuously as possible so that stratigraphic logs were developed of the properties over the sections of interest. Such continuous sampling, coupled with careful handling of the sample material, provides a picture of fracture properties which is more representative of the *in situ* characteristics of the rock than would be obtained by more traditional approaches. Geotechnical and mechanical property data on rocks are usually obtained by testing a few (supposedly representative) samples of material obtained from coring or mining operations. Such limited sampling usually results in the tested material being representative of the more competent and environmentally inert portions of the rock. Zones which are fractured, friable and/or which may suffer degradation once the sample has been taken are often excluded from such selective testing. In the design and analysis of *in situ* explosive fragmentation methods for oil shale it is important that the weaker and more heterogeneous zones and beds be considered also.

7.4.1. NOS 26 Mahogany Zone Shale

All of the direct-pull tensile strength data obtained for the Mahogany Zone shale recovered from NOS 26 (US Naval Oil Shale Reserve no. 1, corehole 26, SE$\frac{1}{4}$ sec. 36, T5S, R95W, Garfield County, Colorado, USA), comprising 909 individual direct-pull tests, are illustrated in Fig. 7.2. The small squares represent the tensile strength values and depth for each test conducted over the interval 265 m (870 ft) to 302 m (990 ft). Several different averaging techniques have been evaluated for providing representative formation values. The continuous curve drawn

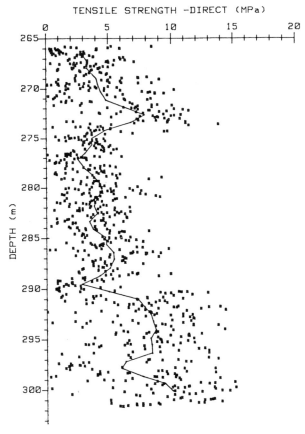

Fig. 7.2. Direct-pull tensile strength data for the 36·6 m Mahogany Zone shale recovered from NOS 26.

through the data in Fig. 7.2 represents a running average value of tensile strength tabulated for 1·52 m (5 ft) intervals calculated and plotted at 0·76 m (2·5 ft) increments of depth. While the raw data appears to be quite scattered the running average of this data clearly reveals zones of high and low average tensile strength. The significance of these variations in average tensile strength both in terms of stratigraphic distribution and of correlations with other shale properties will be discussed in further detail below.

All of the indirect tensile strength data, comprising 1079 data points obtained by the pinch test method, are illustrated in Fig. 7.3. As

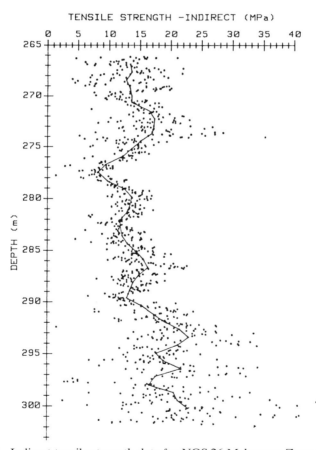

Fig. 7.3. Indirect tensile strength data for NOS 26 Mahogany Zone shale.

previously noted, the indirect tensile strength provides a better measure of average rock tensile properties than does the direct-pull test method as the latter samples preferentially the weakest portions of the rock. This aspect of the indirect tensile strength data is well illustrated by comparing Figs 7.2 and 7.3. The indirect tensile strength data display less scatter than do the direct-pull data and the measured tensile strength values are significantly higher. As discussed below, good statistical correlations exist between direct-pull and indirect tensile strength data with the indirect tensile strength being over twice as great than the direct-pull strength.

Comparable data, comprising 549 data points, were obtained on

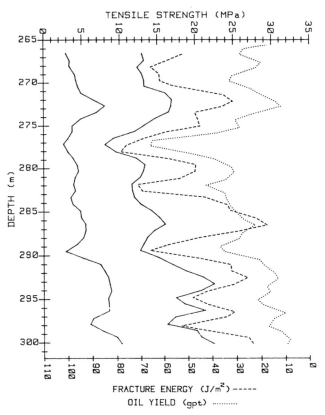

Fig. 7.4. Comparison of oil shale fracture properties and oil yield running averages for NOS 26 Mahogany Zone shale.

fracture energy. While the raw fracture energy data do display a great deal of variability or scatter, the running average reveals zones of distinct high and low fracture energy values. The range of fracture energies measured for the NOS 26 Mahogany Zone, ranging from $8\,\mathrm{J\,m}^{-2}$ to $147\,\mathrm{J\,m}^{-2}$, correspond well with fracture energy values obtained on other rocks (Cooper, 1977). As discussed in greater detail below, the fracture energy has a negative correlation with both the direct-pull and indirect tensile strength properties of the Mahogany Zone shale.

So as to obtain oil yield data with which the shale fracture properties could be directly correlated, Fischer assay oil yields were obtained on the NOS 26 core. The Fischer assays were performed on samples prepared from contiguous $0\cdot3048\,\mathrm{m}$ (1 ft) long sections of the $8\cdot9\,\mathrm{cm}$ diameter core remaining after the fracture property test specimens had been taken. The stratigraphic distribution of oil yield for the NOS 26 core correlates well with other Mahogany Zone oil yields in the southeast corner of the Piceance Basin.

Comparisons between the three fracture property values obtained and the oil yield for the NOS 26 Mahogany Zone are illustrated comparatively in Fig. 7.4. As both fracture energy and oil yield have negative correlations with the direct-pull and indirect tensile strength data, the scale for fracture energy and oil yield (which cover a comparable numerical value range) has been inverted in Fig. 7.4. The running average data presented in Fig. 7.4 suggests strongly that significant correlations should exist between the three fracture property values measured and oil yield.

7.4.2. Tipton Member—Rock Springs Test Site

All of the direct-pull tensile strength data for the rich upper 13 m (43 ft) of the Tipton Member (LETC field site, site 10 core, sec. 15, T18N, R106W, Sweetwater County, Wyoming, USA) of the Green River Formation are illustrated in Fig. 7.5. The continuous curve drawn through the data represents a running-average value of tensile strength over a $0\cdot3\,\mathrm{m}$ (1 ft) interval with the averages calculated at each $0\cdot15\,\mathrm{m}$ ($0\cdot5\,\mathrm{ft}$) of depth. Averages tried over smaller intervals tended to be too 'noisy' and averages over larger intervals (up to $1\cdot5$ m) tended to filter out details of geologic importance. The range in the $0\cdot3$ m running average value from $1\cdot9\,\mathrm{MPa}$ (275 psi) down to $0\cdot05\,\mathrm{MPa}$ (7 psi) is geologically significant and represents the *in situ* variations that should be expected for the upper Tipton. Numerous

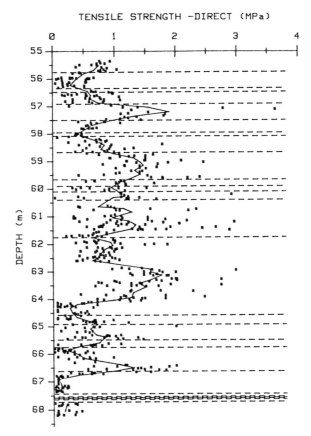

Fig. 7.5. Direct-pull tensile strength values obtained for the 55·3–68·3 m depth interval of the Tipton oil shale.

natural fracture or partings were either existent within the *in situ* formation or developed as the original 15 cm diameter core was being taken. These natural fractures, representing the weakest planes within the formation, are indicated in Fig. 7.5 by the horizontal dashed lines. The frequency of these natural partings suggests that they could play an important role in the fracturing and fragmentation of the Tipton Oil Shale by explosive loading.

Comparable data were also obtained on the indirect tensile strength and fracture energy for the Tipton. As for the NOS 26 Mahogany Zone shale, the indirect tensile strengths were significantly greater than the direct-pull strengths, a maximum value of 12·6 MPa being obtained. The range of fracture energy values, from a minimum of 5 J m^{-2} to a maximum of 87 J m^{-2}, bound the value of 24 J m^{-2} obtained for Solenhofen Limestone by Cooper (1977) and fall within the range expected for a rock like the Tipton oil shale. The correlations between the direct and indirect tensile strength values and the fracture energy values obtained on the upper 13 m of the Tipton oil shale are illustrated in Fig. 7.6 where 0·3 m running average curves for the three

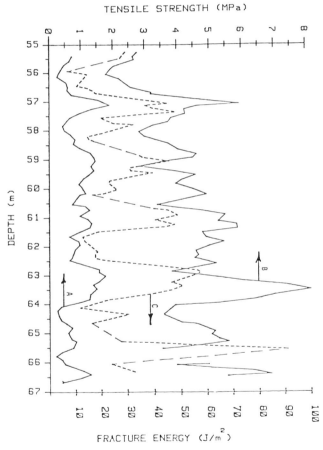

Fig. 7.6. Comparison of the running average curves of direct (A) and indirect (B) tensile strength and fracture energy (C) for the rich upper 11·6 m of the Tipton oil shale.

types of data are superimposed. Although there appears to be a better correlation between the fracture energy and direct-pull tensile strength data than between either of these data groups and the indirect tensile strength data, the statistical analyses discussed below indicate comparable correlations also exist between fracture energy and indirect tensile strength data.

7.4.3. Geokinetics Mahogany Zone Shale

Direct-pull tensile strength data from the third oil shale site studied (sec. 2, T14S, R22E, Uinta County, Utah, USA), the Geokinetics

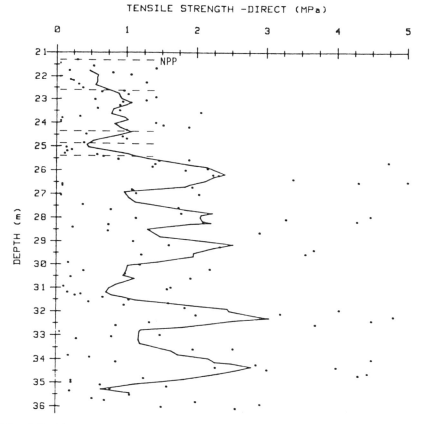

Fig. 7.7. Direct-pull tensile strength data for 14·6 m (48 ft) of the Geokinetics (Utah) Mahogany Zone shale, showing natural parting planes (NPP).

Stratigraphic variations in fracture properties 311

Mahogany Zone shale, is illustrated in Fig. 7.7. As the testing of the Geokinetics core involved only one direct-pull tensile break per 10 cm long test specimen, fewer data points were obtained and a correspondingly larger averaging interval of 0·9 m was employed for the running average curve shown in Fig. 7.7. Although the Geokinetics site encompasses the Mahogany Zone of the Green River Formation, the shale at that location is significantly different from the Mahogany Zone recovered in NOS 26. Due to an increase in clastic material and a proportionate reduction in precipitated carbonates, the Geokinetics shale is significantly more fragile than the NOS 26 Mahogany Zone shale. The tensile strength characteristics of this rock are much more comparable to those measured for the Tipton shale at Rock Springs with tensile strength values ranging from less than 1 MPa and with a high density of low strength planes throughout the section. As noted in Fig. 7.7 there were very few natural parting planes identified in the Geokinetics core.

7.4.4. Statistical Correlations

Comparisons between the different types of fracture property data illustrated in Fig. 7.4 for the Mahogany Zone shale from NOS 26 and in Fig. 7.6 for the Tipton shale indicate that significant correlations exist between the various fracture properties and between them and the oil yield or organic content of the rock. Consequently, a detailed statistical analysis was performed on all the data available from the NOS 26 Mahogany Zone shale and the Tipton shale. As only direct-pull tensile strength data was obtained over the entire section of Geokinetics shale, there was no basis to perform a statistical analysis for this rock.

As neither the fracture property tests nor the Fischer assay for oil yield tests were performed on a lithology controlled basis, it was not possible to group and compare the data for samples with comparable lithology. Also, as the direct-pull and indirect tensile strength tests and the fracture energy tests could not be performed at identical stratigraphic levels, it would be difficult to compare the results on an individual test by test basis. Rather, the statistical analysis was performed utilizing values obtained from the running averages of the various data, with several different averaging intervals being tested. Such an analysis effectively defines the correlations between running average curves, such as illustrated in Figs 7.4 and 7.6.

All types of data obtained on the NOS 26 core display quite good

TABLE 7.1

Best Fit Curve Coefficients and Correlation Coefficients for Rock Fracture Properties and Oil Yield for the NOS 26 Mahogany Zone Shale

Variables[a] (x/y)	Averaging increment	Number of points	Best fit curve	Coefficient of correlation (r)	Coefficient of determination (r^2)
DP/IT	2 m	34	$IT = 5.967 + 1.679DP$	0.96	0.917
	2 m	34	$IT = 5.259 + 1.947DP - 0.022DP^2$	0.96	0.918
	0.5 m	140	$IT = 7.794 + 1.344DP$	0.76	0.576
	0.5 m	140	$IT = 4.087 + 2.777DP - 0.116DP^2$	0.77	0.600
OY/DP	2 m	35	$DP = 9.266 - 0.138OY$	-0.80	0.645
	2 m	35	$DP = 12.625 - 0.385OY + 3.781E - 3OY^2$	-0.90	0.813
	0.5 m	144	$DP = 8.720 - 0.117OY$	-0.68	0.460
	0.5 m	144	$DP = 10.666 - 0.264OY + 2.190E - 3OY^2$	-0.73	0.534
OY/IT	2 m	34	$IT = 22.172 - 0.258OY$	-0.89	0.785
	2 m	34	$IT = 26.492 - 0.568OY + 0.005OY^2$	-0.93	0.871
	0.5 m	140	$IT = 21.348 - 0.226OY$	-0.75	0.563
	0.5 m	140	$IT = 23.683 - 0.400OY + 0.002OY^2$	-0.77	0.596
OY/FE	2 m	34	$FE = 22.025 + 0.895OY$	0.72	0.525
	2 m	34	$FE = 21.901 + 0.904OY + 1.33E - 4OY^2$	0.72	0.525
	0.5 m	142	$FE = 20.084 + 0.954OY$	0.70	0.486
	0.5 m	142	$FE = 19.478 + 0.999OY - 6.677E - 4OY^2$	0.70	0.486
DP/FE	2 m	34	$FE = 73.809 - 5.016DP$	-0.67	0.449
	2 m	34	$FE = 113.273 - 19.942DP + 1.247DP^2$	-0.72	0.521
	0.5 m	142	$FE = 69.779 - 4.300DP$	-0.54	0.295
	0.5 m	142	$FE = 88.802 - 11.606DP + 0.588DP^2$	-0.57	0.326
IT/FE	2 m	34	$FE = 93.835 - 3.132IT$	-0.74	0.542
	2 m	34	$FE = 163.393 - 12.389IT + 0.293IT^2$	-0.79	0.617
	0.5 m	140	$FE = 86.404 - 2.668IT$	-0.58	0.338
	0.5 m	140	$FE = 153.386 - 11.451IT + 0.268IT^2$	-0.66	0.436

[a] DP = direct pull tensile (MPa), IT = indirect tensile (MPa), FE = fracture energy (J m^{-2}), OY = oil yield (gallons ton^{-1}).

correlations over test averaging intervals ranging from 0·5 to 2·0 m (1·6–6·6 ft). Several types of curves were available with the statistical routines employed on the HP 9835 microcomputer used for these analyses, with the resulting correlation coefficients indicating which curves give the best agreement with the data. The best fit curves relating various pairs of data sets and their coefficients of correlation are summarized in Table 7.1.

Comparisons between the direct-pull tensile strength data and the indirect strength pinch test data for two different test averaging intervals are illustrated in Figs 7.8 and 7.9. As seen in the Figures and Table 7.1, these data show quite good correlation with the somewhat unexpected result that the larger averaging intervals give higher correlation coefficients. These better correlation coefficients result because the larger intervals serve to average out the scatter in the test data which is characteristic of mechanical property data on all rocks and especially on oil shales. Also shown in Figs 7.8 and 7.9, but not included in Table 7.1, are third-order polynomial curves to fit the data.

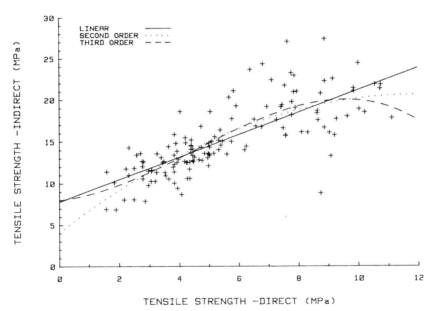

Fig. 7.8. Best fit curves for correlating direct-pull tensile strength with indirect (pinch) tensile strength for the NOS 26 Mahogany Zone shale using a 0·5 m (1·6 ft) sample averaging interval.

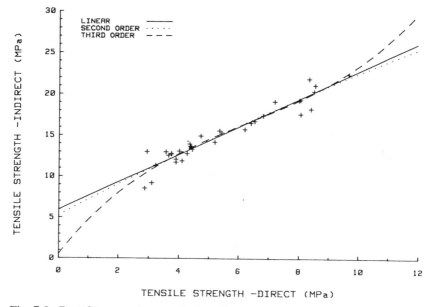

Fig. 7.9. Best fit curves for correlating direct-pull tensile strength with indirect (pinch) tensile strength for the NOS 26 Mahogany Zone shale using a 2·0 m (6·6 ft) sample averaging interval.

The third-order curves give slightly better correlation coefficients than the second-order curves but this improved fit is due to curvature of the fits near the ends of the data sets and does not reflect actual geologically controlled factors. For purposes of predicting various fracture property values from oil yield data or from other fracture properties the second order polynomials provide as much reliability as can be expected.

Both direct-pull and indirect tensile strength data for the NOS 26 Mahogany Zone shale are compared with oil yield in Fig. 7.10. Both types of strength correlate well with oil yield, yielding nearly identical correlation coefficients for both the second degree and third degree polynomial fits. The correlation coefficients of 0·90 for direct-pull data versus oil yield and 0·93/0·94 for indirect tensile strength versus oil yield are consistent with the good correlations noted between other mechanical properties and oil yield (Johnson and Simonson, 1977; Olinger, 1977; Chong et al., 1980). The data and best fit curves in Fig.

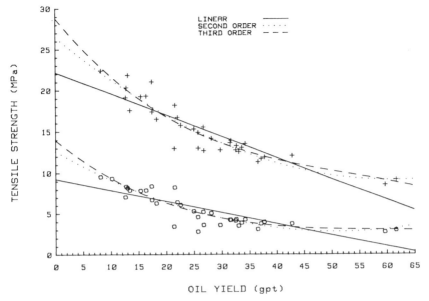

Fig. 7.10. Best fit curves for correlating direct-pull and indirect tensile strength data with oil yield for the NOS 26 Mahogany Zone shale using a 2·0 m (6·6 ft) sample averaging interval.

7.10 also illustrate the characteristically higher values obtained with the indirect tensile strength test as compared to the direct-pull test.

The good correlation coefficients obtained between all of the fracture property data and between them and oil yield, as summarized in Table 1, illustrate that the Piceance Basin Mahogany Zone shale, as represented by NOS 26 data, behaves in a predictable manner. The good correlations between all of the fracture property data with oil yield indicate that oil yields alone may be utilized to predict any of the desired fracture properties to a good first approximation. Site specific and stratigraphically controlled fracture property data would be required only if a detailed analysis of oil shale fracture was required.

The best fit curves, their coefficients and the corresponding correlation coefficients for relating the various Tipton fracture properties and oil yield are summarized in Table 7.2. As indicated in Table 7.2 and as discussed in more detail below, the correlation coefficients for the Tipton data are much lower than for comparable data for the NOS 26

TABLE 7.2
Best Fit Curve Coefficients and Correlation Coefficients for Rock Fracture Properties and Oil Yield for the Tipton Oil Shale

Variables[a] (x/y)	Averaging increment	Number of points	Best fit curve	Coefficient of correlation[b] (r)	Coefficient of determination (r^2)
DP/IT	2 m	10	$IT = 3\cdot312 + 1\cdot076DP$	NM	−0·054
	2 m	10	$IT = 8\cdot595 - 12\cdot187DP + 7\cdot710DP^2$	0·39	0·153
	0·5 m	46	$IT = 3\cdot049 + 1\cdot402DP$	0·41	0·167
	0·5 m	46	$IT = 2\cdot761 + 2\cdot112DP - 0\cdot372DP^2$	0·41	0·169
OY/DP	2 m	10	$DP = 1\cdot036 - 3\cdot058E-3OY$	NM	−0·122
	2 m	10	$DP = 2\cdot071 - 0\cdot105OY + 0\cdot002OY^2$	NM	−0·108
	0·5 m	46	$DP = 0\cdot588 + 1\cdot693OY$	0·20	0·042
	0·5 m	46	$DP = 0\cdot680 + 0\cdot008OY + 1\cdot987E-4OY^2$	0·21	0·042
OY/IT	2 m	10	$IT = 2\cdot957 + 0\cdot067OY$	NM	−0·056
	2 m	10	$IT = -5\cdot693 + 0\cdot923OY - 0\cdot021OY^2$	0·01	0·0002
	0·5 m	46	$IT = 3\cdot585 + 0\cdot037OY$	0·08	0·007
	0·5 m	46	$IT = 4\cdot196 - 0\cdot021OY + 0\cdot001OY^2$	0·09	0·008
OY/FE	2 m	10	$FE = 25\cdot174 + 0\cdot290OY$	NM	−0·071
	2 m	10	$FE = -2\cdot161 + 2\cdot995OY - 0\cdot065OY^2$	NM	−0·047
	0·5 m	44	$FE = 18\cdot020 + 0\cdot541OY$	0·18	0·032
	0·5 m	44	$FE = -23\cdot291 + 4\cdot514OY - 0\cdot089OY^2$	0·28	0·080
IT/FE	2 m	10	$FE = 11\cdot645 + 4\cdot486IT$	0·92	0·838
	2 m	10	$FE = 10\cdot294 + 5\cdot152IT - 0\cdot078IT^2$	0·92	0·838
	0·5 m	44	$FE = 9\cdot253 + 4\cdot682IT$	0·40	0·157
	0·5 m	44	$FE = 6\cdot535 + 6\cdot008IT - 0\cdot150IT^2$	0·40	0·158
DP/FE	2 m	10	$FE = 20\cdot838 + 10\cdot654DP$	0·42	0·172
	2 m	10	$FE = 44\cdot074 - 47\cdot686DP + 33\cdot912DP^2$	0·59	0·342
	0·5 m	44	$FE = 12\cdot224 + 18\cdot064DP$	0·51	0·259
	0·5 m	44	$FE = 38\cdot838 - 47\cdot892DP + 34\cdot606DP^2$	0·65	0·422

[a] DP = direct-pull tensile (MPa), IT = indirect tensile (MPa), FE = fracture energy (J m^{-2}), OY = oil yield (gallons ton^{-1}).
[b] NM = not meaningful.

shale. As for the NOS 26 analysis, running average values obtained over different averaging intervals were employed in establishing the correlation relationships.

Correlations between direct-pull tensile strength and oil yield for the Tipton shale employing a 0·5 m averaging interval are illustrated in Fig. 7.11. As may be seen in this figure and Table 7.2, the correlation coefficients for this data are extremely poor. As indicated in Table 7.2, only slightly better correlation coefficients were obtained between indirect tensile strength and oil yield and between fracture energy and oil yield. No significant improvements in the correlations were obtained by using larger averaging intervals. For many of the cases tested the coefficients of determination were negative indicating no statistically significant correlation between the data sets. Both the direct-pull and indirect tensile strength data appear to be largely controlled by mineralogical variations which are not well related to the oil yield of the rock.

Only marginal correlations exist between the fracture properties and the oil yield of the Tipton oil shale. Knowledge of oil yield can only

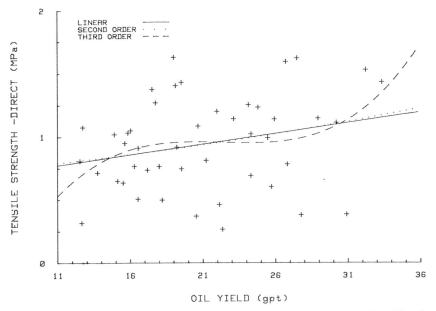

Fig. 7.11. Best fit curves for correlating direct-pull tensile strength with oil yield for the Tipton shale using a 0·5 m (1·6 ft) sample averaging interval.

provide a very rough estimate of the fracture properties that might be expected for this rock. Consequently, knowledge of site and stratigraphic specific fracture properties can only be crudely inferred from oil yield data and additional mechanical property testing would be required to provide reasonably accurate representations of the *in situ* fracture properties. These marginal correlations are attributed to the influence of the clay mineralogy of the rock upon its fracture and the fact that this clay mineralogy has little correlation with the organic content of the shale.

7.5. MODELING RESULTS

One method to generate void volume for *in situ* retorting is to heave the earth's surface by detonating an explosive charge within a relatively shallow oil-shale thickness. This technique has been tested in the field by Geokinetics Inc. in Uinta County, Utah (Lekas, 1979). The void space distributions resulting from the Geokinetics blasting tests have been evaluated empirically, usually by burning the retorts.

This section presents a general explosive fragmentation model incorporating oil shale properties specific to the Geokinetics site. Modeling their blasting program using mechanical properties of the oil shale deposit as primary input can provide an inexpensive tool for optimizing blast results. Although an actual field experiment may involve the timed detonation of numerous explosive boreholes, understanding the mechanics and implications of a single shot hole is important as much of the phenomenology of single-hole breakage is involved in multi-hole fragmentation. Also, the overall environmental impact and program cost may be significantly reduced due to more efficient rubblization.

7.5.1. Rock Fracture Model

The STEALTH sub-model which provides the details of the fracturing process is called CAVS (Crack And Void Strain). The model is general and has various applications (Barbour *et al.*, 1980; Maxwell and Reaugh, 1980). The unique aspect of the CAVS model is that the crack aperture (void strain) changes are coupled precisely to the three-dimensional stress tensor adjustments during crack opening and closing. Due to the extensive book-keeping system within CAVS, it is capable of determining the extent of cracking next to an explosive borehole. It has been proposed that cracks immediately adjacent to the shot hole will become filled with hot explosive reaction products and subsequently open and propagate more readily than if these gases were not present. This phenomenon has been observed and studied quantitatively in the calculations. While the concept of crack communication with the borehole (and explosive gases) is relatively straightforward, in practice the logic could be quite sophisticated since the ultimate pathway of gas travel cannot be determined *a priori*.

7.5.2. Fracture Internal Pressurization Model

Several one-dimensional borehole calculations have been carried out which have allowed the crack internal pressurization logic to be tested. The fluid flow and pressurization model when used in conjunction with the CAVS tensile fracture model essentially works as follows:

1. Explosive detonation within the borehole initiates a compressive wave which is followed by a tensile rarefaction which initiates or propagates fractures in the rock.
2. As fractures open up adjacent to the explosive borehole, hot reactant products are free to enter the fractures at a velocity which is dependent upon the crack's width and length, zone-to-zone pressure gradients throughout the borehole/fracture system and the viscosity of the flowing fluid. The viscosity can be defined as temperature dependent to incorporate the effect of quenching as the reactants cool as they contact the rock fracture faces.
3. Fractures developed in calculational zones which communicate with the high-pressure fluids in the borehole are free to accept fluid.
4. The effect of fluid in a fracture is analogous to pore pressure effects in effective stress analysis in that the normal stress is reduced, i.e. appears less compressive or more tensile, thereby enhancing the fracturing process.
5. As gas penetrates into the fracture system the pressure in the borehole is reduced until equilibrium is achieved, i.e. the excess volume of gas products is exhausted. Mass balance is maintained between the time-dependent fluid available from the borehole and the fluid penetrating the fracture system.

A generalized view of four two-dimensional calculational zones are illustrated in Fig. 7.12. Notice the three possible crack orientations and

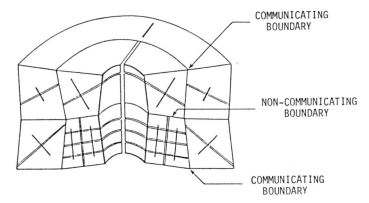

Fig. 7.12. Representation of 4 calculational zones around a borehole illustrating the logic employed for zone to zone fracture communication.

those fractures which are open and communicate with the explosive borehole.

The term 'void strain' refers to the actual opening of a crack within a calculational zone. A value of -0.01 in each of the three principal directions yields a volumetric void strain of 3%, indicating that the calculational zone has expanded by 3% due to crack opening (elastic or plastic volume changes are not reflected in this value). The most complicated aspect of CAVS is in the void closure logic since a crack must not be allowed to over-close and should display self-propping. Closure times for cracks in the three directions are generally different and closure of one crack affects crack void strains in the other directions. Sub-cycling and rearranging of closing sequences solves this problem. The addition of gas pressure to the model is trivial since the crack opening/closing logic is unchanged and only the effective stress is modified. After a crack has opened, perfect closure is not possible due to asperity mismatch. A jumbling or propping logic has been included to incorporate this effect.

The effect of gas pressurization is to allow communication cracks to open up substantially due to the gas pressure. In addition, more cracks are actually generated due to the presence of gas pressure. Immediately adjacent to the borehole wall the effects of gas pressure appear to suppress cracking; however, at distances greater than 4 m the extent of cracking is significantly enhanced due to the influence of gas pressure. There is an intermediate zone where both calculations yield similar crack patterns.

7.5.3. Experimental Data on Oil Shale

Extensive mechanical testing of oil shale has taken place over the past several years. As a result, sufficient information is available to formulate constitutive models for computer codes. Of particular interest in this analysis was the influence of the relatively weak oil shale bedding plane partings and the influence of explosive gas pressurization on initiating and propagating fractures along the weak bedding planes.

Although significant information has been provided for stratigraphic variations in oil shale fracture energy, the fracture model used here, CAVS, at the time of these calculations, only provided tensile failure as the basis for crack initiation and propagation. More recently, shear crack phenomenology has been incorporated into the code (Young *et al.*, 1981) and the inclusion of fracture mechanics concepts, such as fracture energy or toughness, could readily be inserted into the explicit CAVS logic.

The only site specific experimental data, therefore, to be used in this analysis are direct-pull tensile strengths normal to bedding and oil yield. This oil yield was then applied to various empirical formulae as described below to obtain inferred site specific properties.

7.5.3.1. Oil Yield–Density–Organic Volume Relationships

The modified Fischer assay method was used to determine the oil content of core taken at the Geokinetics site (Trudell, 1979). The samples represented approximately $0\cdot3$ m sections each and varied from 5 to 65 gallons shale oil per ton of rock. The amount of organic matter in oil shale has been related to various mechanical properties, as indicated below.

Since oil yields in the Geokinetics blast section varied from approximately 10 gallons ton^{-1} to over 60 gallons ton^{-1}, significant impedance differences existed from layer to layer. These impedance mismatches are especially important to stress wave interaction from layer to layer. The densities of different oil shale grades may be readily estimated from oil yield–organic volume relationships. Rock density has been found to vary with oil shale according to the following relationship (Smith, 1976):

$$OY = 31\cdot563(D_T)^2 - 205\cdot998D_T + 326\cdot624 \qquad (7.2)$$

where OY is the oil yield in gallons ton^{-1} D_T is the rock density in g cm^{-3}. Once this expression is solved for density, the percent organic volume may be found by (Smith, 1976):

$$V_O = 164 \cdot 85 - 60 \cdot 61 D_T \qquad (7.3)$$

where V_O is the organic volume in percent and D_T is the rock density in $g\, cm^{-3}$.

7.5.3.2. Tensile Strength

Tensile strengths parallel to the bedding planes of Green River Formation oil shales similar to the Geokinetics materials have been related to organic volume. Consequently, tensile strengths at this orientation can be evaluated through the blast section. However, oil shale is significantly weaker in the direction normal to the bedding planes. To incorporate this characteristic into the model, direct-pull tensile strength tests were performed on cores taken from the Geokinetics site. Direct measurements of tensile strength have the advantage over indirect methods in that the weakest link in the section tested are always measured. Indirect methods determine strengths at selected depths and therefore do not necessarily test the weakest plane. Tests were performed by pulling apart $0 \cdot 0365\, m$ samples which were $0 \cdot 102\, m$ in length (see Section 7.4.3). Each sample was tested once. Approximately 14 m oil shale were tested. The results are illustrated in Fig. 7.7.

The tensile strength parallel to the bedding planes was provided by an empirical formula based on Brazilian split cylinder tests. Strengths for three types of Green River oil shales are given by the following (Chong et al., 1979, 1982):

$$\text{Colorado:} \quad \sigma_t = 14 \cdot 78 - 0 \cdot 0928 V_O \qquad (7.4)$$

$$\text{Utah:} \quad \sigma_t = 13 \cdot 64 - 0 \cdot 1211 V_O \qquad (7.5)$$

$$\text{Wyoming:} \quad \sigma_t = 23 \cdot 16 - 0 \cdot 3516 V_O \qquad (7.6)$$

where σ_t is the tensile strength in MPa, V_O is organic content in percent. For this study, the representation given for the Utah oil shale was used.

7.5.3.3. Elastic Constants

Elastic constants from stress–strain measurements on Green River Formation oil shales have been shown to vary with organic matter in the rock. The following prediction equations developed from mechanical tests permit evaluation of elastic constants with depth (Chong et al.,

1980):

$$E_x = 10 \cdot 45 - 0 \cdot 1735 V_O + 0 \cdot 3841 S - 0 \cdot 00519 V_O S - 1 \cdot 883 \times 10^{-7} S^4 \tag{7.7}$$

$$E_z = 12 \cdot 34 - 0 \cdot 2196 V_O + 0 \cdot 07461 S - 6 \cdot 82 \times 10^{-5} V_O S - 9 \cdot 869 \times 10^{-8} S^4 \tag{7.8}$$

$$\nu_{zx} = -0 \cdot 04419 + 0 \cdot 00385 V_O + 0 \cdot 00645 S \tag{7.9}$$

$$\nu_{xy} = -0 \cdot 03307 + 0 \cdot 00333 V_O + 0 \cdot 00480 S \tag{7.10}$$

where z is the direction perpendicular to bedding, x and y are directions parallel to bedding, E is the elastic modulus in GPa, ν is Poisson's ratio, V_O is organic volume (%), S is the stress level (actual stress/ultimate stress $\times 100\%$).

Under compressive loading the behavior of these oil shales is nearly isotropic, in spite of their laminar structure. Consequently, the elastic constants can be averaged and applied to three dimensions. The equations contain a stress level function as a percent of ultimate load. This concept may be useful for uniaxial loading, but there is some ambiguity when the stress field is fully three-dimensional. In addition, since within the model the material was allowed to yield under sufficient deviatoric stress, it was assumed that the stress level was 40%, yielding mid-range values for the elastic constants.

7.5.3.4. Plasticity Model

The elastic/plastic yield surface is strongly dependent upon the density (oil yield) of the oil shale. The yield surfaces are similar up to grades of 104 liters tonne^{-1} (25 gallons ton^{-1}) but drop dramatically until approximately 188 liters tonne^{-1} (45 gallons ton^{-1}) at which point yielding is essentially governed by the strength of kerogen which is independent of mean stress. The following formula for yielding has been developed (Johnson, 1979):

$$Y = Y_0 + \Delta Y (1 - \exp{(-aP)}) \tag{7.11}$$

where Y is the yield stress, Y_0, ΔY and a are parameters which depend on density and P is the mean stress. A different yield surface was incorporated into the code for each different density.

7.5.3.5. Input to the Code

The two-dimensional axisymmetric calculational grid utilized is shown in Fig. 7.13. The top boundary condition is a free surface and all other

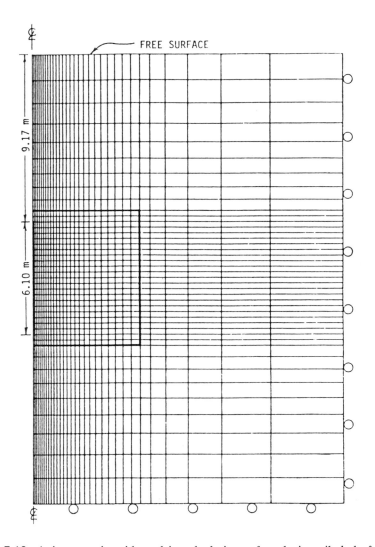

Fig. 7.13. Axisymmetric grid used in calculations of explosive oil shale fragmentation (subgrid used for fracture and contour plots is outlined).

boundaries interact with the wall (roller). The top of the explosive charge is located 9·14 m below the surface and is 6·10 m long. The explosive borehole was initially 0·1524 m in diameter and the explosive was ANFO, which was described by a JWL equation-of-state (Lee et al., 1973).

The vertical zoning within the oil shale blast section is spaced 0·3048 m. A careful correlation was performed and it was determined that all 20 horizons had oil yields that fitted into 8 groups, within ±5%. Figure 7.14 illustrates the stratigraphic distribution of the various layers. Above and below the explosive charge, the rock was assumed to have the parameters of the leanest oil shale (material 1).

The tensile strengths shown in Fig. 7.7 were used for the fracture parallel to the bedding planes. The weakest strength within a given

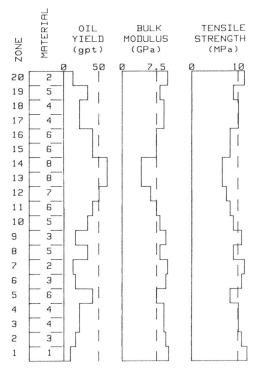

Fig. 7.14. Stratigraphic distribution of the 8 material models used in the calculations.

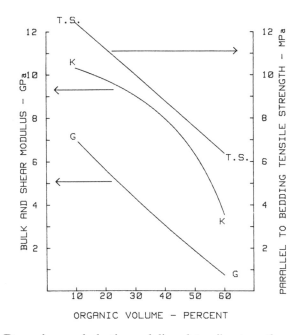

Fig. 7.15. Dependence of elastic moduli and tensile strength upon oil shale organic volume.

layer indicated the stress level to initiate the first crack in a calculational zone. The next higher strength had to be exceeded for the second crack, and so on. Strengths above the third crack value increase arbitrarily by 1·05 times the last crack strength. Tensile strengths in the other two directions were determined from the organic volume relationship presented earlier and are shown in Fig. 7.14. The dependence of tensile strength for fracture across the bedding and of bulk and shear moduli upon organic volume are illustrated in Fig. 7.15.

From the elastic–plastic yield data of Johnson (1979), mentioned earlier, yield coefficients were established for the 8 materials. The yield surfaces are plotted in Fig. 7.16. Note that since materials 7 and 8 are so rich, their yielding is not taken from the equation but rather is a constant of 0·1 GPa based on yield strengths of polymeric materials similar to kerogen (Johnson and Simonson, 1977).

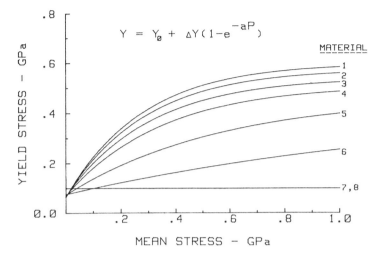

Fig. 7.16. Yield strength curves for the 8 material models employed in the calculations.

7.5.4. Calculational Results

The baseline calculation was performed utilizing the stratigraphic layering and tensile strength criteria presented earlier. The explosive, ANFO, has a relatively long rise time and maintains its peak pressure for longer times than other high explosives. Approximately 44 mol reactant products kg^{-1} charge are produced (Edl, 1980).

A second calculation was run in which there was a uniform material model, i.e. no stratigraphic variations in material properties. For this calculation all properties were defined by the formulae for material 4 94 liters tonne^{-1} (22·6 gallons ton^{-1}).

Maximum gas pressure in a shale zone adjacent to the borehole was 120 MPa. For both calculations, initial overburden stresses were used as the initial condition and the 6·10 m long explosive borehole was top detonated. Gas pressures further from the borehole were significantly reduced because of narrower crack widths.

After 5 ms for the baseline case, certain horizons had little or no gas pressure while others showed a tendency to channel due to the non-uniform void strain distribution. Also the borehole boundary was uneven due to the material layering. The uniform material calculation

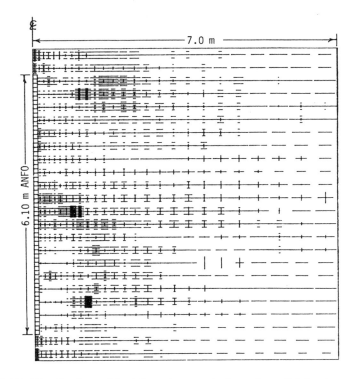

Fig. 7.17. Distribution of bedding-plane and circumferential fractures calculated for the stratigraphically layered (baseline) model at 5 ms.

gave a relatively constant gas profile with depth and the uniform expansion of the borehole wall.

The in-plane crack patterns for the baseline and uniform cases are shown in Figs 7.17 and 7.18, respectively. The influence of the weak bedding planes is obvious. Although relatively little tensile fracturing takes place near the borehole, shear-induced cataclysmic failure is typically present in this area, providing finely fragmented rock. Radial or 'out of plane' crack patterns are illustrated in Fig. 7.19. Several different strengths are utilized for each layer corresponding to the grade of shale in question. Despite the large variation in strength (104 kPa, 15 psi for layer no. 24 and 11·1 MPa, 1615 psi for layer no. 27) only modest differences are noted in the crack patterns.

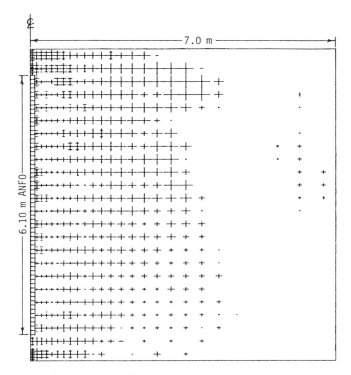

Fig. 7.18. Distribution of bedding-plane and circumferential fractures calculated for the homogeneous model at 5 ms.

7.5.5. Quasi-Static Analysis

In order to investigate the influence of gas pressure at later times and to keep the computational costs to a reasonable limit, a time-step scaling technique was used. It is similar to density scaling which is an artificial increase in density resulting in a lower sound speed and a larger time step. The momentum effects are assumed to be trivial. Instead of increasing the density, a pseudo time-step was defined which was used in the equations of motion which allowed a more efficient variation in the effective scale factor (Maxwell *et al.*, 1978).

The baseline calculation was run in a fully dynamic mode for 5 ms until virtually all shock effects had dissipated from the region of interest. Then the pseudo time-step technique was applied and the

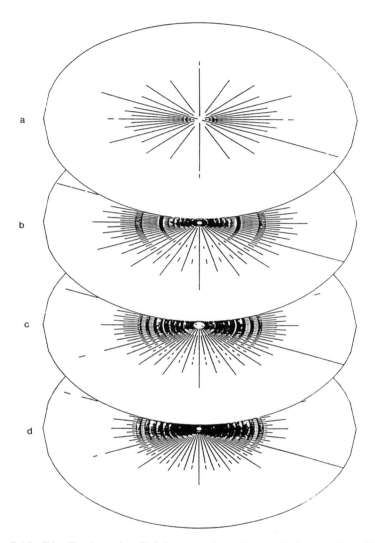

Fig. 7.19. Distribution of radial fractures in 4 layers of the stratigraphically layered model at 2·5 ms (radius of each disk is 3 m). (a) Layer no. 27, 1615 psi; (b) layer no. 26, 55 psi; (c) layer no. 25, 45 psi; (d) layer no. 24, 15 psi.

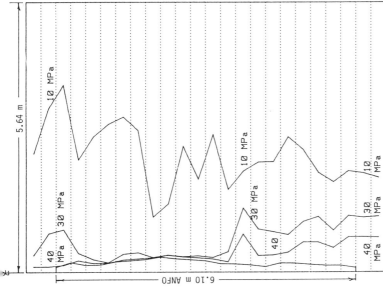

Fig. 7.21. Contours of the explosive gas pressure acting within the fracture system of the layered model at 1 s.

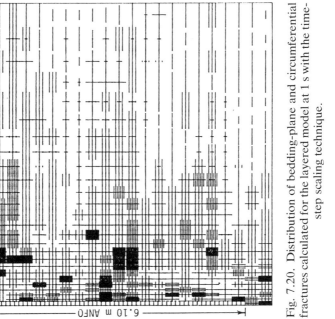

Fig. 7.20. Distribution of bedding-plane and circumferential fractures calculated for the layered model at 1 s with the time-step scaling technique.

calculation proceeded to 1 s. The in-plane fracture pattern is shown in Fig. 7.20. Contours of gas pressure after one second are shown in Fig. 7.21. While the bedding plane fracturing is enhanced, very little increase in the radial cracking is observed. Notice the non-uniform different gas penetration as a result of the stratigraphic variations in material properties.

7.6. CONCLUSIONS

The development of specialized testing techniques combined with the careful handling of core material provide the means to obtain detailed stratigraphic logs of the tensile strength and fracture toughness properties of oil shales and other rocks. Stratigraphic logs of fracture properties for the NOS 26 Mahogany Zone shale, the rich upper 13 m of the Tipton shale and 14.6 m of the Geokinetics Mahogany Zone shale demonstrate the degree of stratigraphic variability in fracture properties that can be expected for oil shales.

The detailed statistical analyses performed on the fracture property data reported here indicate that good correlations may or may not exist between the fracture properties and oil yield depending upon the site specific mineralogy of the rock. The NOS 26 Mahogany Zone shale, with a predominate carbonate/kerogen mineralogy, gives good correlations between the fracture properties and between them and oil yield. Somewhat surprisingly the NOS 26 data give a negative correlation coefficient between tensile strength and fracture energy. This negative correlation has been noted by previous workers (Schmidt, 1977) and may be attributed to the control of tensile strength more by flaw size rather than fracture energy. The good correlation between fracture energy and tensile strength and between these properties and oil yield requires that the inherent rock flaws controlling tensile strength be related to a kerogen related mineralogy of the rock. The intrinsically fine-grained nature of the carbonate/kerogen mineralogy of the NOS 26 shale would be expected to yield increases in ductility and thus fracture energy with increases in kerogen content and decreases in inherent flaw size with increases in kerogen content. The much lower correlation coefficients found between fracture energy, tensile strength, and oil yield of the Tipton shale can be attributed to the control of tensile failure by a clay mineralogy not well related to kerogen content.

Numerical simulations of explosive oil shale fragmentation incorporating stratigraphically varying mechanical properties, such as density, elastic constants, yielding and tensile strengths, demonstrate the significant affect these variations can have upon model results. When bedding plane fracture is significantly weaker than across the bedding fracture this aspect must be incorporated into the model.

The explosive reaction products have been shown to alter the crack pattern by decreasing the amount of cracking near the shothole and increasing the amount and extent of cracking further away. No gas enters this fractured rock until about $1 \cdot 0$ ms since the stress field is still highly compressive until this point. The effect of gas pressure at later times appears to enhance the fracturing process for bedding plane partings; however, the influence on radial crack extension seems to be minor. Gas pressure effects at greater depths could be significantly different.

A most important conclusion to be drawn from the results of this study relates to the extreme variability of oil shale fracture properties with variations in stratigraphic position, mineralogy and field location. Variations in the rock fracture energy values by a factor of 20 for the NOS 26 shale and a factor of 17 for the Tipton shale and the variations in tensile strength from essentially 0 to over 40 MPa for the NOS 26 shale attest to the variability in rock fracture properties that should be expected in the natural *in situ* environment. The numerical simulation of explosive fragmentation further demonstrates the importance of the stratigraphic variability of oil shale. This variability underscores the risks that are taken when a rock such as oil shale is treated as a homogeneous isotropic engineering material in the design and analysis of field experiments. The natural and inevitable variation in rock fracture properties must be properly taken into account in the design, execution and analysis of any *in situ* fracturing and fragmentation experiments to be carried out in the development of oil shale processing methods.

ACKNOWLEDGMENTS

This research was supported by the Laramie Energy Technology Center, Office of Resource Characterization, Laramie, Wyoming, USA. We thank John Ward Smith for his interest in and many contributions to both the experimental and numerical portions of this study. We would like to credit Donald E. Maxwell of Science Applications Inc., San Leandro, California, for the initial development of the CAVS failure model. Timothy G. Barbour of Science Applications Inc., Golden, Colorado and Krishan K. Wahi of Science Applications Inc., Albuquerque, New Mexico are acknowledged for their contributions to refinement of the CAVS model, to the fracture internal pressurization logic and/or implementation of the time-step scaling technique.

REFERENCES

ASTM (1972). *Standard Method of Test for Direct Tensile Strength of Rock Core Specimens*, ASTM D-2936. American Society for Testing and Materials, Philadelphia pp. 862–4.

Barbour, T. G., Wahi, K. K. and Maxwell, D. E. (1980). Prediction of fragmentation using CAVS, *Society of Experimental Stress Analysis, Fall Meeting*, Fort Lauderdale, Florida.

Barbour, T. G. and Young, C. (1980). Numerical analysis of multiple fracturing in a wellbore, *21st US Symposium on Rock Mechanics*, Rolla, Missouri.

Barker, L. M. (1977). A simplified method for measuring plane strain fracture toughness, *Eng. Frac. Mech.*, **9**, 361.

Beasley, R. R. and Boade, R. R. (1980). Instrumentation and evaluation of the Talley Energy Systems, Inc. oil shale project, part II, *Sandia National Laboratories Report, SAND 78-1886*, Albuquerque, New Mexico.

Chong, K. P., Smith, J. W., Chang, B. and Roine, S. (1979). Oil shale properties by split cylinder method, *J. Geotechnical Eng. Div.*, ASCE, **105**(GT5), proc. paper 14567, 595–611.

Chong, K. P., Uenishi, K., Smith, J. W. and Munari, A. C. (1980). Nonlinear three-dimensional mechanical characterization of Colorado oil shale, *Int. J. Rock Mech. Min. Sci.*, **17**, 339–47.

Chong, K. P., Smith, J. W. and Borgman, E. S. (1982). Tensile strengths of Colorado and Utah oil shales, *J. Energy*, AIAA, **6**(2), 81–5.

Clifton, R. J., Simonson, E. R., Jones, A. H. and Green, S. J. (1976). Determination of the critical-stress-intensity factor, K_{IC}, from internally pressurized thick-walled vessels, *Expl. Mech.*, **16**(6), 233–8.

Cooper, G. A. (1977). Optimization of the three-point bend test for fracture energy measurement, *J. Mat. Sci.*, **12**, 277–89.

Edl, J. (1980). Private communication.

Friedman, M., Handin, J. and Alani, G. (1972). Fracture surface energy of rocks, *Int. J. Rock. Mech. and Min. Sci.*, **9**, 757–66.

Hofmann, R. (1976). STEALTH, A Lagrange explicit finite difference code for solids, structural and thermohydraulic analysis, *EPRI NP-260 Vol. I, User's Manual*, prepared by Science Applications Inc. for Electric Power Research Institute, Palo Alto, California.

Jaeger, J. C. and Cook, N. G. W. (1969). *Fundamentals of Rock Mechanics*, London, UK.

Johnson, J. N. (1979). Calculation of explosive rock breakage: oil shale, *Proc. 20th US Symposium on Rock Mechanics*, Austin, Texas, pp. 109–18.

Johnson, J. N. and Simonson, E. R. (1977). Analytic failure surfaces for oil shales of varying kerogen content, *Los Alamos Scientific Laboratory, Report LA-UR-77-1005*, Los Alamos, New Mexico.

Lee, E. L., Finger, M. and Collins, W. (1973). JWL equation of state coefficients for high explosives, *Lawrence Livermore Laboratory Report UCID 16190*.

Lekas, M. A. (1979). Progress report on the Geokinetics horizontal *in situ* retorting process, *Proc. 12th Oil Shale Symposium*, Golden, Colorado, pp. 228–36.

Maxwell, D. E. and Reaugh, J. E. (1980). A continuum fracture model called CAVS, *Symposium on Conventional Methods in Nonlinear Structural and Solid Mechanics*, George Washington University/NASA-Langley Research Center, Washington DC.

Maxwell, D. E., Hofmann, R. and Wahi, K. K. (1978). An optimization study of the explicit finite difference method for quasi-static thermo-mechanical simulations, *SAI-FR-821-3*, prepared by Science Applications Inc. for Union Carbide Corp. (*Contract W-7405-ENG-26*).

Olinger, B. (1977). Elastic constants of oil shale, *Los Alamos Scientific Laboratory, Report LA-6817-PR*, Los Alamos, New Mexico, pp. 24–6.

Parrish, R. L., Boade, R. R., Stevens, A. L. and Long, A. (1980). The Rock Springs site 12 hydraulic/explosive true *in situ* oil shale fracturing experiment, *Sandia National Laboratories Report, SAND 80-0836*, Albuquerque, New Mexico.

Reichmuth, D. R. (1963). Correlations of force-displacement data with physical properties of rock for percussive drilling systems, *Proc. 5th Symposium on Rock Mechanics*, Pergamon, New York, pp. 33–59.

Ridley, R. D. (1978). Progress in Occidental's shale oil activities, *Proc. 11th Oil Shale Symposium*, Golden, Colorado, pp. 169–75.

Schmidt, R. A. (1976). Fracture-toughness testing of limestone, *Expl. Mech.*, **16**(5), 161–7.

Schmidt, R. A. (1977). Fracture mechanics of oil shale—unconfined fracture toughness, stress corrosion cracking and tension test results, *Proc. 18th US Symposium on Rock Mechanics*, Colorado School of Mines, 2A2-1–2A2-6.

Smith, J. W. (1976). Relationship between rock density and volume of organic matter in oil shales, *Report of Investigations, LETC/RI-76/6*, Laramie Energy Technology Center, Wyoming.

Smith, J. W., Beard, T. N. and Trudell, L. G. (1979). Oil shale resources of the Naval Oil Shale reserve No. 1, Colorado, *Report of Investigations, LETC/RI-79/2*, Laramie Energy Technology Center, Wyoming.

Stevens, A. L., Lysne, P. C. and Griswold, G. B. (1975). Rock Springs oil shale fracturization experiment: experimental results and concept evaluation, *Sandia National Laboratories Report, SAND 74-0372*, Albuquerque, New Mexico.

Trent, B. C. and Langland, R. T. (1983). Subsidence modeling for underground coal gasification, *In Situ*, **7**(1), 53–85.

Trudell, L. (1979). Lithologic description of samples submitted for assay,

Illustration No. SBR-5029P, Laramie Energy Technology Center, Wyoming, 16 April 1979. Oil shale assays by modified Fischer retort method—Geokinetics Inc. corehole W-14, *Illustration No. SBR-5016P*, Laramie Energy Technology Center, Wyoming, 16 April 1979.

Wang, K. A. and Chong, K. P. (1980). Diametrical compression of transversely isotropic disks, *Proc. 21st US Symposium on Rock Mechanics*, Rolla, Missouri, pp. 243–8.

Young, C., Barbour, T. G. and Trent, B. C. (1981). Geologic control of oil shale fragmentation, *Society of Experimental Stress Analysis, Fall Meeting*, Keystone, Colorado.

Young, C., Patti, N. C. and Trent, B. C. (1982). Stratigraphic variations in oil shale fracture properties, *Report of Investigations, DOE/LC/RI-82-5*, Laramie Energy Technology Center, Wyoming.

Chapter 8

Model Studies of Fragmentation

W. L. FOURNEY, D. C. HOLLOWAY and D. B. BARKER

*Mechanical Engineering Department, University of Maryland,
College Park, USA*

SUMMARY

*This chapter describes model tests conducted to learn more about the
process of fragmentation in hard rock. Earlier testing as described in the
open literature was conducted with homogeneous materials and ignored
the presence of flaws in most rock structure. Those tests in our opinion
have yielded results which underestimated the role of stress waves in
creating fragments in a blasting operation.*

*Results from two series of tests will be presented. In the first series
dynamic photoelasticity was used to examine models made of transparent polymeric materials which had been artificially flawed. These models
were examined closely while being subjected to explosive loading to
determine the effects of the presence of the flaws on the fragmentation
process and made it possible to identify mechanisms responsible for crack
initiation and growth. Tests were conducted in models containing large
as well as small flaws and the effect of these flaws on optimum delay
times between multiple hole charges was also determined.*

*The second series of tests involved testing small rock plates. Plates
made of granite were used to examine fragmentation of 'fault free' rock
structure. Tests with granite plates with artificially induced flaws and
naturally flawed limestone plates were used to examine fragmentation in
flawed rock structure. Areas of interest as identified from the polymeric
testing were examined closely in the rock plate tests by using various
methods of high speed photography. In this way it was possible to*

338 *W. L. Fourney, D. C. Holloway and D. B. Barker*

determine if the mechanisms identified in the transparent model testing were operative in actual rock blasting situations. The chapter is concluded with a description of small scale single hole blasting in a marlestone formation.

8.1. INTRODUCTION

This chapter describes tests conducted with laboratory models made of brittle transparent polymers and tests conducted on small rock plates. These tests were used to study the process of fragmentation and in particular to assess the effects of the presence of flaws on fragmentation.

There presently exists disagreement with regard to the mechanism that is responsible for the fragmentation that occurs in quarry blasting operations. The particular contribution of gas pressurization as opposed to stress wave propagation has not been clearly established. The various schools of thought range from the notion that gas pressurization plays the dominant role (Langefors and Kihlstrom, 1963; Persson et al., 1970; Porter and Fairhurst, 1970; Ash, 1973; Hagen and Just, 1974) to the idea that combinations of the stress waves and gas pressurization are responsible for the major amount of damage (Kutter and Fairhurst, 1971; Johansson and Persson, 1974; Coursen, 1979) to the thought that stress alone (Duvall and Atchison, 1957; Starfield, 1966; Bhandari, 1977) is responsible for the resulting fragmentation.

Nearly all previous model studies (laboratory as well as small boulder tests) have been conducted primarily in non-flawed plastics or fault free rock. If the mechanism of fragmentation is stress wave related then testing in flaw free models would skew the experimental results towards the theory that gas pressurization played the dominant role in fragmentation.

In this study, flaws found in a typical limestone quarry were artificially simulated in birefringent models made of Homalite 100. By using birefringent models and the technique of dynamic photoelasticity it was possible to determine how the various stress waves generated by the detonation of an explosive charge interacted with the various artificial flaws. It was also possible to view the initiation and growth of both pressurized and non-pressurized cracks. In this manner the mechanisms responsible for crack initiation and propagation were determined. Similar tests were then conducted in small rock plates of granite and limestone. These plates were viewed during explosive loading by means of high speed photography and special attention was focused on areas demonstrated to be of importance in the previously conducted polymer testing.

8.2. MODEL STUDIES OF FRAGMENTATION

In this particular section testing will be reviewed which defines the type of fragmentation observed in flaw free models and what mechanisms are responsible for the fracturing observed in the post test observations.

8.2.1. Experimental Procedure

The experimental program involved the testing of many small two-dimensional polymeric models. High speed photography in conjunction with dynamic photoelasticity was used to visualize the dynamic fracture process. The tests were conducted in a material sold commercially as Homalite 100 which is available from SGL Industries of Wilmington, Delaware. The initiation toughness (K_{IC}) is around $0\cdot40$ MPa$\sqrt{m}$ and is less than that found for most rock. It is therefore more brittle in behavior than rock. At high crack driving energy levels, crack propagation behavior is felt to be similar to that in rock. Since the crack velocities expected in a typical fragmentation round are large (300–400 m s^{-1}), it is felt that the details of the fragmentation process can be studied in Homalite 100 and applied qualitatively to fragmentation in rock.

All models were $6\cdot4$ mm thick and were typically 300 mm square or 300×380 mm depending upon the type of test being conducted. 200–250 mg PETN (pentaerythritoltetranitrate) was tightly packed into $6\cdot4$ mm diameter boreholes. An attempt was made to contain the detonation gases by tightly capping the boreholes with O-ring seals and steel caps.

All of the experiments used dynamic photoelasticity to identify the wave type and shear stress magnitude occurring at a given location. Dynamic photoelasticity (Dally, 1971) is an optical method of stress analysis which permits a full-field visualization of the state of stress in a transparent birefringent material. The dynamic fringe patterns obtained provide a way of simultaneously observing the interaction between the propagating cracks and the local stresses that drive these

340   *W. L. Fourney, D. C. Holloway and D. B. Barker*

cracks. Each fringe order, N, is related to the difference in the principal stresses, σ_1 and σ_2, according to the stress optic law

$$\sigma_1 - \sigma_2 = \frac{Nf_\sigma}{t} \tag{1}$$

where f_σ is the material fringe value and t is the model thickness.

The use of the Cranz–Schardin multiple spark gap camera to record dynamic stress waves and crack propagation has previously been described in detail (see, for example, Dally and Riley, 1967; Riley and Dally, 1969). The camera can record 16 independently controlled frames of dynamic information at a rate of 30 000–1 000 000 frames s^{-1}.

8.2.2. Explosive Loading and Wave Systems Produced

Upon detonation of the explosive in a borehole an intense pressure pulse is applied to the borehole wall. A quasi-static solution to the stresses about the borehole can be derived from the classical solution for a pressurized thick-walled cylinder (Timoshenko and Goodier, 1951). This solution shows that, on the borehole wall, the tangential stress, σ_θ, is tensile and equal in magnitude to the compressive radial stress, σ_r, and that σ_r equals the applied pressure.

Using numerical techniques Bligh (1972) has developed a dynamic solution. The quasi-static and dynamic solutions are schematically shown in Fig. 8.1 for an assumed pressure pulse $P(t)$. Of primary importance in the explosive process is the tangential stress since this is the stress that will initiate borehole cracks. In the dynamic analysis the tangential component starts as a compressive stress and after a few microseconds changes to tension. The tangential component ultimately reaches a maximum value typically 50% greater than that predicted from a quasi-static analysis. This knowledge of the stress field about the borehole helps explain the physically observed crack initiation behavior.

This dynamic pressure pulse applied to an interior point in a half space results in two elastic body waves and four reflected elastic wave systems being generated as shown schematically in Fig. 8.2. The faster of the two waves which propagate within an elastic body is called the primary or P wave. It is also referred to as the dilatational wave because the deformation associated with this wave includes a volume change. Detailed examination of the stress state in this wave from a typical detonation (Barker *et al.*, 1978) has indicated that in the

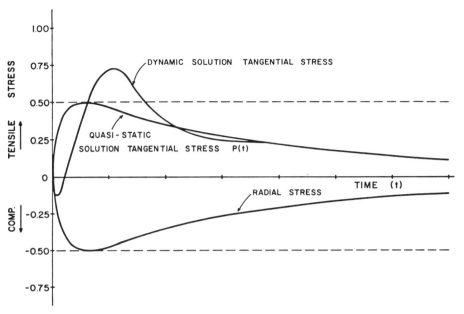

Fig. 8.1. Quasi-static and dynamic borehole stresses for an assumed pressure pulse $P(t)$.

leading edge both σ_r and σ_θ are compressive with σ_r being about 2·5 times larger than σ_θ. We have observed experimentally a small tensile tail for the radial component of stress but usually no tensile tail for σ_θ. It is felt that the tensile tail of σ_θ is consumed in borehole crushing. The typical speed of the P wave in Homalite 100 plates is 2000 m s^{-1}.

The slower of the two elastic body waves is called the secondary or shear wave. A theoretical analysis of a point charge does not predict the presence of a shear wave due to the axisymmetry of the problem. In practice, however, crushing and crack initiation at the borehole walls destroys the axisymmetry and thus accounts for the existence of a

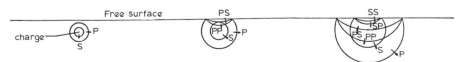

Fig. 8.2. Waves generated from an explosive source near a free boundary.

shear wave. The disturbance caused by this shear wave involves no dilatation and the typical speed of propagation in Homalite 100 is found to be $1100 \, \mathrm{m \, s^{-1}}$. The stress state for the shear wave is pure shear. The magnitude of $\sigma_{r\theta}$ varies with angular position around the borehole.

Reflected wave systems are similarly named. When striking a material interface the P wave generates a reflected primary wave, the PP wave, and a reflected secondary, PS wave. Similarly the S wave generates an SP and SS wave system. The reflection of a P wave of normal incidence to a free surface is the cause of classical spalling or slabbing. Upon reflection the compressive front of the incident P wave undergoes a phase shift and is reflected as a tensile wavefront. Spalling occurs when the reflected tensile component constructively interferes with the tensile tail of the incident wave. The exact location of this spall fracture is a function of the wavelength of the disturbance and the magnitude of the critical tensile stress required for crack initiation. Since the radial stress component in the P wave is approximately three times the tangential component the orientation of the spall should be almost parallel to the free face.

Very near the free face there is a region of material that is subjected to negligible tensile stresses. This region is known as the shadow zone. As the leading edge of the incident P wave is reflected as a tension wave, it destructively interferes with the remainder of the outwardly traveling compressive wave resulting in a lower effective stress in this region. As the reflection process continues the tensile wave constructively interferes with the outwardly directed tensile tail. The region where this occurs is subjected to a high tensile stress.

In the more general case of wave reflection the P wave is not normally incident to the free boundary and a PS wave is also generated. At the shadow zone boundary the PP wave has not yet outdistanced the PS wave and the additional driving energy of the PS wave will branch cracks initiated by the PP wave. Some of these branches will be driven towards the free surface breaking up the shadow zone region.

For a thorough understanding of the wave systems being encountered in this study it is recommended that a good reference on wave propagation such as Graff (1975) be consulted. Additional details on stress separation and wave speed as determined from dynamic photoelasticity can be obtained from Barker and Fourney (1978) and Barker et al. (1978).

8.2.3. Borehole Crack Network and Free Face Effects

8.2.3.1. Creation of Borehole Cracks

After detonation, crushing begins to occur in the immediate vicinity of the borehole. This crushing is caused by the intense shear stresses generated. Prior to the tangential stress wave becoming tensile, the maximum shear stress is small since all three principal stresses are compressive and similar in magnitude hence very little crushing will occur. When the tangential stress changes sign the maximum shear stress becomes significant. The resulting crushed zone propagates radially outward until the maximum shear stress decreases below some critical value required to cause fracturing. Outside the crushed zone, cracks initiate due to tensile loading and not shear.

In this case existing flaws are opened by the tensile stress component normal to the crack faces. When this stress component produces a stress intensity factor (K) greater than the materials fracture toughness (K_{IC}) the crack will start to propagate. Once moving, the crack will align itself perpendicular to the maximum tensile stress so as to maximize the instantaneous dynamic stress intensity factor (K_D). This leads to the general conclusion that cracks always propagate perpendicular to the maximum tensile stress.

Initiated cracks in the borehole vicinity will propagate in a radial direction perpendicular to the maximum tensile stress. These cracks will continue to propagate in a radial direction until the local stress field at the crack tip is altered. When the stress field is altered, the crack will adjust its propagation direction so that it is perpendicular to the maximum tensile stress.

The dynamic stress field about the borehole is due to the generation of the P wave by the impacting of the rapidly expanding explosive gases with the borehole wall. As explained earlier, this stress wave travels outward through the surrounding media with velocity c_P. The destruction of rotational symmetry due to non-uniform crushing and crack initiation about the borehole generates the S wave which travels outward with velocity c_S. The pure shear stresses contained in the S wave are significant in magnitude and the shear disturbance is emitted for a significant duration in time as the radial cracks continue to propagate.

These two stress waves rapidly outdistance the propagating cracks that were initiated at the crushed zone boundary. Theoretically the maximum possible crack velocity is the Rayleigh wave velocity, c_R, approximately $0\cdot9c_S$, but cracks have not been observed to travel at much over half this velocity. Typically a crack will bifurcate at a stress level that corresponds to a much lower velocity. In these fragmentation tests with Homalite 100 individual cracks were measured with velocities in the borehole region of 600 m s^{-1}. This velocity is slightly greater than half c_R and over 40% higher than velocities corresponding to a stress intensity factor required for successful branching. These high velocities occur in the borehole region and are thought to be elevated due to stress field interference from neighboring cracks that inhibits branching.

With the outdistancing of the crack tips by both the P wave and the S wave, a quasi-static situation exists about the borehole. The dense radial crack pattern is driven by borehole pressurization with gases partially penetrating into the cracks and pressurizing their faces. Ouchterlony (1974) has shown that uniform growth of all cracks in a radial system is virtually impossible. Dominant cracks will emerge from the dense radial crack pattern and grow at the expense of the shorter ones. From experimental results Persson et al. (1970) have found the number of dominant cracks to be about 6. Langefors and Kihlstrom (1963) put the number of dominant cracks between 8 and 12, a number we tend to confirm in our experiments.

A typical explosively generated radial crack pattern in a homogeneous material is shown in Fig. 8.3. This figure is a post mortem model made of Homalite 100 which contained a small charge. 11 dominant cracks labeled B and D through M are clearly evident emerging out of the dense radial borehole crack pattern. These 11 cracks propagated an average distance of 19 times the borehole diameter. These long radial cracks succeed in breaking the material into large pie-shaped segments, instead of fragments. More fragmentation would occur if these dominant radial cracks would bifurcate so as to increase the number of active radial cracks.

If the dominant radial cracks are going to branch, they will do so as soon as they emerge from the dense radial borehole crack pattern. Ouchterlony (1974) has shown that for the case of partial pressurization of radial cracks, the crack network is propagating into a decreasing K field after an extension of less than half the borehole radius. For the case of complete crack pressurization K reaches a constant maximum. Thus, borehole cracks reach a maximum K while within the region of dense radial cracking. Thus higher than normal crack velocities can be expected, as previously noted, since the stress field due

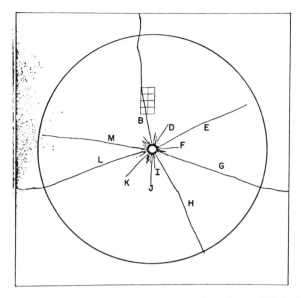

Fig. 8.3. Post mortem crack pattern from a lightly charged hole (test 0).

to neighboring cracks inhibits early branching. As the dominant cracks emerge from the dense radial pattern, the influence of the shorter neighboring cracks decreases and branching will occur. Thus branching due to explosive gas pressurization alone will not occur at distances far outside the dense radial cracking zone.

An example of dominant radial borehole crack branching is shown in the post mortem crack network photograph in Fig. 8.4. In this model and in all cases except when a very small charge is used, the dense radial crack pattern in the immediate vicinity of the borehole was so severe that it was impossible to piece the fragments back together. 5 dominant crack systems are labeled A through E in the photograph. The feathery appearance of the cracks as they propagate out of the dense radial cracked region indicates that the K_D is only marginal for successful branching.

8.2.3.2. Influence of Reflected Stress Waves on Borehole Cracks
Compare the degree of fragmentation in the model shown in Fig. 8.5 with that of Fig. 8.4. The added fragmentation is due to the development of a circumferential crack pattern. The difference between the

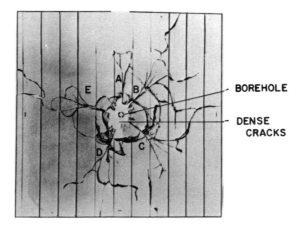

Fig. 8.4. Post mortem of model containing a larger charge (test 26).

two models is the type of charge used. In Fig. 8.4 the charge consisted of 1 part PETN to 5 parts smokeless pistol powder, and in Fig. 8.5 the charge was all PETN. For the model in Fig. 8.4 the detonation was less intense and so was the magnitude of the P wave. The more developed circumferential crack pattern seen in Fig. 8.5 is due to the interaction of the reflected stress waves with the outwardly traveling radial cracks. The outgoing P wave, upon striking a free boundary, is returned as an

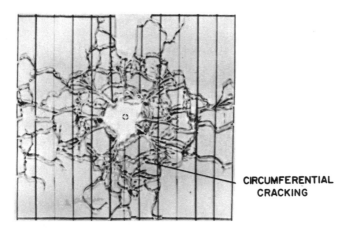

Fig. 8.5. Fracture pattern resulting from a fully coupled charge of PETN (test 28).

image of itself but with a change in the sign of the stresses. The reflected wave (PP) consists primarily of biaxial tension with the radial tensile component approximately 3 times the tangential tensile component. When this stress field is superimposed on the local stress field at the crack tips, the local maximum tensile stress changes direction from tangential to radial. The propagating cracks correspondingly adjust their direction and run in the circumferential direction. Multiple bifurcation will also take place due to the increase in K_D, but it will take place in the new preferred circumferential direction. As the tensile leading edge of the PP wave passes, the maximum tensile stress returns to the tangential direction and the cracks turn and travel once again in the radial direction.

This intersection of the outgoing radial cracks with the PP wave and the resultant change in crack direction and multiple bifurcation has been termed barrier branching. The outgoing radial cracks appear to run into a barrier, branch and then continue in the radial direction. This phenomenon of barrier branching is very important for overall fragmentation. First, it initiates circumferential cracks that tend to break up the large pie-shaped segments between the dominant cracks. Second, due to extensive bifurcation of these circumferential cracks under the influence of the reflected stress wave, more radial cracks are created when the branched circumferential cracks turn and start to propagate again in the radial direction. Many of these newly initiated radial cracks arrest quickly due to the limited driving capability of the explosive gas pressure, but they can be easily reinitiated with the passage of the next reflected stress wave.

Barrier branching was discussed in detail for the case of the PP wave. In the general case of an explosive detonated in the vicinity of a free surface, the result is not so straightforward. Consider first the region along a radial line perpendicular to the free face. There will be barrier branching due to both the PP wave and the SS wave. The location of the SS barrier branching region will not be as distinct as the PP because the radial cracks emerging from the PP barrier will not be traveling at uniform velocities and the SS barrier cracks will not be initiated in the true circumferential direction (the maximum tensile stress in the shear wave is at $45°$ with respect to a borehole radial line and the superposition of this stress field on the existing local field results in a local maximum tensile stress slightly askew of the tangential direction). Thus, shear wave barrier branching does not change the direction of the radial cracks as significantly as the PP wave.

348   *W. L. Fourney, D. C. Holloway and D. B. Barker*

Regions further removed from the borehole are also influenced by the *PS* and *SP* reflected waves. The further the area is from the borehole the more energy one finds in the *PS* and *SP* waves. There are now 4 distinct wave systems, *PP, PS, SS* and *SP*, that will interfere with the outwardly directed radial cracks. In general each one of these waves will momentarily increase the K_D of the radial cracks. This increase in the K_D will cause either branching or will reinitiate already arrested cracks and drive them for a short period of time. The exact preferential crack growth direction during the passing of these reflected wave systems cannot be stated in general, but each of these 4 wave systems will result in crack growth in directions other than the radial one.

8.2.3.3. Summary of Fragmentation in a Flaw Free Media

The detonation of an explosive in a homogeneous flaw free infinite plate produces a very simple radial fragmentation pattern. Only 8–12 dominant cracks emerge from a very dense radial network around the borehole. These 8–12 dominant cracks can travel significant distances, splitting the borehole region into large pie-shaped segments. Branching of the dominant cracks only produces long thin slivers of material along the crack's path and does not greatly succeed in breaking up the material between the dominant cracks.

When the explosive is detonated near a free face, the fragmentation pattern becomes quite complex. Additional fragmentation is caused by reflection of the *P* and *S* wave components back into the medium. These reflected wave components drastically influence the propagation behavior of the borehole radial cracks. These reflected waves break up the pie-shaped segments by causing radial cracks to be reoriented in a circumferential direction. The large tensile stresses in the reflected waves cause the circumferential cracks to multiply (branch). Upon passage of the reflected waves these branched cracks are reoriented back to a radial direction. The combination of circumferential cracking and the presence of a large number of radial cracks results in a much improved fragmentation pattern over that obtained in an infinite medium.

8.2.4. Fragmentation Resulting From Small Flaws Subjected to Intense Stress Waves

In order to study the effect of small flaws on fragmentation, models were used which had flaws artificially routed into 1 surface. These flaws were made with a sharp 60° router bit and were 6 mm in length and 1 mm deep. The flaw depth as well as charge size were selected so as to have crack initiation occur primarily at the artificial flaw locations and not in 'flaw free' regions of the model.

The model depicted schematically in Fig. 8.6 was one of many which have been used to study flaw initiation. The specimen was fabricated from 6·4 mm thick Homalite 100. 8 flaws were routed into the front surface of the model as shown in the figure. Flaws labeled A through F were placed 133 mm from the borehole with flaws D and C also being parallel to and 19 mm from the lower free surface. 2 additional flaws, G and H were placed parallel to and 19 mm from the left free surface.

8.2.4.1. Small Flaws Initiated by Outgoing Stress Waves

The early phases of crack initiation are seen in the sequence of pictures in Fig. 8.7. Each picture is an enlargement of the dynamic isochromatic fringe pattern around flaws E and F. The borehole is outside the picture frame on the right as marked by the black cross. Flaw E is the lower of the 2 flaws shown.

80 ms after detonation the compressive front of the P wave has just passed over the 2 flaws as shown in Fig. 8.7(a). The compressive peak can be seen as the broad light fringe passing through the vertical grid line, just to the left of the flaws. The tensile region of the P wave is just approaching the flaws.

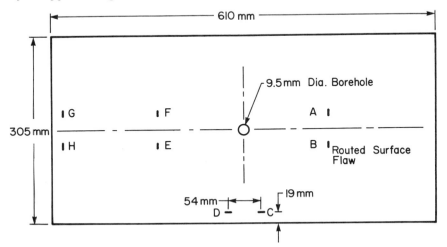

Fig. 8.6. Model used in artificially flawed tests in transparent plastics.

350 W. L. Fourney, D. C. Holloway and D. B. Barker

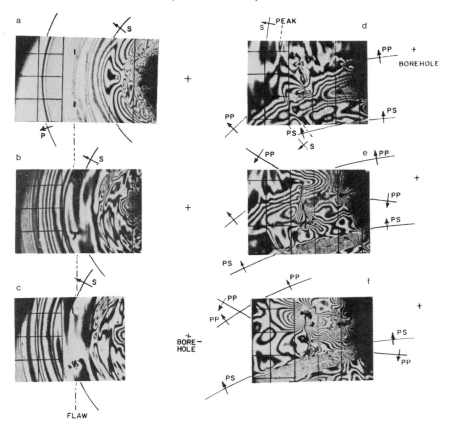

Fig. 8.7. Crack initiation at artificial flaw by stress waves. Lapsed time: (a) 80 μs; (b) 103 μs; (c) 116 μs; (d) 137 μs; (e) 155 μs; (f) 170 μs.

In Fig. 8.7(b) taken 103 μs after detonation, a building stress intensity factor is visible at the lower flaw due to the tensile radial component in the *P* wave tail. Since the flaw is not oriented in an exact borehole circumferential direction, mixed mode loading exists as confirmed by careful observation of the isochromatics. Mixed mode loading means that the crack tip stresses are a combination of tensile and shear components. This mixed mode loading is more obvious in Fig. 8.7(c) where the isochromatic loops are seen skewed in a counterclockwise direction. At this time cracks have already initiated at the

Model studies of fragmentation 351

upper and lower corner of flaw E. This initiation occurred somewhere between 103 and 116 μs after detonation. At 116 μs the S wavefront has just reached the flaw and its full effect will not be felt until later when the peak of the S wave arrives. Note that no initiation occurred at the upper flaw. The lower flaw was routed slightly deeper and since the stress intensity factor is proportional to the square root of the crack depth, a greater stress intensity factor occurred at the lower flaw.

Figures 8.7(d), (e) and (f) are enlargements of the region around this P wave initiated flaw (flaw E) showing its later propagation behavior. The two cracks that were initiated at the top and bottom of the flaw started to propagate in a circumferential direction, perpendicular to the σ_r tensile component of the P wave tail. In Fig. 8.7(d) the cracks are propagating in a direction approximately 22° off the circumferential. This new propagation direction occurred abruptly and can be seen as a kink in the crack path located about halfway between the crack tips and the flaw. This abrupt change in direction occurred as the peak of the shear wave passed over the crack. The shear wave peak can be seen as the circumferential fringe located halfway between the flaw and the S wavefront. Thus, the propagation direction in Fig. 8.7(e) is due to the combined influence of the decaying P wave tail and the building S wave.

The PP wave component from the upper free surface is entering the field of view in Fig. 8.7(e) and has just reached the upper crack tip as has the PP wave from the lower free surface. Since K_D was nearly equal to K_b already (the stress intensity factor necessary for branching) the added energy contained in the PP wavefront tends to explode the crack tip as seen in Fig. 8.7(f). The preferential branching direction is parallel to the PP wavefront, since the tensile stress component perpendicular to the wavefront is approximately 3 times the parallel tensile component.

The initial velocity of the initiated cracks was about 280 m s^{-1} but increased to approximately 350 m s^{-1} before branching occurred under the influence of these reflected waves. The PP wave supplied the energy to branch the original crack but with its passage the majority of the resulting branches arrested.

The PS wave is a pure shear wave containing a resolved tensile stress of significant magnitude as evidenced by the closely spaced isochromatic fringes shown approaching the lower branches in Fig. 8.7(f). Unfortunately, the dynamic recording ended at this time. It is possible to determine further crack extension by comparison of Fig.

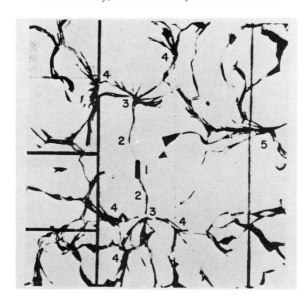

Fig. 8.8. Post mortem photograph showing the development of P wave initiated cracks. 1. Flaw initiation by P wave tail; 2. reorientation with S wave; 3. PP branching; 4. PS branching; 5. PP branching of radial crack.

8.7(f) with the final post mortem photograph (Fig. 8.8). The PS wave apparently caused reinitiation and branching of these lower branches as indicated in the post mortem picture. Further crack extension occurred with the passage of the SP and SS waves which have yet to come into the field of view in Fig. 8.7.

Comparison of Figs 8.7 and 8.8 also shows that most of the branches at the upper main crack arrested with the passage of the two PP waves. Additional branching did occur on the left and right main branches due to the combined influence of the two PS waves. Again the SP and SS waves served as additional sources of energy to extend the existing cracks.

8.2.4.2. Initiation by Reflected Waves
In this section results of tests conducted with fragmentation models to demonstrate flaw initiation by reflected wave systems will be examined. Examples will be presented showing initiation of routed flaws by individual waves as well as by wave system combinations. The exam-

Model studies of fragmentation 353

ples presented do not include all initiation sources encountered but do give a good representation of the major initiating sources encountered.

In the fragmentation tests conducted in Homalite 100, classical spalling as described in Section 8.2.3 was rarely seen since the tensile tail of the P wave was negligible, having been consumed in the creation of the dense radial borehole crack pattern. Flaws were initiated outside the shadow zone, but not at a spacing as expected in classical spalling. Cracks were initiated by the PP wave at various distances from the model boundaries depending upon whether or not the tensile stress component in the wave was large enough to produce a local critical K value. Due to the natural attenuation of the reflected wave, more flaws were initiated near the free surfaces than at points located in the interior of the model.

Figure 8.9 is a sequence of photos showing initiation at 2 routed surface flaws located just inside of a straight free face. A schematic of this model is shown in Fig. 8.6. The flaws shown in the photographs in Fig. 8.9 are labeled flaws C and D in the schematic. The borehole is out of the field of view, but the wavefronts are marked in the photographs.

The first picture, taken $80 \mu s$ after detonation, shows the P wavefront just after it strikes the lower free surface. The S wave is following behind. No disturbances are seen around the flaws due to the compressive stresses in the P wave.

Initiation at the routed flaws has occurred in the next frame taken $93 \mu s$ after detonation. The PP and PS wavefronts have passed over the flaws while the S wavefront has yet to arrive. Due to the high K_D, characteristic branching is occurring at the ends of the routed flaws. The third photograph taken $105 \mu s$ into the dynamic event shows this branching better. In this photograph the S wavefront has passed over the flaws, but its maximum influence has not yet been experienced. The most intense stresses in the S wave can be seen following the wavefront as indicated by the discontinuity in the PP wavefront directly above the two flaws.

Figure 8.9(d), taken $170 \mu s$ after detonation, shows the 2 initiated flaws after the reflected S wave components have passed. The branching at both ends of the flaws was in an H pattern. The major branches are driven to the free surface or radially inward.

The H-like appearance of the initiated flaws in Fig. 8.9(d) was found to be typical of spall type fracture. The crossbar of the H corresponds to the original orientation of the flaw. The maximum tensile stress in

354 W. L. Fourney, D. C. Holloway and D. B. Barker

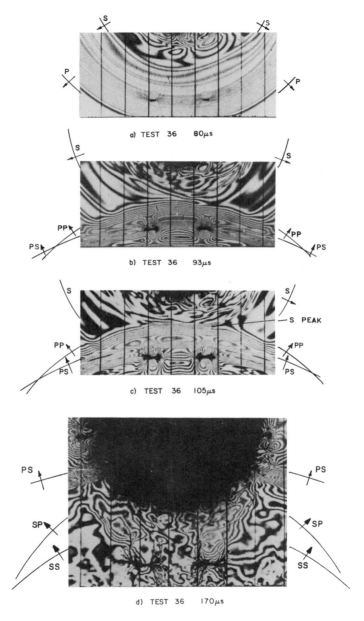

Fig. 8.9. Crack initiation due to the initial action of the PP and PS waves.

the *PP* wave is perpendicular to the wavefront and hence flaws oriented parallel to the wavefront are most likely the ones to be initiated. Note that the H crossbars are all parallel to the free surface in this figure. The multiple crack branching or brooming at either end of the original flaw forms the legs of the H. This multiple branching is due to the extremely large K_D values generated. The cracks are originally initiated when the local stress field around the flaw produces a K greater than the fracture toughness of the material. As the cracks grow, K_D increases with the square root of the crack length and in direct proportion to the tensile stress magnitude. Thus as the maximum peak of the *PP* wave passes over the already initiated cracks, extremely large K_D values can be generated producing multiple branching. From the last frame in Fig. 8.9(d) there appears to be no preferred branching direction. The biaxial tension in the *PP* wave seems to explode the crack tips.

8.2.4.3. Summary of Fragmentation as Influenced by Small Flaws

Fragmentation in a flawed model was found to be quite different from that observed in a flaw free media. The greatest effect of small flaws on fragmentation is caused by cracks initiated at flaw sites remote from the borehole region by the combined action of the *P* wave tail and the shear wavefront. Flaws initiated in the immediate borehole vicinity by these waves have only a minor effect since they generally propagate in a radial direction and rapidly coalesce with the borehole cracks.

Remotely initiated flaws located further from the borehole generally do not propagate in a radial direction. This non-radial propagation succeeds in breaking up the large pie-shaped segments created by the dominant radial cracks. The maximum distance from the borehole that the combined *P* and *S* waves can initiate flaws is a direct function of the attenuation of the wave and the severity of the flaw. It is easily possible to initiate and propagate cracks at distances further from the borehole than the borehole generated cracks propagate. Once a flaw is initiated by the *P* wave tail and the leading front of the *S* wave the remainder of the *S* wave has sufficient energy to keep the crack from arresting.

The outwardly directed *P* and *S* waves can initiate flaws anywhere independent of the presence of a free surface. Outside the immediate borehole vicinity, only severe flaws oriented approximately parallel to the wavefront are initiated. As was the case for flaw free material the reflection of the *P* and *S* waves from a free boundary causes additional fragmentation. The strongest flaw initiating mechanism observed in the models containing small flaws was the combined action of the *PP* and *PS* wave near a free surface. This mechanism is not necessarily the most important for overall fragmentation, but it has been observed to initiate flaws near free surfaces at extreme distances from the borehole. Due to the natural attenuation of the reflected *P* wave components, the majority of flaws seem to be initiated near the shadow zone boundary.

As the reflected *P* wave travels back into the medium it meets the outward moving *S* wave. Constructive interference between these two wave systems produces 2 distinct zones for further crack initiation. These 2 zones are indicated in Fig. 8.10 which is a schematic of the region between the borehole and the free face showing the major regions where cracks are initiated. The left side of the figure shows the principal initiating wave systems and the zones where initiation can take place.

On the right side of Fig. 8.10 the locations of barrier branching for a radial crack with an assumed constant radial velocity are shown. These locations are included in the figure to show all primary zones where non-radial cracks can be initiated both in the flaw initiated network

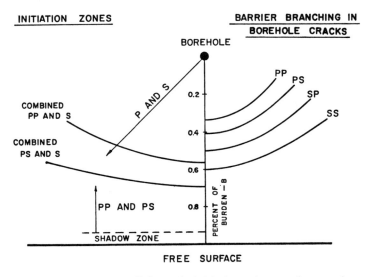

Fig. 8.10. Zones of non-radial crack initiation. Assumptions: $c_P/c_S = 0.60$; radial crack $= 0.2c_P$; P wavelength $= 0.5B$; S peak lags front by $0.3B$.

Model studies of fragmentation 357

and in the borehole crack network. Once flaws are initiated, the reflected *SP* and *SS* wave systems serve as a source of driving energy to continue propagation and further branching of the cracks. Secondary reflections of all the stress waves can also further the crack extension process.

Figure 8.10 was constructed for the particular parameters of wave speed and crack velocity encountered in Homalite 100, but these parameters are similar to those expected in rock blasting. From the figure it is seen that there is an almost uniform distribution of primary initiation zones for non-radial cracks between the borehole and a free surface. This uniform distribution helps explain why large pie-shaped segments are not found in typical fragmentation blasting.

8.2.5. Fragmentation Resulting from Interaction of Stress Waves with Large Joints and Bedding Planes

Figure 8.11 is a photograph of a working bench face in the Pinesburg Limestone Quarry located near Hagerstown, Maryland. The fractures seen in Fig. 8.11 are typical for those found in quarry faces. The fracture structure consists of the natural joint sets and blast induced damage from prior explosive shots. These detonations loosen up the joints and faults that are present in the formation, but not to such an extent that the material can be removed.

The spacing of these joint sets and bedding planes varies from rock type to rock type and even from point to point in the same rock formation. The orientation of these fault sets also varies, but the assumption is that the joint sets and the bedding planes make up an orthogonal triad of planes. Rock joints in this quarry range from open sets filled with mud to very tightly bonded calcite joints with a strength nearly equal to the tensile value found in the adjoining rock masses.

In an effort to study fragmentation in bedded and jointed rock masses jointed models were constructed of Homalite 100. All models were 6·4 mm thick and were made by bonding together 50 mm wide strips of Homalite 100. The strips were rough cut on a band saw and then routed to final size to provide smooth edges. One very important parameter in the model is the bond formed between the 2 photoelastic strips. During the experimental program variations in bonding strengths were utilized from grease filled joints to very tough epoxy glues. The particular bonding agent used throughout the test series described in this chapter is a product sold under the trade name M Bond 200. M Bond 200 is marketed by Micro Measurements Inc. and

Fig. 8.11. Bench face showing bedding and joint sets at Pinesburg Maryland Quarry.

contains cyanoacrylate ester which sets rapidly under slight pressure. The adhesive was spread uniformly with a resulting bond that was approximately 0·075 mm in thickness. The joint could be classed as medium in strength in that it could normally be handled without separating, but in some instances did fail to remain intact when the samples were being prepared for testing. Tests conducted to determine bond strength showed a tensile failure in the bond at about 6·9 MPa and a shear strength that exceeded the shear strength of the Homalite 100. That is to say, when a model was loaded in tension at 45° to the bond line, the Homalite 100 failed before the bond did. Most models which separated during handling did so as a result of bending loads being applied.

The 6·4 mm thick two-dimensional models were 305 × 381 mm with two boreholes. The geometry of the models tested is given in Fig. 8.12. Each borehole was loaded with 200 mg PETN and the gases contained in the manner of the previous tests.

The first 4 of 16 frames taken during one test are shown in Fig. 8.13. This test had an elapsed time of 400 μs between the detonation of the first borehole and the second. Figure 8.13(A) was taken 34 μs after detonation of the first borehole. The fronts of the outgoing P and S

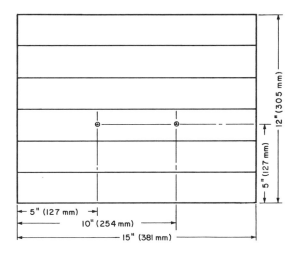

Fig. 8.12. Geometry of jointed photoelastic models. (Notes: 1. Boreholes are 0·25 in (6·35 mm) in diameter; 2. model consists of six 2 in (51 mm) wide, 0·25 in (6·35 mm) thick H-100 strips.)

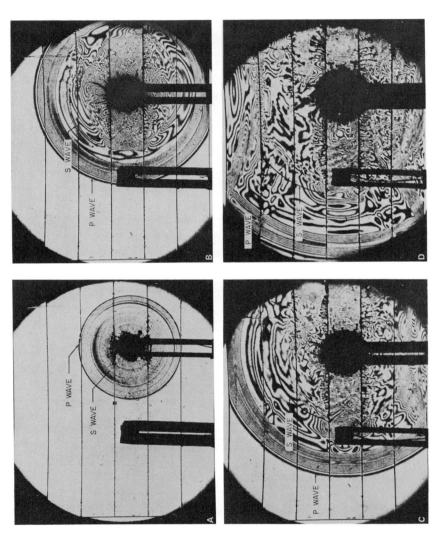

Fig. 8.13. Photoelastic patterns obtained from a jointed model test. Time after detonation: (A) 34 μs; (B) 62 μs; (C)

Model studies of fragmentation

waves have been marked. The shadows of the fractures seen in this frame are radial in direction. Although it cannot be proven conclusively, some appear to have originated at the borehole wall and some at the bond lines between layers. All of these fractures were, most likely, produced by the outgoing P wave. The second frame (Fig. 8.13(B)) was taken 62 μs into the dynamic event. The long radial fractures are due to bondline initiated fractures that coalesce with the borehole radial crack pattern. A large number of non-radial cracks are evident along the joint line. These fractures were initiated at the bond lines adjacent to the borehole and are traveling *away* from the borehole layer at angles of 60–80° to the bond line. These joint initiated cracks were the dominant fragmentation mechanism in the layered polymeric models. These fractures are considered to be the result of high shear loading on the bonded interface. In earlier layered model testing described elsewhere (Fourney *et al.*, 1979), crack initiation has clearly been linked with the passage of the compressive peak of the P wave. In this particular test also, there is evidence that these cracks are initiated by the compressive peak of the outgoing P wave. Observe from Fig. 8.13(A) the area just in front of the S wavefront and behind the P wave on the joint line below and to the right of the charge. Small dark areas along this joint can be barely discerned. These are initiation sites of the intense joint initiated cracking seen in Fig. 8.13(B). Once these sites are initiated by the resolved shear stress in the outgoing P wave, they grow in a direction roughly perpendicular to the bond line as a result of the tensile stress in the outgoing shear wave. The stress state of the P waves responsible for this mechanism of fragmentation are sketched in Fig. 8.14. Notice from Fig. 8.14 that the maximum fragmentation due to this mechanism should occur where a borehole radial line intersects a joint line at 45°. This observation is confirmed from the photographs in Fig. 8.13.

Figure 8.13(C), taken at 86 μs, shows the further development of fragmentation due to the first detonation. Notice the relatively strong shear wave that is traveling along the bond line above and to the left of the borehole in Fig. 8.13(C). Note also that the radial cracks in the borehole layer have turned and are propagating parallel to the bond line. A number of cracks have also been initiated at the second interface directly above the borehole. Unlike the earlier joint initiated cracks these propagate both toward and away from the borehole. These cracks have been initiated and driven by the shear wave since the region of maximum shear in that wave system is directly above and below the borehole.

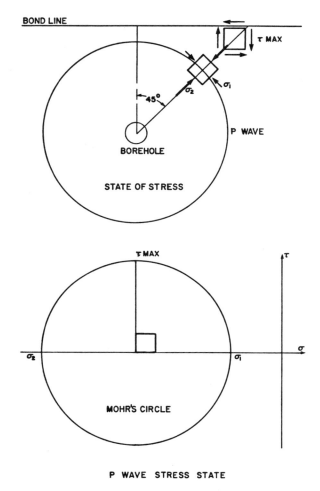

Fig. 8.14. Stress state at bond line due to outgoing P wave.

Cracks initiated from a slight imperfection on the bond line (whether the imperfection is due to small debonded areas or to small areas debonded by the passage of the P wave) by pure shear loading should initiate in a direction of 70° from the bond line (Erdogan and Sih, 1963). All of the joint initiated cracking seems to follow this behavior.

At 116 μs (Fig. 8.13(D)) the fragmentation in the borehole layer is beginning to develop due to the arrival from the right of the reflected P

Model studies of fragmentation 363

wave, which is now predominantly biaxial tension. Some initiations have occurred near the second borehole due to reflections from the undetonated hole. The reflected waves from the right free surface have caused extensive branching when passing over the propagating joint initiated cracks. This is very obvious in the layer just above and to the right of the first borehole.

The photographs presented in Fig. 8.15 were selected to show the additional damage done to the model by the second borehole detonation. In Fig. 8.15(I), taken $419\,\mu$s into the event and $19\,\mu$s after the second detonation, the fragmentation is due to the first detonation except for the region inside the P wavefront emanating from the second borehole. Notice the intense fragmentation that has occurred in all four corners of the model due to the reinforcement of reflected stress waves. The small 'butterfly-like' fringe loops at the tips of cracks imply that many are still active from the first detonation and continue to propagate. The lower layer has separated from the rest of the model over most of its length, yet many cracks continue to propagate in it. Of particular interest is the complete or near complete fragmentation which has occurred in the borehole layer and the two layers abutting it.

Figure 8.15(II), taken $47\,\mu$s after detonation of the second borehole, shows the continued growth of the region affected by the wave systems from the second borehole detonation. Many of the cracks within this affected region have been reinitiated and have changed propagation direction due to the sudden change in stress state created by the passing of the stress waves.

The increased fragmentation caused by the second detonation is very evident in Fig. 8.15(III), taken $108\,\mu$s after the second detonation. Considerable crack growth and additional joint initiations have occurred to the left of the second borehole. Notice in particular the bundle of joint initiated cracks labeled A in Fig. 8.15(III). From Figs 8.15(I) and (II) it is evident that these were not initiated by the outgoing P wave but, rather, occurred when the reflected waves from the left free boundary passed over already existing cracks. Additional cracks have been initiated from the bond lines (cracks labeled B and C in Fig. 8.15(III), for example). Many other cracks have become highly active even to the point where much multiple branching has occurred (D and E).

Figure 8.16 shows the fragmentation pattern of the model at about $700\,\mu$s or nearly $300\,\mu$s after the second detonation has occurred. The fragmentation of the model is intense, especially in the lower 5 layers.

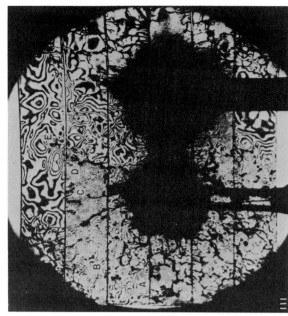

Fig. 8.15. Dynamic photoelastic photographs taken after detonation of second borehole. Time after detonation: (I) 19 μs; (II) 47 μs; (III) 108 μs.

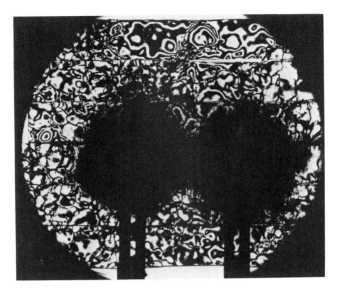

Fig. 8.16. Frame taken very late in the fragmentation event (298 μs after second detonation).

The only sizable fragments are in the uppermost layer, but some cracks in that layer are still active. By this point in time the model has separated along most of the bond lines and large rigid body motions are occurring. Keep in mind that almost all the fragmentation that has occurred is due to stress wave action—not crack pressurization.

From close examination of the results from approximately 20 tests similar to the one discussed it appears that the large number of cracks initiated at the bond lines are the result of a shear loading on the layer interface. This conclusion is based upon several different observations:

A. The speed at which the initiation front appears to travel along bond lines agrees within experimental error to the speed of the P wave. The resolved shear stresses as shown in Fig. 8.14 must therefore initiate the cracks.

B. The appearance of the small black dots at initiation sites corresponds in time with the arrival of a high shear loading. The dots, also known as caustics, are an optical effect caused by the localized stress state. A miniature lens is formed when the stresses are high and light from the light source is refracted out of the optical system. Directly above the borehole this shear loading is associated with the arrival of the shear wave while along the layer interface at other locations it is associated with the arrival of the compressive peak of the P wave.

C. The most intense cracking and the cracks which appear to accelerate the fastest are in locations where the highest shear loading is found.

D. Finally, the direction of propagation of the initiated cracks is between 70 and 90° to the layer interface and directed away from the borehole layer. This agrees well with analytical solutions for shear loading of an existing crack. This would imply that the layer interfaces appear to the outgoing stress wave to be a very long partially bonded crack. When the shear loading is applied by the stress waves, cracks at about 70–90° to the interfaces are initiated wherever the shear stress in the outgoing wave exceeds the level necessary for cracking to occur.

For lack of a more descriptive title we have named this type of fracture joint initiated fracture or JI cracking. The typical speed of these cracks during the early part of the dynamic event is around 400 m s^{-1} and in many cases the cracks are being driven with sufficient energy to result in branching.

The tests described would represent blasting conducted in a material which had one predominant joint set present. The same mechanism of fragmentation was observed when checker board like models were fabricated to represent perpendicular joint sets. Figure 8.17 presents a photograph from a test which used a line charge ignited from the bottom to represent a two-dimensional approximation for a column charge in a heavily bedded rock formation. Note from the figure the presence of joint initiated cracks which occur as a result of the PETN detonation.

To the authors' knowledge, the fragmentation mechanism observed in the layered tests has not been identified previously. The resulting fragmentation is intense when compared to homogeneous models at comparable load levels. In layered tests with two boreholes, for example, approximately 54% of the model weight was reduced to particles with a maximum dimension of less than 25 mm. For similar tests with homogeneous models of comparable size and loading levels, only 28% of the model weight resulted in particles with a maximum dimension less than 25 mm.

Model studies of fragmentation 367

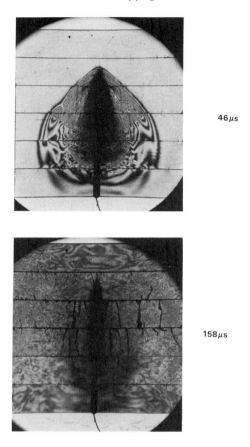

Fig. 8.17. Fragmentation in a model simulating bedding planes.

Results from production quarry blasting seem to support the type of fragmentation that has been identified in these tests. Very few fragments from the muck pile after the blast are pie-shaped. They are blocky and their size seems to increase slightly as the free face is approached. The pattern of fragmentation, as determined from observation of high speed photographs of quarry blasts by scientists from Martin Marietta Laboratories, Baltimore, Maryland, is presented in Fig. 8.18 (Winzer et al., 1979). Note the blocky nature and the decrease in fragment size as the borehole region is approached. This

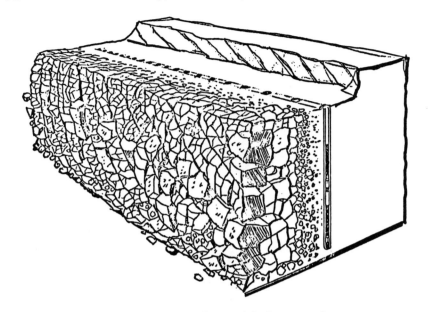

Fig. 8.18. Fragmentation model of a quarry face.

pattern is easily explained on the basis of the fragmentation mechanism observed in the model tests described in this chapter.

8.2.6. Fragmentation Enhancement from Interaction Between Multiple Boreholes

Several tests were conducted with jointed models with two boreholes. The delay time between the two boreholes was varied and the resulting fragmentation pattern was studied. This section describes the results of those tests.

After each test the fragments were collected and sieved to determine the size distribution of particles. All model weights were identical because of identical geometry and material. The weight distribution of particles was used to construct histograms for graphically illustrating the distribution of fragments for each delay. Figures 8.19(a)–8.19(e) are these results: 8.19(a), one hole only; 8.19(b), instantaneous detonation; 8.19(c), 200 μs; 8.19(d), 400 μs; and 8.19(e), 600 μs delay. Plotted vertically is the weight of fragments not passing through a given sieve size while the horizontal coordinate gives the sizes of the holes in the sieves utilized. A leftward skew implies large fragments.

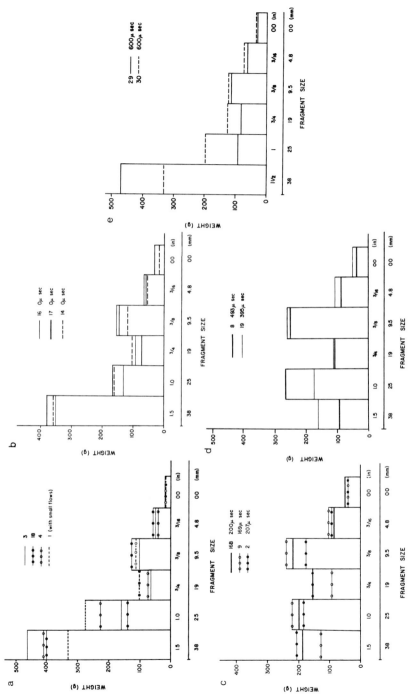

Fig. 8.19. Histograms of fragment sizes for dual borehole tests. (a) One hole only; (b) instantaneous detonation; (c) 200 μs delay; (d) 400 μs delay; (e) 600 μs delay.

The histograms show a leftward skew for 1 hole and instantaneous detonations while a more uniform distribution is shown for 200 and 400 μs delays. An extreme left skewness is shown again for the 600 μs delay tests.

The 200 μs delay histogram shows an almost equal distribution of fragments in the over 38 mm: between 19 and 25 mm, and between 4·8 and 19 mm categories. The 400 μs delay appears to result in even better fragmentation than the 200 μs tests. The amount of fragmentation greater than 38 mm has been reduced while the number of fragments in the 19–25 mm and 4·8–19 mm ranges have increased.

Increasing the time delay to 600 μs resulted in poorer fragmentation rather than better. Close examination of Figs 8.19(e) and (b) indicates that the longest delay yielded fragmentation about equal to that obtained from instantaneous detonation of the two boreholes. A frame from a dual borehole layered model test is presented in Fig. 8.20. This was a 600 μs detonation delay test and this particular frame was taken

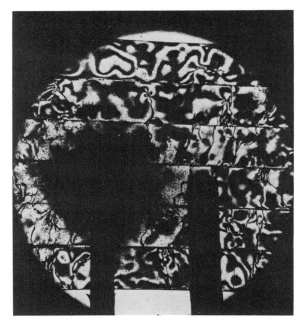

Fig. 8.20. Photograph of model taken just prior to 600 μs delayed borehole detonation.

at 593 μs. At this time there is complete separation of the layers. As a result, the stress waves generated from the second detonation are very ineffective. Some additional fragmentation occurs, but by this time the model is too broken up to transmit the stresses to regions away from the borehole.

In one of the jointed single hole detonations, small flaws 6·2 mm in length and 1·6 mm deep were routed into the front surface of the model on 25 mm centers. This was done to assess the effect of the combination of small flaws and large flaws on blast induced fragmentation. As can be observed from the histograms in Fig. 8.19(a), a small increase in fragment size resulted. However, this was not large enough to be considered significant in that the results fall within the scatter range of the other data for the single hole experiments.

From the histograms presented in Fig. 8.19, a weighted average fragment size was computed. This average size gives an idea as to the size of a typical fragment. The following formula was utilized for calculating F_{av}:

$$F_{av} = \frac{\sum_{i=1}^{n} X_i W_i}{\sum_{i=1}^{n} W_i}$$

F_{av} is the weighted average, X_i is the mean of the ith and $(i-1)$th sieve dimensions, W_i is the total weight of fragments retained by the ith sieve. When more than 1 test per delay was available an average of the weights retained by a given sieve was used; n is the total number of sieve sizes used.

This same calculation was used by Singh (1979) for homogeneous dual borehole models made from Homalite 100. The results from Singh's tests are presented in Fig. 8.21(a). The average fragment size for homogeneous models was still decreasing at a 600 μs delay. Unfortunately Singh had no data between 600 μs and an infinite delay but the decrease in average fragment size between 520 and 600 μs was not as great as the decrease from 380 to 530 μs. An estimate of a minimum fragment size of about 30·5–33 mm is felt to be accurate in spite of the extrapolation necessary. Fragmentation results for 1 hole detonation are also shown on the figure.

Similar results for the layered models are shown in Fig. 8.21(b). The model sizes of the homogeneous and jointed models were the same except for thickness. The charge sizes were scaled in the same ratio as the thicknesses. Single hole detonation resulted in an average fragment size of a little under 25 mm for the jointed media and over 76 mm for

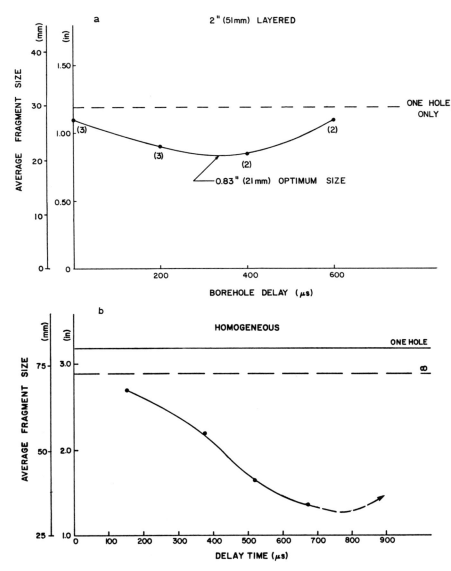

Fig. 8.21. Particle size as a function of delay.

Model studies of fragmentation 373

the homogeneous models. The minimum average fragment size for the jointed media was about 20·3 mm and occurred at a delay of 360 μs. For the homogeneous models the value extrapolated from Fig. 8.21(a) is a minimum average fragment size of 31·8 mm at about 775 μs delay.

For a jointed medium the fragmentation mechanism of joint initiated cracking yields a much smaller average fragment size than would be obtained in a homogeneous medium. This reduction in fragment size is at least 1·5 times. For some delays it can be as high as 2 times.

The delay needed to optimize the fragment size is much more critical with a jointed medium than for a homogeneous one. The optimum delay time is much shorter for the jointed media by about a factor of 2 over that of the homogeneous models. The increase in fragment size with increased delay time once the optimum is exceeded is more drastic with a jointed medium. This observation is not surprising. After a given amount of time the joints separate. The only additional damage done by the second detonation is restricted to the area in the immediate vicinity of the second borehole.

8.3. STUDIES OF FRAGMENTATION IN ROCK

The natural progression in scaled laboratory investigations is to move towards testing materials of real interest once the phenomenological sequences have been studied in transparent plastics. The immediate and obvious problems that are encountered when this is attempted is that there is a much greater variability in the properties of the rock samples, and that the stress wave and crack propagation in rock is much harder to detect. In this section results obtained in testing rock specimens will be described.

8.3.1. Experimental Procedure

The basic testing parameters were similar to those of the Homalite tests. Key tests were selected to confirm that the mechanisms active in the fragmentation of the Homalite were also present in the rock. Descriptions will be given of tests conducted in homogeneous granite plates as well as naturally flawed limestone plates. The granite plates were loaded with centrally located boreholes as well as boreholes offset towards the plate corners. In addition tests will be described which were conducted in limestone plates that had been separated and

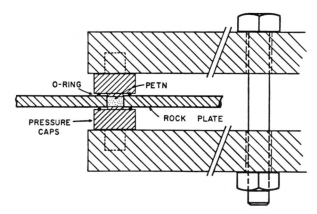

Fig. 8.22. Pressure containment device.

recemented. These latter tests were similar in geometry to the jointed Homalite models described earlier.

For most tests the single boreholes were 4·7–6·3 mm in diameter and the PETN explosive was fully coupled to the borehole with a density of $0·7$–$1·3$ g cm^{-3}. A pressure cap arrangement similar to the one used in Homalite testing was used to contain the explosive gases and is shown in Fig. 8.22. The specimens were nominally $20 \times 20 \times 1·3$ cm and were cut with a diamond saw from blocks of Westerly granite, St Paul (Stone River Formation) limestone and Chambersburg limestone. The St Paul limestone was nearly 100% carbonate, with about 3% dolomite and less than 1% quartz and graphite. The Chambersburg was 83% carbonate, with 11% dolomite and 16% quartz and pyrite. These compositions were obtained by Montenyohl (1978). The granite served as the nearly homogeneous material and the limestones served as the naturally flawed models. In most tests the specimens were photographed during the dynamic event by 1 of 3 high speed photographic methods. Each method has its own particular advantages and disadvantages. The purpose of the photographic instrumentation was to determine typical crack velocities and when the cracks formed. Wave speeds in the rocks were computed from information gathered during earlier testing.

8.3.1.1. Conventional Rotating Drum Camera

A Cordin Dynafax 350 framing camera was operated at 35 000 frames s^{-1}. There are available with this camera 224 frames of 16 mm image width. The film used for these tests was Kodak's Tri-X. The best photographs were obtained when the rock surface was lapped flat and then lightly coated with silver by vacuum deposition. The lighting system consisted of 2 banks of 6 electronic flash units (guide number of 80 at ASA 25) located at an angle of ±45° with respect to the normal to the specimen surface.

The camera has an exposure time of 0·75 μs but does not have the spatial resolution to record crack initiation. The time between pictures is about 25 μs which is not fast enough to follow in detail the crack growth. What this camera does provide is an extended recording period.

8.3.1.2. Cranz Schardin Spark Gap Camera

A 16 gap Cranz Schardin camera was used in a reflection mode as shown in Fig. 8.23. Light from a 4×4 array of spark gaps passed through a 50-50 beam splitter and was collimated by the field lens. It then reflected off the polished and silvered rock surface back through the lens to the beam splitter where it was projected onto the 4×4 array of camera lenses. Both Tri-X and Gravure Positive film were used, with the best results coming from the Gravure Positive film

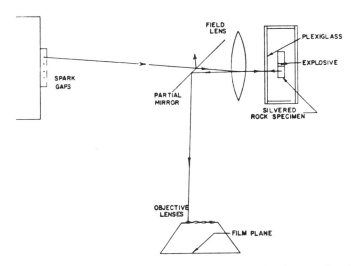

Fig. 8.23. Cranz Schardin camera arrangement for reflection mode of operation.

developed for 15 min in D-11, a developer manufactured by Kodak Inc.

The rock samples were polished using an overarm lapping machine with a cast iron head and carborundum slurries. The final polish was achieved with a felt buffing head and $0.06\,\mu$g alumina powder. The rock was placed in a vacuum oven for several hours at 60°C and then put in a vacuum deposition chamber where a layer of silver was applied. Accurate measurements of the layer thickness were not made, but it was assumed that the layer thickness was of the order of a few wavelengths of light.

The reflection method for photographing the exploding rock has several advantages and disadvantages. Since there are no moving components to the camera, the resolution of the image is limited only by the duration of the light source and the quality of the optical components. The other attractive feature is that there are 16 pictures taken during the event which allow the development of the cracks to be studied. Some problems were encountered when the rock was not well polished. In this case there was not enough reflected light to expose the film. In addition, the image smear at the camera lens board allowed light from neighboring spark gaps to overlap. This effect caused a ghosting of the image and made crack identification difficult. Even with excellent specimens, there was some ghosting of the image. This latter effect results from the fact that, even though the exposure time of the camera is approximately $1.0\,\mu$s, there exists a long but low level glow at the spark gap. This after glow continues to expose the film which results in a ghosted image. For reasons not understood, not all spark gaps exhibited this behavior to the same degree. The net effect of this problem is that not all 16 pictures were useable for data reduction. A final difficulty encountered was that of focus. For transmission work, as in dynamic photoelasticity, a slight misfocus of the image is not detrimental. For the reflection work, if the image is slightly misfocused or soft, the sharp crack line in the rock and coating will disappear. This problem and the ones mentioned above reduced the useable number of pictures from 16 to typically 6.

8.3.1.3. Holographic Interferometry

The technique of holographic interferometry was applied to the study of rock fracture by modifying the normal holographic procedure. The standard method of taking 1 holographic image before detonation of the explosive and then the second at some time after the explosion could not be used because the surface deformations are so great for this problem that all interferometric information would be lost.

Instead, both holographic images were taken after the detonation with the time between holograms reduced to 800 ns. In this way the stress waves served as a qualitative indicator of the cracks that existed in the rock at the time the holograms were taken. The stress waves create surface displacements that are in turn measured by the holographic technique. Where there is a crack the surface displacements are not continuous across the crack and thus the cracks show up as discontinuities in the holographic fringe pattern.

The holographic arrangement was the standard off-axis reference beam method with a slight diverging reference beam and a diverging object beam. The holographic plates were Agfa 10E75 developed in D-19 for 8 min. The specimens of limestone and granite were spray painted white on the front face, and the explosive was loaded and contained in the same manner as with the other tests.

The advantage of this method is that it permits an excellent visualization of the developing crack pattern at the instant the 2 holograms are taken. It is assumed that the cracks do not travel very far in the time interval between pulses. In fact, taking the highest observed crack velocity and the interpulse time yields an upperbound movement of $0·14$ mm. The drawbacks of this technique include the cost of the laser system and the fact that only 1 picture of the event is obtained. Some additional details of the 3 techniques can be found in Holloway *et al.* (1979, 1980).

8.3.2. Results of Rock Plate Testing

Figure 8.24 shows a reassembled granite plate ($30·5 \times 30·5 \times 5$ cm) that was loaded with 8 g PETN. The borehole was located in the upper right corner. The features of the crack pattern are very similar to those described earlier for the non-flawed Homalite. There is an intensely damaged zone near the explosive, then radial and circumferential cracks. The circumferential cracks are a result of the reflected waves from the free boundary. In some cases the circumferential crack has formed before the radial crack, for example at the junction of squares C-9 and D-9. At squares F-9 and G-9, the reflected waves reorientated the radial crack. Multiple bifurcation of cracks as seen with the Homalite is not as evident in the rock, the problem being one of detecting branched but arrested cracks of small length.

A post mortem assembly of a Chambersburg plate is shown in Fig.

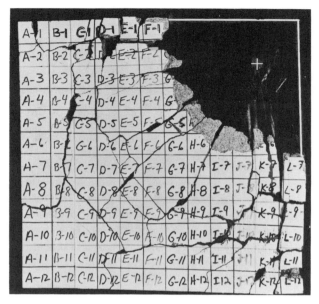

Fig. 8.24. Post mortem of granite plate.

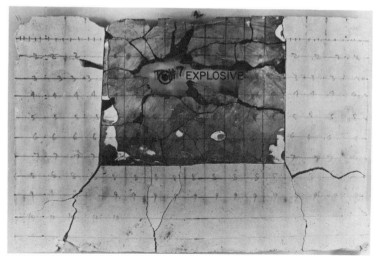

Fig. 8.25. Post mortem of Chambersburg limestone plate.

8.25. For this test Hydro-stone (a high strength gypsum product made by US Gypsum) was used on 3 sides of the model. The acoustic impedances of the limestone and Hydro-stone (ρc_1) are, respectively, $11 \cdot 1 \times 10^6 \text{ kg s}^{-1} \text{ m}^{-2}$ and $6 \cdot 79 \times 10^6 \text{ kg s}^{-1} \text{ m}^{-2}$. This resulted in about 75% of the incident wave being transmitted into the Hydro-stone and 25% being reflected back into the limestone with a change in sign. As can be seen from the photograph there are many radial cracks in the limestone but none penetrated into the Hydro-stone. Those in the Hydro-stone are believed to have occurred upon reflection of the outgoing P and S wave from the free boundary of the Hydro-stone. The Y shaped cracks at the 90° corners of the Hydro-stone rock boundary were observed in other similar tests.

Since many fractures in the previous test were felt to be the result of reflections of the wave systems, the use of high speed photography was necessary to isolate the fracture development. Figure 8.26 shows 1 frame from 16 taken with the Cranz Schardin camera operating in a reflection mode. The specimen was St Paul limestone which was

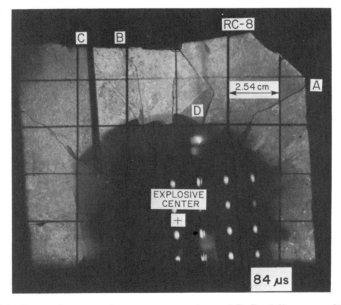

Fig. 8.26. Frame from spark gap camera taken of St Paul limestone block at 84 μs.

polished and then silvered by vacuum deposition. The specimen therefore acts like a mirror and light reflected from it is captured by one of the camera lenses. Should a portion of the sample rotate, then light reflected from this segment will not reach the camera and the image will appear black. At the time shown in the photograph several radial cracks labeled A, B and C have reached the boundaries of the model. Crack velocities for these cracks ranged from 28 to 39% of the P wave velocity. Other fractures started at internal flaws or at the boundaries and were observed to be traveling at 14% of the P wave velocity. The well defined black region in the center of the model corresponds to segments that have fractured and rotated. From the sharp changes in light intensity on the boundary of this pattern, it is clear that the radial fractures have preceded the circumferential ones. A region labeled D has turned grey showing that it also is fractured. The region is bounded by a clearly open fracture to the right, a circumferential fracture at the bottom and a grey/white transition. This transition is a fracture but it is not open enough to be detected by normal means. This particular method of photographing fractures shows the deficiencies of standard high speed photography in locating fractures.

Figures 8.27(a) and (b) are 2 photographs from another limestone test. 2 plates of limestone were bonded together with epoxy cement in the middle of the photograph. The bond line was interrupted with 2 pieces of tape. The dark features parallel to the bond line are structural characteristics of the limestone. The explosive was located in the upper half, and at $31 \, \mu$s (Fig. 8.27(a)) there are numerous radial fractures in this section. At $51 \, \mu$s (Fig. 8.27(b)) 2 prominent fractures have appeared in the lower plate. Both of them have initiated at the outer transition of tape and epoxy. These fractures are analogous to the joint initiated fractures described in the Homalite tests.

Similar testing was conducted with continuous bond lines in granite plates, and the specimens were photographed with the Cordin 350 camera. Figures 8.28(a) and (b) show 2 frames from a typical test. The specimen consisted of 2 plates bonded together at the indicated seam with epoxy glue. An aluminum plate shown as the dark central band was taped to the rock surface and served to deflect the combustion products away from the lower plate. In Fig. 8.28(a), taken at $105 \, \mu$s after detonation of the explosive, numerous cracks have appeared in the image. There are radial borehole cracks present as well as cracks that have initiated at the bond lines and adjacent to the boundaries of the model. The cracks nearly parallel to the edges of the sample are

Model studies of fragmentation

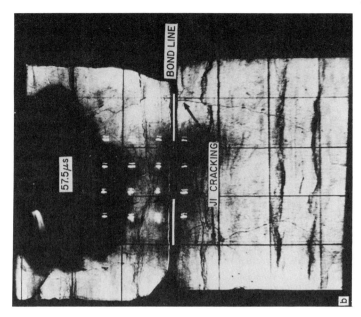
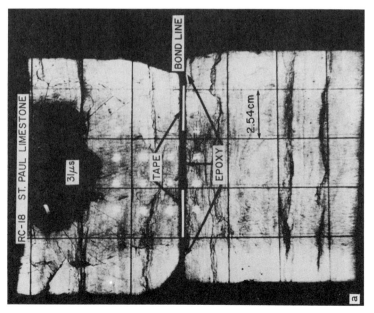

Fig. 8.27. Pictures of fragmentation in a jointed limestone plate.

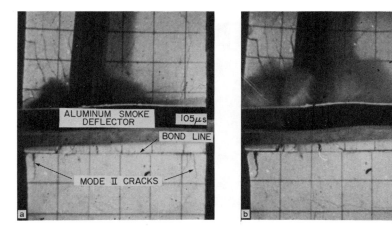

Fig. 8.28. Pictures of fragmentation in a jointed granite plate.

the familiar spall type failure resulting from the reflected P wave. The cracks initiated at the bond line are believed to have occurred as a result of the shear stress component of the P wave in the direction of the bond. This would correspond to the joint initiated fractures in the Homalite tests. In Fig. 8.28(b), taken at 180 μs, there is extensive movement of the rock fragments. It is interesting to note that the radial borehole cracks did not penetrate the glue layer or propagate into the lower plate.

A photograph of the reconstructed image of a doubly exposed

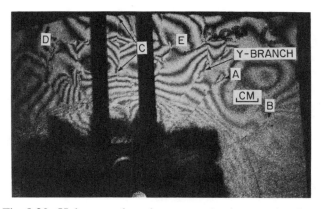

Fig. 8.29. Hologram of cracks propagating in a granite plate.

hologram of a granite model is shown in Fig. 8.29. The time of the first light pulse was 61 μs after detonation and the interpulse time was 800 ns. The 2 vertical stripes in the photograph are the steel arms of the pressure containing device. The reconstructed holographic image of the specimen contains interference fringes related to the change in the surface displacements between the 2 exposures. Clearly seen are the radial cracks emanating from the borehole region. Two of the more prominent cracks have been labeled A and B. The resolution is also sufficient to locate the open crack line behind the crack tip. In addition, typical crack behavior near the upper boundary is also visible. Here a radially oriented crack, labeled C, has started at the free boundary and is propagating inwards. This crack is believed to have originated from the circumferential tensile stress in the tail of the P wave or from the tensile stresses in the P wave upon reflection. Cracks like this one have been observed by Winzer and Ritter (1980) in the testing of 20 ton limestone boulders. There is also a spall crack whose ends have been labeled D and E.

Figure 8.30 is a post mortem photograph of the reassembled specimen. It is apparent that the radial cracks A and B propagated to the boundary whereas crack C has arrested at the location shown in Fig. 8.29.

From Fig. 8.29, the true tips of cracks A and B appear to be about 8 mm ahead of the optically resolved tip. The other cracks are not open sufficiently for them to be seen in a normal photograph.

From this and other similar tests, the borehole radial cracks in granite were observed to be traveling at about 44% of the P wave velocity.

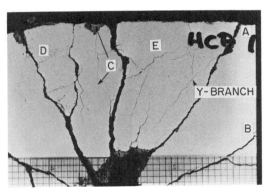

Fig. 8.30. Post mortem photograph of granite plate shown in Fig. 8.29.

8.3.3. Field Testing.

Whenever possible field tests are conducted to confirm the observations of the polymer and rock plate tests. These tests usually take place in large boulders, small benches or in quarry or mine floors. One such test took place in an underground mine. The experimental arrangement consisted of a single 38 mm diameter hole, 914 mm deep that was filled with one-third of a fully tamped and coupled 66% unigel dynamite, one-third of drill fines and one-third of a gypsum cement stem. The hole was located in the floor of the mine 406 mm from a clearly open vertical joint. Figure 8.31 shows the excavated test site after the test. It is clear that the presence of an open joint dominated the fracture results. On the near side of the joint the fragment size was very small as a result of the reflected waves. There were essentially no fractures or damage to the shale on the far side of the joint. The important effects of an open joint suggest that a dynamic pre-split or smooth blasting technique may be very useful for wall control in room and pillar mining methods.

Figure 8.32 shows joint initiating fractures in a test similar to the one described in Fig. 8.31. The tape measure in the figure is 50 mm across. Several new fractures running perpendicular to the joints have been initiated. These fractures are in every way similar to the joint initiated fractures described in detail in the previous sections.

Fig. 8.31. Excavated test site showing effect of an open joint.

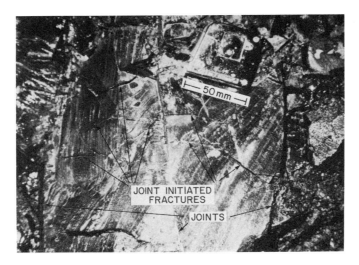

Fig. 8.32. Joint initiated fracture on mine floor test.

8.3.4. Summary of Rock Testing

Testing in rock has shown that the fracture behavior of the rock is very similar to that of the Homalite 100 specimens. For nearly homogeneous models the radial and circumferential crack networks bear strong similarity. The observed crack to wave speed ratios are higher in rock than those of the Homalite. The borehole cracks in granite were traveling at about 44% of the P wave velocity and those in the limestone at between 28 and 39% of the P wave velocity. Cracks initiated remote from the borehole traveled at much smaller velocities. The fact that the crack speed ratio is higher for rocks would mean that barrier branching would occur at larger distances from the borehole.

The crack branching behavior is different in rock than it is in the Homalite. In fact, no information is available on the dynamic stress intensity factors necessary for branching or the relationship between K_D and crack velocity for rocks. This is an area where additional research should be conducted.

The number of initiated fractures at flaws remote from the borehole were less in rock. This is attributed to the fact that severe flaws were not present in the rock samples and to the fact that the static fracture toughness of Homalite is less than that of the rocks tested. Fractures originating at the boundaries of the rock specimens and propagating inwards were observed more often in rock than in the Homalite. In almost all tests there was at least 1 such fracture starting from the free faces nearest the borehole.

The effects of the presence of joints in a rock formation were found to play a role in fragmentation results. At cemented joints or naturally occurring large flaws, cracks were observed to originate with the outward passage of stress waves. These cracks were similar to those observed in the testing of transparent plastic models but were not as numerous and did not result in intense fragmentation. When the joints were open the effect was the same as the presence of a free face. The outward directed stress waves were reflected back into the rock creating an intensely damaged area between the borehole and the joint.

8.4. CONCLUDING REMARKS ON TESTING

The testing described has attempted to examine in detail the role of flaws in the fragmentation process. In particular the interaction of these flaws with outgoing and reflected stress waves have been documented. Those mechanisms observed to be of importance in fragmentation of plastic transparent models have been observed to be active in the fragmentation process in rock. They appear, however, to be less important in the rock than in the plastics. This is felt to be due to the higher value of stress intensity required to initiate flaws in rock along with the relatively light charges utilized in the tests. The fact that qualitatively the fragmentation mechanisms in rock are in agreement with those observed in the transparent models is encouraging. With increases in charge size it is quite possible that the joint induced cracking could be of great importance in the overall fragmentation results. This could have significant implications for the proper timing of multiple hole, multiple row blasting.

The role of stress waves has been found to be of importance in fracture initiation and growth. The intent of this chapter, however, has been by no means to claim that gas pressurization is of no importance. It is in fact believed that once the fractures are initiated and grown to a given length by stress waves that additional growth and coalesence occurs at later times by the gas pressurization. This later growth is of importance and in particular the later movement of fragments is entirely determined by gas pressurization.

REFERENCES

Ash, R. L. (1973). The influence of geological discontinuities on rock blasting. PhD Dissertation, University of Minnesota.

Barker, D. B. and Fourney, W. L. (1978). Photoelastic investigation of fragmentation mechanisms, part II—flaw initiated network, *Report to National Science Foundation*, University of Maryland.

Barker, D. B., Fourney, W. L. and Dally, J. W. (1978). Photoelastic investigation of fragmentation mechanisms, part I—borehole crack network, *Report to National Science Foundation*, University of Maryland.

Bhandari, S. (1977). On the role of quasi-static gas pressurized stress waves in rock fragmentation by explosives, *Proc. 6th International Colloquium on Gas Dynamics of Explosives and Reactive Systems*, Stockholm, Sweden.

Bligh, T. P. (1972). Gaseous detonations of very high pressures and their applications to a rock-breaking device. PhD Thesis, University of Witwatersrand, Johannesburg.

Coursen, D. L. (1979). Cavities and gas penetrations from blasts in stressed rock with flooded joints, *Acta Astronautica*, **6**, 341–63.

Dally, J. W. (1971). Applications of photoelasticity to elastodynamics, *Proc. Symposium on Dynamic Response of Solids and Structures*, Stanford University.

Dally, J. W. and Riley, W. F. (1967). Stress wave propagation in a half plane due to a transient point load. In: *Developments in Theoretical and Applied Mechanics*, Vol. 3, Pergamon Press, New York.

Duvall, W. I. and Atchison, T. C. (1957). Rock breakage by explosives, *US Bureau of Mines, RE 5356*.

Erdogan, F. and Sih, G. C. (1963). On the crack extension in plates under plane loading and transverse shear, *J. Basic Engineering, ASME*, **IV**, 8514.

Fourney, W. L., Barker, D. B. and Holloway, D. C. (1979). Mechanism of fragmentation in a jointed formation, *Report to National Science Foundation*, Photomechanics Laboratory.

Graff, K. F. (1975). *Wave Motion in Elastic Solids*, Ohio State University Press, Columbus.

Hagen, T. N. and Just, G. D. (1974). Rock breakage by explosives—theory, practice and optimization, *Proc. 3rd Congr. Int. Soc. Rock Mechanics*, Denver, Colorado.

Holloway, D. C., Kobayashi, T. and Barker, D. B. (1979). High speed photography of dynamic crack propagation in rock, *IUTAM Symposium Optical Methods on Mechanics of Solids*.

Holloway, D. C., Barker, D. B. and Fourney, W. L. (1980). Dynamic crack propagation in rock plates, *Proc. 21st US Rock Symposium*, 1980.

Johansson, C. H. and Persson, P. A. (1974). Fragmentation systems, *Proc. 3rd Congr. Int. Soc. Rock Mechanics*, Denver, Colorado.

Kutter, H. K. and Fairhurst, C. (1971). On the fracture process in blasting, *Int. J. Rock Mech. Min. Sci. and Geomech. Abstr.*, **8**, 181–202.

Langefors, U. and Kihlstrom, B. (1963). *The Modern Technique of Rock Blasting*, Wiley, New York.

388   W. L. Fourney, D. C. Holloway and D. B. Barker

Montenyohl, V. I. (1978). Defect structures of the Pinsburg Station quarries, *Martin Marietta Laboratory Report TR 78-22c*.

Ouchterlony, F. (1974). Fracture mechanics applied to rock blasting, *Proc. 3rd Congr. Int. Soc. Rock Mechanics*, Denver, Colorado.

Persson, P. A., Lundborg, N. and Johansson, C. H. (1970). The basic mechanism in rock blasting, *Proc. 2nd Int. Soc. Rock Mechanics*, Belgrade.

Porter, D. D. and Fairhurst, C. (1970). A study of crack propagation produced by the sustained borehole pressure in blasting, *Proc. 12th Symp. Rock Mechanics*, University of Missouri, Rolla.

Riley, W. F. and Dally, J. W. (1969). Recording dynamic fringe patterns with a Cranz Schardin camera, *Exp. Mech.*, **9**, 27N–33N.

Singh, B. (1979). Experimental investigation of the effect of time delays on fragmentation in homogeneous models, *Report to National Science Foundation*, Photomechanics Laboratory.

Starfield, A. M. (1966). Strain wave theory in rock blasting, *Proc. 8th Symposium on Rock Mechanics*, University of Minnesota.

Timoshenko, S. and Goodier, J. N. (1951). *Theory of Elasticity*, McGraw-Hill, New York.

Winzer, S. R. and Ritter, A. P. (1980). Role of stress and discontinuities in rock fragmentation: a study of fragmentation on large limestone blocks, *Proc. 21st US Rock Symposium*, 1980.

Winzer, S. R., Montenyohl, V. I. and Ritter, A. (1979). The science of blasting, *Proc. 5th Conf. on Blasting and Explosive Techniques*, Society of Explosive Engineering, St. Louis.

Chapter 9

Surface Uplift Blasting for Oil Shale Production

KEITH BRITTON

President, BlastMasters Inc., San Leandro, California, USA

SUMMARY

True in situ *retorting, utilizing explosively rubblized retort beds, has long been a goal of those involved in oil shale recovery. Potentially, substantial mining costs and environmental disruption may be avoided and otherwise sub-economic resources processed. To date, progress in the area has been dismal, with the notable exception of the Geokinetics Lofreco process. Cursory examination of fundamentals of explosives systems and the unavoidable geometrical constraints suffices to explain the failures and to identify the principles, essential to success, embodied in Lofreco blast designs. Basically, space must be provided for rubblization to be effective and it is provided in the Lofreco process by the raising of a barren overburden. The unique demands of this step have necessitated methods novel to blasting science.*

Once space is available, the intended retort zone is then similar to confined retort beds of the modified in situ *kind and rubblized similarly. Such blasting introduces matter new to the science and the Lofreco process also differs from other* in situ *processes in utilizing and therefore blasting for horizontal burns. Blasting theories are reviewed in light of fragmentation processes observed to be of practical significance to the Lofreco process. The rubble condition is discussed and the factors affecting and methods to control its utility as a retort bed considered. Blasting practice for the unique overburden raising step is treated in detail, as is the subsequent extension of the rubbled zone by blasting methods which,*

390 *Keith Britton*

though developed for the Geokinetics process, may be more generally relevant to confined blastings.

Geokinetics' success has not only rested on the theoretical soundness of its process—it has also rested on, and required, substantial expertise regarding the problems of practice. Particularly noteworthy are the constraints imposed by the limitations of delay detonators and initiating systems.

9.1. INTRODUCTION

The literature mostly covers theory and small scale experiments. Hopefully, this brief review will prove useful in identifying that which has been observed or used in practice within the scale and scope of a successful *in situ* process. As rapidity of progress and limited resources have afforded little opportunity for repetition or confirmatory experiment, the reader is cautioned that the following necessarily lacks scientific rigor.

Conceptually, *in situ* retorts using blast holes drilled from the surface have long been attractive. Capital cost is potentially low and spent shale disposal avoided. In practice, apparent simplicity has proved deceptive, resulting in numerous expensive failures. Unsuccessful approaches attempted prior and contemporary to the development of Geokinetics' (GKI) Lofreco process fall into two categories: use of explosively loaded hydrofracs by the US Government, Talley Industries and Dow Chemical and use of a columnar charge array, as in US Government work in Wyoming. Fragmentation was achieved but permeability proved inadequate for retorting.

The hydrofracture approach is technically interesting and illuminating. In its most sophisticated form, it consisted of a vertical access hole drilled from the surface, from which were extended two planes of fracture along the nearly horizontal bedding (see Fig. 9.1). The fractures were driven, filled and propped by a slurried explosive which was then detonated from the central borehole, the intention being to rubblize and render permeable the rock between the fractures.

With simultaneous detonation, stress waves interact within the target mass and reflect from the acoustic mismatch presented by the gas (explosion product) filled fractures. Predictably, the material was intensely damaged but, being clamped between two planar high pressure zones, remained essentially monolithic and impermeable, though micro-

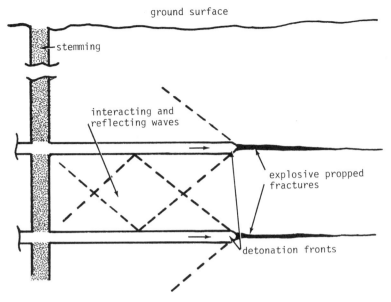

Fig. 9.1. Explosive loaded hydrofracs.

fractured. Any material will fail if its strength is exceeded, by definition, but it need not thereby become permeable. Bulk motion will not occur in the absence of a pressure gradient. Differential motions will not occur with uniform loading. Obviously, the geometry/timing of the approach was necessarily fatally defective.

Less obviously, it was also unsuitable from the point of view of gas penetration of the broken rock. Gas penetration is effected by the available pressure differential and hence was inhibited by the isostatic condition between the pressurized fractures. Planar rather than sharply curved stress fields further inhibited gas entrainment by suppression of gross fracture propagation, the effect being compounded by sealing resulting from excessive explosive detonation pressure plastically deforming the fracture surfaces.

The columnar charge array attempts of the US Government were more promising but also failed. Fundamentally, in the absence of a free face and especially where columnar charges are arrayed normal to the plane of available relief (Johansson and Persson, 1970), blasting processes are extraordinarily inefficient. Thus, the US Government venture failed through inadequate specific charge while the almost equally naive short column array design of Geokinetics' retort 1 partially succeeded through successful rubblization of a limited zone where the specific charge approached $5\,\text{kg}\,\text{m}^{-3}$. Since such specific charge was economically infeasible and the effectiveness of the design would drastically diminish with increased scale, a new approach was needed.

The Lofreco blasting process is unique in being fundamentally a two-step process. The overburden overlying the intended retort zone is first raised to provide void space and a free face. Subsequent detonation of fragmenting charges may then be efficient in both fragmenting the oil shale and redistributing the void to render the retort zone rubble highly permeable.

The overburden raising step is achieved by the application of gaseous pressure from explosion products under a sufficient area of the overburden that the edge fails and the overburden is raised. Pressurization of the interface between the overburden and the intended retort zone is geometrically the most efficient approach. It is awkward to achieve at small scale and results in an horizontal free face, which may be awkward or inefficient for subsequent blasting. Accordingly, the Lofreco process initially utilized the V cut arrangement of Fig. 9.2. It provided a free face the full thickness of the intended retort zone to allow efficient use of vertical blast holes and, of great importance for research, a geometrically primitive and hence analytically tractable arrangement.

Lifting may also be accomplished by pressurizing a zone of rubble rather than a discrete fracture plane. This may be readily achieved by the simultaneous firing of a charge array as in Figs 9.3 and 9.4. Lifting is comparatively inefficient and, if the columns are of significant length, permeability degrades rapidly with depth due to lack of differential motions.

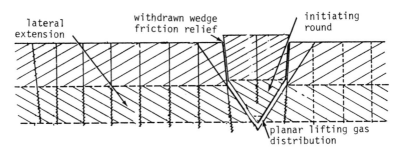

Fig. 9.2. V cut and lateral extension (side view).

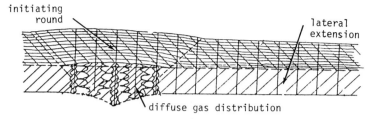

Fig. 9.3. Vertical column initiating round and walking W design lateral extension (side view).

However the overburden is raised, void space and a free face become available. A comparatively conventional blast arrangement then suffices with raising of additional overburden. Basically, rows of blast holes are sequentially fired, each moving its burden horizontally and vertically to fill the void space left by the preceding charges.

Theoretical knowledge is not sufficient for success. Numerous practical problems and limitations must also be thoroughly understood to avoid impracticable designs or field failures of sound designs.

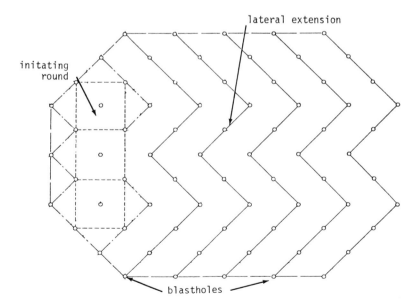

Fig. 9.4. Initiating round and walking W extension (plan view).

394   *Keith Britton*

Notable in this respect are the characteristics of commercial explosives and initiators and allowance for rock motions.

9.2. GENERAL PRINCIPLES

The Lofreco process cycle is in principle extremely simple: on the basis of surface and subsurface surveys, a round is designed, drilled, loaded and blasted to rubblize and render permeable the desired retort bed. Re-entry drilling is then used to emplace air-in and air-out ducting, pumps and thermocouple instrumentation, and any needed recontouring or surface sealing is carried out. Concurrently, data are recovered from post blast surveys, coring, drill logs and flow testing. The retort is then ignited at the air-in end and the fire front driven to the air-out end by gas flow, the retort being maintained at a slightly negative pressure to control leakage. As the fire front advances, shale oil is recovered by pumping from sumps. Eventually, the fire front penetrates to the final air outlet or pump position and the retort is shut in and abandoned. As of 1982, retorts are typically nearly square, ranging from around 70×70 m to 100×100 m in plan and rubblizing a nominal 10 m thickness of almost horizontally bedded oil shale under an overburden ranging in thickness from 11 to 30 m. Some 70–150 tonnes explosive are used. Thus the geometry of the process differs radically from that found in VMIS (vertical modified *in situ*) approaches.

Geometrical considerations touch the lithology, blast design and eventual burn and must be thoroughly understood as their influence is often profound in subtle as well as gross ways. From the retorting point of view, the Lofreco burns proceed parallel to the bedding, utilizing the anisotropic void distribution resulting from failure on beddings and permitting, therefore, efficient retorting with as little as 7% void. The burn is horizontal, the burn front being primarily vertical. Both geometries interact positively as the structurally competent lean oil shales provide support and permeability through the otherwise incompetent hot high-grade materials. For large retorts, lateral edge conditions affect only a small percentage of the total volume. A high degree of control is available on flow due to the wide transverse span on the ducting and access is excellent for the required process instrumentation.

Little leverage exists for vertical control of the retort front, a major concern of early workers in the field, but in practice even an order of magnitude difference in permeability between the top and bottom of the retort zone proves nearly inconsequential. The overall geometry necessitates use of an array of vertical columnar charges for fragmentation of the retort zone. The remnants of the blast holes and the major vertical fracture systems of their primary radial breakage establish substantial structures within the permeability system which provide the vertical transport necessary to continually renormalize the fire front to a vertical geometry.

The ability to lift overburden is proportional to the area affected by a lifting pressure, inversely proportional to the lifted mass and, in regard both to the initial loosening and to subsequent energy losses through deformation processes, inversely proportional to some function of the periphery. Any increase in areal scale therefore improves blast efficiency. The geometrical relationship between the retort width and the overburden thickness thus assumes considerable importance as an index not only of blast efficiency but, as the overburden thickness approaches or exceeds the retort width, of process feasibility.

Less obviously, there are advantages and constraints placed upon the blast design by the overall geometry. Where process economies prohibit a mining approach, all drilling must be done from the surface, and in practice it must be done with vertical blast holes, as inclined drilling is effectively limited to the shallowest overburdens. It follows, therefore, that the general case for both lifting and fragmenting charges is a long or squat, vertical, columnar geometry. Such blast holes are fundamentally unsuitable for breaking and moving rock for lifting purposes, as is evident from comparison of the blasting efficiency of charges emplaced by blast holes drilled normal and parallel to their effective free face (Johansson and Persson, 1970), the difference being an order of magnitude. The lifting efficiency of present designs is in consequence only around 2% and is the process limiting factor both technically and economically. For combined lifting and fragmenting charges, fragmentation and void introduction are also inefficient, increasingly so with increased blast column length, where poor bottom breakage may become a limiting factor. In contrast, once void space and a free face of significant vertical extent have been developed, the processes contributing to retort rubblization and further overburden lifting become comparatively efficient and more akin to conventional bench blasting with vertical blast holes.

Numerous approaches have been taken in attempts to erect a

satisfactory theoretical basis to explain observed blasting phenomena. For what may be termed 'classical' breakage in hard uniform rock, the Swedish work which culminated in the definitive monograph by Langefors and Kihlstrom (1963), advanced the art from a skilled craft to a comparatively well controlled science. Basically, their approach consisted of analyzing blast behavior in terms of geometrical primitives so simple as to be amenable to well understood scaling laws. Constants were then easy to obtain as the simple geometries were experimentally tractable. Geometric primitives were aggregated as needed for analysis of the more complex geometries of practical rounds. Where appropriately applied, the approach has successfully withstood, both qualitatively and quantitatively, the test of 20 years practical use.

Unfortunately, the Swedish work was only of limited value for the Lofreco process. Most obviously it suffered from that classic difficulty of empirical approaches, having its constants generated from a database radically dissimilar to the required application. This was particularly the case in regard to the material being broken, which exhibits properties unusual amongst rocks and very different from those of massive hard rock. Further, the Swedish work has little need to consider in detail the processes operative within the first few radii of the blast hole, nor cause to study in detail the processes of fragmentation within the broken material, both matters of some concern for blasting oil shale. 'Classical' breakage, moreover, though present and of importance, as evidenced by cores, is not the source of most of the new surface developed during a Lofreco blast.

'Classical' breakage proceeds as follows and as indicated in Fig. 9.5. The borehole wall is initially stressed by the passage of the detonation front. The pressure peak at the reaction zone, though brief, is sufficiently intense that a shock wave is formed in the material of the borehole wall. This may be elastic but is usually intense enough to commence as a plastic wave which permanently alters and translates the material it passes through, losing intensity as it does so by expansion and by performing work upon its medium. Shortly, the intensity drops to a level at which the medium is competent to propagate elastic waves, which thereafter principally decay by divergence. This first zone is identifiable by permanent blast hole dilation and reworking of the material. Typically, as viewed from inside, the explosive worked surface is strikingly reticulate, but closer examination or sectioning identifies the surface pattern as superficial, the reworked zone being nearly fracture free and forming, therefore, a seal to inhibit the entry of the

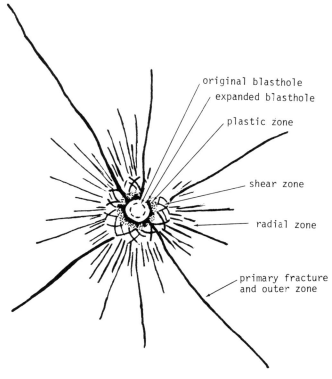

Fig. 9.5. 'Classical' rock breakage.

explosion gases into the fracture systems of the outer concentric breakage zones.

At the point where the rock is just competent to withstand the pressure wave, an annular zone of shear failure may be observed, followed by a zone of intense and principally radial fracturing which decays with a more indefinite boundary than the preceding zones to a much larger outer zone characterized by a small number of unforked radial fractures. Probably only these fractures extend inwardly to eventually relieve the explosion pressure from the blast hole (staining from reaction products being frequently evident in same and normally absent elsewhere). Certainly, at some point the reaction gases escape into the formation, probably initially at a pressure of less than 5 kbars, reflecting the drop from detonation pressure (in the 50–60 kbar range), through the explosion pressure, to a pressure reflecting the blast hole dilation.

Some further energy from gas pressurization is thus available to extend the primary fractures of the outer zone and more energy may become available from a returning tensile wave generated by reflection of the initial compressive wave at a free face. Neither the former nor (for an isotropic medium and the burdens and specific charges commonly employed) the latter should contribute materially to fragmentation except in the extension of the primary fractures approaching the free face to permit the detaching of a nearly monolithic prism of burden rock. The fragmentation which does occur within the burden mass is, with this scenario, then principally attributed to secondary mechanisms involved with activation of pre-existing weaknesses and collisions between fragments.

While this model works well for hard, massive rock, it has long been known to be inadequate to describe phenomena readily observable during the blasting of soft sedimentary rocks. Too much fragmentation occurs, too early, the fragments tend to be rectangular rather than the pie-shaped products of radial fracturing and the burden appears to detach in advance of gas pressurization to the free face. Early photography showed a striking and well developed reticulate failure of the burden block preceding visible venting from the primary fractures (Patterson, 1957). Recent studies (Fourney et al., 1979) have elucidated the mechanisms involved, flaw activation and spalling at resulting fractures, but more work is needed on breakout mechanisms and pressurization. Notably, the photography of bench-blasted oil shale has shown a breakage pattern dominated by fractures positioned to the side of and below the blast hole (Wright et al., 1953).

GKI blasted retorts have been extensively cored, frequently utilizing 45° inclined cores positioned to closely approach blast hole positions and intersect probable primary fractures. Classical radial fracturing is certainly identifiable, but only in the immediate vicinity of a blast hole, to some 1.5 m from the blast hole, or around 15 radii. Primary fractures almost certainly exist outside this zone, but the evidence is not unequivocal, consisting typically of core loss and recovery of finely diced material where such were expected. Beyond the immediate blast hole vicinity and ignoring comminuted material from movement planes, the balance of the material is extraordinarily uniformly broken, appearing to be far more affected by variations in materials properties with grade than by blast hole proximity. Failure is typically that of an impulsively overloaded laminate.

Surface uplift blasting for oil shale production   399

Beddings delaminate at material property boundaries, typically with propagation of fractures at right angles to the failed plane and with noticeably increased severity of damage to thin beds of markedly dissimilar properties to the surrounding parent, and in zones of pronounced alternating beds of sharply differing properties. However, while Fourney's recent work provides a qualitatively excellent description of the mechanisms producing the rubble types, both the classical approach and a simple model after Fourney, allowing exponential stress level decay, require a more pronounced diminution in fragmentation with distance from the shot point than is observed in practice. Nor, moreover, is the gradient found to differ significantly in the tight areas underlying the initially lifted overburden, where no lateral motion or interaction with a free face occurs. This phenomenon is, however, fully consistent with the recently published breakage theory due to Coursen (1977, 1980).

In the Coursen model, breakage in the immediate vicinity of the blast hole proceeds classically until gas pressurization of the primary radial fracture system occurs. The pressurization, involving but small volumetric increase, thus raises to the then borehole pressure a nearly circular (cuspate between fracture tips) zone of considerable extent, the isostatically stressed zone extending to nearly the tips of the primary fractures, as indicated in Fig. 9.6. Fragmentation mechanisms appropriate to this isostatic zone would necessarily produce uniform fragmentation, as in Fig. 9.7, between the outer limit of the classical breakage and a fairly sharply defined outer perimeter where fragmentation would comparatively abruptly cease. Existence of the latter has not been demonstrated and experimental difficulties have precluded rigorous quantitative analysis of the fragment sizes but, subjectively, distribution of breakage best follows the Coursen model.

Two additional factors may be involved. From overburden velocity studies at the GKI field site it is inferred that pressure decay is very rapid, despite a substantial degree of confinement. This may lead to breakage by release of load, as suggested by Cook (1958). Pressurization may also occur by bedding invasion resulting from the extraordinary strains exhibited by high grade strata.

The purpose of the blasting is not primarily the disruption of the material but the production of a retort bed. For the present scale of the Lofreco process and the lithology at the Kamp Kerogen site, the actual mechanisms of fragmentation are not of great importance as any blast design sufficiently energetic to achieve adequate overburden motion for void introduction (or muck motion for void redistribution) typically

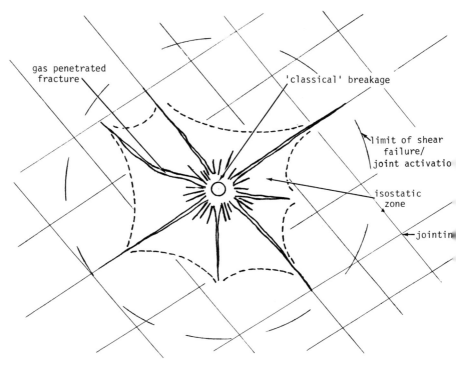

Fig. 9.6. Rock breaking processes (after Coursen).

substantially overshoots the rubble in terms of mean particle size. Present concern is in fact to reduce rather than increase the degree of fragmentation. Obtaining desirable permeability is of much greater importance than control of fragmentation.

Successful retorting has been demonstrated with void volumes varying from around 7% to around 50%. As pointed out by Simon (1979), randomized beds of reasonably uniform sized fragments must necessarily contain a substantial void percentage. Since practical fragmentation by blasting does not remotely approach Fuller's curve, retort beds at the lower void limit must necessarily have remained basically ordered whilst at the other extreme only complete randomization of somewhat oblong particles suffices to support the observed void percent. Observation from coring confirms this and reveals that between these extremes the retort bed is invariably bimodal, consisting of zones of disorder and high void fraction interspersed with zones of order and characteristically low void fraction. The upper limit for void in an ordered bed and the lower limit for a randomized bed appear to neither coincide nor overlap. (As a rule of thumb, random beds have 25–50% void, ordered <10%.) The significance of this to permeability, and hence blast design, is very great. Significant lateral motion of the rubble must be associated with randomization and hence rapid void consumption. Gas flow through such material is tortuous and increasingly inefficient with decrease in mean particle size, despite maintenance of a comparatively high percentage void. Flow may be very much less tortuous and hence more efficient in ordered bed material, very largely compensating for sharply lower void percentage. Note, however, that this is extremely sensitive to the lithology. The GKI site offers a nearly optimum case, a laminate of sharply differing properties where the desired flow is parallel to the separated laminations and vertical jointing, at approximately 0·5 m spacing, limits slab size.

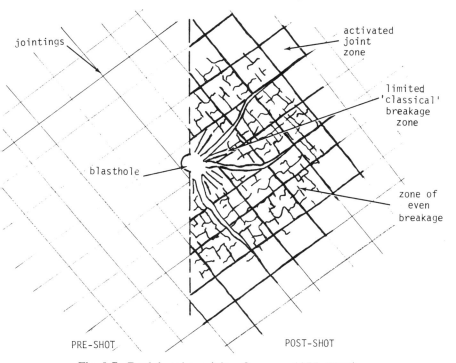

Fig. 9.7. Rock breakage (after Coursen, 1977, 1980).

Permanent separation of the laminations is required to maintain permeability, but, to judge from cores, is achieved on only a minor proportion of the fractures. Propping mechanisms include relative motion, small fragment insertion and rotation or displacement of indigenous fragments. It is enhanced by measures which increase the volume of gas entrained into the formation and pressure differentials to promote rapid flow in desired directions. Inert decks positioned to control axial gas flow within the blast hole are particularly useful in this respect.

9.3. INITIAL OVERBURDEN RAISING

The basic principle of all fragmentation systems for the Lofreco process is to force an overburden into motion, await development of a free face and void space, and complete fragmentation with propping of the raised overburden to limit fallback. In the original embodiment (Britton and Lekas, 1979) gaseous pressure from detonating explosive charges arrayed on inclined planes raised the overburden (see Fig. 9.2). The arrangement is both experimentally tractable and highly efficient since, from pre-splitting practice, the blast hole array may be expected to form an opened gas-filled plane with charges in the $0 \cdot 5{-}1 \, \text{kg m}^{-2}$ range and little energy needs to dissipated in fragmenting processes. (Lifting efficiency becomes maximal with the use of low explosive and a horizontal planar array, a geometry treated cursorily in early work.) Scales for this early work were quite small, ranging from around $7 \, \text{m}^2$ and breaking $3 \cdot 4 \, \text{m}$ oil shale under $5 \, \text{m}$ overburden to nearly $13 \, \text{m}^2$ and breaking about $7 \, \text{m}$ of oil shale. Maximum vertical velocities of the raised overburden block varied in the range $5{-}11 \, \text{m s}^{-1}$. High speed photography showed that the overburden motion was not ballistic, but heavily braked by edge effects despite considerable effort in edge preparation by use of inclined pre-splits to provide withdrawn wedge relief of friction. Without such preparation, most of these rounds would have failed.

Breakage was found in practice to be dominated by the site jointing system rather than the imposed pre-split failure planes. Accordingly the latter were dispensed with in preference to orientation of the retorts to accurately conform to the jointing, with the result that future $1:1$ width: overburden ratio shots, though still much edge affected, became reasonably predictable.

Surface uplift blasting for oil shale production 403

Production drill and blast advantages then caused emphasis to shift to the use of vertical holes and simple columnar charges, the advantages outweighing the sharp loss in efficiency and the very great increase in experimental complexity.

This led to the initiating round and walking W geometry of Fig. 9.3 (Zerga, 1980). Overburden was raised by gaseous pressure from the explosion products entrained into the bulk of the oil shale formation rather than confined to a discrete open plane. Detonating explosive was then essential, to ensure predictable entrainment, involving losses from deformation of the blast hole walls and, of course, the fragmentation processes within the rubble.

The projected area of the blast holes is quite small, so the initial lifting force is not large even with detonation pressure enhancement from end effect. Pressure rapidly falls to reflect the explosion pressure, borehole expansion simultaneously occurring with increase in volume and further pressure drop. From theoretical studies (Johansson and Persson, 1970) the borehole expansion may be expected to occur within a few tens of microseconds. This is supported by the results of practical experimentation at the Kamp Kerogen site, springing holes with stemmed and unstemmed column charges. While some dissimilarities were noted, overall volumetric expansion was negligibly affected by the confinement, demonstrating that it must occur within at least the low millisecond regime. Volumetric blast hole expansions were measured within the range $2 \cdot 5 : 1$ to $3 \cdot 5 : 1$, for blast holes of 114 and 140 mm and two explosives of high and low detonation pressure.

Within very few milliseconds of detonation, the borehole pressure would then have diminished from the 60 kbar detonation pressure typical of IRECO slurry explosives to something more of the order of 4–8 kbars. Assuming no loss and simple cylindrical expansion, any lifting force would have remained essentially constant through this short period as increased projected area compensates for the pressure drop. Considering a typical blast hole of the initial lifting array for a large retort with around 20 m overburden, the area of action of the 4 bar lithostatic pressure is some 1500–2000 times greater than that of the original blast hole. Taking the overburden as an ideal Newtonian body, and neglecting edge restraint, no more than about 3 gravities acceleration is then initially available. Initial peak accelerations of 150–200 gravities are, however, recorded by accelerometers placed in the overburden and optical tracking confirms mean accelerations in the 10–20 gravity range for some milliseconds. This can only occur as a consequence of gas entrainment into the formation increasing the lifting force by extending the projected area of action, which would initially cause but small volumetric increase. Areal increase limits when the space between the blast holes is gas flooded and peak lifting force is then developed. Analysis is complicated by gradients away from the blast hole pressure reservoir, but is initially insensitive to whether pressurization is by invasion of a bedding separation or by the Coursen model.

This point is critical. There must exist a sufficient force on the lower surface of the overburden to cause both failure of its periphery and its upward acceleration, or the round will fail. Assuming the edge fails, the driving force will then rapidly diminish as the gas volume increases due to the raising of the overburden. For the example given, about 2 cm of overburden lift suffices to double the volume.

Neglecting acceleration, for a not untypical initial velocity of 3 m s^{-1}, 7 ms or so would be required for the necessary overburden motion. In practice, shown in Fig. 9.8 (a plot from an experimental shot with 10 m overburden thickness) accelerations are an order of magnitude lower than would be expected with this model and the decay of pressure very much faster. The gases cool and lose the bulk of their energy within about 10 ms. Similar results are obtained from optical tracking of overburden motions. Even for overburdens exceeding 20 m in thickness, the driving force is clearly impulsive rather than sustained, firing times corresponding to marked inflections on a displacement–time plot. Acceleration appears to commence and fall to a small residual positive value within 6–8 ms, but these data are only approximate, being derived from 16 mm photography at 500 frames s^{-1}.

It is not known how the gas is partitioned into vertical fractures or bedding separations but the best evidence is that entrainment is diffuse rather than primarily in a plane at the discontinuity between the retort zone and the overburden, though it does appear rather strongly concentrated towards the upper portion of the blast holes. For reasons not presented here, it is believed that as the overburden recedes the damaged rock dilates concertina-fashion. Without further intervention, the overburden would fall back and settle to almost its original position, with very little resulting permeability. Accordingly, a second array of charges is fired after a suitable delay to both continue the lifting process, fragment, prop and establish desirable conditions for a third and final array.

Assuming the delay has been chosen to allow both significant

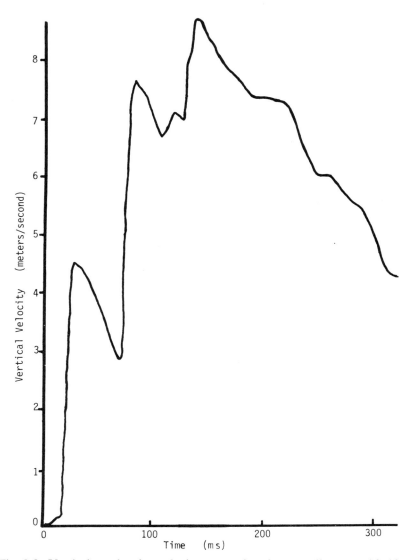

Fig. 9.8. Vertical overburden velocity versus time for a small retort with 10 m overburden.

pressure decay and development of significant void volume, the second firing array has very different initial conditions from the first. For the former, radiated waves are essentially lost to the formation, performing little work as they expand through unflawed material. For the second firing charges, a high proportion of this radiated energy will be retained within the rock damaged by the first array, and hence be available to promote fragmentation. As, from the literature, radiated energy is estimated to contain around 9% of the total explosive energy, this assumes considerable importance. Of equal importance, the void zones developed by the earlier array constitute a substantial pressure sink relative to the explosion pressures of the second array. In sharp contrast to the more isostatic condition of the earlier array, pressure gradients are considerable, so lateral forces and significant gas flows become major factors. Gas is again rapidly entrained into the formation and a further, but similarly short, lifting force is applied to the rising overburden.

Overburden motion continues, rapidly increasing the void fraction. A third charge array is then typically fired, primarily to provide the bulk of the propping necessary to resist overburden fallback and so maintain permanent permeability, but with further major contribution to overall fragmentation and minor additional contribution to overburden lift.

Timing is of crucial importance. As the bulk of the lifting force available from a blast hole is delivered within some 10 ms, common firing precision becomes extremely important. Scattered charges co-operate to only a limited extent. Much energy is then wasted, deforming the overburden through differential loading, and unnecessary retort leakage may result. A marked improvement in lifting ability has been demonstrated where carefully measured detonating cord initiating trains were utilized to render the initial firings precise to the sub-millisecond regime. Firing precision for the first charges has the additional benefit of increasing by superposition the peak stresses and complexity of waves experienced by material which would otherwise be the least fragmented within the retort. For this reason also, it is desirable that a high degree of precision be maintained for the firing of the second firing array, but high precision is actually detrimental to the third firing charges. Propping and fragmentation will be adequate without cooperation which could cause undesirable lineation in final permeability.

The timing between charge arrays is of even greater importance than the precision within them. It has been demonstrated in large scale experimental shots that a simple timing change can halve overburden lift, specific charge remaining constant. Results from well instrumented small research shots, Geokinetics' 20 series, showed that the maximum specific force was generated by the first firing array, but the second firing array, while generating a lower specific force, contributes about twice as much specific energy transfer to the overburden. Specific force is again reduced for the case of the third firing array, with even greater reduction in specific energy transferred.

The maximum force observed corresponds to the minimum void, i.e. unbroken material, equivalent to a zero delay period. Energy transfer, however, passes through a maximum and then decreases with increasing delay. The physical reasons for these observations are probably as follows: for the force peak in unbroken material there are alternative explanations. If the Coursen breakage model is correct, one would expect the maximum product of pressure and area when radial fracture propagation ceases and the pressurized zone tends to isostatic conditions. The void volume would then represent the dilated blast hole, the compressive strain of the material within the isostatic zone and some compression of that adjacent to the isostatic zone. Pressure drop would then principally reflect cooling until the semi-static situation altered with the failure of the overburden edge and the introduction of fresh volume by overburden motion. Further work is then possible, cooling surfaces exposed and continued gas expansion enabled. Alternatively, one may assume a model where there is a substantial pressure gradient from the blast hole. For the latter model, the quasistatic stress situation should be much less uniform within the lower part of the overburden and the void volumetric increase would more than likely lead to an increase in lifting force as the area of action should initially increase much faster than the associated volume.

To date, the second firing array has only been timed at or subsequent to inception of overburden motion, which is unfortunate as the earlier period offers opportunities to distinguish between alternative breakage models and, depending on which mechanisms actually obtain, possibly an optimum situation for momentum transfer.

For second firing array charges timed subsequent to inception of overburden motion, pressure gradients (to the pressure sinks of the initial fracture system) are available to facilitate rapid pressurization of the intervening damaged rock. The shorter the time interval, the shallower the gradient due to the more limited volume and the higher residual pressure from the preceding charge array. Unsurprisingly, the shorter the timing, the greater the lifting force and, since it is acting upon a moving overburden without suffering pressure decay prior to inception of motion, the more energy is transferred to the overburden, though both fragmentation and propping suffer.

The third firing array faces a situation akin to that of the second with a long delay period. Void from significant overburden motion dominates subsequent behavior. For overburden motions of the order of 30 cm or so, a not uncommon situation, it is probable that the overburden bridges and a discrete separation occurs at the interface between the receding overburden and the intended retort zone, providing an horizontal free face registered at the top of the explosive column. Alternatively, the upper few feet may have been loosened and penetrated on beddings. In either case, one would expect that upon detonation lateral motions of the rubble would be pronounced and upward motions, with pressurization of a loose rubble, would occur for at least the upper portion of the broken zone. Energy usage would then go primarily to fragmentation and lateral motion though some additional lift would be expected from pressurization and the kinetic energy of the upwardly dilated rubble, with effects consistent with those observed in practice.

Manifestly, the periphery of the attacked overburden must break or the round will fail. There must be energy loss in edge deformation during the lifting process. Some further losses must occur from overburden deformation within the raised block itself. As directly or indirectly these effects limit process feasibility and dominate drill and blast design and economics, they have been studied intensively, both analytically and by computer modeling (Schamaun, 1981). Results from the computer code IJUMP utilizing the GKI database suggest that less than 10% of the explosive's energy is applied to overburden lifting and distortion, energy absorption exceeds lifting energy by more than 5 times and more than one-third of the absorbed energy is taken through edge effects. The motion energy of the overburden may be readily estimated and 2% is a fair representative figure, not unsurprising in view of the inherent inefficiencies in lifting with vertical blast holes. Absorptive losses may, however, be much smaller than estimated from the model.

For smaller rounds, with a geometrically disadvantageous overburden thickness to width ratio, braking is demonstrably severe, beautifully illustrated by high speed photography of an early small shot which showed masses of loose material falling upward from the surface of the rock overburden as its deceleration sharply exceeded that due to gravity. Detailed analysis of motions is unequivocal, however, in demonstrating that at least some of the overburden for shots of high width to overburden thickness ratio achieves or approximates ballistic motion (see Fig. 9.9). By definition there is negligible net energy change within the system and, therefore, without a balancing force, negligible edge or internal energy loss. (The presence of a significant lifting force exactly counterbalancing edge losses may be ruled out, at least at the time of greatest elevation, since no significant inflection occurs in the rate of change of velocity as the direction of motion reverses.)

Once the geometrically primitive V cut design had been abandoned, analytical complexities precluded useful progress until continuing scale-up reached areal dimensions large enough to include a central area unaffected by edge effects, approximating a unidimensional case which, from mining overburden subsidence theory, would be expected when the width to overburden thickness ratio exceeded about 1·4.

Subsidence theory and experience provides a useful approach to both the energetics and mechanisms of failure. Considering overburden raising as the inverse of subsidence, it would be surprising if the strain energy absorption were in fact to be as high as predicted. For subsiding spans comparable to retort dimensions, it is axiomatic that full subsidence can require no more energy input than the original potential energy of the mass. Despite the very great difference in strain rates, it would be surprising if the resulting increase in energy absorption was as much as in the model. Material flow during subsidence, moreover, tends to be convergent, increasing friction, while overburden lifting is primarily divergent, reducing friction.

Surface expression of failure modes has varied from sigmoidal to a near scarp, the most common profile being a gentle uplift of the surrounding material, typically with stepped fractures from compressive strain, terminating inwardly in a fairly sharp break of slope to a steeper incline of constant slope which again terminates in a perceptible though typically less marked break of slope to a gently domed or mesa-like central zone, about two-thirds of the total change in elevation being typically between the two breaks of slope and most of the balance above. Presently, data is inadequate for unequivocal identification and description of failure modes. It is known, however, that, analogous to subsidence behavior, high strain conditions (particularly

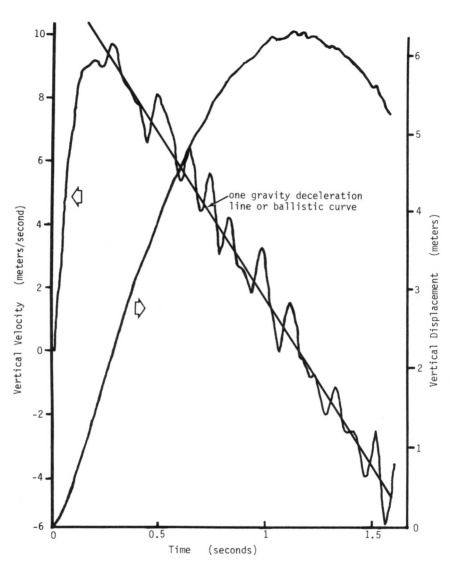

Fig. 9.9. Vertical overburden velocity and displacement versus time for the central zone of a large initiating round with 17 m overburden.

Surface uplift blasting for oil shale production 411

where associated with the shallower overburdens) tend to produce a scarp-like failure. Taking the included angle between the upper extremity of the edge charges and the two breaks of slope, typically one-third of the edge effect will occur without the retort and two-thirds within, for an included angle of around 45°. As the charge is increased or the overburden reduced, the angle tends to narrow and the breaks of slope be displaced outwardly to form, for instance, a 30° edge-centered angle. Conversely, as the overburden is increased or the specific charge dropped to more closely approach marginal conditions, the angle broadens and both breaks of slope move inwardly. This is to be expected. As the round becomes marginal, contributions to edge failure become increasingly important from the interior blast holes. Breakage energy is known to be a non-linear function of overburden thickness, but insufficient data presently exists to determine its exact form.

Fortunately, edge effects can in practice be ignored for production rounds to at least 25 m overburden thickness, so long as the required specific charge is acceptable, as retort dimensions convenient from the retorting point of view are then large enough ($70-150$ m^2) to ensure the favorable width to overburden thickness ratio. For the smallest scale research shots or production shots at greater overburden thicknesses, edge restraint may be improved by explosive fracturing of the periphery, ideally utilizing an inclined pre-split to provide withdrawn wedge relief of friction and the specific charge may be increased locally at the edges. The pre-splitting rendered practicable the earliest shots but was found of limited effectiveness compared to the natural anisotropy of the area jointing pattern. Re-orientation of retort sides to align with the jointing pattern eliminated the need for artificially generated weaknesses and precision orientation to jointing has been an important feature of all subsequent designs. It is known that linear charge arrays cooperate in fracture plane generation over extraordinary distances when precisely aligned to jointing, with a very sharp reduction in effectiveness at only a few degrees deviation from such conformity. This must be borne in mind when increased overburden thickness justifies introduction of artificial weaknesses.

Results are naturally sensitive to both specific charge and the type of explosive employed. Lifting with low explosive and gas-filled discrete planes can probably be made around 20% efficient, or roughly double the efficiency calculated for the hydrofrac geometry with detonating explosive (Chazin *et al.*, 1979). A variety of methods have been described to approach this situation (Britton and Lekas, 1979; Britton, 1980). Significant further lift may be obtained by displacing material downwards from the bottom of the raised block, as demonstrated by fragmenting charges emplaced within the V cut block of early retorts. The fragmentation requirement can hardly be less than 0.25 kg m^{-3}, which requires a well positioned good free face and adequate void. For optimum explosives usage it is in principle best, therefore, to partition the explosives into purpose designed lifting and fragmenting charges. For production blasting in the shallower overburdens, however, the overall simplification of combined lifting and fragmenting vertical charges has much to commend it, despite the penalty in specific charge. Appropriate specific charges for the variety of designs and dimensions addressed to date, therefore, span the rather wide range from less than 0.5 to more than 1.5 kg m^{-3} gross specific charge. Similarly, blast hole sizes have ranged from 64 to 311 mm.

In the main, once scales were adequate to overcome random perturbations, laws of verisimilitude could be effectively applied in the manner taught by Langefors and Kihlstrom (1963), to facilitate reasonably predictable extrapolation from Geokinetics' expanding database. Scaling has also been discussed by Clark (1967) and, specifically the case of column charge arrays, by Redpath (1977). For the geometries involved, the bulk of the motion is uniaxial and the specific charge required for verisimilitude in lifting should therefore be a fourth power function of the linear dimension, due to gravitational effects. This must ultimately limit the process. It must also, as a practical matter, determine the technical or economic limits of lossy approaches such as multipurpose vertical column charge arrays. For the latter, a question arises as to the effective specific charge.

Downward lengthening of the charge columns must approach some limiting value for overburden lifting effect. Even with the comparatively short charge columns in current use, Redpath's analysis, using the entire column weight of explosive, produces results which are low compared to that seen in practice. The question has been addressed experimentally by GKI. Explosive specific energy was reduced in the upper portion of the charge columns, with a corresponding increase in the lower portion to maintain the overall specific charge constant on an energy basis. A reduction in overburden motion resulted, very closely paralleling the reduction in energy for the upper portion of the charge. Following the Coursen breakage model one would expect, and finds, fair agreement between computed and observed results utilizing the

Surface uplift blasting for oil shale production 413

parameters for only that explosive contained in the upper portion of the column to a depth corresponding to one-half the spacing/burden.

For vertical column lifting charges, investigation typically shows that intense fragmentation begins abruptly at the interface between the overburden and the charge array defined retort zone. Material is diced and extensively randomized for 3 or 4 m, becoming tighter below, with inclusion of ordered zones which dominate to exclusion, typically for the lower third of the retort zone. Excessive fragmentation in the upper portion may be reduced by use of a low detonation pressure explosive and, as might be expected, fragmentation may be increased at the toe with use of high energy slurry explosives. The latter does not, however, necessarily result in improved permeability since very little more void is introduced and, with many more particles, the mean separations become smaller and paths more tortuous. A noticeable improvement can, however, be obtained by use of GKI's 'Fast Leader' technique, described in Section 9.4. After the firing of the last charge array, for retorts of large area to overburden thickness ratio, overburden motion continues basically ballistically, neither edge restraint nor residual gas pressure being significant. As extensometers within the overburden show, it does not dilate until the fallback phase, when dilation is comparatively small. Fallback (corrected for dilation) is not quite to the elevation at the firing time for the last firing charges, so some bulking of the rubble may continue to occur during the subsequent motion time of the overburden. This may result from gas detraining from the lower portions of the rubble under the influence of the pressure gradient generated by the void left by the rising overburden. Alternatively, loose rubble from the underside of the overburden slab may rotate and prop.

The times and dimensions for overburden motion can be considerable. For the large retort producing the motion curve of Fig. 9.9, the driving impulses, smeared by photographic instrumentation, are quite short and give a clearly impulsive character to the driving force. The primary driving force then diminishes to a nearly constant low level force which approximately balances gravitation (the details again being lost through instrumentation and processing degradation) but from as little as 235 ms after inception of motion the motion is ballistic and remains so through apogee some 900 ms later until fallback is limited by the rubble pile more than 1·5 s into the shot. The maximum displacement for this shot exceeded 6 m, above an original 10 m retort zone.

Fallback is objectionable as it implies some measure of waste and some unnecessary damage (and hence retort leakage) at the edges of the raised mass. In principle it could be minimized by the use of decked charges to fire around the time of maximum elevation. Unfortunately, this presents practical difficulties in that the available precision from commercial pyrotechnic delays is inadequate. There is sufficient time for slow processes (which are not a problem in more normal blasting) to penetrate stemming, dead pressing the explosive or damaging the initiation train, and charges are liable to dislocation through rock motions. It has not yet been necessary for much work to be done in this area, as to date dimensions have been sufficiently small for the desired propping to be accomplished externally.

Once the overburden has been put into ballistic motion and sufficient space has developed (1·5–3 m), fairly heavily charged blast holes arrayed at the periphery of the raised overburden may be fired, resulting in large lateral motions which effectively bulk the newly broken and previously damaged material into firm contact with the rising overburden for as much as 20 m into the rubble zone. The void created by the lateral motion of the burden on these charges becomes the void and free face required for lateral extension of the damaged area, thus this has hitherto been the preferred method of finally propping the initially raised overburden. Practical difficulties with decking cutoff, dead pressing, etc., are avoided, but the method is similarly limited to only about half its potential by the limitations of pyrotechnic delays. A further limitation occurs for geometrical reasons as scale increases. If the initially attacked area is too large, since propping proceeds only a limited distance from the edge inwardly, the middle subsides, an effect which becomes increasingly apparent as the dimensions exceed around 50 m. This becomes a limiting factor as the overburden thickness increases and large areal dimensions become essential for lifting feasibility.

9.4. LATERAL EXTENSION

Once space and a free face have been developed by the firing of an initiating round, the damaged zone may be efficiently laterally extended by the firing of more conventionally designed charges. Granted adequate performance from the first firing row, each succeeding row will be presented with a free face geometrically similar to that of the

Surface uplift blasting for oil shale production 415

preceding and with a void to accommodate bulk motion equal to that available to the preceding less the volume consumed by bulking of the preceding charge's burden but plus the volume contributed by any additional overburden raising.

Edge effects from the initially raised overburden tend to propagate an open plane fracture at the overburden/retort zone interface, providing an horizontal free face for breakage processes. Further breakage may then proceed in a fashion similar to quarry bench blasting with toe restriction from the muckpile. For continued blasting and comparatively long delays between rows (of the order of 50 ms), there is time for each successive row to continue the process of overburden raising as each increment of dilating burden contributes further lift.

The process is subtle and complex. The broken material bulks to fill the available vertical space, adjusting itself to fit by adopting the appropriate ratio of randomized and ordered bed modes. Post-blast coring reveals that the vertical mode distribution is surprisingly stable, implying that once the overall geometry has been established by the early firing rows, subsequent rows will tend to have lateral burden motion inhibited where prior rows left ordered material. For this reason, use of inert decks as charge separators requires caution as the uncharged zone may initiate a persistent ordered layer. At least for the shallower overburdens, to 20 m, the process will typically be initially unstable. With inadequate specific charge the bed will revert to an ordered bed by rising hard bottom and may freeze. With excess specific charge, breakage conditions will typically improve and stabilize but diminish and stabilize where a substantial excess of void was initially provided. Lateral motions within the randomized material may be quite pronounced, 10 m or so, despite the restraint imposed by the necessity to continually raise the overburden.

Similar timing effects are observed to those within the initiating round areas. As the interrow delay increases, lateral motions increase due to the development of pressure gradients and the lifting force then developed probably primarily represents the conversion of kinetic energy of the vertically bulking rubble. Necessarily though, energy must be lost to absorption within the continually flexed overburden. For very short delays, less than $0\cdot3\,\mathrm{ms\,m^{-1}}$ burden, the resulting deformation within the overburden can be minimized and the lifting force reach high values, but at the cost of leaving a bed which, because of the small pressure differentials developed, then becomes basically ordered. Since these ordered beds have been demonstrated to retort

416   *Keith Britton*

with high efficiency despite quite small void volumes, rapid timing may be the preferred course. It is particularly attractive, for instance, where the lithology encourages the separation of charges with a central deck, as one charge array may be fired in rapid sequence to perform lifting, with subsequent fragmentation of the balance and the decked zone utilizing a much slower sequence for the second array.

Thicker overburden involves greater deformation losses and the round must eventually freeze when void originally provided by the initiating round has been consumed and insufficient new void can be sourced from the surface by uplift. Limits have not yet been established but, at scales of 20 m overburden and 50 m retreat, the main problem at present seems to be ensuring that adequate rubble is left to support the overburden.

Fragmentation and permeability are affected by too many factors for explicit computation, a situation which will persist pending significant advance in the extent of the available database and possibly also computer science. Computer aided design proves most useful but control is readily achieved in practice by corrective adjustment based on informed judgement. The properties of particular patterns have been discussed elsewhere (Britton and Lekas, 1979; Britton, 1980) and are beyond the space limitations of this chapter. Lithology has a perceptible influence, but the properties of explosives may be locally varied to control particle size, as in initiating rounds. Timing and geometry dominate, however, the latter being especially troublesome regarding bottom breakage and void. Inert decks are desirable to stop axial venting up the blast hole, but decks of conventional dimensions restrict lateral motions and critically reduce the local specific charge. As in the initiating rounds, improvement results from the 'Fast Leader' technique, which also offers a convenient example of subtleties which distinguish blasting for *in situ* bed preparation from more conventional blasting.

In the 'Fast Leader' technique (see Fig. 9.10) the column charge is initiated by a mid or top primer, which initiates both a main charge and a high velocity, low coreload detonating cord. Cord detonation outruns that of the main charge, traverses a short ($\simeq 1$ m) inert deck (with minimum disruption) and fires a primer at the hole bottom to initiate a high energy toeload in the reverse direction. System dimensions and explosives are selected to precisely position the resulting wave superposition, normally at a point corresponding to the lower extremity of the inert deck. The stress front from the main charge,

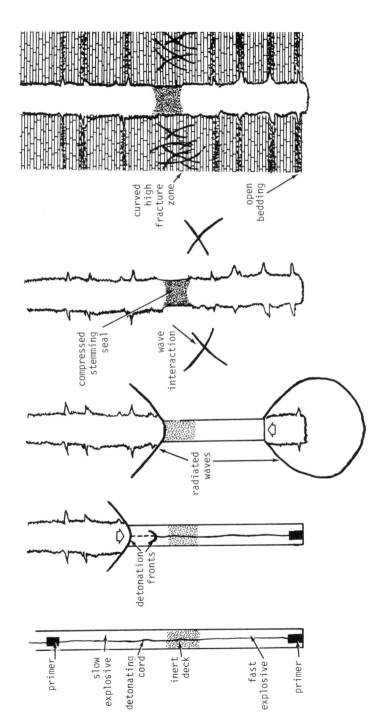

Fig. 9.10. The 'Fast Leader' technique.

which would normally be largely lost beneath the retort, then combines in a complex interaction with waves from the toe charge to fragment the uncharged material adjacent the inert deck. The deck is sintered to high integrity by the balanced pressure from the end effects of the two mutually opposing detonation heads. When pressure differentials develop, reaction products from the toe charge are axially strongly contained within a severely damaged medium and therefore vent through the rock. The result, from coring, is typically a tightish but broken zone (typically exhibiting curved fractures) aligned with the inert deck and underlain by coarse, even, ordered breakage with noticeable open flow channels. Desirable permeability is thus obtained at the lowest levels of the retort for fluid drainage and retorting. Lateral motion at the toe is also facilitated, as evidenced by fragment comminution on very low level slip planes.

9.5. ADDITIONAL ASPECTS

The necessity of allowing for practical limitations is well illustrated by consideration of commercial delay detonators. Ideally, charges of the lateral extension would begin to fire when the initially raised overburden reached its maximum elevation. Thereafter, the interrow delay should be minimized, but should exceed 5 standard errors in precision to avoid cut-off holes from out of sequence firing. Delay detonators are so inaccurate that use of more than one period is intolerable. Poor precision likewise rules out all electric caps and non-electric delays exceeding 600 ms. Typical GKI practice therefore uses a common period downhole delay, with a common period surface delay to control interrow accuracy and monotonicity. (Timing is adjusted to avoid downline cutoffs from the extensive edge effects.) Care is taken to select the product by ensuring that all caps within a period are of the same batch. All incoming caps are characterized to screen out bad batches and provide data for Monte Carlo modeling. Caps may be paralleled to improve scatter and reliability. Measurements on 600 ms delays reveal thermal coefficients of $\simeq 1 \, ms \, °C^{-1}$. Time is allowed for the explosive, which may be emplaced at over $70 °C$, to normalize to rock ambient. Because of round dimensions, systematic and random propagation delays from the initiating train are significant and allowed for.

Despite this care, the achievable coefficient of variation is little

Surface uplift blasting for oil shale production 419

better than 1·5%, a matter of considerable seriousness. It limits the process to around 0·5 s for initial overburden motion delay and to ≃40–50 ms for interrow delay. About half the overburden motion is wasted (or ≃30% of the drill and blast cost). The effective duration of gas pressure from a charge is comparable to or shorter than one standard error in long delay precision, so intrarow sequence and hence firing time geometry are unpredictable. Options are therefore limited for practicable designs, and adequate tolerance is usually at the expense of efficiency.

System dimensions, coupled with the need to fire up to 25 tonnes explosive on a single and precise delay, result in powerful and unusually low frequency seismic radiation, which may also exhibit marked directionality. Superposition appears to occur in the direction of the surface timing wave, which exceeds the rock longitudinal velocity by a ratio of 3:2, the match being almost exact for the worst case of the diagonal of a corner initiated rectangular array. For the latter, the early part of a wave train plot contained a striking, near sinusoidal, 5 Hz half cycle (the instrument limit) of high amplitude. From limited studies using Fourier analysis, some reduction in vibration seems realizable by careful selection of delay periods to avoid driving resonances (as also anticipated (Langefors and Kihlstrom, 1963)).

Presently, there seems no bar to extending the process from its present 30 m overburden to at least 50 m, utilizing decking, timing changes, attention to scale and possibly some edge preparation. Increasing the retort zone thickness, to 25 m+, appears even more straightforward, but actually involves more risk since the retorting characteristics and permeability requirements are unknown. This, however, applies to GKI's Kamp Kerogen site. Blast designs developed for a particular scale and lithology may be totally inappropriate to another. Edge restraint may be very different—ductile, for instance, with clay or permafrost. Soft oil shales will tend to compact rather than fragment with blasting. Uniform oil shales will not exhibit 'destroyed laminate' failures. Very large scales will require mining to obtain the access needed for emplacement of massive charges and to achieve the necessary charge geometries.

Lifting at great scale will require the use of low explosives or propellants for reasonable efficiency. Even at present scales, detonation pressures of common commercial explosives are excessive for blasting high grade oil shales, deforming the borehole wall rather than fracturing and pressurizing the rock. Low detonation pressure explo-

420 *Keith Britton*

sive has been found to invade the higher grade beddings, resulting in a notably ragged rather than cylindrical profile, possibly facilitating formation gas penetration. Rapid gas cooling appears to occur, suggesting that better results may be obtained with explosives generating permanent gases. For ANFO, 6 of 11 mol condenses to water and the situation worsens for aluminized explosives unless the aluminum content is sufficient to produce hydrogen as in the work by Coursen (1977). New explosives and digital electronic delay detonators are required.

ACKNOWLEDGMENTS

The author is indebted to Geokinetics Inc. for permission to publish this material, which derives from experience and observation during development of their Lofreco process, work in part supported by the US Department of Energy and instrumented by Sandia National Laboratories.

REFERENCES

Britton, K. C. (1980). Principles of blast design developed for *in situ* retorts of the Geokinetics surface uplift type, *13th Oil Shale Symposium Proceedings*, Colorado School of Mines, pp. 169–180.

Britton, K. C. and Lekas, M. A. (1979). *US Patent 4,175,490.*

Chazin, D., French, G., Hanna, K., Heydari, M., McCarthy, H., Moore, E., Nelms, R. and Wing, E. (1979). Commercial feasibility evaluation of selected oil shale recovery systems, *Report to Laramie Energy Technology Center*, Science Applications Inc., p. B2.

Clark, G. B. (1967). Blasting and dynamic rock mechanics, *Failure and Breakage of Rock, Proceedings of the 8th Symposium on Rock Mechanics*, ed. C. Fairhurst, American Institute of Mining, Metallurgical and Petroleum Engineers Inc., New York, pp. 463–99.

Cook, M. A. (1958). *The Science of High Explosives*, ACS Monograph 139, Krieger, New York, pp. 344–6.

Coursen, D. L. (1977). Cavities and gas penetrations from blasts in stressed rock with flooded joints, *Acta Astronautica*, **6**, 341–63.

Coursen, D. L. (1980). A gas penetration model of fragmentation, *Society for Experimental Stress Analysis Annual Meeting*, Fort Lauderdale, Florida, 12–15 October.

Fourney, W. L., Barker, D. B. and Hollaway, D. C. (1979). Mechanism of fragmentation in a jointed formation, *Report to NSF grant no. DAR. 77–05171.*

Surface uplift blasting for oil shale production

Johansson, C. H. and Persson, P. A. (1970). *Detonics of High Explosives*, Academic Press, London and New York, pp. 243–8, 272–4.

Langefors, U. and Kihlstrom, B. (1963). *The Modern Technique of Rock Blasting*, John Wiley, New York; Almqvist and Wiksell, Stockholm, pp. 281–4.

Patterson, E. M. (1957). Photography applied to the study of rock blasting, *J. Photographic Sci.*, **5**, 140, 142.

Redpath, B. B. (1977). Application of cratering characteristics to conventional blast design. *Monograph 1 on Rock Mechanics Applications in Mining*, eds W. S. Brown, S. J. Green and W. A. Hustrulid, American Institute of Mining, Metallurgical and Petroleum Engineers Inc., New York, pp. 213–20.

Schamaun, J. T. (1981). Lumped mass modeling of overburden motion during explosive blasting, *Sandia National Labs Report SAND 80–2413*.

Simon, R. (1979). Oil shale *in situ* retorting: effect of porosity on yield, *12th Oil Shale Symposium Proceedings*, Colorado School of Mines, pp. 278–82.

Wright, F. D., Burgh, E. and Brown, B. C. (1953). Blasting research at the Bureau of Mines oil shale mine, *BuMines, RI 4956*.

Zerga, D. P. (1980). *US Patent 4,205,610*.

Chapter 10

A Strain-Rate Sensitive
Rock Fragmentation Model

E. P. CHEN, MARLIN E. KIPP and D. E. GRADY

Sandia National Laboratories, Albuquerque, New Mexico, USA

SUMMARY

The dependence of rock fracture and fragmentation on strain-rate is examined in this chapter. The dependence is first derived within the context of dynamic fracture mechanics. This feature is then incorporated explicitly into a dynamic rock fragmentation model in which the fracture process is modeled as a continuous accrual of damage in the rock mass, leading to a state of fragmentation. The damage is defined to be the volume fraction of material that has been stress relieved by multiple crack growth. The model is calibrated on data over the strain rate range important to explosive blasting. Application of the model to oil shale has been made by using a two-dimensional finite difference wave propagation code. Correlation of computational results with available experimental data from both laboratory and field scale tests is made and an example of a blasting scheme suggested from the calculations is included.

10.1. INTRODUCTION

Fracture and fragmentation of rock lies at the heart of suitable bed preparation for resource recovery techniques as applied to oil shale formations. Effective fracture and fragmentation may be accomplished by proper selection of explosive charges, geometries and timing. This selection process could be significantly aided by interaction with accurate simulations of particular choices of these variables. To develop computational predictive capabilities, it is necessary to determine the response of rock media to dynamic stress-wave loading. In particular, it is necessary to understand the influence of transient tensile waves on crack growth in the rock. Since many rocks, including oil shale, have an existing flaw structure (McHugh *et al.*, 1977), it is expected that these flaws will activate and grow under tensile pulse loading until such time as interaction between cracks occurs, or the load is removed.

Explosive loading configurations that would be used in field applications generally result in a compressive spherical or cylindrical wave emanating from the source. The divergent particle motion induced in the rock pulls the rock into tension in the circumferential direction (and usually in the radial direction as well). The strain rate at which the tension develops in divergent geometries is calculated to range between $10^1\,\mathrm{s}^{-1}$ and $10^3\,\mathrm{s}^{-1}$. If, however, these same waves are incident on free surfaces, tension can develop at strain rates of $10^4\,\mathrm{s}^{-1}$ to $10^5\,\mathrm{s}^{-1}$ as the result of interacting release waves. In both of these regions, the tension formation at early times can be characterized by constant strain-rate loading.

In the strain-rate regime of $10^1-10^5\,\mathrm{s}^{-1}$, inertia may have a major influence on the response of a material containing flaws. Such effects have been observed by Green and Perkins (1968) and Birkimer (1971) in several types of geologic media in the form of a strain-rate dependent fracture stress. More recently, Kalthoff and Shockey (1977) have analyzed the role of inertia under short impulse loading conditions and demonstrated that cracks in polycarbonate respond as predicted to those loads, while Grady and Lipkin (1980) have shown that inertial effects influence the fracture behavior of several rock types.

At much lower strain rates $(10^{-7}-10^0\,\mathrm{s}^{-1})$, the fracture stress has also been noted to be a function of the loading rate (e.g. John, 1974; Sinha, 1979). However, at these strain rates, mechanisms of creep are likely to be more important than the dynamic wave propagation effects discussed in this chapter.

The range of fracture stresses over the strain-rate interval $10^1-10^5\,\mathrm{s}^{-1}$ may be as large as an order of magnitude. For example, in oil shale, the fracture stress at strain rates of $10^4-10^5\,\mathrm{s}^{-1}$, obtained in spallation experiments, is of the order of 100 MPa, and is insensitive to orientation (Grady and Hollenbach, 1979). In contrast, the static fracture stress is of the order of 5–20 MPa (for competent material) and is quite sensitive to the loading orientation relative to the bedding planes (Schmidt, 1977). Fracture stresses ranging from 30 to 50 MPa have been obtained at intermediate strain rates by torsional split Hopkinson bar techniques (Lipkin and Jones, 1979).

The mechanical properties of oil shale have been under investigation in a program at Sandia National Laboratories that has involved laboratory and field experimentation along with complementary analytical activities. Through this program a numerical fragmentation model has been developed for use in finite difference wave codes to simulate the response of oil shale to dynamic loads of the type experienced in explosive blasting. The model incorporates the analytical and experimental observations that fracture thresholds and fragment dimensions depend strongly on the rate of load application or on the rate at which material is strained. Description of the fragmentation model, the analytical and experimental studies upon which it is based and the verification of its applicability to blasting problems have been presented in several reports (Grady and Kipp, 1980; Grady *et al.*, 1980; Kipp and Grady; 1980; Kipp *et al.*, 1980; Boade *et al.*, 1981), and a review of this work will be given here.

Section 10.2 of this chapter considers the response of an isolated crack to constant strain-rate tensile loading in order to gain insights into the basis for the variation in dynamic fracture strength with strain rate. Analytical observations from this study are incorporated into the development of the rock fragmentation model, which is described in Section 10.3. Verification of the model through correlation with field experimental data is given in Section 10.4 and a few concluding remarks in Section 10.5 complete this chapter.

10.2. ELASTODYNAMIC FRACTURE MECHANICS

The strain-rate effect on the dynamic fracture strength of rock material is studied through the response of an isolated crack subjected to the action of constant strain-rate tensile loads. The impact response of an elastic solid containing a crack and subjected to tensile loading normal to the crack surface has been well characterized, and is extensively discussed by Chen and Sih (1977) in terms of the dynamic stress intensity factor. The response of the solid to other loading functions applied to the crack has been stated in general form by Freund (1973), but no explicit characterization exists for other than the impact loading

426   *E. P. Chen, Marlin E. Kipp and D. E. Grady*

case. For the present discussion, we will limit the topic to the events leading to fracture initiation. The subsequent material bulk strength degradation will be discussed in the next section.

Note that the material properties for Anvil Points shale (McHugh *et al.*, 1977; Schmidt, 1977; Grady and Hollenbach, 1979; Lipkin and Jones, 1979) have been used in the numerical calculations in this study.

10.2.1. The Dynamic Stress Intensity Factor

A study of the response of isolated flaws, or cracks, to transient tensile loads is aided by the well developed theory of linear elastic fracture mechanics applied to the dynamic loading of an isolated crack. The response of cracks to various types of Heaviside loading has been outlined by Chen and Sih (1977). The Heaviside loading response function may then be readily employed as a Green's function for other dynamic pulse shapes (Freund, 1973).

Specifically, if a Heaviside load of magnitude σ_0 is applied to a crack with a characteristic dimension a, the functional form of the stress intensity factor, K_1, at the crack tip, is

$$K_1(a, t) = \sigma_0 \sqrt{a\pi} f(c_2 t/a), \tag{10.1}$$

where c_2 is the shear wave velocity. The response to an arbitrary stress loading function, $\sigma(t)$, may then be expressed as (Freund, 1973)

$$K_1(a, t) = \sqrt{a\pi} \int_0^t \sigma'(s) f(c_2(t-s)/a) \, ds \tag{10.2}$$

The loading function which provides a convenient kinematic parameter for evaluating various explosive loading geometries is that of constant strain-rate loading. For an elastic material, the elastic modulus relates the stress rate ($\dot{\sigma}_0$) and strain-rate ($\dot{\varepsilon}_0$), so for constant strain-rate the stress rate is constant and $\sigma'(s) = \dot{\sigma}_0$. Equation 10.2 may then be rewritten as

$$K_1(a, t) = \dot{\sigma}_0 \sqrt{a\pi} \int_0^t f(c_2(t-s)/a) \, ds \tag{10.3}$$

Through a simple change of variable, eqn 10.3 further simplifies to

$$K_1(a, t) = \dot{\sigma}_0 \sqrt{a\pi} \int_0^t f(c_2 s/a) \, ds \tag{10.4}$$

At some time, t_c, the stress intensity factor calculated by eqn 10.4 may become large enough to exceed the critical stress intensity factor, K_{1C}, at which time crack growth initiates. At this critical time, t_c, the applied stress level will be the crack activation, or fracture stress σ_c, where

$$\sigma_c = \dot{\sigma}_0 t_c \qquad (10.5)$$

Equation 10.4, when equated to K_{1C} at time t_c, is then written as

$$K_{1C} = \dot{\sigma}_0 \sqrt{a\pi} \int_0^{\sigma_c/\dot{\sigma}_0} f(c_2 s/a) \, dc \qquad (10.6)$$

Equation 10.6 is now an implicit equation in the parameters a and σ_c, where K_{1C} and c_2 are material properties, and $\dot{\sigma}_0$ is the loading rate parameter. As an alternative, eqn 10.6 could be considered an expression relating $\dot{\sigma}_0$ (or $\dot{\varepsilon}_0$) and σ_c, in which, for constant crack size, a, the dependence of the activation stress on strain rate is defined.

Before discussing the relationship of crack size, strain rate and activation stress obtained from eqn 10.6, it is important to remain cognizant of the nature of the function $f(c_2 s/a)$. In the elasticity analysis for the Heaviside loading, as discussed by Chen and Sih (1977), a Laplace transform is applied to the wave equations. At the crack tip the specific response of the stress intensity factor may then be

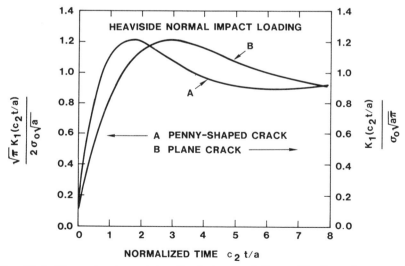

Fig. 10.1. Dynamic stress intensity factor variation with time for normal Heaviside loading.

428   E. P. Chen, Marlin E. Kipp and D. E. Grady

formulated in terms of a Fredholm integral. This integral is numerically evaluated to obtain the stress intensity factor in Laplace transform space. Numerical inversion of this (numerical) function is required to finally obtain the function $f(c_2 s/a)$ in eqn 10.6. The resulting dynamic stress intensity factors for a plane crack of length $2a$, and for a penny-shaped crack of radius a, are plotted in Fig. 10.1. In each case the dynamic stress intensity factor has been normalized by the corresponding static stress intensity factor.

To complete the present analysis, it is necessary to perform a numerical integration of these functions as prescribed by eqn (10.6), bringing the total numerical operations to three for the results to be discussed in the following section.

10.2.2. Penny-shaped Crack

Consider first the crack size dependence of the fracture activation stress for a penny-shaped crack. K_{1C} was chosen to be $1\,\text{MPa}\sqrt{\text{m}}$, which is a value within the observed range for K_{1C} obtained by Schmidt (1977) for Anvil Points oil shale. K_{1C} for this rock varies between $0\cdot3$ and $1\cdot1\,\text{MPa}\sqrt{\text{m}}$, depending upon grade and layer orientation relative to crack direction. The shear wave velocity is approximately $1\cdot74\,\text{km}\,\text{s}^{-1}$ for oil shale with a density of $2\cdot0\,\text{Mg}\,\text{m}^{-3}$. For a given strain rate, we may then determine the relationship of crack size to the fracture stress. (The stress rate and strain rate were assumed to be related by a modulus of $15\cdot6\,\text{GPa}$.)

In Fig. 10.2, the fracture stress is plotted as a function of crack size for several representative strain rates. The range of crack sizes is similar to that considered in previous characterizations of oil shale (McHugh et al., 1977). For reference, the static stress-crack radius relationship is also plotted in Fig. 10.2 ($\dot{\varepsilon}_0 = 0$). As the strain rate increases, the point of departure from the static solution moves toward smaller crack radii and correspondingly higher fracture stress levels. We also observe that for each strain rate (greater than zero), there is an intermediate crack length for which the fracture stress is a minimum, i.e. a preferential crack size. Although some particular intermediate crack size may have a slightly lower fracture stress, suggesting pulse tailoring to optimize the strain rate, the significant difference occurs between static and dynamic loading. When a solid with an array of cracks is loaded statically, the largest flaw will dominate the response of the solid, limiting the maximum load that can be applied. If a preferred orientation of the largest flaws exists, the

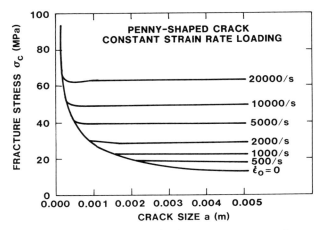

Fig. 10.2. Fracture stress versus crack size at constant strain rate loading. (Penny-shaped crack.)

material will also show an orientation dependence for the fracture stress. In the dynamic case, however, the largest crack no longer dominates; rather, cracks with a wide range of sizes are clearly activated nearly simultaneously, so failure occurs by fracturing the solid through multiple crack growth. Even with some preferred flaw orientation, the dynamic fracture stress tends to be independent of orientation.

The insensitivity of the fracture stress over a large range of crack sizes for constant strain-rate loading suggests that the inherent flaws in the rock are the basis for the strain-rate dependent fracture stress, i.e. it is a geometric and not a material effect. For crack radii larger than 2·0 mm (Fig. 10.2), the static fracture stress, which is governed by the largest flaw in the solid, is less than 20 MPa for the oil shale. The fracture stress is plotted as a function of strain rate for a crack size of 5·0 mm in Fig. 10.3 (solid line). A fracture stress of 100 MPa is calculated at strain rates in excess of $10^4 \, \text{s}^{-1}$. This dependence of fracture stress on strain rate is consistent with experimental results on oil shale (Grady and Kipp, 1979a) and holds over a large range of crack sizes.

An approximate analysis has been suggested (Grady and Kipp, 1979b) that describes the strain-rate dependent fracture stress observed in Fig. 10.3. Assume that the penny-shaped crack response to

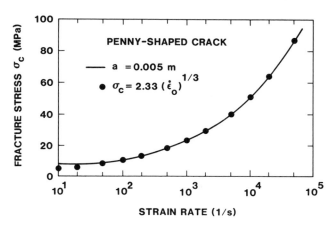

Fig. 10.3. Fracture stress versus strain rate for fixed crack radius. (Penny-shaped crack.)

Heaviside loading may be described at small normalized times by

$$K_1(t) = N \frac{2}{\sqrt{\pi}} \sigma_0 \sqrt{a} \sqrt{c_2 t/a}, \qquad (10.7)$$

where N is a geometry coefficent obtained from the early portion of the numerical curve in Fig. 10.1. As before, using eqn 10.7 as a Green's function for an arbitrary loading $\sigma_0(t)$ of the crack,

$$K_1(t) = N \frac{2}{\sqrt{\pi}} \sqrt{c_2} \int_0^t \sigma_0'(s) \sqrt{t-s}\, ds \qquad (10.8)$$

For constant stress rate,

$$\sigma_0'(s) = \dot\sigma_0, \qquad (10.9)$$

Equation 10.8 becomes

$$K_1(t) = N \frac{2}{\sqrt{\pi}} \sqrt{c_2}\, \dot\sigma_0 \frac{2}{3} t^{3/2} \qquad (10.10)$$

Using the results in eqns 10.5 and 10.6,

$$K_{1C} = N \frac{4}{3\sqrt{\pi}} \frac{\sqrt{c_2}}{\sqrt{\dot\sigma_0}} \sigma_c^{3/2} \qquad (10.11)$$

Relating $\dot\sigma_0$ to the strain rate $\dot\varepsilon_0$ by the modulus E, and solving eqn

10.11 for σ_c provides the strain-rate dependence,

$$\sigma_c = \sqrt[3]{\frac{9\pi E K_{1C}^2}{16 N^2 c^2}} \dot{\varepsilon}_0^{1/3} \qquad (10.12)$$

The coefficient N for the penny-shaped crack in Fig. 10.1 is 1·12. The remaining physical parameters are unchanged ($K_{1C} = 1$ MPa$\sqrt{\text{m}}$, $E = 15\cdot 6$ GPa, $c_2 = 1\cdot 74$ km s^{-1}) and eqn 10.12 becomes

$$\sigma_0 = 2\cdot 33 \dot{\varepsilon}_0^{1/3} \qquad (10.13)$$

This equation is plotted in Fig. 10.3 as the solid circles. Except at strain rates below 50 s^{-1} where crack size is becoming important, the cube root function is indistinguishable from the numerically integrated solution.

10.2.3. Plane Crack

For the same material parameters, the fracture stress has been determined as a function of crack size for the plane crack. In Fig. 10.4, the static solution is again plotted for reference. Then for the same strain rates as in Fig. 10.2 for the penny-shaped crack, the fracture stress-crack size dependence is plotted. The general behavior is similar to that discussed for the penny-shaped crack, including the shallow minimum indicating a preferential crack size at a given strain rate.

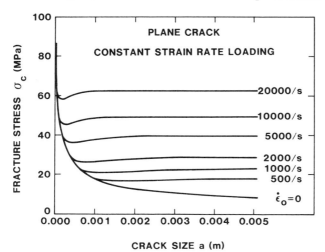

Fig. 10.4. Fracture stress versus crack size at constant strain rate loading. (Plane crack.)

But in the plane crack case, there is an analytic solution to the stress intensity factor up to the time of arrival of a signal from the other crack tip. This is the semi-infinite crack solution (Chen and Sih, 1977), which is characterized by the square root of the time,

$$K_1(t) \sim \sqrt{t} \tag{10.14}$$

Since this is of the same functional form as eqn 10.7, considered for the penny-shaped crack, the fracture stress will be dependent on the cube root of the strain rate, and a figure similar to Fig. 10.3 would be obtained for the plane crack.

10.2.4. Crack Shape Dependence

We now briefly compare the plane and penny-shaped crack responses to constant strain-rate loading. In Fig. 10.5, we have plotted on the same figure the results of Figs 10.2 and 10.4. Observe that for static loading, a solid containing a penny-shaped crack has a larger fracture stress than one with a plane crack of the same characteristic dimension.

As the strain rate becomes non-zero, we note that beyond some crack size, the fracture stresses for both the penny-shaped and plane cracks are identical. That is, beyond a certain crack size, the solid

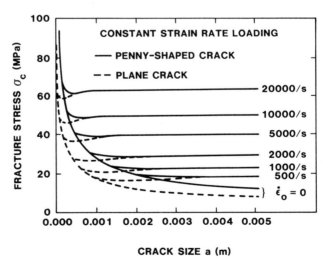

Fig. 10.5. Comparison of plane and penny-shaped crack response to constant strain rate loading.

cannot discriminate the geometry of the flaw being loaded. This merging of fracture stress for the plane and penny-shaped cracks suggests that the crack front curvature is no longer important, and that for a characteristic crack dimension, the shape is of no consequence to the constant strain-rate response.

Evidence has been presented for strain-rate dependent fracture response of an elastic solid containing a crack. Certainly, the analysis shows that the fracture initiation stress is a strong function of the strain rate for a given crack size. This result appears to provide a good basis for explaining the strain-rate dependent fracture response of rocks with existing flaw distributions. We have also observed that in dynamic loading, the fracture stress becomes independent of the crack geometry.

The preferential crack initiation that was shown to exist for a given stress rate is a very weak phenomenon. The calculation is complicated by the severe difficulties that exist in determining the Heaviside loading response from the numerical inversions required to form solutions to the dynamic elastic problem. Within the accuracies of the currently available solution, however, the calculated fracture stress minimum appears to be a real phenomenon.

10.3. ROCK FRAGMENTATION MODEL DEVELOPMENT

The development of a rock fragmentation model that is strain rate sensitive is presented in this section. The model is based on observations from laboratory fracturing and fragmentation experiments and analytical observations from the previous section which indicate that fracture thresholds and fragment dimensions depend strongly on the rate of load application, or on the rate at which the material is strained (see Fig. 10.6). This type of behavior can be explained by hypothesizing that the material in its virgin state contains a distribution of flaws with various sizes and orientations. The flaws become active, i.e. grow, when subjected to tensile loads of various magnitudes. The active flaws grow during the period of load application and eventually coalesce, causing material failure and termination of the flaw growth. When a load is applied slowly, only those flaws that become active at low stress levels actually contribute to the fragmentation process because these flaws grow and coalesce before the applied load reaches a level of stress high enough to activate other flaws. This results in a low

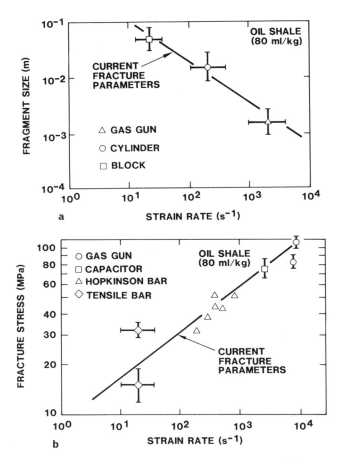

Fig. 10.6.(a) Fracture stress dependence on strain rate; (b) fragment size dependence on strain rate.

apparent threshold for material failure, and comparatively large fragments, because the number of contributing flaws is relatively small. When the load is applied more rapidly, a higher level of stress can be reached before flaw coalescence occurs; thus, a greater number of flaws become active, causing the fragment dimensions to be smaller and the apparent threshold for material failure to be higher.

The dependence of the fragmentation phenomenon on strain rate is important in explosive applications because strain rates realized in

A strain-rate sensitive rock fragmentation model 435

typical blasting events extend over a range comparable to that depicted in Fig. 10.6. For instance, near a cylindrical borehole charge, strain rates may be in the $10^2–10^3\,\mathrm{s}^{-1}$ range, while at larger radii, they may drop to the $10^0–10^1\,\mathrm{s}^{-1}$ range, or lower. If a free face is present, strain rates in the $10^3–10^4\,\mathrm{s}^{-1}$ range may be reached when tensions develop after a compressive wave reflects from the free surface.

10.3.1. Continuum Damage Concept

The above behavior for dynamic material failure was assumed in developing the numerical formulation of the fragmentation model for oil shale. The instantaneous state of the material with regard to structural integrity is described in the model by a damage parameter, D, which is defined as the volume fraction of material that has lost its load-carrying ability. Damage is considered a scalar parameter,

$$0 \leqslant D \leqslant 1 \tag{10.15}$$

such that $D = 0$ corresponds to the intact, undamaged rock, $D = 1$ corresponds to the inability to transmit tensile stress (full fragmentation) and intermediate values of D correspond to intermediate values of fracture damage.

A fractured material subject to a tensile stress will undergo a strain determined by a modulus K_f which is less than the intrinsic elastic modulus of the unfractured materials, K. The damage D is regarded as the parameter which determines the reduced modulus of the fractured material, $K_f = K(1 - D)$. The elastic energy expression is then,

$$\mathscr{E} = \tfrac{1}{2}K(1 - D)\varepsilon^2 \tag{10.16}$$

where ε is a one-dimensional tensile strain. For simplicity we will initially only consider the one-dimensional problem. A more complete description of the strain state will be provided later. If the damage remains the same, D is simply a constant that characterizes the reduced elastic modulus of the material. If, however, under the applied tensile load, fracture is initiated in the material, then D is no longer constant but may be regarded as an internal state variable that characterizes the instantaneous level of fracture damage (Davison and Stevens, 1973). A complete material description would then require a law governing the time-dependent growth of damage. Thermodynamic restrictions require that the tensile stress be given by

$$\left(\frac{\partial \mathscr{E}}{\partial \varepsilon}\right)_D = \sigma = K(1 - D)\varepsilon \tag{10.17}$$

The derivative

$$\left(\frac{\partial \mathcal{E}}{\partial D}\right)_\varepsilon = \Gamma, \tag{10.18}$$

can be regarded as a strain energy release rate density analogous to the expression for the extension of a single crack.

In microstructural theories of the elastic properties of fractured rock, the elastic modulus is found to be approximately reduced by a factor

$$D = Nv, \tag{10.19}$$

where N is the number of idealized penny-shaped flaws per unit volume and $v = \frac{4}{3}\pi r^3$ is the spherical region surrounding the penny-shaped flaw of radius r which approximates the stress-relieved volume due to the traction-free boundary of the crack. Consequently, the damage, D, can be regarded as the volume fraction of the material which has lost its load carrying capability.

10.3.1.1. Damage Activation and Growth

The activation of flaws could be based on a Griffith criterion or, due to rapid loading, on more recently developed dynamic fracture crieria, as discussed in Section 10.2. Recourse to explicit flaw geometry is not required to evaluate the damage, however. Studies of the brittle fracture of rock and other solids have shown that a Weibull function provides a satisfactory description of the inherent flaws leading to tensile fracture (Jaeger and Cook, 1969). This concept will be pursued and the flaw distribution will be described by a two parameter Weibull function,

$$n = k\varepsilon^m, \tag{10.20}$$

where n is the number of flaws which can activate at or below a tensile strain level of ε. The constants k and m will be regarded as material properties characterizing fracture activation.

When the tensile strain is increased by the increment $\delta\varepsilon$,

$$\delta n = n'(\varepsilon)\,\delta\varepsilon \tag{10.21}$$

new flaws become available for activation. Because of previous damage, however, a volume fraction D of the material will have been stress relieved, thus the number of flaws that will actually activate is given by

$$\delta N = \delta n (1 - D) \tag{10.22}$$

A strain-rate sensitive rock fragmentation model 437

and the rate of flaw activation is

$$\dot{N} = n'(\varepsilon)\dot{\varepsilon}(1-D) \tag{10.23}$$

The total damage at time t will be the superposition of the spherical volumes, $v(t-\tau)$, of all the cracks that have activated at past times τ,

$$D = \int_0^t \dot{N}(t)v(t-\tau)\, d\tau \tag{10.24}$$

The current damage D will be governed by a microstructural law for $v(t-\tau)$ which will determine the growth of cracks that activated at past time, τ. In general, crack growth will depend on the time interval, $t-\tau$, and the strain, $\varepsilon(t)$. We will assume that cracks quickly approach a constant fracture growth velocity, C_g, after activation; thus, invoking a spherical region of crack influence,

$$v(t-\tau) = \tfrac{4}{3}\pi C_g^3 (t-\tau)^3 \tag{10.25}$$

C_g can be regarded as an additional tensile fracture property that governs the rate of damage growth during dynamic fracture. Equation 10.24 may finally be written as an integral equation in the damage, D,

$$D(t) = \tfrac{4}{3}\pi C_g^3 \int_0^t n'(\varepsilon)\dot{\varepsilon}(1-D)(t-\tau)^3 \, d\tau, \tag{10.26}$$

or, in terms of the rate of damage growth,

$$\dot{D}(t) = 4\pi km C_g^3 \int_0^t \varepsilon^{m-1}\dot{\varepsilon}(t-\tau)^2 \, d\tau. \tag{10.27}$$

10.3.1.2. Fragmentation

To address the question of fragment size, or the amount of fragment surface area created in a fragmentation event we assume that the surface area of a flaw at time t is given by $a(t) = 2\pi(C_g t)^2$ (two flat surfaces of a circular flaw) and that the total surface area per unit volume is given by

$$A(t) = \int_0^t \dot{N}(\tau)a(t-\tau)\, d\tau \tag{10.28}$$

$$= 2\pi km C_g^2 \int_0^t \varepsilon^{m-1}\dot{\varepsilon}(1-D)(t-\tau)^2 \, d\tau \tag{10.29}$$

The rate of growth of fracture surface area is

$$\dot{A}(t) = 4\pi k m C_g^2 \int_0^t \varepsilon^{m-1} \dot{\varepsilon}(1-D)(t-\tau)\, d\tau \qquad (10.30)$$

Equations 10.24 and 10.28 can be readily transformed to integrals over the crack radius, r, by setting $r = C_g(t-\tau)$, giving,

$$D(t) = \int_0^{C_g t} \mathcal{N}(r, t)\tfrac{4}{3}\pi r^3\, dr \qquad (10.31)$$

and

$$A(t) = \int_0^{C_g t} \mathcal{N}(r, t)2\pi r^2\, dr \qquad (10.32)$$

where

$$\mathcal{N}(r, t) = \frac{1}{C_g}\,\dot{N}(t - r/C_g) \qquad (10.33)$$

The damage and fracture surface area distribution functions over crack radius are then,

$$\mathcal{D}(r, t) = \mathcal{N}(r, t)\tfrac{4}{3}\pi r^3 \qquad (10.34)$$

and

$$\mathcal{A}(r, t) = \mathcal{N}(r, t)2\pi r^2 \qquad (10.35)$$

Fragmentation is assumed to be completed when the damage,

$$D(t_f) \equiv 1 \qquad (10.36)$$

which corresponds to fracture coalescence at time t_f. At fracture coalescence we will assume that the fracture faces form fragment sides and that a fragment size dimension is given by $L = 2r$. The distributions of fragment volume and surface area over fragment size are then given by

$$\mathcal{F}(L) = \tfrac{1}{2}\mathcal{D}(L/2, t_f) \qquad (10.37)$$

and

$$\mathcal{A}_f(L) = \tfrac{1}{2}\mathcal{A}(L/2, t_f) \qquad (10.38)$$

Fragment size distributions measured in blasting events are typically broader and more skewed toward fine fragments than is predicted by eqn 10.38. We currently believe that this is due to further or secondary breakage of initial fragments due to kinetic energy still available in the impulse-loaded rock mass. Consequently, average fragment sizes would be expected to be somewhat smaller than calculated.

10.3.1.3. Model Behavior Under Specified Loading Conditions

The model provides a description of fracture and fragmentation that is dependent on the transient tensile strain. It is convenient, both experimentally and for purposes of interpretation, to characterize a rate-sensitive phenomenon by dependence on the loading strain rate. Accordingly, the preceding model will be specialized to a tensile strain history

$$\varepsilon(t) = \dot{\varepsilon}_0 t \tag{10.39}$$

where $\dot{\varepsilon}_0$ is a constant strain rate.

Under constant strain-rate loading, eqn 10.26 may be written in the form,

$$D(t) = \tfrac{4}{3}\pi k m C_g^3 \dot{\varepsilon}_0^m \int_0^t \tau^{m-1}(t-\tau)^3(1-D(\tau))\,\mathrm{d}\tau. \tag{10.40}$$

It has been shown (Grady and Kipp, 1980) that for D appreciably less than unity, for constant strain rate loading, the damage growth in eqn 10.40 depends, to a good approximation, on time according to

$$D(t) = \alpha \dot{\varepsilon}_0^m t^{m+3} \tag{10.41}$$

where

$$\alpha = \frac{8\pi C_g^3 k}{(m+1)(m+2)(m+3)}, \tag{10.42}$$

is a constant function of the three fracture parameters, k, m, and C_g.

The dynamic fracture stress is defined as the maximum stress a material can support before failure by fracture occurs. From eqn 10.17, the stress

$$\sigma(t) = K(1-D)\varepsilon(t), \tag{10.43}$$

specialized, using eqn 10.41, to constant strain rate loading is

$$\sigma(t) = K\dot{\varepsilon}_0 t(1 - \alpha \dot{\varepsilon}_0^m t^{m+3}) \tag{10.44}$$

This expression exhibits constant tensile stress rate loading at early time, followed by rapid stress relaxation as damage accumulates. Equation 10.44 can be maximized with respect to time, providing a maximum stress,

$$\sigma_M = K(m+3)(m+4)^{-(m+4)/(m+3)}\alpha^{-1/(m+3)}\dot{\varepsilon}_0^{3/(m+3)} \tag{10.45}$$

occurring at a time,

$$t_M = (m+4)^{-1/(m+3)}\alpha^{-1/(m+3)}\dot{\varepsilon}_0^{-m/(m+3)} \tag{10.46}$$

Dependence of the fracture stress on strain rate is provided by eqn 10.45.

Fragmentation occurs at a time t_f corresponding to $D(t_f) = 1$. Using the damage expression, eqn 10.34, results in,

$$t_f = \alpha^{-1/(m+3)} \dot{\varepsilon}_0^{-m/(m+3)} \qquad (10.47)$$

The total fracture surface area per unit volume is,

$$A(t) = \int_0^{C_g t} \mathscr{A}(r, t)\, dr, \qquad (10.48)$$

and, at $t = t_f$, corresponds to the total fragment surface area A_f. For constant strain rate loading,

$$A_f = \frac{m+3}{2C_g} \alpha^{1/(m+3)} \dot{\varepsilon}_0^{m/(m+3)} \qquad (10.49)$$

and also increases with strain rate as expected.

The fragment size corresponding to the peak of the fragment distribution curve in eqn 10.36 is given by

$$L_M = \frac{6C_g}{m+2} \alpha^{-1/(m+3)} \dot{\varepsilon}_0^{-m/(m+3)} \qquad (10.50)$$

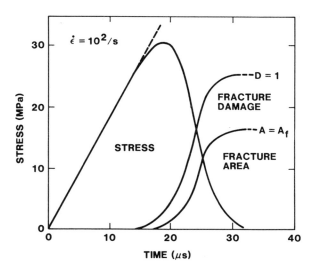

Fig. 10.7. Material response for constant strain rate loading.

A strain-rate sensitive rock fragmentation model 441

Note that within a small factor it is just the inverse of the fracture surface area in eqn 10.49.

We have anticipated fracture parameters to be obtained for oil shale and approximate computational relations to be described in the next section to show the behavior of the fracture model under a constant strain rate loading of $10^2 \, s^{-1}$ in Fig. 10.7. Note that loading is essentially linear elastic until a stress and time correspond to the onset of damage growth. At this time fracture damage grows rapidly and a corresponding catastrophic loss of strength occurs as integrity of the rock body is lost.

In Fig. 10.8 the behavior is shown for several values of strain rate. Note in particular the sensitive dependence of fracture stress, fracture time and the total amount of fracture surface area on the loading strain rate, consistent with the observed behavior shown in Fig. 10.6.

10.3.2. Computational Model

The expressed objectives of the present fracture and fragmentation modeling is to develop a computational capability for predicting explosive rubblization in oil shale. Specifically, it is desired to perform calculations to predict the extent and degree of fracture and fragmentation for *in situ* or modified *in situ* explosive events that encompass geometries not part of previous mining experience. The calculations will be used to scope explosive type, placement and energy release rate, proximity of free faces or interfaces, sequencing of charge detonations, and other features critical to optimizing rubblization.

The geometries of explosive rubblization methods that have been tried or proposed are complex. Consequently, the calculational model must be streamlined to provide useful results within current computer storage and running time capabilities. The model should predict the essential features of stress wave propagation, fracture and fragmentation but forego extensive calculations of microstructural features of the fracture process when possible. Toward this end the preceding fragmentation description has been generalized to three dimensions and coupled with the laws governing stress-wave propagation.

10.3.2.1. Stress-wave Propagation and Fracture

In addition to the fracture and fragmentation description, when dynamic loading involves the propagation of stress waves, the basic conservation laws of momentum and energy must be incorporated.

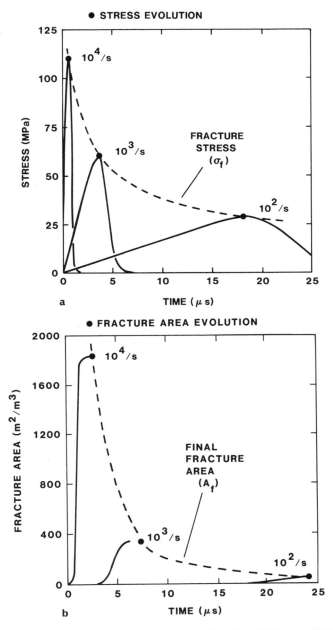

Fig. 10.8. Material response for several rates of loading. (a) Stress; (b) fracture area.

A strain-rate sensitive rock fragmentation model

Neglecting body forces the balance of linear momentum requires

$$\frac{\partial \sigma_{ij}}{\partial X_j} - \rho \frac{\partial U_i}{\partial t} = 0 \tag{10.51}$$

where ρ is the mass density and σ_{ij} and U_i are the stress tensor and particle velocity vector, respectively. Neglecting thermal and radiative sources, energy conservation yields

$$\frac{\partial \mathscr{E}}{\partial t} - \sigma_{ij} \frac{\partial U_i}{\partial X_j} = 0 \tag{10.52}$$

where $\mathscr{E}$ is the internal energy.

For states of tensile stress and strain, we assume material response will be governed by an isotropic elastic energy function.

$$\mathscr{E} = \tfrac{1}{2} K(1-D)\varepsilon^2 + 2G(1-D)(\varepsilon_{ij}\varepsilon_{ij} - \tfrac{1}{3}\varepsilon^2) \tag{10.53}$$

where ε_{ij} is the strain tensor, $\varepsilon = \varepsilon_{kk}$, the volume strain, and K and G are the unfractured bulk and shear moduli, respectively. D is the scalar damage parameter defined previously. Note that the description is isotropic and remains isotropic during damage growth. Also both the bulk and shear moduli are assumed to degrade with damage.

Thermodynamic restrictions require the stress to be given by

$$\left(\frac{\partial \mathscr{E}}{\partial \varepsilon_{ij}}\right)_D = \sigma_{ij} = K(1-D)\varepsilon\,\delta_{ij} + 2G(1-D)(\varepsilon_{ij} - \tfrac{1}{3}\varepsilon\,\delta_{ij}) \tag{10.54}$$

and further,

$$\left(\frac{\partial \mathscr{E}}{\partial D}\right)_{\varepsilon_{ij}} \dot{D} \leq 0 \tag{10.55}$$

to satisfy energy dissipation.

To complete the constitutive description in tension, differential equations governing the growth of fracture damage and fracture surface area have been determined from eqns 10.27 and 10.30

$$\dot{D} = (36\pi)^{1/3}[n(\varepsilon)]^{1/3}D^{2/3}(1-D)C_g \tag{10.56}$$

$$\dot{A} = (48\pi^2)^{1/3}[n(\varepsilon)]^{2/3}D^{1/3}(1-D)C_g \tag{10.57}$$

As rate laws they are computationally simpler and have been shown to agree well with the exact expressions over strain histories expected in dynamic tensile fracture.

10.3.2.2. Fragment size calculation

In computation the question of fragment size is addressed by defining an average value for the flaw radius at time t, $\bar{r}(t)$, together with average values for flaw area, $\bar{a}(t)$, and volume of influence, $\bar{v}(t)$. Equations 10.24 and 10.28 can be rewritten as

$$D(t) = \int_0^t \dot{N}(\tau)\bar{v}(t)\,\mathrm{d}\tau = N(t)\bar{v}(t) = \tfrac{4}{3}\pi[\bar{r}(t)]^3 N(t) \qquad (10.58)$$

and

$$A(t) = \int_0^t \dot{N}(\tau)\bar{a}(t)\,\mathrm{d}\tau = N(t)\bar{a}(t) = 2\pi[\bar{r}(t)]^2 N(t) \qquad (10.59)$$

Eliminating $\bar{r}(t)$ from eqns 10.58 and 10.59 leads to an equation for the average fragment size

$$L(t) = [1/N(t)]^{1/3} = (9\pi/2)^{1/3}[D(t)]^{2/3}/A(t) \qquad (10.60)$$

The logic for identifying $L(t)$ as the average fragment size follows from the definition that $N(t)$ is the density of active flaws, so its reciprocal is the average volume per active flaw; $L(t)$ then, being the cube root of this volume, can be considered to be the average distance between flaws or an approximate measure of fragment size. When fracture damage goes to completion $(D = 1)$ eqn 10.60 provides a relation between the average fragment size L_f and the fracture surface area A_f.

10.3.2.3. Model Parameters

The parameters used for the computations made to date have been based on experimental data for an oil shale with a Fischer assay grade of about 20 gallons ton^{-1} (80 ml kg^{-1}). For the elastic-plastic constitutive relation, the quantities of interest are initial density, bulk sound speed, Poisson's ratio and yield strength, which were given values of $2{\cdot}27\ \mathrm{Mg\ m^{-3}}$, $3000\ \mathrm{m\ s^{-1}}$, $0{\cdot}4$ and $200\ \mathrm{MPa}$, respectively.

The constants in the fragmentation model (k, m and C_g) were evaluated by using the constant strain rate expression for fracture stress, eqn 10.45, and fragment size, eqn 10.50, in conjunction with the experimental data in Fig. 10.6. The results, which are represented by the straight lines in Fig. 10.6 are $k = 1{\cdot}7\ (10^{-27})\ \mathrm{m^{-3}}$, $m = 8$ and $C_\mathrm{g} = 1300\ \mathrm{m\ s^{-1}}$. It is significant to note here that the constants in the fragmentation model were determined on the basis of data from laboratory experiments that in general were performed on samples having very small characteristic dimensions in comparison to those associated with explosive events performed in the field. Yet, the predictive capability of the model appears to be satisfactory for large-scale experiments as well as for small-scale experiments, as will be demonstrated in the following section.

10.4. NUMERICAL SIMULATIONS OF FRACTURE AND FRAGMENTATION

The complexity of the waves that are generated by explosive charges and the appearance of relief waves from regions that have fractured, necessitate the use of wave-propagation codes to address realistic geometries. The codes numerically integrate the conservation equations of mass, momentum and energy, along with the constitutive equations for the materials. A principal assumption made is that the initial stress wave emanating from the explosive creates the majority of fractures.

The model described in the previous section has been incorporated into the Lagrangian two-dimensional wave propagation code, TOODY-IV (Swegle, 1978). The damage and fracture surface area differential eqns 10.56 and 10.57 are integrated in conjunction with the expressions for the stresses, eqn 10.54. The strain measure ε used to drive the damage was chosen to be the residual elastic bulk strain

$$\varepsilon = p/K \qquad (10.61)$$

where $p = \sigma_{kk}/3$ is the mean tensile stress. The explosives used to drive the rock to a state of fragmentation are modeled as energy sources, with isentropic release of that energy as a gas from the Chapman–Jouguet point. The timing of the energy release is based upon prescribed distances from the detonation point, and the characteristic detonation wave velocity.

One of the purposes of using wave-propagation codes in blasting simulations is to obtain insight into the early stress wave behavior that leads to fracture of the rock. Such studies may lead to more efficient use of the available explosive energy. Five examples will be considered in this section: (a) simulation of the detonation of a small explosive charge in a meter-sized block; (b) damage variation with explosive burial depth (burden); (c) damage variation with charge shape; (d) simulation of an experiment in the floor of a mine; and (e) construction of a retort blasting technique.

446   *E. P. Chen, Marlin E. Kipp and D. E. Grady*

10.4.1. Block Experiment

One of the experiments (see Fig. 10.9(a)) used to evaluate the model involved the detonation of 80 g C-4 explosive in a block of oil shale having dimensions of about 1 m on each side. The oil shale block was mounted with its bottom surface, which was flat, on a slab of low-density foam so that, insofar as stress-wave interactions are concerned, this surface would be free. The remaining portions of the block were encased in a rock-matching grout to prevent interactions of stress waves at the lateral and top surfaces, thereby reducing the probability for extraneous rock damage. After the detonation, the block was sectioned to reveal the effects of the blast (Fig. 10.9(b)). Measurements of fragment dimensions were made in the sectioned block and compared to predictions based on the fragmentation model. As would be expected, results of the computations (Fig. 10.9(c)) and the experiments both indicate that damage is heavy in the region immediately around the charge as well as near the free face; the latter occurs as the compressive wave from the charge reflects as a tensile wave. There is also evidence from the sectioned block that preexisting faults or other discrete features in the rock can have a strong influence on the resulting fracture state. In this case, the computations indicate (Fig. 10.9(c)) that the spherical wave, diverging from the charge, is followed quite closely by the formation of a damaged region. When the wave reflects from the surface, an extensive region of damage is created, a result of the interaction of the release wave from the free surface and the release wave trailing behind the outward moving spherical shock wave.

10.4.2. Influence of Burial Depth for Spherical Charges

The configuration in the previous example was taken as the starting point for a series of calculations to examine the effect of varying the burial depth of the explosive (burden, B). Figure 10.10(a) illustrates the region of oil shale that experienced damage levels greater than $0\cdot2$ from the 80 g spherical charge. The volume encompassed by this damage contour was chosen to characterize the fractured region, which when normalized with the mass of the explosive is called the powder factor. As the burden is changed, the volume of material that is damaged is expected to change, since the relative size of the regions subjected to divergent and spall tensions will change. In particular, there should be some burial depth which optimizes the amount of fractured rock.

A strain-rate sensitive rock fragmentation model

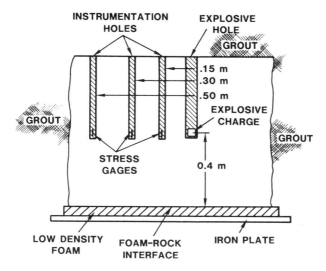

Fig. 10.9.(a) Experimental configuration of oil shale block.

Calculations were performed for burdens ranging from 10 cm to 90 cm. The results are summarized in Fig. 10.10(b). As the burden is increased, the volume of damaged material approaches a constant level, since the presence of a free surface no longer has any influence. There is clearly a broad minimum, 20–24 cm burden, in which the

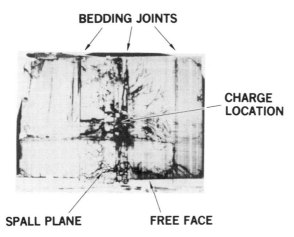

Fig. 10.9.(b) Surface of block section through the explosive charge.

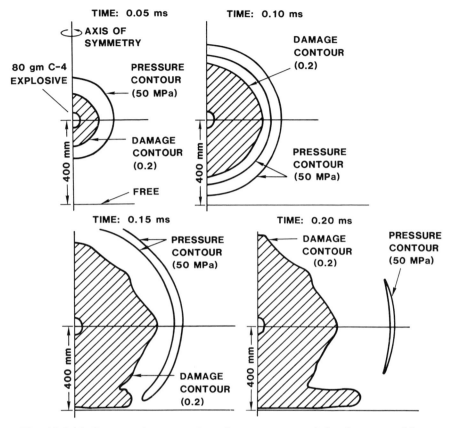

Fig. 10.9.(c) Computed propagation of stress wave and development of fractured rock.

energy of the explosive provides the optimum fracture damage. When the charge is near the surface, there is a significant decrease in the volume of fractured oil shale. The behavior reflected in Fig. 10.10(b) is consistent with blasting tradition.

10.4.3. Influence of Charge Shape

In blasting operations, spherical charges are rarely of interest, so the charge in the previous example was changed to a column with a length-to-diameter ratio of 8. The center of the 80 g charge was located 40 cm from the free surface, and detonation initiated at the point most

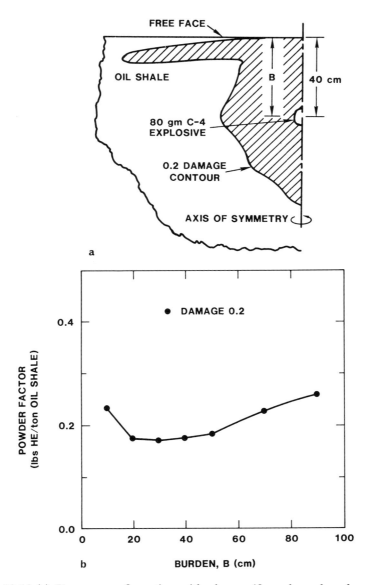

Fig. 10.10.(a) Damage configuration with charge 40 cm from free face; (b) calculated powder factors for a spherical charge.

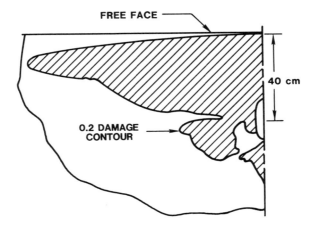

Fig. 10.11. Column charge: bottom detonation.

distant from the free surface. As the detonation proceeds towards the free surface, a divergent wave is formed that again causes fracturing to occur. After interaction with the free surface, the stress wave generates a large volume of fractured material (Fig. 10.11), extending to nearly twice the volume of the spherical charge case. The fractured oil shale is most heavily concentrated in a region between the top of the column and the free surface, with the characteristic cone-shaped region of rock that would be expected to be freed by this configuration.

10.4.4. Mine Simulation—Single Hole

Another source of data for evaluating the fragmentation model has been the cratering experiments (Harper and Ray, 1981) conducted by the Los Alamos National Laboratory in the Colony oil shale mine near Grand Junction, Colorado. A comparison between model calculations and measurements for one of the cratering experiments is shown in Fig. 10.12. This experiment involved the detonation of a 5·2 kg, 0·75 m long charge of ANFO at the bottom of a 0·1 m diameter, 2·0 m deep borehole drilled into the floor of the mine. After the shot, fragmented rock not ejected from the crater as a direct result of the blast was excavated. Profiles of the cleared crater were then measured at 90° intervals around the axis of the borehole; these profiles are shown as the dashed curves in Fig. 10.12(a). The solid curves in the figure represent the results of the numerical simulation of the blast.

A strain-rate sensitive rock fragmentation model 451

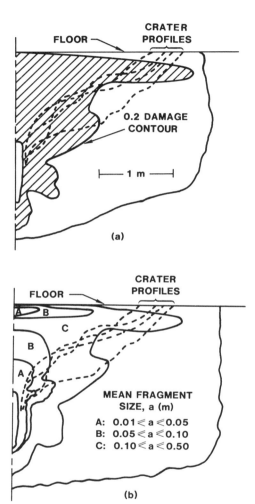

Fig. 10.12. Calculated damage profile, fragment dimension contours and measured, crater profiles (at 90° intervals around charge) for one of Los Alamos National Laboratories' cratering experiments at the Colony Mine.

The solid curve in Fig. 10.12(a) is a contour along which the damage parameter, D, was determined to have a value of 0·2; inside the contour the damage is more severe while outside the contour the damage is less severe. In Fig. 10.12(b), the solid curves delineate regions in which mean fragment dimensions were determined to be in the ranges indicated.

Based on a comparison between calculations and measurements, it has been concluded, for this model, that the damage contour of $D = 0·2$ represents a good criterion for determining whether or not material has been damaged sufficiently to be ejected. This is true except in the region near the charge, in particular near the lower portion of the charge. In the remaining regions, discrepancies between measurements and calculations generally have magnitudes comparable to fragment dimensions. The largest of the fragments, which are near

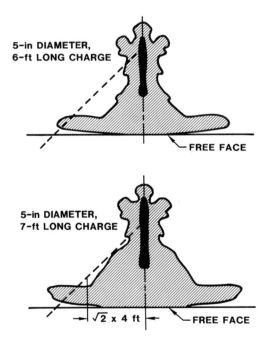

Fig. 10.13. Calculated damage patterns for two deck-1 charges having different lengths.

the boundary of the crater, may in fact be only marginally formed and thus may remain in place.

10.4.5. Application to a Retort Blasting Scheme

Although the model has only been incorporated into a two-dimensional code, calculations have been applied to approximate retort blasting schemes (Boade *et al.*, 1981). The demand for blasting a large quantity of rock into a mined cavity is a non-traditional configuration, and one of the uses of codes, as mentioned above, is to obtain insight into various geometries, in an attempt to narrow the possible choices of charge configuration, timing, etc.

Without considering wave interactions among charges, the fragmentation from a first deck of charges in a square array was constructed from single borehole fractured geometries (Fig. 10.13) to determine the spacing required to obtain good coverage from this first layer of charges (Fig. 10.14(a)). By displacing the subsequent deck so that a charge in this deck is centered with respect to four charges in the previous deck, maximum use can be made of free surfaces (Figs 10.14(b) and (c)). Appropriate timing of the decks should create good overall fragmentation.

10.5. CONCLUSIONS

Essential physical processes of the dynamic fracture and fragmentation of rock have been included in the model outlined in this chapter. The overall agreement of the calculations with experiments on different scales indicates that the model correctly accounts for the scale of the event, even though the parameters are based upon laboratory-scale experiments. One of the most useful aspects of the model, when coupled into a wave-propagation code, is the ease with which changes in significant parameters, such as explosive type or detonation position, can be evaluated. Hence, a thorough understanding of the wave propagation in a blasting geometry can be obtained, and this results in an approximate survey of the expected fragment sizes. Although complex geometries involving multiple boreholes cannot yet be calculated explicitly, individual borehole studies can reduce the number of uncertainties involved in such designs. This has the potential of greatly enhancing the efficient use of explosives to fragment rock.

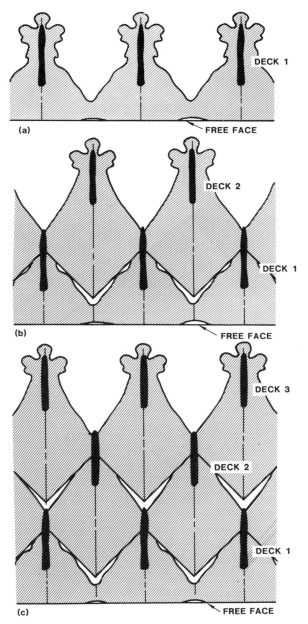

Fig. 10.14. Calculated damage regions for a three-deck charge array showing that the blasting concept proposed here gives good coverage in terms of rock damage. (a) Deck-1 charges detonated; (b) deck-2 charges detonated; (c) deck-3 charges detonated.

ACKNOWLEDGMENTS

This work was performed at Sandia National Laboratories and was supported by the US Department of Energy under contract no. DE-AC04-76DP00789.

REFERENCES

Birkimer, D. L. (1971). A possible fracture criterion for the dynamic strength of rock, *Proc. 12th Symp. Rock Mech.*, Society of Mining Engineers, New York, pp. 573–90.

Boade, R. R., Kipp, M. E. and Grady, D. E. (1981). A blasting concept for preparing vertical modified *in situ* oil shale retorts, *SAND81-1255, Sandia National Laboratories*, Albuquerque, New Mexico.

Chen, E. P. and Sih, G. C. (1977). *Elastodynamic Crack Problems*, ed. G. C. Sih, Noordhoff International Publishing, Amsterdam.

Davison, L. and Stevens, A. L. (1973). Thermomechanical constitution of spalling elastic bodies, *J. Appl. Phys.*, **44**, 668–74.

Freund, L. B. (1973). *J. Mech. Phys. Sol.*, **21**, 47–61.

Grady, D. E. and Hollenbach, R. E. (1979). Dynamic fracture strength of rock, *Geophys. Res. Letters*, **6**, 73–6.

Grady, D. E. and Kipp, M. E. (1979*a*). The micromechanics of impact fracture of rock, *Int. J. Rock. Mech. Min. Sci.*, **16**, 293–302.

Grady, D. E. and Kipp, M. E. (1979*b*). Oil shale fracture and fragmentation at higher rates of loading, *Proc. 20th US Symp. Rock Mech.*, Austin, Texas, pp. 403–6.

Grady, D. E. and Kipp, M. E. (1980). Continuum modelling of explosive fracture in oil shale, *Int. J. Rock Mech. Min. Sci. and Geomech. Abstr.*, **17**, 147.

Grady, D. E. and Lipkin, J. (1980). Criteria for impulsive rock fracture, *Geophys. Res. Letters*, **7**, 255–8.

Grady, D. E., Kipp, M. E. and Smith, C. W. (1980). Explosive fracture studies in oil shale, *Soc. Pet. Eng. J.*, **20**, 349.

Green, S. J. and Perkins, R. D. (1968). Uniaxial compression tests at varying strain rates on three geologic materials, *Proc. 10th Symp. Rock Mech.*, ed. K. E. Gray, American Institute of Mining, Metallurgical and Petroleum Engineers, Austin, Texas, pp. 35–54.

Harper, M. D. and Ray, J. M. (1981). Experimental design and crater profiles of intermediate scale experiments in oil shale, *Explosively Produced Fracture in Oil Shale: Quarterly Report, LA-8553-PR*, April–June 1980, Los Alamos National Laboratory, New Mexico.

Jaeger, J. C. and Cook, N. G. W. (1969). *Fundamentals of Rock Mechanics*, Chapman & Hall, New York.

John, M. (1974). *Proc. 3rd Int. Cong. Rock Mechanics*, Vol. IIA, ISRM, Denver, pp. 330–5.

456   *E. P. Chen, Marlin E. Kipp and D. E. Grady*

Kalthoff, J. F. and Shockey, D. A. (1977). Instability of cracks under impulse loads, *J. Appl. Phys.*, **48**, 986–93.

Kipp, M. E. and Grady, D. E. (1980). Numerical studies of rock fragmentation, *SAND79-1582, Sandia National Laboratories*, Albuquerque, New Mexico.

Kipp, M. E., Grady, D. E. and Chen, E. P. (1980). Strain-rate dependent fracture initiation, *Int. J. Frac.*, **16**(5), 471–8.

Lipkin, J. and Jones, A. K. (1979). Dynamic fracture strength of oil shale under torsional loading, *Proc. 20th Symp. Rock Mech.*, Austin, Texas, pp. 601–6.

McHugh, S. L., Seaman, L., Murri, W. J., Tokheim, R. E. and Curran, D. R. (1977). Fracture and fragmentation of oil shale, *Stanford Research Institute Final Report.*

Schmidt, R. A. (1977). Fracture mechanics of oil shale, *Proc. 18th Symp. Rock Mech.*, Keystone, Colorado.

Sinha, N. K. (1979). Rate sensitivity of compressive strength of columnar-grained ice, *SESA Spring Meeting*, San Francisco, paper no. R79-164.

Swegle, J. W. (1978). TOODY IV—A computer program for two-dimensional wave propagation, *SAND78-0522, Sandia National Laboratories*, Albuquerque, New Mexico.

Chapter 11

Mining and Fragmentation Oil Shale Research

WILLIAM HUSTRULID

Professor, Department of Mining, Colorado School of Mines, Golden, USA

ROGER HOLMBERG

Swedish Detonic Research Foundation, Stockholm, Sweden

and

ERNESTO PESCE

Montevideo, Uruguay

SUMMARY

Although there has been interest in the exploitation of oil shale resources since the early 1900s, preparations for large scale development began with the establishment of the Anvil Points research facility (Rifle, Colorado) by the US Bureau of Mines in the mid 1940s. Since that time true in situ, *open pit, underground and modified* in situ *methods have been applied under large-scale field situations in attempts to recover the kerogen from the host rock. This chapter presents the mining aspects of these methods with particular emphasis on drilling and blasting practices. The discussion of true* in situ *experience includes the explosive fracturing procedures employed by the US Bureau of Mines in the shallow beds of the Green River Formation in Wyoming and the bulk bed rubblization with lifting of the overburden used by Geokinetics in Utah. Open pit mining of oil shale is discussed with respect to the Sao Mateus do Sul property of Petrobras (Brazil). Experience in the underground mining of oil shale includes the activities of the US Bureau of Mines, Mobil and Paraho at the Anvil Points facility, Colony (Parachute Creek, Colorado), Union (Parachute Creek and Long Ridge, Colorado), and White River Oil Shale (Vernal, Utah). The modified* in situ *experience* of Occidental (*Logan Wash, Colorado*) has been described in great detail with a first summarization of much of the publicly available information. Details of the Rio Blanco (*Piceance Creek, Colorado*) modified in situ *process based upon the vertical crater retreat system are* included.

11.1. INTRODUCTION

Oil shale deposits, widely distributed throughout the world, occur at different depths in various thicknesses, grades, orientations, etc. This has led to the consideration of a number of different approaches for the recovery of the organic matter from the host rock. The more important ones are:

(i) borehole access, layer rubblization and *in situ* recovery (true *in situ*);
(ii) surface mining with surface retorting (surface);
(iii) underground extraction with surface retorting (underground);
(iv) underground partial extraction, bulk rubblization and *in situ* recovery (modified *in situ*).

Some of the approaches are still in the conceptual (paper design study) stage, others have been tried in the laboratory or on a bench scale, others have progressed to mini-field experiments. This chapter will focus on a detailed discussion of the mining and, in particular, of the fragmentation aspects of some major projects which have progressed to the large scale field testing/commercial prototype stage. For several of the operations, the amount of readily available information is sketchy due to its proprietary nature. Where possible, the authors have tried to supplement this information with (a) other data within the public realm; (b) engineering estimates; and (c) in some cases educated conjecture, with the thought of presenting a better overall and hopefully accurate presentation. Care has been taken to accurately document the information source so that the interested reader can repeat the analysis, if desired.

11.2. TRUE *IN SITU* RETORTING

11.2.1. Introduction

In true *in situ* retorting, the only access to the oil shale formations is through boreholes. Ideally, one would like to increase uniformly the permeability of the entire vertical section of the pay formation over a large area while maintaining the integrity of adjacent layers so that the combustion zone is confined. Through the use of carefully placed and controlled injection and extraction wells one should be able to control the retorting of the formation and the recovery of the kerogen. From an economic viewpoint, the wells used both for fracturing and extraction should be as widely spaced as possible. For creating the fractures within the formation, several possible techniques are available. These include explosive fracturing and hydraulic fracturing. Both have been used. Inherent in any fracturing process is the fact that for new surfaces to be formed a source of expansion volume must be present. For shallow beds, expansion can occur by lifting the overlying layers toward the free surface. For deep beds one is largely restricted to utilizing the volume of the borehole itself and formation porosities. Both tend to be extremely small. Even 250 mm diameter holes drilled in a 3×3 m pattern (a very unlikely situation from an economic viewpoint) would yield an expansion volume of 0·5% of the total rock mass. The discussion in this section will be limited to shallow beds in which lifting of the surface to provide the required void is possible.

11.2.2. Liquid Explosive Fracturing Tests

The US Bureau of Mines (Hay and Scott, 1965; Campbell *et al.*, 1970; Thomas *et al.*, 1972; Burwell *et al.*, 1973; Miller *et al.*, 1974) has run a number of laboratory and field experiments to determine the feasibility of using liquid explosives to fracture oil shale. Laboratory experiments (Hay and Scott, 1965) were done to study whether a dry porous rock would imbibe a sufficient amount of a nitroglycerine (NG)–ethylene glycol-dinitrate (EGDN) mixture to yield a detonable charge and a high detonation velocity. Tests with a dry Berea sandstone having a density of $2·2$ g cm^{-3} showed that this rock could imbibe about 12% of its own volume of liquid. With a booster of 73 g of NG–EGDN Miller *et al.* (1974) succeeded in initiating detonation in a rock sample ($51 \times 51 \times 152$ mm) that had absorbed 8·2% NG–EGDN. The detonation velocity was measured as 4700 m s^{-1}. 15% gelled NG (density, $\gamma = 1·37$ g cm^{-3}) absorbed in a sample of sodium chloride (particle diameter $= 287$ mm) had a detonation velocity of 1550 m s^{-1}.

The Bureau of Mines also showed that NG–EGDN could be detonated in cracks. A detonation velocity of 7500 m s^{-1} was recorded for a 1·6 mm open crack. NG–EGDN poured into a sand-filled crack detonated with a detonation velocity of 2100 m s^{-1}. In a field test (Miller *et al.*, 1974) conducted in a limestone quarry, about 5·5 liters desensitized NG was poured into a presplit crack having an average

460   *William Hustrulid, Roger Holmberg and Ernesto Pesce*

width of 3 mm. The detonation extended the fracture about 40 m and the crack width increased to about 70 mm as the limestone displaced horizontally towards the vertical face. Tests run in oil shale (Rock Springs and Green River, Wyoming sites) indicated that NG will detonate and the explosion will propagate in water and sand-filled natural and hydraulic fractures. The oil shale was fragmented to such extent that retorting was indeed possible. However, the difficulties in controlling the NG flow pattern were such that its use is not recommended. Pelletized TNT was shot in wells and an extensive fracturing out to a radius of 15 m was disclosed by seismic methods. Air flow measurements between wells indicated the presence of fractures but the evaluation techniques did not indicate the extent of rock fragmentation.

Prototype oil shale fracturing tests were conducted by the Bureau of Mines (Campbell *et al.*, 1970) in the Green River Formation near Rock Springs, Wyoming. Five wells were drilled in a square pattern (hole distance = 7·6 m) with one well in the center to a depth of 15–27 m. At this test site the Bureau of Mines tested electrolinking, hydraulic fracturing without and with sand propping and explosive fracturing. Liquid NG was used for the explosive fracturing.

Electrolinking and hydraulic fracturing without sand propping were found to be relatively ineffective. Hydraulic fracturing with sand propping created horizontal fractures with desirable flow capacity. Two wells having depths of 22–24 m and 24–26 m, respectively, were used for the tests. Almost 300 liters of a desensitized NG was poured into each well and was allowed to migrate into the hydrofractured rock. After the detonation, the maximum surface elevation had increased about 50 mm directly above the wells. To evaluate the fracturing, airflows between selected wells were measured. Campbell *et al.* (1970) indicate that there was a significant increase in fracture permeability when an adequate NG shot was detonated (i.e. 300 liters). A 100 liter NG shot resulted in a lower permeability. The 300 liter shots increased the injection capacity up to eight fold. The report does not mention anything about the degree of vertical fracturing that occurred after the blasts. However, Burwell *et al.* (1973) suggest that the detonation of the NG explosive undoubtedly created breakage of the shale on both sides of the fractures which resulted in a large surface area. The shale was apparently fractured to allow the *in situ* retorting process to proceed and the oxygen utilization and rate of burning were improving steadily at the time the test was terminated.

Mining and fragmentation oil shale research 461

Burwell *et al.* (1973) also report their experience with another test (test site 7) where only hydraulic fracturing first took place before ignition. Two ignition attempts were made but they were terminated because the injection rates could not be maintained. When 160 kg pelletized TNT was detonated in the wellbore the permeability increased and it became no problem to ignite the oil shale.

Coursen (1977) and McNamara *et al.* (1979) have shown that true *in situ* blasting can increase the permeability and explosion cavities if the depth is not too large and the tectonic stresses are not too high. However, a very large specific charge must be used (or several reshots), if such a permeability should be established, which allows a high degree of recovery of oil.

11.2.3. The Geokinetics Horizontal Retorting Process

Geokinetics began operations at Kamp Kerogen located 70 miles south of Vernal, Utah on land owned by the State of Utah, in April 1975. The oil shale beds strike in an east–west direction and dip to the north at about 120 ft mile^{-1}. Overburden ranges from zero at places where the shale forms the ground surface to a maximum of 110 ft. The Mahogany Zone is approximately 30 ft thick and has an average grade of 23 gallons ton^{-1}. Their process is designed specifically for areas where oil shale beds are relatively close to the surface. To date, the maximum overburden depth has been 55 ft (Table 11.1).

In the process as described by Lekas (1981), a pattern of blastholes is drilled (using a CP 650 track mounted blasthole rig) from the surface through the overburden, and into the oil shale bed:

'The holes are loaded with explosives and fired using a carefully planned blast system. The blast results in a fragmented mass of oil shale, with a high permeability. The void space in the fragmented zone comes from lifting the overburden and producing a small uplift of the surface. The fragmented zone constitutes an *in situ* retort. The bottom of the retort is sloped to provide drainage for the oil to a sump where it is lifted to the surface by a number of oil production wells. Air injection holes are drilled at one end of the retort and off gas holes are drilled at the other end. The oil shale is ignited at the air injection wells, and air is injected to establish and maintain a burning front that occupies the full thickness of the fragmented zone. The front is moved in a horizontal direction through the fractured shale towards the off gas

TABLE 11.1
Summary of Data From Geokinetics *in situ* Retorts 1–25.

Retort number	Date blasted (month/year)	Date ignited (month/year)	Thickness of shale (ft)	Average overburden thickness (ft)	Width (ft)	Length (ft)	Barrels of shale oil recovered
1	7/75	9/76	10	0	10	50	56
2	7/75	3/76	3	10	10	30	28
3	1/76	7/76	10	17	20	40	82
4	2/76	2/77	10	16	20	40	146
5	2/76	5/77	11	19	20	81	354
6	Abandoned—not blasted						
7	11/76	—	10	15	20	50	—
8	11/76	—	23	22	20	83	—
9	12/76	9/77	22	22	40	83	1 007
10	12/76	1/79	11	14	20	50	445
11	3/77	4/77	12	14	20	45	272
12	3/77	—	11	31	30	50	—
13	6/77	—	11	31	30	50	—
14	6/77	2/78	12	29	40	70	384
15	7/77	5/78	20	31	50	75	1 003
16	8/77	8/78	20	41	62	87	2 067
17	5/78	6/79	17	26	72	156	3 700
18	7/78	11/79	17	27	108	156	5 500
19	12/78	—	30	50	126	182	—
20	4/79	—	24	36	40	100	—
21	6/79	—	23	35	40	100	—
22	6/79	—	23	34	50	100	—
23	9/79	—	24	36	50	100	1 762[a]
24	11/79	12/80	28	45	217	230	12 741[a]
25	7/80	—	28	55	217	230	21 000[a]

All data from Lekas (1981) except [a] from Shaler (1982).

wells at the far end of the retort. The hot combustion gases from the burning front heat the shale ahead of the front, driving out the oil, which drains to the bottom of the retort, where it flows along the sloping bottom to the oil production wells. As the burn front moves from the air-in to the off gas wells, it burns the residual coke in the retorted shale as fuel. The combustion gases are recovered at the off gas wells.'

The principles (Britton and Lekas, 1979; Lekas, 1979; Britton, 1980; Lekas, 1981) are summarized below:

1. The permeability in the oil shale layer is obtained by lifting the overburden. The containment for the retorting process is achieved by keeping the overlying layers intact. Therefore the

application depends on achieving lift while maintaining containment.

Since the lift is obtained by increasing the volume of the retortable oil shale, if the layer is thin compared to the thickness of the overburden it will be difficult to lift the overburden without developing a very large area. On the other hand, if the overburden thickness is small compared to the thickness of the shale layer, it will be difficult to maintain a seal at the edges.

2. There are several ways of creating a second free face to which the shale can expand. One way is to create a wedge (see Fig. 11.1) which is lifted vertically. The time required for the block to start to rise is termed the release time. The release time for a block varies with the overburden depth. For an overburden thickness of 15–30 ft it is of the order of 24–32 ms. The charges freeing the

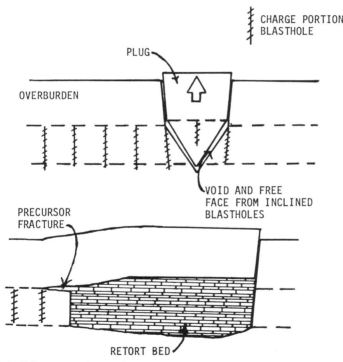

Fig. 11.1. Diagrammatic vertical section showing the rubblizing procedure. (Britton, 1980.)

block should all be fired within a period of one-third that of the release time of the block. Shorter times are better and it is desirable for all lifting charges to be fired within 10% of the total release time.

3. The delay times for charges immediately adjacent to the cut block should be set so that they explode when the block reaches about half to three-quarters of the unaided rise height of the block.
4. The timing of sequential firing fragmentation charges significantly affects the degree of fragmentation, total lift, differential lift and subsequent permeability. The timing is a compromise between the demands for lift and optimal fragmentation, as well as minimizing flexure of the overburden. Satisfactory results have been achieved for intervals of about one-third to 5 ms ft^{-1} of effective burden. The optimum time interval is about one-third to 1·5 ms ft^{-1}.
5. A staggered pattern of blastholes is to be preferred. The ratio of effective burden to effective spacing should have 1:1·5 as the lower limit and 1:4 as an upper limit. The firing sequence is extremely important as it controls the direction of movement of the rubble. One arrangement is shown in Fig. 11.2.
6. Low explosives (ANFO for example) are used as opposed to high explosives (TNT, dynamite) to reduce the intensity of the peak stress experienced by the overburden. The longer duration also affects the breakage. Initiation of the charges is done at the top of each charge thereby effecting propagation of the explosion in a downward direction and preventing the waves generated from

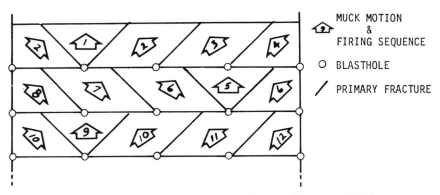

Fig. 11.2. Firing sequence for rubblization. (Britton, 1980.)

disrupting the overburden. Appropriate powder factors appear to be of the order of 2–3 lbs ANFO yard^{-3}.

7. For retort widths of the order of the overburden thickness, edge effects prevail and careful and expensive edge preparation is necessary. For ratios of 7 or greater, edge effects can be ignored.

From their results, Geokinetics (Lekas, 1981) have concluded that:

—It is possible to drill a pattern of blastholes from the surface into the oil shale and fracture the shale with explosives to establish a zone of high permeability with a relatively impermeable zone between the fragmented shale and the surface.

—It is possible to drill through the rubblized material and construct the various wells for the operation.

—A point ignition can be made in the rubblized shale and expanded into a burn front that covers the cross-section of the retort.

—The burn front can be moved down the length of the retort as a cohesive temperature front with satisfactory sweep efficiency.

—Produced oil can be recovered from a well drilled to the bottom of the rubblized zone.

—Recovery of inplace oil of up to 50% can be achieved.

11.3. SURFACE MINING OPERATIONS

There are a number of oil shale deposits located around the world which are amenable to surface mining techniques. Petrobras (Petroleo Brasileiros SA), Brazil's state oil company (Dayton, 1981f) is the first to start construction of an oil shale module (25 000 barrels day^{-1}) to be supplied from a strip mine. This first of two units is scheduled for initial production in 1985. Eventually 112 000 tonnes day^{-1} of mine run oil shale will be required to produce a daily output of 54 000 barrels of shale oil.

The Irati prototype mine and plant are located at Sao Mateus do Sul about 150 km southwest of Curitiba, Brazil. The site covers a north-westerly striking outcrop zone. The beds dip at 1·5% to the southwest under a topography of rolling hills and gentle slopes. Two shale beds separated by barren layers of shale and limestone (see Fig. 11.3) wil be mined. The shales are dark brown and grey to black in color, very fine-grained, finely laminated and fissile. The kerogen content ranges

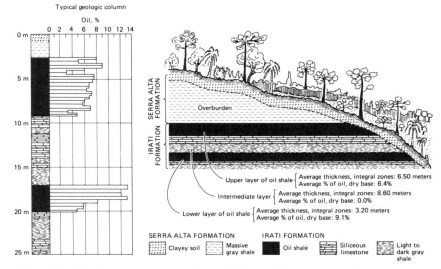

Fig. 11.3. Typical cross-section through the Irati formation in Sao Mateus do Sul, Parana, Brazil. (Dayton, 1981f.)

from 2 to 14% with the density decreasing from about 2·45 to 1·80 g cm^{-3} with increasing oil content.

A 30 m overburden depth establishes the practical mining limits. Within the 64·5 km^2 area so defined is contained 560 million barrels oil, 8·7 million tonnes sulfur, 3·9 million tonnes LPG and 19×10^9 m^3 light combustible gas. The overburden and intermediate beds will be blasted and then loaded by a fleet of draglines (two of 105 yard^3, two of 65 yard^3 and two of 18 yard^3). The two shale beds will be blasted and loaded by four 30 yard^3 shovels. 25 large trucks will be used to haul the output to the plant and return the retorted shale to the mine.

A series of blasting experiments have been conducted (Petrobras, 1982) to establish patterns and explosives that yield a satisfactory fragmentation while minimizing the amounts of fines. This latter aspect is important since fines (less than $\frac{1}{4}$ in) are not acceptable as feed to the vertical gravity flow retort. Undersize will be returned to the mine. The Bragel CB(IRECO) aluminized slurry explosive (density = 1·12 g cm^{-3}, detonation velocity = 3500 m s^{-1}) was selected for the following reasons:

—safe;
—water resistant;

—velocity matches rock characteristics;
—low density;
—good sensitivity;
—easy to handle and store.

The patterns selected (Fig. 11.4, Tables 11.2, 11.3) have provided the desired fragmentation with a minimum of fines and only 3% boulders

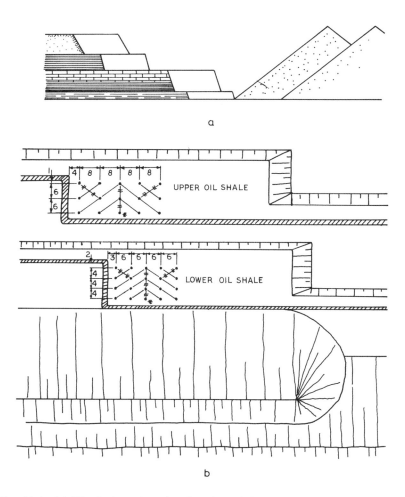

Fig. 11.4. (a) Vertical section showing the mining sequence of the two shale layers. (b) Plan view showing the blasting pattern: dimensions in m; ≈, delay; * point of initiation; scale, 1:1500. (Petrobras, 1982).

468　William Hustrulid, Roger Holmberg and Ernesto Pesce

TABLE 11.2
Blasting Parameters for the Upper Seam
(Petrobras, 1982)

Blasting pattern	8×6 m
Hole diameter	$4\frac{1}{2}$ in
Hole depth	6·7 m
Stemming	3 m
Explosive	Bragel CB
Cartridge size	$3\frac{1}{2} \times 24$ in
Connection	V
Delays	30 ms
Sub drill	0·3 m
Powder factor	0·090 kg m^{-3}

TABLE 11.3
Blasting Parameters for the Lower Seam
(Petrobras, 1982)

Blasting pattern	6×4 m
Hole diameter	$4\frac{1}{2}$ in
Hole depth	3·7 m
Stemming	1·8 m
Explosive	Bragel CB
Cartridge size	$3\frac{1}{2} \times 24$ in
Connection	V
Delays	50 ms
Sub drill	0·4 m
Powder factor	0·180 kg m^{-3}

(pieces with at least one dimension longer than 80 cm), no vibration damage to surrounding structures and no fly rock problems.

11.4. UNDERGROUND MINING OPERATIONS

11.4.1. Background

Large scale room and pillar mining methods for oil shale have been under development since the 1940s when the US Bureau of Mines opened their Anvil Points facility at Rifle, Colorado (East and Gardner, 1964). In addition to the Bureau, Mobil (Sellers *et al.*, 1971; Zambas *et al.*, 1972; Crookston, 1976; Crookston and Weiss, 1980)

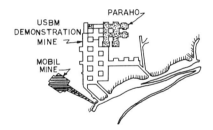

Fig. 11.5. Plan view of the Anvil Points Mine. (Modified from Crookston and Weiss, 1980.)

and Paraho (Crookston, 1976; Gauna, 1978; Crookston and Weiss, 1980) have conducted mining operations at Anvil Points (Fig. 11.5). Union (Crookston, 1976; Crookston and Weiss, 1980) and Colony (Agapito, 1972; Crookston, 1976; Crookston and Weiss, 1980; Dayton, 1981c) operated their own sites. The work was primarily confined to the Mahogany Zone which has an approximate thickness of 60 ft. By 1982, the development of commercial operations was underway by Union (Mitten, 1981; Randle and McGunegle, 1982) at their Long Ridge site, Parachute, Colorado and White River Oil Shale Corp. (Parrish, 1982) at sites U-a and U-b, Bonanza, Utah.

11.4.2. Anvil Points Mine
The early work (1944–1956) by the Bureau of Mines at the Anvil Points facility (Fig. 11.5) provided both extremely useful information regarding the design and stability of large openings (Merrill, 1954; Obert and Merrill, 1948) in oil shale and a proving ground for equipment. Fragmentation involved the use of cartridge explosives (dynamites) in small diameter holes and hence this experience is not as relevant to the large diameter ANFO-based systems planned today.

The tests conducted by Mobil (Sellers et al., 1971; Zambas et al., 1972) during 1967 and 1968 involved driving four rooms 78 ft high and 60 ft wide using a top heading (40×60 ft) and bench (38×60 ft) arrangement. These rooms were separated by rib pillars 40 ft in width. After completing the rooms the ribs were cross cut full height with a single blast. Drilling the headings and cross cuts was performed by a single boom version of the two boom rotary drill developed by Colony over the period 1965–67. The drill had the capability of drilling holes 3–5 in diameter up to 32 ft deep at any required angle in a face 30 ft

470   *William Hustrulid, Roger Holmberg and Ernesto Pesce*

wide and 40 ft high from one set up. Penetration rates (and bit life) for the rotary drills were dependent upon the grade of oil shale. Typical values for penetration rate were:

Headings
Richer shale zones: $8\,\text{ft}\,\text{min}^{-1}$
Harder: $4\,\text{ft}\,\text{min}^{-1}$
Average: $5\,\text{ft}\,\text{min}^{-1}$
Benching $>8\,\text{ft}\,\text{min}^{-1}$

ANFO primed by 70% dynamite was detonated with electric millisecond delay caps. For the headings, the powder factor was 0·6–0·7 lbs ton^{-1} whereas for the benches it was 0·35 lbs ton^{-1}. Both standard (Fig. 11.6) and presplit (Fig. 11.7) heading rounds were used. The standard advance per round was 25 ft. The bench round initially used is shown in Fig. 11.8. This produced an unacceptable amount of spalling from the ribs and the presplit pattern shown in Fig. 11.9 was devised.

The Paraho Development Corp. (Crookston, 1976; Crookston and Weiss, 1980) conducted field operations at Anvil Points beginning in 1974. Rooms were nominally driven by full headings 42 ft high and 60 ft wide. Pillars were 60 × 80 ft by 42 ft high. The drill jumbo leased from Colony was used to drill the 27 holes required. The pattern (Gauna, 1978) is as shown in Fig. 11.10. Blasting using ANFO resulted in a powder factor of 0·85 lbs ton^{-1}. The number of holes and the relatively high powder factor (compared to Mobil) reflected the desire to maximize fragmentation.

A summary of some of the important data is presented in Tables 11.4–11.6.

11.4.3. Colony Oil Shale Project

The original Colony oil shale mine was begun in 1964 by a group that included TOSCO, Sohio Petroleum Co. and Cleveland Cliffs Iron Co. (Dayton, 1981c). Atlantic Richfield Co. (ARCO) later joined the project and Sohio and Cliffs dropped out in 1971 making the ownership 40% TOSCO and 60% ARCO. In May 1980, Exxon purchased ARCO's share. It is located on the Middle Fork of Parachute Creek, 16 miles upstream from the town of Parachute, Colorado and about 3 miles north of Union Oil's Long Ridge development..

Over the period of 1965–72, more than 1·25 million tons oil shale

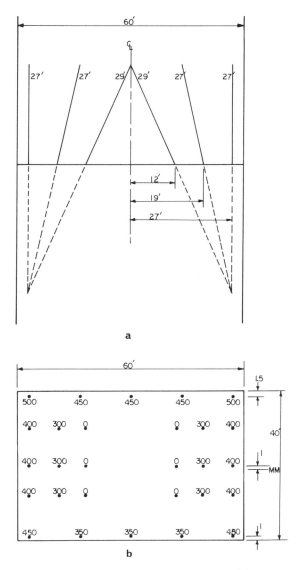

Fig. 11.6. Standard heading round employed by Mobil. (a) Plan view; (b) face view (4 in diameter holes; ms delays; MM, Mahogany marker). (After Zambas et al., 1972.)

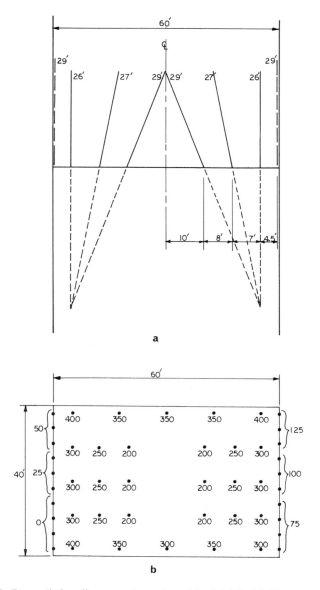

Fig. 11.7. Pre-split heading round employed by Mobil. (a) Plan view; (b) face view (4 in diameter holes, ms delays). (After Zambas et al., 1972.)

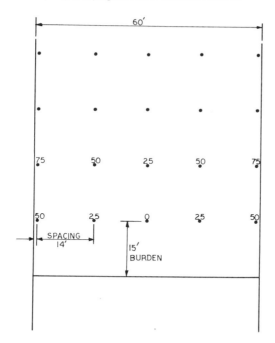

Fig. 11.8. Standard bench round used at Mobil (ms delays). (After Zambas et al., 1972.)

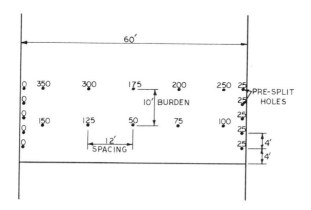

Fig. 11.9. Pre-split bench round used by Mobil (ms delays). (After Zambas et al., 1972.)

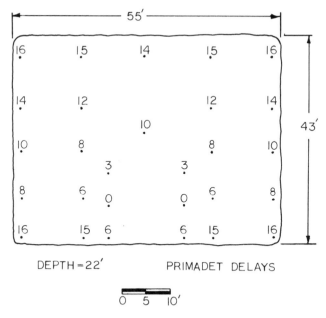

Fig. 11.10. Heading round employed by Paraho. (After Gauna, 1978.)

TABLE 11.4
Oil Shale Blasting—Powder Factor (Gauna, 1979)

Operation	Type	Explosive	Powder factor ($lb\ ton^{-1}$)
Bureau of Mines Anvil Points (East and Gardner, 1964)	V cut	45% Semi gelatin	0·41, 0·47
Union Oil Mine (Crookston, 1976)	No standard	ANFO	0·50
Mobil Anvil Points, Stage II Zambas et al., 1972)	V cut	ANFO	0·6–0·7
	Presplit and V cut	ANFO	?
	Crosscut, presplit and V cut	ANFO	0·56
	Bench	ANFO	0·35
	Presplit, bench	ANFO	0·30
Paraho (Gauna, 1978)	V cut	ANFO	0·7

TABLE 11.5
Oil Shale Blasting—Drilling Data (Gauna, 1979)

Operation	Round	Heading Height (ft)	Width (ft)	Drilling Number of holes	Diameter (in)	Depth (ft)	Production ($ton\,ft^{-1}$ of hole)
Bureau of Mines Anvil Points (East and Gardner, 1964)	V cut	27	60	72	$2\frac{1}{8}$	15	1·4
Union Oil Mine (Crookston, 1976)	No standard	28	30	40–55	$2\frac{1}{2}$–$2\frac{3}{4}$	17	1·0–1·25
Mobil Anvil Points Mine, Stage II (Zambas *et al.*, 1972)	V cut	40	60	28	4	25	5·7
	Presplit and V cut	40	60	25 + 20 presplit	4	25	3·56
	Crosscut, presplit and V cut	78	40	86 + 40 presplit	4	20	3·57
	Bench	38 spacing = 14 burden = 15	60	20	$4\frac{1}{2}$	36	6·8
	Bench	38 spacing = 12 burden = 10	60	10 + 10 presplit	$4\frac{1}{2}$	36	3·4
Paraho (Gauna, 1978)	V cut	40	55	28	$4\frac{1}{4}$	22	4·9

TABLE 11.6
Oil Shale Blasting—Burden (Gauna, 1979)

Operation	Round	Cut burden (ft)	Remaining[a] burden (ft)
Bureau of Mines Anvil Point			
(East and Gardner, 1964)	V cut	11	5
Mobil Anvil Points Mine,			
Stage II (Zambas et al., 1972)	V cut	13	10
	Presplit		
	and V cut	13	10
	Crosscut,		
	presplit		
	and V cut	10	10
	Bench	—	15
	Presplit		
	and bench	—	10
Paraho (Gauna, 1978)	V cut	9	9

[a] For the V cuts, represents the weighted average of relievers, slab, lifter, rib and crown holes. For the benches, represents burden between rows.

averaging 34–35 gallons ton^{-1} was mined from the Pilot Mine (Fig. 11.11) using room and pillar methods (Agapito, 1972; Agapito, 1974; Crookston, 1976; Hardy et al., 1978; Crookston and Weiss, 1980; Weakley, 1982). The depth of cover varies from about 600–800 ft. For the most part, top heading (30 ft high × 60 ft wide × 30 ft deep) and bench rounds were used. Some headings were drilled with a Twin-Boom Gardner-Denver mobile jumbo ($4\frac{1}{4}$ in blastholes up to 21 ft deep).

Rotary drill development resulted in production rates of more than $4 \cdot 8$ tons ft^{-1} of hole and average penetration rates of more than 7 ft min^{-1} (Crookston and Weiss, 1980)

Heading rounds 60 ft high × 55 ft wide × 30 ft deep were tested as a possible alternative. The advantage was a reduction in move-and-set up times and cost. The disadvantages were the size and cost of the drilling, charging, bolting and scaling equipment. The nominal pillar dimension was to be 60 × 60 ft. However, due to heavy blasting, strength deterioration with time and patterns of jointing, the result was assorted pillar dimensions and several different roof spans. The measured extraction was 75%.

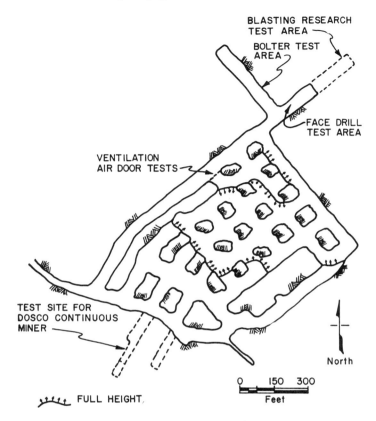

Fig. 11.11. Plan view of Colony pilot mine. (Room size: 55–80 ft spans, 30–60 ft high. Pillars: 60×60 ft. Extraction ratio: 75%. Overburden height: 350–850 ft.) (After Weakley, 1982.)

In preparation for commercial mining on a large scale, a number of equipment types were tested. The DOSCO Twin Boom 600 miner (Fig. 11.12) was one such device (McKinlay, 1982). The shale was very hard (10 000–17 000 psi) and very tough to cut. It was not abrasive in the normal sense. The peak cutting rate was about 160 tons h^{-1}. Although pick consumption was relatively high (1 per 20 tons), the economics regarding the use of the miner in development headings appears attractive.

Although presently in a holding pattern, the proposed commercial

Fig. 11.12. The Dosco TB600 (twin boom) mining machine. (McKinlay, 1982.)

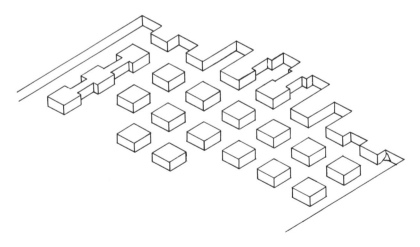

Fig. 11.13. Proposed system for mining the upper heading, Colony Mine. (Weakley, 1982.)

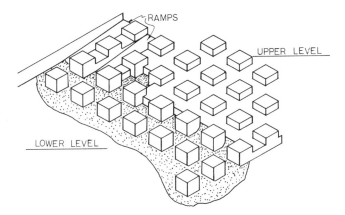

Fig. 11.14. Proposed mining system of the bench, Colony Mine. (Weakley, 1982.)

scale development (Dayton, 1981c) would employ (Figs 11.13 and 11.14) room headings 30 ft high and no more than 60 ft wide advanced at the level of the upper mining horizon. It is planned to drill 30 ft rounds ($4\frac{1}{4}$ in holes) on this level using wedge blasting cuts. The blastholes will be primed with dynamite and non-electric blasting caps which will initiate an ANFO charge. The broken rounds will be mucked with 10–15 yard^3 loaders. A 30-unit fleet of 80-stone off-highway haulers will transport the oil shale to the crusher. Scaling of the face and walls will be done with a backhole type of machine suitably modified for reaching the floor and the roof from a position on top of the bench.

After the upper headings have been advanced and secured, the lower benches will be quarry drilled slightly below grade and blasted. Trimming of the lower bench blast will be done from the upper bench using the scaling rig.

11.4.4. Union Test Adit

During the period 1955–1958 the Union Oil Co. of California operated a 1200 ton day^{-1} experimental oil shale retort on Parachute Creek, Garfield County, Colorado (Crookston, 1976; Crookston and Weiss, 1980). Part of the supply for the retort was provided by an adit driven full heading 28 ft high × 30 ft wide. Initially pneumatically powered percussion drilling machines were used to drill holes 17 ft deep

480   *William Hustrulid, Roger Holmberg and Ernesto Pesce*

and up to $2\frac{3}{4}$ in in diameter. Later rotary drills were used. Penetration rates of $13 \, \text{ft} \, \text{min}^{-1}$ were achieved in the rich oil shale. A variety of spacings and cut-hole arrangements were tried. Toward the end of the program production was $1 \cdot 0 - 1 \cdot 25$ tons shale ft^{-1} drill hole. 40–55 holes were used with a burden of 15–$20 \, \text{ft}^2$ per hole. A powder factor of $0 \cdot 5 \, \text{lbs} \, \text{ton}^{-1}$ was experienced using pneumatically loaded ANFO and millisecond delays. The data from this test are provided in Tables 11.4–11.6.

11.4.5. Union Oil Co.'s Long Ridge Operation

Union announced plans in early 1978 to build the first phase of a project to develop the first commercial shale oil complex in the USA. Phase I included a mine, retort and upgrading facility that processed $12\,500$ tons day^{-1} ore, yielding $10\,000$ barrels high quality syncrude (Dayton, 1981*b*). On-site construction of the retort and upgrading facility began late in 1981 with completion of the project planned for mid-1983.

Conventional room and pillar mining was employed (Mitten, 1981; Randle and McGunegle, 1982) recovering up to 70% of the shale in the mine zone and leaving pillars to support the mine roof. Union owns nearly $20\,000$ acres of oil shale lands in fee. The Mahogany Zone outcrops approximately 1000 ft above the valley floor on a cliff face approximately 2000 ft high. The access road was first developed between 1955 and 1958 when the test adits were developed supplying shale for Union Oil's experimental retort.

The basic equipment selected for Phase I mining was as follows:

—50 ton off-highway construction haul trucks;
—12 yard^3 articulated front end loaders;
—automatic or semi-atuomatic roof bolters;
—two boom Atlas Copco Boomer R276 hydraulic jumbos with face coverage capabilities of 35 ft high and 30–50 ft wide (Fig. 11.15);
—5 yard^3 loaders for face cleanup;
—hydraulic high reach scaling units and prill charging units capable of 35 ft high reach.

The maximum plan dimension of this particular orebody lies in a N70°E direction and is the orientation of the main haulage entires. (The strike of the primary joint set is N70°W.) The haulage layout is one of long straight drifts. Each heading is 27 ft high and 40 ft wide. The two hydraulic boom jumbos drilled out the face. In 1981 (Mitten,

Fig. 11.15. One of the Atlas Copco Boomer R276 hydraulic drill rigs used at Union. (Mattila, 1982.)

1981), the optimum pattern was a V cut consisting of 23 holes, $4\frac{1}{4}$ in diameter, 24 ft deep. Today (Baloo, 1984) 31 holes are being used. The holes are charged with ANFO and initiated by nonelectric blasting caps. The blasted rock is loaded by 15 yard^3 loaders and hauled by 50 ton trucks. Scaling is by high reach hydraulic operated scalers, followed by automatic roof bolters. The bolting pattern is dictated by ground conditions but the minimum pattern is 8 ft long bolts on 8 ft centers. Panels 950 ft wide and 2150 ft long are to be developed. The rooms are 50 ft wide and 27 ft high producing 90 tons per linear foot. The remaining pillars are 50 ft wide and 100 ft long. Temporary stopings will be erected as the mining front advances to maintain sufficient air flow across the front. After the panels have been developed completely at the 27 ft height, the mining front then reverses and extracts another 33 ft from each room floor. Operating experience dictates if this second cut will be vertically bench drilled or horizontally face drilled. As this bottom cut retreats, ramps will be established up to the top level haulage routes at regular intervals. The overall extraction ratio is expected to be 75%.

11.4.6. White River Shale Project

The White River shale project is a consolidation of Utah Tracts U-a and U-b. The project has commenced (Parrish, 1982) with the sinking of a 30 ft diameter air intake shaft and a production decline. Required station, entry and panel development will be accomplished by mid-1988 after which the mine will be capable of continuous production at the Phase I level of 27 330 tons day^{-1} (mid-1989). During the development program, two more 30 ft diameter shafts will be sunk, one for service and one for ventilation exhaust. The proven room and pillar mining method will be used. Phases II and III will follow with a full production capability of 176 740 tons day^{-1} in mid-1998.

The mining will take place in the Mahogany Zone of the Green River Formation. The mining depth at the service shaft is 1025 ft below the surface. The mining zone is 55 ft thick averaging 28 gallons ton^{-1}. The upper bench will be 25 ft high and the lower bench 30 ft high. A typical configuration is shown in Fig. 11.16. The

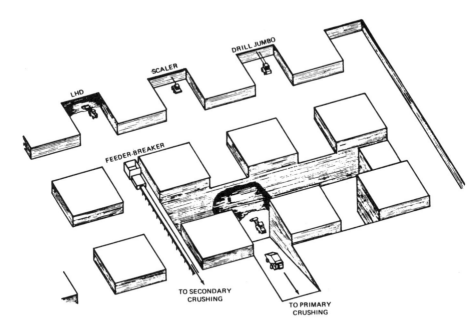

Fig. 11.16. Room and pillar two-bench mining system used by White River Oil Shale Co. (Parrish, 1982.)

Mining and fragmentation oil shale research 483

mine will be divided into panels consisting of 9 rooms and 15 or more crosscuts. Rooms and crosscuts will be 55 ft wide and pillars will be 60 × 70 ft. Resource recovery will be approximately 72% within production panels and 55% in the development entries, averaging 63% overall.

A three-entry system will provide access to mining panels. The entries, 30 ft high by 45 ft wide, providing air intake, haulage and service activities, will be driven in the lower portion of the mining zone. Entry pillars will be 70 ft wide, the length determined by panel access requirements.

Based on gassy mine design, all mine equipment to entry development and upper bench levels beyond the last open crosscuts will be Schedule 31 certified.

Drilling: Specialized drill jumbos capable of drilling 3 in diameter or larger horizontal holes up to 22 ft deep. Lower bench drilling will be by vertical blasthole drills.

Blasting: Each hole will be primed with a high-strength primer and an electric delay blasting cap. ANFO will be pneumatically placed in the drill holes. Approximately 0·5–0·6 lbs explosive will be used for each ton of shale broken, charging carriers with pressure pots.

Loading and hauling: Diesel powered front-end loaders will load haulage trucks or LHD units.

Rock bolting: Resin type or split-set bolts 7 ft in length will be installed on approximate 5-ft centers to support the roof spans.

The excavation of the decline is presently underway using a Paurat Series E134 heavy duty heading machine.

11.5. MODIFIED *IN SITU*

11.5.1. Introduction

The modified *in situ* process is an underground method in which a portion of the oil shale in a particular stope is mined providing void space for the rubblization of the remainder. The kerogen can be removed from the rubblized mass by a variety of retorting processes. In practice, it is desired to create a mass with vertical and horizontal uniformity of permeability containing fragments of a size that can be processed by the hot gases. The presence of a high permeability zone can cause channeling of the retort gases and poor recoveries. The rubblization is achieved by a careful arrangement of void spaces, drill

484 *William Hustrulid, Roger Holmberg and Ernesto Pesce*

holes, explosive arrays and blast sequencing. Although paper designs for horizontal retorts have been made for application to relatively thin beds, the commercial prototypes presently under development by Occidental Oil Shale and Rio Blanco are for vertical retorts in thick sequences of thick layers. Details regarding the very important fragmentation process are in general not easily obtainable. The authors have attempted to assemble as much of the appropriate public information as possible in this chapter.

11.5.2. Occidental Oil Shale Inc.'s Experience (Retorts 1–4)

Occidental Oil Shale Inc. (OOSI) has had the greatest amount of experience, by far, in the mining and *in situ* retorting of oil shale. Because of (a) the general importance of their contribution to the mining of oil shale (as well as specifically to the evaluation of modified *in situ* (MIS) retorting) and (b) the fact that retorts 5–8 were developed under partial sponsorship of the US Department of Energy (DOE), the discussion has been divided into two parts. This section will cover retorts 1–4 and the following section retorts 5–8. All the retorts were constructed at the company's Logan Wash Mine located outside Grand Junction, Colorado. The technology was developed for eventual application at the Cathedral Bluffs (C-b) site. Site development work at C-b, shaft sinking, etc., is currently underway. Much of the work done to date, including most of that pertaining to fragmentation, is considered proprietary. Hence, the general literature is devoid of details. The authors have tried to present a somewhat more complete picture by assembling information from a variety of sources.

11.5.2.1. Retort 1E

The first retort of this series (Ridley and Chew, 1975; McCarthy and Cha, 1976; Ridley, 1978; Lumpkin, 1980; Shaler, 1982) was prepared in 1972 with retorting commencing in July 1973. It was about 31 ft on each side and 72 ft high (Fig. 11.17). The void volume (about 25%) was in the form of a small room at the base of the retort and a vertical and circular center raise (8 ft diameter). The blastholes were vertical and paralleled this center raise. Verification drilling revealed that the rock broke to within 1–2 ft of the top, bulking fully. The retort created contained about 4000 tons of broken shale. A hole was drilled from the surface down to the top of the retort for an entry way for air and recycle gas and a burner was lowered into position. The fire started at the top of the retort was worked down through the rubblized shale under controlled conditions. This first retort produced 1250 barrels shale oil or a yield of 53%, based upon Fischer assays.

11.5.2.2. Retort 2E
Retort 2E was prepared in 1973 with retorting commencing in 1974 (Ridley and Chew, 1975; McCarthy and Cha, 1976; Ridley, 1978). It employed the same basic rubbling design (vertical and circular center raise, 9 ft diameter) and plan area (32 × 32 ft) as retort 1E. The changes were an increase in retort height to 94 ft, a reduction in void volume (21%) and a change in blasting pattern (Fig. 11.17). It produced 1424 barrels shale oil and a yield of 54% (Shaler, 1982). Retort 2E operated with a greater pressure drop and lower flow rate than retort 1E.

11.5.2.3. Retort 3E
Retort 3E was a small scale 32 × 32 × 113 ft multiple level horizontal free face experimental retort (Ridley and Chew, 1975; McCarthy and Cha, 1976; Ridley, 1978). The void volume was of the order of 25% and consisted of two sublevels plus the undercut (Fig. 11.17). These sublevels were mined from a short vertical raise outside the retort.

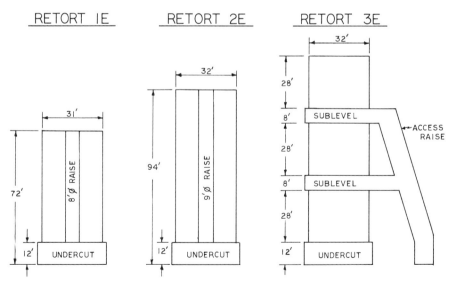

Fig. 11.17. Vertical sections through retorts 1E, 2E and 3E. (Modified after Shaler, 1982 and Lumpkin, 1980.)

486   *William Hustrulid, Roger Holmberg and Ernesto Pesce*

Blastholes were drilled perpendicular to the free faces. Because of the different fragmentation system, this retort had significantly different operating characteristics than the prior retorts. It produced 1616 barrels shale oil for a yield of 59% (Shaler, 1982). Retort 6 which will be discussed later is essentially a scaled up version of retort 3E.

11.5.2.4. Retort 4

Retort 4, the first large scale retort, was 120 ft square in plan and about 298 ft high (Ridley, 1978; Dayton 1981*d*; Shaler, 1982). The scale up to retort 4 size was done to evaluate

- —geologic and rock mechanics factors;
- —enlargement of blasting patterns;
- —retort flow control over a 14 000 ft² cross-sectional area

(all factors that might not be critical to and hence illustrated by smaller retorts) as well as to demonstrate the process on a near commercial scale. It was felt that while blasting pattern effects could be treated theoretically as they related to tons of rock broken, they could not project with any accuracy particle size distributions or the effect of limited void volume. Furthermore, the effect of geologic variables could only be determined by these large scale tests. Retort 4 had two parallel vertical void slots which extended across the cross-section (Fig. 11.18). Vertically these slots extended from the floor of the retort to a mined out room at the top. Although the actual retort dimensions have not been published, those included on Fig. 11.18 by the present authors are probably representative. From these, the calculated void volume is about 30–35%. Large diameter (estimated to be 8–9 in) blast holes were drilled parallel to the slots from the upper level. The burden and spacing appear to be of the order of 16 ft. Although originally designed as a 50-fold scale up of the retort 1E design, it was subsequently changed by substituting two vertical slots for the central raise concept.

Retort 4 was ignited in December 1975 through holes drilled from the surface and retorting continued until June 1976. Occidental recovered 27 500 barrels oil from retort 4 providing a yield of 44% (Shaler, 1982). Ridley (1978) and Ricketts (1982*a*) indicate that rock mechanics and geologic conditions aggravated by mining techniques that prevented adequate rubblization of a specific section of the retort played a major part in this reduction in yield.

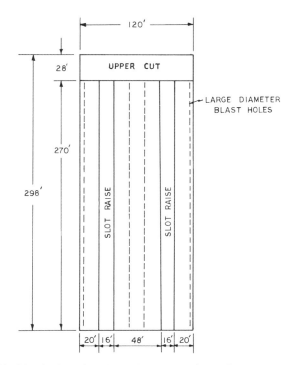

Fig. 11.18. Vertical section through retort 4. (Modified after Shaler, 1982.)

11.5.3. Occidental Oil Shale Inc.'s Experience (Retorts 5–8)

11.5.3.1. Introduction
The US Department of Energy and Occidental Oil Shale Inc. entered a two-phase cooperative agreement consisting of engineering development of the Occidental modified *in situ* process as the first phase and a technical feasibility demonstration as the second phase. The primary objective of the first phase was to evaluate two specific retort designs (vertical free face retort system (retort 5) versus horizontal free face retort system (retort 6)) for the purpose of selecting a particular retort design for the technical feasibility demonstration (retorts 7 and 8).

11.5.3.2. Explosive Types and Hole Sizes
The fragmentation part of the program was considered to be outside of the cooperative agreement and no direct information regarding explo-

488 William Hustrulid, Roger Holmberg and Ernesto Pesce

sive types, hole sizes and blasting patterns used in any of the retorts has been presented by OOSI. Published information in the form of patents, symposium papers, progress reports, etc., combined with some engineering calculations and judgement has provided a basis for conjecture at least for retorts 5–8. In this section, the information presented should be considered in orders of magnitude only. The blasting patterns discussed with respect to individual retorts have basically been scaled off published drawings and are presented to aid the reader in understanding the phenomena.

Ricketts (1982a) has indicated that the same bagged explosive was used in retorts 5–8. In Ricketts (1982b), dealing with downhole delay assemblies, it was indicated that 'the delay explosive package contains a slurry explosive such as aluminized TNT slurry'. It will be assumed that the explosive used in all rubblizing and pillar holes was such an aluminized TNT slurry. A review of the literature for such explosives would suggest a density range of $1 \cdot 4$–$1 \cdot 6$ g cm^{-3} with the most common value being about $1 \cdot 4$. The weight strength, S_w, of the explosive in comparison to ANFO ($S_w = 1$) has been derived (Loucks, 1979; Ricketts, 1980a) to be $1 \cdot 293$. The bulk strength (B_s) for such an explosive would be about $2 \cdot 13$ times that of ANFO (density = $0 \cdot 85$ g cm^{-3}).

The large diameter vertical holes for retort 5 were drilled using a small crawler-mounted blast hole rig (Ridley and Chew, 1975; Loucks, 1979). A pneumatically driven down-the-hole hammer was employed. For retort 6, the large diameter holes were drilled with a rotary machine. The pillar holes (estimated by the authors to be of the order of 5–6 in diameter) were drilled using a rotary-percussive (air-track type) drill. For retorts 7 and 8 most of the large diameter vertical holes were drilled with the rotary drill. The remainder (about 20%) were drilled with the down-the-hole hammer. A hydraulic rotary-percussive drill jumbo was used to drill the pillar holes at the intermediate level with the air track type drill for the remainder (Ricketts, 1983). The hole diameters are estimated by the authors to be of the order of 2–4 in.

11.5.3.3. Design Equations for Retorts 6–8

The deep cratering design philosophy which appears to have been used for retorts 6–8 is described in various patents (Ricketts, 1981; Ricketts, 1982d; Ricketts and Redpath, 1982). The present authors have

Mining and fragmentation oil shale research

attempted to summarize the appropriate equations:

Charge length (l):

$$l = 0 \cdot 5L \qquad (11.1)$$

Hole diameter (D):

$$D \leqslant \frac{l}{20} \qquad (11.2)$$

Depth of burial (DOB):

$$DOB = 0 \cdot 75L \qquad (11.3)$$

Hole spacing (S):

$$S < \tfrac{7}{8}DOB \text{ for lean } (<20 \text{ gallons ton}^{-1}) \text{ oil shale}$$
$$S < \tfrac{2}{3}DOB \text{ for rich } (>20 \text{ gallons ton}^{-1}) \text{ oil shale} \qquad (11.4)$$

Scaled point charge depth of burial (S_{dob}):

	Oil shale		
S_{dob} $(mm\ cal^{-1/3})$	Grade $(gallons\ ton^{-1})$	Designation	Comments
<6	—	—	Not recommended as the explosion may be too energetic. Over expansion
6–9	>20	rich	
9–12	<20	lean	
>12	—	—	Not recommended as there may not be sufficient energy to expand formation
$8 \cdot 5$	$15 \cdot 7$–$19 \cdot 3$	lean	Used for design of retorts 6–8?

Explosive energy:

$$W_c = \frac{DOB(S^2 \times 10^9)}{(S_{dob})^3} \qquad (11.5)$$

Explosive volume:

$$V = \frac{\pi D^2}{4} l \times 10^2 \qquad (11.6)$$

490   *William Hustrulid, Roger Holmberg and Ernesto Pesce*

Explosive bulk energy content:

$$B_s = \frac{W_c}{V} \tag{11.7}$$

Explosive weight:

$$W_{kp} = \frac{W_c \times 10^{-3}}{C} \tag{11.8}$$

where: L = thickness of unfragmented formation (m); DOB = depth from the horizontal free face to the center of mass of the explosive charge (m); D = hole diameter (cm); W_c = energy of the individual charge (cal); l = charge length (m); S_{dob} = scaled point charge depth of burial (mm cal$^{-1/3}$); S = hole spacing (m); V = volume of explosive (cm^3); B_s = explosive bulk energy content (cal cm^{-3}); W_{kp} = explosive weight per hole (kg force); C = explosive weight strength (cal g^{-1}).

The following example will serve to illustrate the application of the equations for designing the sill holes of retort 6:

Determine the blasting parameters that apply under the following conditions: L = 10·67 m (35 ft); D = 26·67 cm (10·5 in); S_{dob} = 8·5 mm cal$^{-1/3}$; B_s = 1300 cal cm^{-3}.

Charge length:

$$l = 0.5 \times 10.67 = 5.34 \text{ m}$$

Depth of burial:

$$DOB = 0.75 \times 10.67 = 8 \text{ m}$$

Hole diameter:

$$\frac{5.34}{20} = 0.267 \text{ m (okay)}$$

Explosive volume:

$$V = \frac{\pi}{4} (26.67)^2 (5.34 \times 10^2) = 2.98 \times 10^5 \text{ cm}^3$$

Explosive energy:

$$W_c = (2.98 \times 10^5)1300 = 3.88 \times 10^8 \text{ cal}$$

Spacing:

$$S = [(3.88 \times 10^8)8.5^3]/(8 \times 10^9)^{1/2} = 5.46 \text{ m}$$

(For lean oil shale, $S = 5.46 < \frac{7}{8}DOB = 7$ m (okay))

Mining and fragmentation oil shale research 491

It is noted that the cratering scaling equations (S_{dob}) are expressed in terms of the energy (calorie) content of the explosive as opposed to the weight of the explosive:

$$\frac{mm}{cal^{1/3}} \quad versus \quad \frac{ft}{lb^{1/3}}$$

This is an apparent attempt to generalize the application of cratering experiments done with one explosive to all explosives. There are two problems with this approach. First, and most importantly, the breaking ability of an explosive in a given rock is not necessarily directly related to the energy content and one cannot expect 100% of the energy to be used in rock breakage. To utilize all of the energy as expansion work, it must be possible for the gaseous products to expand to a very low pressure. Rock breakage and primary fragmentation (except fragmentation due to later collisions between blasted rock) are already completed when the detonations have expanded to a pressure of about 1500–15 000 psi. Some rocks break better when explosives with a high gas volume release are used, others break better by use of an explosive with a high shock wave energy. Secondly, few manufacturers provide energy content data for their explosives. They usually only provide the weight strength (S_{ANFO}) with respect to ANFO, the density (P_{exp}) and possibly the velocity of detonation. Ireco is one exception and specifications for several of their explosives (Ireco, 1982a, b) are given in Table 11.7. An approximate relationship is given by:

$$B_s = 700 \times S_{ANFO} \times \frac{P_{exp}}{P_{ANFO}} \tag{11.9}$$

where: B_s = bulk energy content (cal cm^{-3}); S_{ANFO} = weight strength relative to ANFO; P_{exp} = density of the explosive; P_{ANFO} = density of *ANFO*.

11.5.3.4. Discussion of Retort 5 (Loucks, 1977, 1978a,b,c, 1979)
An isometric drawing of the configuration of retort 5 is given in Fig. 11.19. Calculations revealed that the rubblized shale contained approximately 53 000 barrels oil in place prior to retort ignition. The overall retort dimensions were $118 \times 118 \times 200$ ft high (including a 40 ft thick sill pillar at the top of the rubble). Retort 5 used a single vertical slot to provide the void volume and the free face necessary for the rubblization blast. Retort 5 was developed from two mine levels. It

TABLE 11.7
Properties for Ireco Explosives (Ireco, 1982a, b)

Explosive series	Density $(g\,cm^{-3})$	Energy $(cal\,g^{-1})$	Weight strength	Energy $(cal\,cm^{-3})$	Bulk strength
Iregel A	1·10	720	0·84	792	1·13
	1·10	815	0·95	905	1·29
	1·12	901	1·04	1 009	1·42
	1·13	986	1·14	1 114	1·57
	1·14	1 114	1·28	1 270	1·78
	1·15	1 322	1·50	1 520	2·10
	1·16	1 488	1·68	1 726	2·38
Iregel U	1·20	657	0·78	788	1·14
Iregel C	1·25	745	0·87	931	1·33
	1·25	808	0·94	1 010	1·43
	1·25	887	1·02	1 109	1·55
	1·25	970	1·13	1 212	1·77
	1·25	1 054	1·21	1 317	1·84
Iregel E	1·25	630	0·78	788	1·20
	1·25	713	0·87	891	1·33
	1·25	796	0·94	995	1·43
	1·25	878	1·02	1 098	1·55
	1·25	960	1·13	1 200	1·72
	1·25	1 042	1·21	1 303	1·84
ANFO	0·82	858	1·00	704	1·00

was accessed from the lower mine (product) level (elevation 7561 ft) and the U-500-S drift at the upper mine level (elevation 7830 ft). At the air level (Fig. 11.20), two rib pillars approximately 20 ft wide and 16 ft high were left to provide support for the roof.

The vertical slot for retort 5 was mined by first drilling a 4 ft diameter raise and then blasting toward the raise forming a slot 117 ft long and averaging 21 ft in width. The slot had been tapered into a trapezoidal shape at the bottom to facilitate mucking from a single drawpoint. The void volume in the slot represented 17·4% of the total volume in the rubblized portion of the retort. Two rows of vertical blast holes were drilled on each side of the slot from the air level. The 20 rubblization holes (Petersen, 1979) were loaded with 170, 150 lbs explosive (each hole contained about 8500 lbs explosive) (Loucks, 1979). The explosive column was 160 ft in length with the upper 40 ft of hole (that in the future sill) being stemmed.

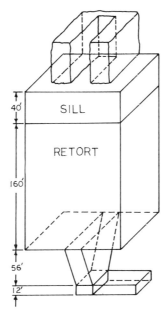

Fig. 11.19. Isometric representation of retort 5. (Loucks, 1977.)

Each hole was multiprimed, to provide a high degree of reliability and equipped with a blast detector to give a positive indication of explosive detonation. No primadet lines remained in any of the blast-holes thus indicating positive activation of each hole. The planned delay sequence for the 20 holes is indicated in Table 11.8.

Following rubblization, an inspection (Loucks, 1977) revealed some blast damage. On the lower level: (a) cracks were observed in the area where the retort 5 bulkhead was fastened to the walls; (b) a 20×50 ft section of wire and roof bolts had come down at a drift intersection about 500 ft from the center of the retort leaving a large slab for scaling; and (c) minor rock spalling appeared on the sides and roof of the access drifts. On the air level (at the top of retort 5) about 2 ft of rock came down from the center sections of the roof and some rock fell from the sides. Pressure tests run before and after the blast revealed that the sill pillar had retained enough integrity to be used as a barrier pillar during retorting. The results of ground motion measurements made during the rubblization blast will be discussed in Section 11.5.3.7.

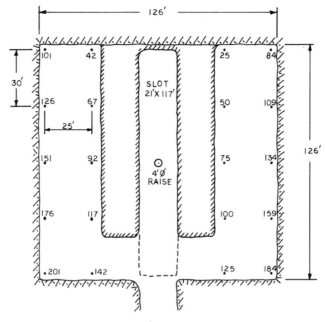

Fig. 11.20. Best estimate of fragmentation scheme for retort 5. Plan view of the air level. The numbers represent the planned delay (ms).

TABLE 11.8.
Delay Sequence for the Rubblizing Blast of Retort 5 (Petersen, 1979)

Hole number	Planned delay (ms)	Hole number	Planned delay (ms)
101	201	111	125
102	176	112	100
103	151	113	75
104	126	114	50
105	101	115[a]	25
106	142	116	184
107	117	117	159
108	92	118	134
109	67	119	109
110[a]	42	120	84

[a] Corner holes.

Tracer tests conducted prior to retorting revealed that:

—The rubble underlying the front center of the middle finger (position of blast initiation) contained more void than was found in any other area.
—In some sections, there was a high resistance to gas flow between adjacent holes indicating non-uniform breakage in the radial direction.
—Some of the void spaces near the outer edge of the retort were inactive.

The retorting procedure was modified to try and overcome the expected non-uniformity in permeability. In spite of the fact that no burners were used in the front center area, the combustion front moved rapidly through this zone anyway. As a result, the quantity of oil produced (11 047 barrels or 21% oil-in-place) was substantially less than anticipated.

11.5.3.5. Discussion of Retort 6 (Loucks, 1978b,c, 1979; Ricketts, 1980a)

Retort 6 (basically a scaled up version of retort 3) having a plan area of 165 × 165 ft utilized a horizontal free face retort system. It is shown in isometric view in Fig. 11.21 and in cross-section in Fig. 11.22. Void

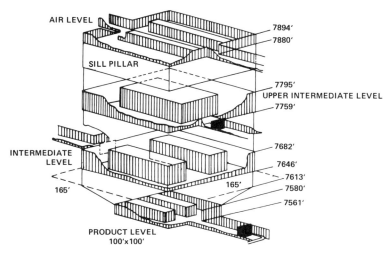

Fig. 11.21. Isometric view of retort 6. (Ricketts, 1980a.)

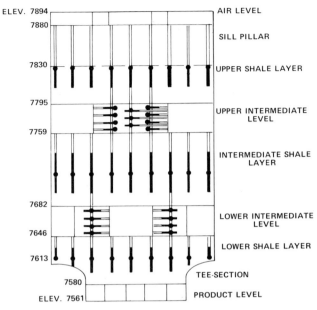

Fig. 11.22. Section view of retort 6 blast design (elevation in feet). (Ricketts, 1980a.)

TABLE 11.9
Dimensions for Retort 6

Item	Number	Width	Length	Height/thickness
		Dimensions (ft)		
Air level		165	165	15
Air level pillars	2	16	145	15
Sill				50
Upper shale layer				85
Upper intermediate void level		165	165	36
Upper intermediate level pillar	1	70	116	36
Intermediate shale layer				77
Lower intermediate void level		165	165	36
Lower intermediate void level pillar	2	36	112	36
Lower shale layer				33–36
Product level		100	100	19
Product level pillars	2	20	19	60

TABLE 11.10
Drilling Statistics for Retort 6

	Hole diameter (in)	Pattern (ft)	Total number of holes
Rubblization blast	10–11	20×20	218
Upper level pillar (short)	5–6	9×9	2×52
Upper level pillar (long)	5–6	9×9	36
Lower level pillars	5–6	9×9	2×48

space was located on the upper and lower.intermediate levels as well as on the product level. The symmetric blasting concept used in retort 6 consisted of expanding equal rock into equal void on each of the void levels. The planned void volume was 23%. The dimensions of retort 6 are given in Table 11.9. Above the retort was an air level containing two permanent pillars for supporting the roof. The pillars on the other levels are only temporary and will be explosively removed prior to the main rubblizing blast.

The large diameter rubblizing holes are drilled from the air level, the upper and lower intermediate void levels. In those holes drilled from the air level, the upper 50 ft was stemmed so that a 50 ft thick sill pillar would remain. Drilling statistics for retort 6 are given in Table 11.10. The blast was designed using the principles described earlier in Section 11.5.3.3. Some of the relevant information is presented in Tables 11.10 and 11.11.

Figures 11.23–11.28 are sketches which show vertical section views of retort 6 at various times in the blasting sequence. The sequencing condition required that the three horizontal shale layers be detonated at the same time so that each layer could only expand into its portion of the void. At time $t = 0$, the electric blasting caps detonated, initiating the non-electric primacord trunk line networks on all three levels.

TABLE 11.11
Explosive Statistics for Retort 6

Explosive type	Bagged, TNT sensitized, aluminized slurry
Total explosive	499 205 lbs slurry in ANFO equivalent 386 000 lbs explosive
Overall average ANFO powder factor	$1 \cdot 45$ lbs ton^{-1}

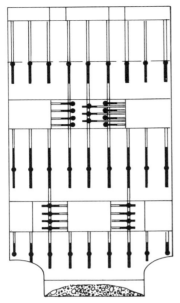

Fig. 11.23. Retort 6 detonation sequence at $t_0 = 0$ ms. (Ricketts, 1980a.)

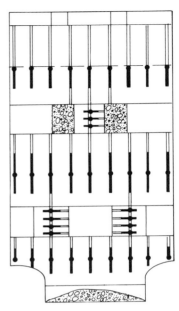

Fig. 11.24. Retort 6 detonation sequence at time t_1. (Ricketts, 1980a.)

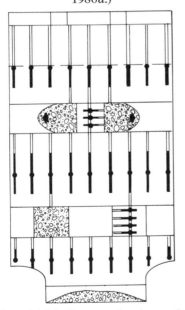

Fig. 11.25. Retort 6 detonation sequence at time t_2. (Ricketts, 1980a.)

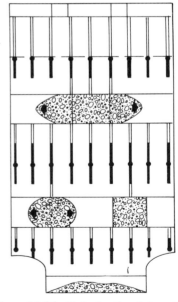

Fig. 11.26. Retort 6 detonation sequence at time t_3. (Ricketts, 1980a.)

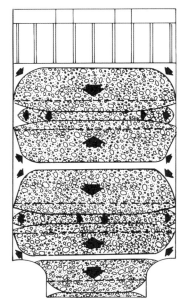 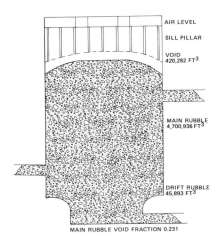

Fig. 11.27. Retort 6 detonation sequence at time t_4, all charges detonated. (Ricketts, 1980a.)

Fig. 11.28. Section view retort 6, post shot. (Ricketts, 1980a.)

These in turn initiated, within several milliseconds, the non-electric downlines for the pillar holes and main vertical blastholes. Delays were introduced through downhole units which sat in the holes 'burning' until their prescribed time for detonation. The downhole delay concept was necessary because ejecta from the pillar blasts could have destroyed any initiating lines for the main rubblizing blast lying on the floor of the void levels. The circles superimposed on each of the holes show the point of initiation. The results of the ground motion survey taken during rubblization are presented later. The following major observations were made after the blast (see Fig. 11.28):

—The retort did not completely fill with rubble. Instead, an open void region was left between the top of the rubble pile and the base of the sill pillar (the gap varied from about 8 ft (2·5 m) at the center to 35 ft (11 m) at the corners. The void volume was 420 300 ft^3.
—The sill pillar varied in thickness from about 50 ft at the corners and walls to 40 ft at the center.

500 William Hustrulid, Roger Holmberg and Ernesto Pesce

—The rubblized pillar was domed rather than flat as had been expected.

—A significant quantity (45 900 ft^3) of material was blown out of the access drifts to the upper and lower intermediate rooms. This appears to have had a significant effect on the permeability distribution (gas flow) near the drifts.

—The particle size appeared to be of the desired range.

—A preliminary analysis (Loucks, 1979) of overcoring data made in the air level (sill) pillars before and after rubblization indicated a decrease in stress from 1400 psi to $\approx$200 psi.

—Tracer tests on the rubble bed showed the permeability along the edges was about twice that in the center which was better than any previous retort.

Occidental and outside consultants (Loucks, 1979) studied whether the sill pillar had adequate mechanical strength to allow it to remain intact during retorting (particularly during the two-week start-up period). It was concluded that the sill would maintain its structural integrity even if it were to spall to a thickness of 20–25 ft. No significant spalling was expected during the scheduled 20-day start-up period. To enhance the mechanical strength of the sill pillar, it was rock bolted on a 10 ft pattern using grouted bolts ranging in length from 30 to 40 ft.

Start-up was on 28 August 1978. By 5 September, a survey indicated that spalling of the sill pillar of 12–23 ft had occurred. Extensometer readings near the center of the sill pillar indicated that the rate of the sag had accelerated by 12 September. Midday on the 12th, a small hole appeared in the sill showing that spalling had been more serious than anticipated. It was cooled with water and grouted. That evening a larger hole developed which was followed later by a partial collapse of the sill pillar. Retort gases were introduced into the air level and it became inaccessible. Major corrective action was required by Occidental. Even with the loss of access to the piping system (and subsequent loss of control of the combustion front) and flow channeling down the north (vent raise) side of the retort, the design produced about a 115% increase in recovery ratio over the vertical void design of retort 5.

Retort 6 pointed out the following technical and operational changes that could offer improvements:

—Better distribution of the void level pillars. This was particularly true of the large pillar on the upper intermediate void level, which was difficult to blast, required a large number of blastholes and decks, and was difficult to distribute evenly over the entire cross section.

—Reduction in the rubble flow out of the void level entry drifts.

—Minimization of the number of deep, multiple-decked vertical blastholes required to pass through void level pillars to reach the shale beneath these pillars. There were 25 such holes in retort 6.

—Loading the retort with a reliable, bulk pumpable explosive product. Retort 6 was loaded by hand with 20–60 lb bags of explosives over a period of several days.

—Reduction in the retort edge effects of higher permeability.

—Reduction in the rubble bed mounding shape at the top. A flat upper profile would be preferred from uniformity as well as processing start-up viewpoints.

11.5.3.6. Discussion of Retorts 7, 8, and 8x (Nelson, 1982; Ricketts, 1982*a*)

Retorts 7, 8 and 8x as well as numerous mini-retorts were developed under Phase II of the cooperative agreement between Occidental Oil Shale Inc. and the US Department of Energy. The horizontal free face concept of retort 6 was utilized with design modifications made to eliminate some of the problems experienced. Additional studies included (a) the fracturing produced in pillars between retorts for simulated full scale operation and (b) the problems/possibilities with bulk pumpable explosives.

The arrangement of the three retorts can be seen in isometric view in Fig. 11.29. All three retorts are of production scale cross-section (165 × 165 ft). The rubblized heights of retorts 7 and 8 are 241 ft whereas retort 8x has a height of only 63 ft. This latter retort was used primarily for practising blasting techniques (prior to rubblizing retorts 7 and 8) as well as creating the 50 ft wide pillar with retort 8. The major focus of this section will be on retorts 7 and 8 (see Figs 11.30 and 11.31) whose geometries are identical. The important dimensions for retorts 7 and 8 are given in Table 11.12. A comparison, between retorts 6 and 7/8 of some design parameters are given in Table 11.13. It is noted that:

—The air level has been eliminated. Holes from the surface are used for initiation and controlling retorting.

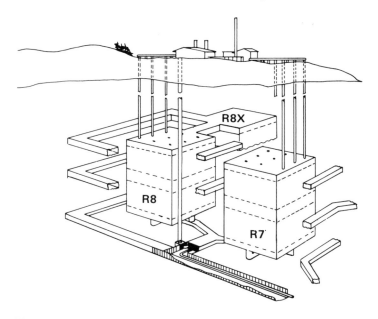

Fig. 11.29. Retorts 7, 8 and partial retort 8x isometric view. (Ricketts, 1982a.)

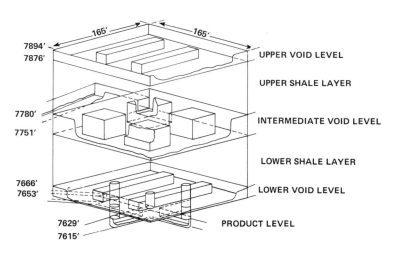

Fig. 11.30. Retorts 7 and 8, isometric view before blasting. (Ricketts, 1982a.)

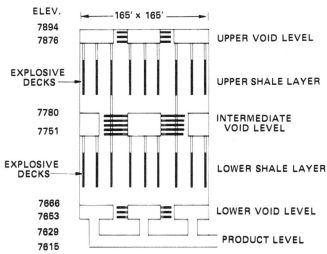

Fig. 11.31. Retorts 7 or 8 blast design in section view. (Total retort height: 241 ft. Total shale: 4 953 756 ft³. Total void: 1 460 462 ft³. All explosive decks center initiated.) (Ricketts, 1982a.)

—The large intermediate level pillar has been replaced by four smaller pillars thus helping to better distribute the pillar rock as well as reducing drilling/blasting problems.
—The number of long vertical-rubblizing holes has been reduced.
—A separate product collection level has been added.

TABLE 11.12
Dimensions of Retorts 7 and 8

Item	Number	Width	Length	Height/thickness
Upper void level		165	165	18
Upper level pillars	2	16	105	18
Upper shale layer				96
Intermediate void level		165	165	29
Intermediate level pillars	4	36	36	29
Lower shale layer				85
Lower void level		165	165	13
Lower level pillars	2	16	105	13
Floor shale layer				24
Product level				14

504 William Hustrulid, Roger Holmberg and Ernesto Pesce

TABLE 11.13
Comparison of Retorts 6 and 7/8

	Retort	
Factor	6	7/8
Plan area (ft)	165×165	165×165
Rubblized column height (ft)	269	241
Mined void ratio (%)	27	23
Number of mined levels	4	4
Air level	Yes	No
Separate product level	No	Yes
Control of retort process	Underground	Surface
Intermediate level pillars size (ft)	$(1-70) \times 116$	$(4-36) \times 36$
Long vertical holes through pillars—number/length (ft)	15/179 10/163	8/184
Multidecked pillars	Yes	No
Number of large diameter vertical holes	~218	168
Length of large diameter vertical hole (ft)	14 700	12 300
Number of pillar holes	~260	371/386
Length of pillar holes (ft)	~6 400	~5 700
Powder factor[a]	1·45	1·55/1·57
Tons rubblized	362 000	348 940/347 750
Average oil content (gal ton^{-1})	15·6	16·9/17·3

[a] ANFO, lbs ton^{-1}, calculated from actual explosive used based upon weight strength.

A best estimate for the blasting parameters is given in Table 11.14. The rubblizing concept is essentially identical to that described for retort 6. The shale layers are expanded upward and downward into equal void volumes. The explosive used for retorts 5 and 6 was also selected for retorts 7, 8 and 8x. A series of tests (Nelson, 1982; Ricketts, 1982a) were performed with two bulk pumpable explosive products (water gel and emulsion slurry) supplied by different manufacturers in 200 ft deep holes with the hope of improving operational aspects of the loading process. A given test series usually consisted of three holes which were loaded and then allowed to sit for 6–7 days in the hole to simulate a full scale retort loading operation. Each product experienced severe problems (misfires, low velocity detonations, etc.) on at least one test series in which it was employed and hence the same explosive used for retorts 5 and 6 was selected for retorts 7, 8 and 8x.

Mining and fragmentation oil shale research 505

TABLE 11.14
Blasting Parameters (Best Estimate)

Large diameter vertical holes
Diameter	10–11 in
Pattern	18×18 ft and 18×24 ft
Explosive	Bagged TNT sensitized aluminized slurry
Booster	3 lb cast primer

Horizontal pillar holes
Diameter	2–3 in
Pattern	5×5 ft
Explosive	Bagged TNT sensitized aluminized slurry

Blast initiation was through the use of electric zero delay seismic blasting caps (a non-electric back-up system was in place as well). These initiated non-electric detonating cord systems consisting of two independent (but crosslinked at various locations) trunkline systems passing over each hole on the three void levels simultaneously. Each trunkline was connected to a different downline leading to a delay cap within the powder column. Each powder column was dual primed. The detonating cord systems initiated the downline cords in all of the pillar holes and vertical holes within a period of several milliseconds. The detonation propagated down to the downhole delay caps which then waited to detonate at their prescribed times. For each of the explosive columns the point of initiation was midway along the charge length. Figures 11.32–11.37 show the progression of the blast. The delay caps used in retorts 7 and 8 showed much less scatter than those used in retort 6. No overlaps were observed between successive delays. The extremely low scatter on these caps presents the possibility for seismic enhancement damage for these large retorting blasts. Ricketts (1982*a*) suggests that 'the excellent powder distribution in these designs separates the energy sufficiently for a specific delay to keep the seismic amplitude to acceptable levels'. A summary of the results of ground motion studies is given in Section 11.5.3.7.

The rubblization of retorts 7 and 8 went very much as planned.

—The tops of the rubble piles were both quite flat and level. The average void height (distance between the tops of the retort and the rubble pile) was 14·9 ft (retort 7) and 14·6 ft (retort 8).

—Pressure drop contours were fairly flat and level showing good

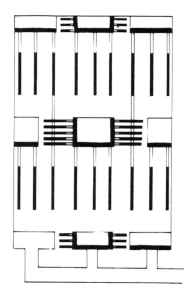 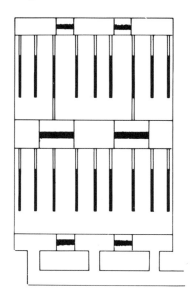

Fig. 11.32. Retort 7 or 8 detonation sequence at time t_0. (All primalines initiated for downhole delays.) (Ricketts, 1982a.)

Fig. 11.33. Retort 7 or 8 detonation sequence at time t_1 when center portions of pillars are detonated. (All pillars blasted simultaneously.) (Ricketts, 1982a.)

lateral uniformity. Vertical profiles revealed a uniform increase with depth suggesting no unblasted or poorly blasted layers.
— In general these retorts appear well rubblized, very uniform and virtually identical to each other.
— The permeability in the rubble was 10^4 times greater than the most damaged region (near the perimeter of the retort) of the 50 ft wide pillar separating retorts 8 and 8x. This suggests that it would be sufficient for retort isolation during retorting.

There were some problems experienced as well:

— For retort 8, the area around the center drawhole (bottom of retort to the product level) was badly fractured during rubblization. It had to be rebuilt.
— Approximately 5 min after the retort 7 blast, detonation products from the blast were ignited by electrical equipment in the mine maintenance shop. A flashfire resulted and followed the fumes to

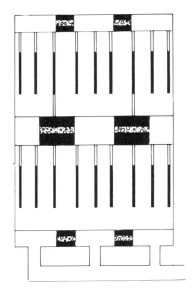 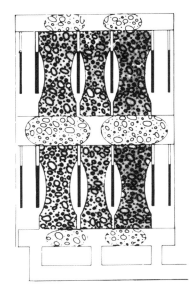

Fig. 11.34. Retort 7 or 8 detonation sequence at time t_2 when upper and lower portions of pillars are detonated. (All pillars blasted simultaneously.) (Ricketts, 1982a.)

Fig. 11.35. Retort 7 or 8 detonation sequence at time t_3 when main vertical blasthole sequence begins. (Main blasthole sequence initiated.) (Ricketts, 1982a.)

the retort 7 upper entry level where several small stable fires occurred in loose shale meant to contain fly rock.

Retorting results available at that time (Anon., 1982) suggested a recovery of more than 190 000 barrels shale oil.

11.5.3.7. Ground Motion and Air Blast Data

The most complete information regarding ground motion studies has been published on retort 5. The preliminary data are given in Table 11.15. URS Blume (Keith, 1979), using the surface data, found that:

$$V_{max} = 566\,428 R^{-1.63} \tag{11.10}$$

where: V_{max} = maximum particle velocity (in s^{-1}); R = distance (ft). Assuming a cube root scaling relationship:

$$V_{max} = K_1 \left(\frac{R}{W^{1/3}}\right)^{-1.63} \tag{11.11}$$

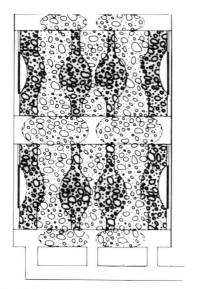 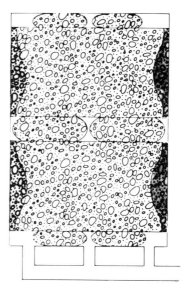

Fig. 11.36. Retort 7 or 8 detonation sequence at time t_4 when main vertical blastholes continue to detonate. (Main blasthole sequence continues to detonate.) (Ricketts, 1982a.)

Fig. 11.37. Retort 7 or 8 detonation sequence at time t_5 when last delay interval detonates. (Detonation sequence completed). (Ricketts, 1982a.)

where: K_1 = constant; W = pounds of explosive per delay. Equation 11.11 can be rewritten as:

$$V_{max} = K_1 W^{0.543} R^{-1.63} \tag{11.12}$$

Keith (1979) suggests that although a cube root scaling is thought to be relatively good for close-in stations, for the more distant surface stations a weight scaling exponent of 0·7 may be more appropriate:

$$V_{max} = K_2 W^{0.7} R^{-1.63} \tag{11.13}$$

where K_2 = constant. Since for retort 5, $W \triangleq 8500$ lbs, eqns 11.12 and 11.13 become, respectively:

$$V_{max} = 4164 W^{0.543} R^{-1.63} \tag{11.14}$$

$$V_{max} = 1006 W^{0.7} R^{-1.63} \tag{11.15}$$

Seismic data collected for retorts 4–8 is presented in 'normalized'

TABLE 11.15
Preliminary Ground Motion Data from the Retort 5 Blast (Explosive Per Delay $\simeq$ 8 500 lbs)

Station	Range (ft)	Peak amplitude (in s^{-1}) Vertical	Longitudinal	Transverse	Comments/location
5[a]	47[c]	—	22	—	On rib, room 5
5[a]	47[c]	—	—	—	On pillar, room 5
5[a]	—	36	—	—	Overhead, room 5
5[a]	45[c]	43, 66, 121 43, 63	—	—	Floor room 5, steps in velocity as individual holes detonate
17[a]	145[c]	—	—	—	Outside of room 5, records not reasonable
18[a]	320[d]	4·0	2·9	—	Next to room 4
19[a]	347[d]	2·0	—	—	Across from entry to room 4
16[a]	400[d]	7·3	5·7	7·3	In above near recording room
8[a]	800[d]	2·2	3·7	3·4	Near upper portal
22[a]	530[d]	4·0	—	—	(Provided most complete match to planned blasting sequence)
Tank[a]	~2 500[d]	1·9	1·1	1·1	Near tank, upstream from heater treater plant
Stack[b]	556[e]	5·7	5·6	5·0	On concrete collar at base of stack (bedrock)
Microwave[b]	687[e]	10·8	5·2	6·7	On fill
Transformer[b]	1 243[e]	7·2	5·5	3·1	On fill
Mountain[b]	3 540[e]	0·71	0·82	0·82	Surficial soil and weathered rock?

[a] S^3 measurements (Petersen, 1979).
[b] URS Blume measurements on surface (Keith, 1979).
[c] Range is the distance to the closest explosive for room 5 stations.
[d] Range is the horizontal distance to center of room 5.
[e] Radial distance from center of room 5.

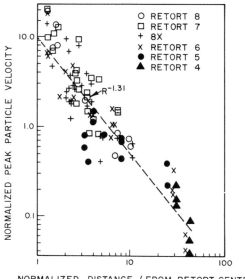

Fig. 11.38. Normalized peak particle velocity versus normalized distance for retort rubblization blasts. (Ricketts, 1982a.)

form in Fig. 11.38. The motion at stations distant from the blast appear to respond as suggested by:

$$V_{max} \propto R^{-1.31}$$

Closer to the blast, the relationship may be more like:

$$V_{max} \propto R^{-1.6} \quad \text{(fitted by eye)}$$

Assuming that cube root scaling applies at far distances and square root scaling applies at near distances then:

far:

$$V_{max} = K_1 W^{0.44} R^{-1.31} \qquad (11.16)$$

near:

$$V_{max} = K_3 W^{0.8} R^{-1.6} \qquad (11.17)$$

Fitting both equations to the data for the stack (retort 5) yields:

far:

$$V_{max} = 417 W^{0.44} R^{-1.31} \qquad (11.18)$$

TABLE 11.16
Comparison of Peak Particle Velocities Using Cube Root and Square Root Scaling

Distance (ft)	V_{max} (in/s^{-1})	
	Eqn 11.18	Eqn 11.19
50	133	269
100	54	89
200	22	29
300	13	15
400	8·7	9·7
500	6·5	6·8
1 000	2·6	2·2

TABLE 11.17
Particle Velocity Results for Retorts 7 and 8 (Peterson and Brown, 1982)

Distance[a] (ft)	Peak velocity[b] $(in\ s^{-1})$	Distance[a] (ft)	Peak velocity[b] $(in\ s^{-1})$
133	98	323	11
134	49	328	9
135	112	343	12
154	52	347	4
158	35	382	17
161	70	400	6
163	37	462	4
170	27	472	4
188	39	701	2
240	45	718	5
262	20	747	7
269	16	759	10
277	21	1 066	4
305	13	1 079	3

[a] Horizontal distance from the retort center (ft).
[b] Main blast peak velocity (in s^{-1}).

TABLE 11.18
Peak Air Pressures as a Function of Distance for Retorts 7, 8 and 8x
(Peterson and Brown, 1982)

Retort	Distance[a] (ft)	Peak pressure (psi)	Retort	Distance[a] (ft)	Peak pressure (psi)
8x	235	1·3	7	765	0·5–0·6
	240	1·5		850	0·13–0·16
	570	1·04		1 400	0·27–0·43
	1 250	0·35		1 700	0·12
7	180	0·35	8	360	2·1
	215	0·9–1·4		1 030	0·24
	235	0·4		1 330	0·02
	235	0·35		1 450	0·76
	590	0·73			

[a] Drift distance from retort center.

near:

$$V_{max} = 101 W^{0·8} R^{-1·6} \qquad (11.19)$$

A comparison of the results of eqns 11.18 and 11.19 is given in Table 11.16. The results of ground motion measurements have been reported as a function of position for retorts 7, 8 and 8x (Peterson and Brown, 1982). Table 11.17 presents the results of main blast peak velocity as a function of horizontal distance from the retort center. As can be seen, the velocities vary from about 100 in s^{-1} at 130 ft to 3 in s^{-1} at 1000 ft. A series of air pressure measurements have been reported as well. These are summarized in Table 11.18.

11.5.4. Rio Blanco Experience
The Rio Blanco Oil Shale Company (a 50/50 general partnership of Gulf Oil Company and Standard Oil Company (Indiana)) has designed, constructed and processed two modified *in situ* retorts (MIS) at Tract C-a (Piceance Creek, Colorado) using newly developed techniques for rubbling, ignition and operation through surface drill holes (Berry, 1978, 1982; Berry *et al.*, 1982; Rutledge, 1982). The MIS program cost \$132 000 000 over a period of $4\frac{1}{2}$ years (Dayton, 1981*e*). It concluded with the successful demonstration of two retorts (retort 0 and retort 1).

An isometric view of the mine and retort development plan is shown

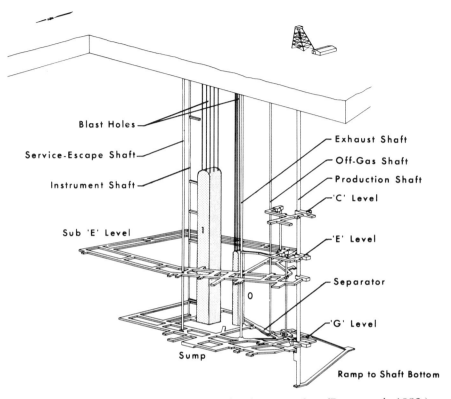

Fig. 11.39. Rio Blanco Mine and retort development plan. (Berry et al., 1982.)

in Fig. 11.39. The main production level ('G' level) is at a depth of 840 ft below the surface. Level 'E' was used for the dewatering of the area in the vicinity of the retorts. A summary of the two retorts is given in Table 11.19. Figures 11.40–11.42 show conceptually Rio Blanco's proprietary procedure for constructing the retorts.

The blastholes for rubblizing the retorts can be drilled from the surface by either truck or trailer mounted drill rigs. Five blast holes were planned for retort 0 and 13 for retort 1 (Crookston and Weiss, 1980). After drilling the blastholes to the full depth of the planned retort, an undercut is excavated across the retort cross-section. The blastholes are loaded with explosives and a segment of the roof is blasted down into the undercut using an approach similar to vertical

TABLE 11.19
Comparison of Retorts 0 and 1 (Berry et al., 1982)

Factor	Retort 0	Retort 1
Plan dimensions (ft)	30×30	60×60
Height (ft)	165	400
Fischer assay (gallons ton^{-1})	17·3	21·6
Actual yield (barrels)	1 876	24 444
Recovery (%)	68	68
Average front advance (ft day^{-1})	2·7	3·0

crater retreat mining. A portion of the material is then loaded out through the access drift. The process is repeated until the desired void space for rubblizing the remaining volume without further mucking has been created. The exact void space used is not known but some probable ranges are given in Table 11.20. The loading of explosives

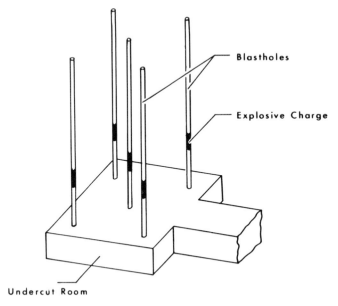

Fig. 11.40. Step 1 in the high void retort development. (Berry et al., 1982.)

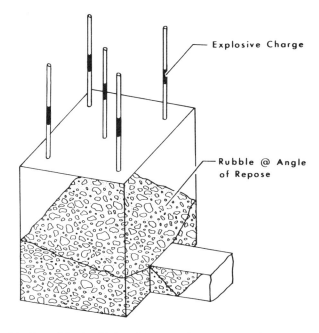

Fig. 11.41. Step 2 in the high void retort development. (Berry *et al.*, 1982.)

and blasting down of the roof continued until the retort is filled with broken rubble. If desired, an attic space may be left at the top of the retort to promote air distribution. Size analysis of the withdrawn muck indicated that a small average particle size was produced.

Assuming that Figs 11.40–11.42 accurately represent retort 0, then:

Charge length ≅ 5 ft
Hole diameter = 9–10 in
Spacing of holes ≅ 20 ft
Depth to center of charge ≅ 12–13 ft

It is surmised that the explosive used was probably aluminized slurry due to the high energy requirements and wet hole conditions. Retort 1 can be formed by repeating the pattern of retort 0.

The ignition and control of the retorts can be performed from the surface through the blastholes.

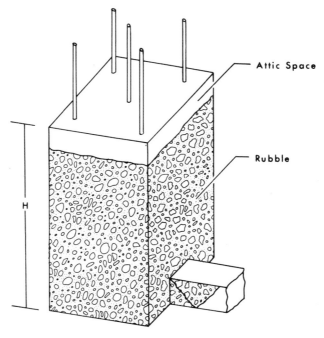

Fig. 11.42. Step 3 in the high void retort development. (Berry et al., 1982.)

TABLE 11.20
Void Space for the Rio Blanco Process

Reference	Approximate void volume
Crookston and Weiss (1980)	40%
Berry (1982)	15–40%
Table 11.18	33% retort 0
	28% retort 1

11.6. CONCLUSIONS

A considerable amount of applied research and development has occurred in the mining and fragmentation of oil shale since the US Bureau of Mines began its pioneering studies at the Anvil Points Mine in 1944. The industry has generated a number of technically advanced

Mining and fragmentation oil shale research 517

variations of four exploitation methods:

—near surface *in situ*;
—surface mining;
—underground mining;
—modified *in situ*

which are available for application to suitable oil shale deposits. This chapter has focused on the technical details of these approaches. Economic and other factors (environmental, for example) have not been included. Whether or not these technically viable approaches actually ever become part of a commercial operation will depend upon the integrated consideration of all the factors involved.

11.7. ACKNOWLEDGMENTS

The authors would like to express their appreciation to the companies who provided information for inclusion in this chapter.

REFERENCES AND BIBLIOGRAPHY

Agapito, J. F. T. (1972). Pillar design in competent bedded formations, PhD Thesis, Colorado School of Mines.

Agapito, J. F. T., (1974). Rock mechanics applications to the design of oil shale pillars, *SME Preprint 74-AIME-26*.

Anon. (1982). *Engineering/Mining J.* December, 75.

Baloo, G. (1984). Personal communication. (Parachute Creek shale oil program, Union Oil Co.)

Bauer, A. and Calder, P. N. (1974). Trends in explosives, drilling and blasting, *Bull. Can. Inst. Min. and Metall.*, **67**(742), 51–7.

Berry, K. L. (1978). Conceptual design of combined *in situ* and surface retorting of oil shale, *Proc. 11th Oil Shale Symposium*, ed. J. H. Gary, Colorado School of Mines Press, Golden, Colorado, pp. 176–83.

Berry, K. L. (1982). *In situ* retorting of oil shale, *US Patent 4,328,863*. May 11.

Berry, K. L., Hutson, R. L., Sterrett, J. S. and Knepper, J. C. (1982). Modified *in situ* retorting results of two field retorts, *Proc. 15th Oil Shale Symposium*, Colorado School of Mines Press, Golden, Colorado, pp. 385–96.

Boade, R. R., Stevens, A. L., Long, A. and Harak, A. E. (1979). True *in situ* processing of oil shale: an evaluation of current bed preparation technology, *SAND 78-2162, Sandia National Laboratories*, Albuquerque, New Mexico.

518   *William Hustrulid, Roger Holmberg and Ernesto Pesce*

Britton, K. (1980). Principles of blast design developed for *in situ* retorts of the geokinetics surface uplift type. *Proc. 13th Oil Shale Symposium*, ed. J. H. Gary, Colorado School of Mines Press, Golden, Colorado, pp. 169–80.

⇢ Britton, K. C. and Lekas, M. A. (1979). Process for producing an underground zone of fragmented and pervious material, *US Patent 4,175,490*.

Burwell, E. L., Sterner, T. E. and Carpenter, H. C. (1973). *In situ* retorting of oil shale, results of two field experiments, *Report of Investigations 7783, US Bureau of Mines*.

→ Campbell, G. G., Scott, W. G. and Miller, J. S. (1970). Evaluation of oil shale fracturing tests near Rocky Springs, Wyo. *Report of Investigations 7397, US Bureau of Mines*.

Canadian Industries Ltd (1968). *Blasters Handbook*, 6th edn.

Cook, M. A. (1956). *The Science of High Explosives*, Reinhold, New York.

Cook, M. A. (1974). *The Science of Industrial Explosives*, Ireco Chemicals.

⇢ Coursen, D. L. (1977). Cavities and gas penetration from blasts in stressed rock with flooded joints. *6th Int. Coll. on Gasdynamics of Explosives and Reactive Systems*, Stockholm.

Crookston, R. B. (1976). Mining oil shale. *Int. Joint Petroleum Engineering and Pressure Vessels and Piping Conf.*, Mexico City.

Crookston, R. B. and Weiss, D. A. (1980). Oil shale mining—plans and practices, *AIME Annula Meeting*, Las Vegas, Nevada, preprint 80–68.

Dayton, S. H. (1981*a*). For oil: a mining alternative, *E/MJ*, June, 61–7.

Dayton, S. H. (1981*b*). Union oil: a fast start on module one, *E/MJ*, June, 68–70.

Dayton, S. H. (1981*c*). Colony mobilizes for 1985 and 47,000 barrels a day, *E/MJ*, June, 71–7.

Dayton, S. H. (1981*d*). Cathedral Bluffs: a plan for 94,000 barrels a day, *E/MJ*, June, 78–84.

Dayton, S. H. (1981*e*). Rio Blanco pursues MIS tests as well as open-pit studies, *E/MJ*, June, 85–9.

Dayton, S. H. (1981*f*). Petrobras: a start on 50,000 barrels a day, *E/MJ*, June, 100–4.

East, J. H. and Gardner, E. D. (1964). Oil shale mining, Rifle, Colorado, 1944–56, *Bulletin 611, US Bureau of Mines*.

E. I. DuPont de Nemours & Co. (1969). *Blasters Handbook*, 15th edn.

Gauna, M. (1978). Personal communication.

→ Gauna, M. (1979). Toughness measurements for fragmentation assessment of oil shale, MS Thesis, Colorado School of Mines.

Haarberg, K. (1982). Private communication.

Hardy, M. P., Agapito, J. F. T. and Page, J. (1978). Roof design considerations in underground oil shale mining. *Proc. 19th US Symp. Rock Mech.*, Stateline, Nevada, pp. 370–7.

Hay, J. E. and Scott, F. H. (1965). Detonability of nitroglycerin contained in porous rock, *Nature*, **208**(5016), 1197.

⇢ Holmberg, R. (1981). Optimum blasting in bedded and jointed oil shale formations, *Final Report Under PL 92216, Laramie Energy Technology Center*, submitted to US Department of Energy.

Ireco Chemicals (1982*a*). Iregel emulsion, *Data Sheet E-1*.

Mining and fragmentation oil shale research

Ireco Chemicals (1982*b*). Iregel packaged slurry blasting agents, *Data Sheet S-1.*

Johansson, C. H. and Persson, P. A. (1970). *Detonics of High Explosives,* Academic Press, New York.

Keith, J. M., (1979). Surface seismic motion from room 5 blast (January 11, 1977), *Final Report for the Period November 1, 1976 through April 30, 1979, TID-28053/1,* appendix B. (Ed. R. A. Loucks.)

Leach, H. J. (1975). Analysis of methods for underground mining of oil shale, *Mining Cong. J.,* May, 33–42.

Lekas, M. A. (1979). Progress report on the Geokinetic horizontal *in situ* retorting process. *Proc. 12th Oil Shale Symposium,* ed. J. H. Gary, Colorado School of Mines Press, Golden, Colorado, pp. 228–36.

Lekas, M. A. (1981). The Geokinetics horizontal *in situ* retorting process. *Proc. 14th Oil Shale Symposium,* ed. J. H. Gary, Colorado School of Mines Press, Golden, Colorado, pp. 146–53.

Loucks, R. A. (1977). *Summary Report for the Period November 1, 1976 to October 31, 1977,* Vols 1 and 2 (Occidental Vertical Modified *In Situ* Process for the Recovery of Oil from Oil Shale—Phase I), *TID 28053,1 and 2.*

Loucks, R. A. (1978*a*). *First Quarterly Progress Report for the Period November 1, 1977 through January 31, 1978* (Occidental Vertical Modified *In Situ* Process for the Recovery of Oil from Oil Shale), *TID 28917.*

Loucks, R. A. (1978*b*). *2nd Quarterly Progress Report for the Period February 1, 1978 through April 30, 1978* (Occidental Vertical Modified *In Situ* Process for the Recovery of Oil from Oil Shale), *TID 28582.*

Loucks, R. A. (1978*c*). *3rd Quarterly Progress Report for the Period May 1, 1978 through July 30, 1978* (Occidental Vertical Modified *In Situ* Process for the Recovery of Oil from Oil Shale—Phase I), *TID 28943.*

Loucks, R. A. (1979). *Final Report for the Period November 1, 1976 through April 30, 1979* (Occidental Vertical Modified *In Situ* Process for the Recovery of Oil from Oil Shale—Phase I), prepared for US Department of Energy (*Contract No. DE-FC20-78LC10036*).

Lumpkin, R. E. (1980). Oxy's development of *in situ* oil shale retorting, In Situ *Energy Recovery Technology Meeting,* Albuquerque, New Mexico.

Mattila, E. (1982). Private communication.

McCarthy, H. E. and Cha, C. Y. (1976). Oxy modified *in situ* process development and update, *Quarterly, Colorado School of Mines (Proc. 9th Oil Shale Symposium),* **71**(4), 85–101.

McKinlay, I. (1982). Twin boom miner progress summary (private communication).

McNamara, P. H., Peil, C. A. and Washington, L. J. (1979). Characterization, fracturing and true *in situ* retorting in the Antrim shale of Michigan, *Proc. 12th Oil Shale Symposium,* ed. J. H. Gary, Colorado School of Mines Press, Golden, Colorado, pp. 353–65.

Merrill, R. H. (1954). Design of underground mine openings—oil shale mine, Rifle, Colorado, *Report of Investigations 5089, US Bureau of Mines.*

Mignogna, R. P. (1979). Conceptual design of a horizontal modified *in situ* oil shale retort, MS Thesis, Colorado School of Mines.

520　　William Hustrulid, Roger Holmberg and Ernesto Pesce

Miller, J. S., Walker, C. J. and Eakin, J. L. (1974). Fracturing oil shale for *in situ* oil recovery. *Report of Investigations 78874, US Bureau of Mines.*

Mitten, R. (1981). Union Oil Co. of California oil shale mining project, *Notes of Oral Presentation at Society of Mining Engineers, Fall Meeting,* Denver.

Nelson, R. M. (1982). *Final Report for the Period June 1, 1979 through August 31, 1981* (Occidental Vertical Modified *In Situ* Process for the Recovery of Oil From Oil Shale—Phase II), prepared for US Department of Energy (*Contract No. DE-FC20-78LC10036*).

Obert, L. and Merrill, R. H. (1948). Oil shale mine, Rifle, Colorado: a review of design factors, *Report of Investigations 5429, US Bureau of Mines.*

Parrish, C. (1982). Private communication. (White River shale project.)

Parrish, R. L., Boade, R. R., Stevens, A. L., Long, A., Jr. and Turner, T. F. (1980). The Rock Springs site 12 hydraulic/explosive true *in-situ* oil shale fracturing experiment, *SAND 80-0836, Sandia National Laboratories,* Albuquerque, New Mexico.

Petersen, C. (1979). Preliminary results of measurements of room 5 blasting (January 12, 1977 and January 28, 1977), *Final Report for the Period November 1, 1976 through April 30, 1979, TID 28053/1,* appendix B. (Ed. R. A. Loucks.)

Peterson, C. F. and Brown, C. S. (1982). Measurements of blasting for retorts 7, 8 and 8x, *Final Report by Systems, Science and Software for Occidental Oil Shale Inc., June 1981.* (Included as appendix B to *Final Report for the Period June 1, 1979 through August 31, 1981* by R. M. Nelson.)

Petrobras Co. (1982). Private communication.

Randle, A. C. and McGunegle, B. F. (1982). Union Oil Company's Parachute Creek shale oil program, *Proc. 15th Oil Shale Symposium,* ed. J. H. Gary, Colorado School of Mines Press, Golden, Colorado, pp. 224–30.

Reynolds, W. J. (1975). Mining considerations for *in situ* oil shale development, *Lawrence Livermore Laboratory, UCRL-51867.*

Ricketts, T. (1980*a*). Occidental's retort 6 rubblizing and rock fragmentation program, *Proc. 13th Oil Shale Symposium,* ed. J. H. Gary, Colorado School of Mines Press, Golden, Colorado, pp. 46–61.

Ricketts, T. E. (1980*b*). Formation of *in situ* oil shale retort with void at the top, *US Patent 4,238,136.*

Ricketts, T. E. (1981). Explosive expansion to a limited void with uniform scaled depth of burial, *US Patent 4,245,865.*

Ricketts, T. E. (1982*a*). Rubblization of Occidental's retorts 7 and 8, *Proc. 15th Oil Shale Symposium,* ed. J. H. Gary, Colorado School of Mines Press, Golden, Colorado, pp. 341–60.

Ricketts, T. E. (1982*b*). Blasting to a horizontal free face with mixing of fragments, *US Patent 4,326,751.*

Ricketts, T. E. (1982*c*). Method for forming an *in situ* oil shale retort, *US Patent 4,326,752.*

Ricketts, T. E. (1982*d*). Cratering in the deep cratering region to form an *in situ* oil shale retort, *US Patent 4,336,966.*

Ricketts, T. E. (1982*e*). Downhole delay assembly for blasting with series delay, *US Patent 4,347,789.*

Ricketts, T. E. (1982*f*). Method of blasting pillars with vertical blastholes, *US Patent 4,353,598.*

Mining and fragmentation oil shale research 521

Ricketts, T. E. (1983). Private communication.

Ricketts, T. E. and Redpath, B. B. (1982). Method of uniform rubblization for limited void volume blasting, *US Patent 4,333,684*.

Ridley, R. D. (1978). Progress in Occidentals' shale oil activities. In *Proc. 11th Oil Shale Symposium*, Colorado School of Mines Press, Golden, Colorado, pp. 169–75.

Ridley, R. D., and Chew, R. T. III (1975). *In situ* oil shale process development, *Quarterly, Colorado School of Mines (Proc. 8th Oil Shale Symposium)*, **70**(3), 123–7.

Rutledge, P. A. (1982). The prototype oil shale program—an update, *Proc. 15th Oil Shale Symposium*, ed. J. H. Gary, Colorado School of Mines Press, Golden, Colorado, pp. 210–23.

Schamaun, J. T. (1981). Lumped mass modeling of overburden motion during explosive blasting, *Proc. 14th Oil Shale Symposium*, ed. J. H. Gary, Colorado School of Mines Press, Golden, Colorado, pp. 53–60.

Sellers, J. B., Haworth, G. R. and Zambas, P. G. (1971). Rock mechanics research on oil shale mining, *SME Preprint 71-AM-232*.

Shaler, J. E. (1982). An update on modified *in situ* retorting of oil shale, *1982 American Mining Congress International Mining Show*, Las Vegas, Nevada.

Snyder, G. B. and Pownall, J. R. (1978). Union Oil Co.'s Long Ridge experimental oil shale project, *Proc. 11th Oil Shale Symposium*, ed. J. H. Gary, Colorado School of Mines Press, Golden, Colorado, pp. 158–68.

Stevens, A. L., Lysne, P. C. and Griswold, G. B. (1975). Rock Springs oil shale fracturization experiment, experimental results and concept evaluation, *SAND 74-0372, Sandia National Laboratories*, Albuquerque, New Mexico.

Thomas, H. E., Carpenter, H. C. and Sterner, T. E. (1972). Hydraulic fracturing of Wyoming Green River oil shale: field experiments, Phase I, *Report of Investigations 7596, US Bureau of Mines*.

Weakley, L. A. (1982). Private communication.

Zambas, P. G., Haworth, G. R., Brackebusch, F. W. and Sellers, J. B. (1972). Large scale experimentation in oil shale, *Trans. SME*, **252**, 283–9.

Chapter 12

Temperature Effects

JOEL DuBow

Professor, Electrical and Computer Engineering Department,
Boston University, Massachusetts, USA

SUMMARY

The static and dynamic mechanical properties of oil shale exhibit consid-
erable temperature dependence. The strength and acoustical velocities of
oil shale decrease with increasing grade and increasing temperature. The
strength and acoustical velocities increase with increasing confining
pressure. Heating rate and ambient gas atmosphere do not significantly
affect the strength of oil shale.

A significant loss of strength occurs between room temperature and
150°C arising from the loss of volatile hydrocarbons from the sample. A
further loss in strength and strength and velocity minimum occurs around
350°C arising from a physical delamination of the shale and plastic
flow. Between 350 and 450°C kerogen composition results in recementa-
tion and a resulting increase in acoustical velocities and mechanical
strength. Between 450 and 500°C, a second order increase also occurs.

The variations in acoustic velocity with temperature, grade, pressure,
void volume and thermal history provide the basis for a novel indirect
diagnostic of oil shale retorts. The variation in strength with temperature
indicates that, for temperatures below 100°C, conventional mine design
techniques apply but at temperatures much exceeding 100°C the shale
weakens considerably and the computations of strength become much
more difficult. At those elevated temperatures, mechanical support from
the shale can only warily be relied upon.

524   *Joel DuBow*

12.1. INTRODUCTION

Shale oil extraction technology will benefit from an empirical and theoretical understanding of the variation with temperature of the mechanical properties of oil shale. All present day extraction technologies are based on the thermal process of the retorting of oil shale. The most significant step in this process is the decomposition of kerogen, the solid carbonaceous material contained in oil shale rock, into oil, gas and residual carbon. This oil and gas must flow through the process bed or retort, and be collected and condensed outside the retort. The retort bed consists of a fractured bed of oil shale rock. Efficient product extraction requires that adequate permeability (10–25%) be maintained throughout the course of processing (Tisot and Sohns, 1971). In order that this degree of permeability be maintained the oil shale rocks comprising the retort rubble must retain enough structural strength during and after retorting to support a load at least equal to the peak pressure exerted by the rubble overburden (Carley, 1976). This pressure can often exceed that expected from the height, diameter and particle size in the retort because the irregular shape of individual rocks leads to excessive stress concentration at points of contact.

Underground or *in situ* retorting commonly involves retorting large rooms separated by pillars. Because of the sedimentary nature of oil shale deposits, these pillars are of necessity composed of material identical to the oil shale being retorted. Heat from the retort bed heats the pillars during retorting. The temperature profile in a pillar, arising from heat conduction from the retort (an exterior, moving boundary value problem), has been solved by Wilhelm *et al.* (1979). This heating can significantly weaken the pillar wall to the point of mechanical failure, with potentially disastrous human, economic and environmental costs. Moreover, resource recovery maximization considerations constrain pillar thickness to the minimum required to retain retort structural shape, both to support the retort roof structurally as well as to prevent the escape from the retort of gases and oils formed during retorting through thermally induced cracks in the pillar wall. Understanding the temperature dependence of the mechanical properties is critical for calculating optimal pillar sizes and support capabilities for the range of conditions likely to be encountered in *in situ* retorts. For above ground retorting, an understanding of the temperature dependence of the mechanical properties is needed to predict the onset of permeability plugging caused by the flow or collapse of the rubble particles (Tisot, 1975), thereby permitting estimation of the maximum vessel size as well as upper and lower limits in rubble size. These properties are also useful for designing retort process diagnostic tools.

The interior of a retort is subject to elevated temperatures, and an atmosphere containing hydrogen, oxygen, CO_2, H_2S, sulfur, nitrogen and combustible gases. These process stream conditions are commonly denoted 'extreme' and, combined with a large process bed and possibly flowing and exfoliating massive rocks, make process diagnostics a rather challenging problem. Direct process diagnostics (DuBow, 1980a), defined as those using sensors in physical contact with the formation, suffer from susceptibility to failure caused by the extreme conditions and unrepresentative data arising from boundary layers near the retort walls or large shale rubble particles. In this regard, the heat flow, mass flow and kinetic equations comprising the process simulation equations are sets of one-dimensional equations and thus require profiles of the process variables. Only these profiles permit process diagnostics to detect lateral non-uniformities and estimate volumetric average quantities. Indirect diagnostic measurements (DuBow, 1980a) are those made external to the material being processed or retort walls. These include electrical, acoustical and seismic measurement techniques and provide volume averages of the observables rather than point estimates, causing difficulty in relating the measurements, which yield profiles of constitutive properties, to the chemical, density and concentration variables used in process simulation equations (Jones, 1980). Research in the author's laboratory (DuBow, 1980b) has focused on providing an experimental and analytical base of constitutive property (electrical, acoustical and thermal impedances (conductivity and capacity)) data as a function of composition, temperature, ambient atmosphere and chemical reaction sequences. The variation of the mechanical and acoustical properties with temperature are needed to model the propagation processes encountered in ray or wave formulations of indirect process diagnostic techniques and for interpreting the data obtained in these measurements.

Since oil shale is a complex, heterogeneous material, the measurement of many different properties over a range of temperatures, pressures and ambient atmospheres is required to predict the behavior of oil shale from a given formation during various aspects of the shale oil extraction process sequence. These types of measurements are commonly called thermophysical characterizations and have been studied extensively in the author's laboratory (DuBow, 1980a, b)

In this chapter a review of the temperature dependence of the mechanical properties of oil shale is presented. Section 12.2 reviews mechanisms underlying the variation of the mechanical properties of rocks with temperature. Section 12.3 discusses the techniques used to measure these properties for oil shale. The next section summarizes the available experimental measurements of the temperature dependence of the mechanical properties. The chapter concludes with a discussion of the use of these data in applications to pillar stability, maximum column height and retort diagnostics.

12.2. MECHANISMS UNDERLYING THE TEMPERATURE DEPENDENCE

The temperature dependence of the mechanical, or elastic, properties of rocks is determined by a combination of intrinsic and extrinsic variables. Intrinsic variables are those based upon the physical and chemical properties inherent to minerals comprising the rock such as bond strength, chemical reaction parameters and the quantity and form of water or other fluids contained in the rock. Extrinsic variables are independent of the physical or chemical properties of the various minerals. They include structural and geological properties such as defects, cracks, porosity, permeability, fluid saturation and the sedimentary or igneous nature of the formation. Generally, the more highly consolidated the rock, the more intrinsic variables dominate the properties.

The temperature dependence of the mechanical properties of oil shale exhibits both intrinsic and extrinsic elements. The low temperature mechanical properties show the influence of thermal expansion of minerals and pore pressure while the high temperature properties are highly influenced by the plastic flow and decomposition of kerogen.

Also, while well defined experimental conditions are important for all mechanical properties measurements, they are particularly important for elevated temperature measurements. For example, differential thermal expansion of oil shale rock constituents and the resulting exfoliation of the rock, the oxidation of kerogen or lack thereof, and the influence of cracks and pores on the mechanical properties of the

Temperature effects 527

material make experimental control especially important in ambient conditions.

The minerals which compose most rocks decompose according to a functional relationship, f, of the Arrhenius type:

$$k = f[A \exp (E_a/KT)] \tag{12.1}$$

where k is the reaction rate constant, A is called the frequency factor or 'infinite temperature' rate constant, E_a is the activation energy, K is Boltzmann's constant and T is the absolute temperature (Eyring *et al.*, 1980).

As the temperature is increased, the various forces binding molecules comprising the rock decrease. This causes an intrinsic decrease in the static and elastic moduli of the rock. However, these intrinsic mechanisms are usually overshadowed by extrinsic and structural factors in experimentally observed variations of rock mechanical properties with temperature.

Included in these extrinsic variables are porosity, fluid or water saturation and attendent pore pressure, structural defects, such as cracks and joints, and stress arising from differences in thermal expansion between various constituents of a rock (Rzhevsky and Novik, 1981; Agapito and Hardy, 1982). Oil shale has additional complications arising from the existence of compounds such as nahcolite and dawsonite which decompose at temperatures below the 450°C temperature commonly used for oil shale processing (Stagg and Zenkiewicz, 1968). Mineral decompositions alter total porosity and pore pressure in a complex manner. The fine-grained sedimentary structure of oil shale, in which kerogen rich sedimentary layers, or varves, are often interleaved with lean layers for a given average richness, complicates the extension of the laboratory data to field conditions. Finally, since practical retorts are fractured packed beds, extending measurements on consolidated cores to heterogeneous media is necessary for practical application of these data in commercial retort technology.

The elastic moduli of inorganic minerals tends to behave ideally elastically, wherein changes in the elastic moduli can be attributed to changes in temperature within the mineral. However, kerogen flows plastically at elevated temperatures. As a result, the temperature dependence of the elastic properties of oil shale tends to exhibit both an elastic component and a plastic component (elasto-plastic) which arise from plastic phenomena occurring at the organic–inorganic

528   *Joel DuBow*

boundaries. The combination results in creep phenomena being superimposed on thermoelastic behavior. These characterizations are discussed in more detail by Dimitryev *et al.* (1972), Jaeger and Cook (1976) and Sladek (1980). Jaeger and Cook (1976) have provided a standard text on the mechanical properties of rocks, which treats elasticity and plasticity in an interesting manner by using electrical analogies. A linearly elastic, or Hookean substance may be modeled as a spring (resistor):

$$T = Ye \qquad (12.2)$$

where T is the applied stress, e the resulting strain and Y the stiffness (Young's modulus in uniaxial compression). A perfectly viscous or Newtonian substance may be treated as a dashpot (capacitor):

$$T = \dot{V}e \qquad (12.3)$$

where V is the viscosity and the dot denotes a derivative with respect to time. Combining these resistors and capacitors in readily analyzed network configurations facilitates the derivation of the most commonly observed linear and even non-linear materials which combine instantaneous strain, transient creep, and steady state creep (Burgers and Bingham substances) (Jaeger and Cook, 1976). An electrical model, based on networks, or their graph theoretical extensions, of long time, elevated temperature and mechanical creep behavior, could be a useful contribution to understanding the mechanical behavior of oil shale.

The multiplicity of influences on the temperature dependent mechanical properties of oil shale are fortunately not equally significant for a specific sample. However the number of variables are indicative of the care required in making measurements or in interpreting reported data. The number of variables which can influence the mechanical properties of a particular oil shale is a source of the often heard cautionary note that these and other properties are 'site specific'.

While it is attractive to try to identify a single mechanism as dominating the mechanical property variations, in practice it is often expedient to make a numerical curve to fit the data, derive equations and estimate the size of the coefficients of the various terms. In a recent study of the mechanical properties of oil shale as a function of temperature (Ozdemir *et al.*, 1981) a multiple linear regression analysis was performed using a number of parameters. While this identified significant variables, the data were not readily amenable to a more refined mechanistic interpretation.

Temperature effects

Despite all the potential complexity, a few general trends are common to most porous sedimentary rocks. Since most rocks contain pores and cracks and since increased confining pressure tends to seal those pores and cracks, the elastic moduli tend to increase with increasing confining pressure (Terzaghi, 1945). The elastic moduli should decrease with increasing porosity, since the pores may be considered to act as 'cracks' in the material.

In raw oil shale most of the pores are either kerogen- or moisture-filled. While extremely lean oil shales (less than 30 liters $tonne^{-1}$) may exhibit porosities of up to 10%, most oil shales exhibit less than 1% measurable porosity (Baughman and Hedrickson, 1978). On heating, various forms of water (pore water, free water, water of hydration) and light hydrocarbons, vaporize and, at higher temperatures, the kerogen swells and ultimately decomposes.

The primary purely mechanical effects of heating are internal stress caused by differing thermal expansion rates of the various constituent minerals and pore pressure increases which act as offset applied stresses caused by moisture and gas expansion. The general form of the thermal stress caused by differential thermal expansion between two materials (P_t) 1 and 2 is of the form:

$$P_t = Y(B_1 - B_2)Tf(s) \qquad (12.4)$$

where Y is the Young's modulus, $(B_1 - B_2)$ is the difference in thermal expansion coefficients, T is the temperature change and $f(s)$ is a function of the contact area between the two media.

The pore pressure effects are more difficult to estimate since they depend upon the permeability of the specimen and act internally in the rock (Spencer and Nur, 1976). These internal pore pressures apply a force in a direction opposite to the applied stress. They are often treated using the concept of effective stress wherein, as suggested by Terzaghi (Petit, 1976):

$$P_e = P_c - P_p \qquad (12.5)$$

where P_e = effective stress, P_c = confining pressure, P_p = pore pressure.

There have been numerous studies of the effect of pore pressure on the mechanical properties of rocks. Useful background to the study of oil shale elastic properties is provided by Spencer and Nur (1976) in which lithium niobate transducers $LiNbO_3$ of the type used in our laboratory by Petit (1976) and later by Mraz (1981) are described in a study of pore pressure induced inversions of expected sonic velocity versus depth profiles in westerly granites.

The next section describes measurements and available data for oil shale. These data indicate the following features of the variation of elastic properties with temperature. As oil shale is heated, the thermal stress will act to reduce the elastic moduli. Further heating will decompose light hydrocarbons (Lonvik et al., 1980) and drive off moisture (typically between 0·4 and 3%) (Closman and Bradley, 1979), increasing the porosity. The increased porosity will further decrease the elastic moduli. At higher temperatures the plastic flow of

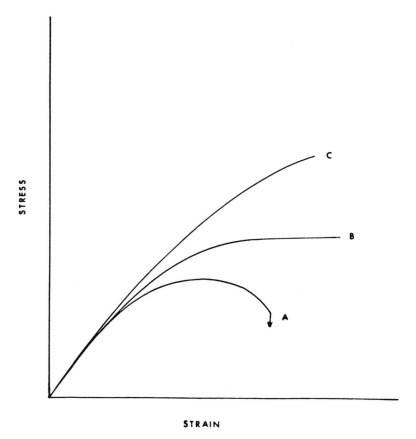

Fig. 12.1. Stress–strain characteristics for a typical rock at low (A), medium (B) and high (C) confining pressure.

Temperature effects   531

kerogen into voids and pores will lead to a more pronounced, almost anomalous, decrease in elastic properties. As the shale is further heated, compaction caused by confining stress will decrease the porosity and stabilize the elastic moduli. This is seen even at confining stresses of 345 kN m^{-2} (50 psi). However, at temperatures exceeding 350°C the kerogen will begin to decompose, leading to further porosity increases and elastic moduli reduction. At temperatures of 500°C and above, or for times long enough for the kerogen to decompose, confining stress will result in compaction, decreased porosity and an increase in elastic moduli. Even without compaction, recementation of the mineral matrix will occur for temperatures high enough for kerogen decomposition to occur.

Confining pressure tends to increase flow before rupture of the material and tends to significantly increase the rupture strength of the material. Thus, in Fig. 12.1, curve A is typical of a rock at low confining pressure. Curve B is characteristic of an elasto-plastic rock and curve C characteristic of a rock under high confining pressure. This latter situation is characterized by poorly defined yield stress and strain hardening.

12.3. STATIC AND DYNAMIC MEASUREMENT TECHNIQUES

The elastic moduli are usually defined to include Young's modulus, shear modulus, Poisson's ratio and, in many cases, the bulk modulus and compressibility. Measurements of these properties for rocks are complicated by the non-linearity of the stress–strain curves and the observation that measured moduli are dependent upon the rate, duration, magnitude and geometry of loading. Figure 12.2 is a typical stress–strain curve for a rock, with two types of Young's modulus defined.

The secant modulus (large deflection modulus) is the value observed in standard testing machines where a sample is stressed until failure and the resulting deformations observed. The tangent modulus (or small deflection modulus) is obtained using ultrasonic (pulse or resonance) techniques. For a highly elastic material the tangent and secant moduli are nearly equal and will be approximately equal even if the tangent value is measured at low or zero load.

Figure 12.2 points up the origin of the noted caveat that elastic

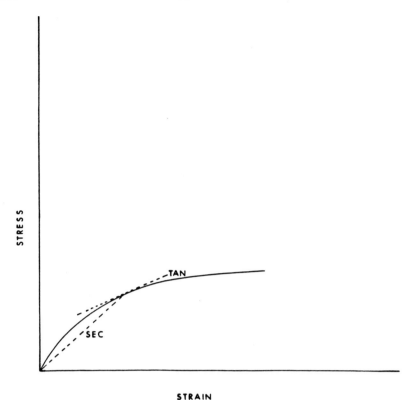

Fig. 12.2. Typical stress–strain curve exhibiting the small deflection (tangent) and large deflection (secant) definitions of Young's modulus.

constants are meaningful only in the context of specified measurement conditions. For example, the ultrasonic technique at low load will overestimate the Young's modulus compared to an estimate obtained from the secant modulus. However, at high applied stress values the tangent value will be lower than the secant value.

Furthermore, most rocks exhibit a reduced modulus if the stress rate (stress increase per second) is increased substantially. If an application is concerned with the load bearing ability of a rock, the static high deformation measurements described in this chapter, or ultrasonic method, are used to characterize the material. For applications to fracture, blasting and rubblization, shock wave derived moduli are used.

At low applied stress values, pore spaces and cracks exert a larger effect on sonic measurements (where they act as scattering centers (or discontinuities with high acoustic impedance)) than they do on large deformation measurements. The ultrasonic velocity data on oil shale were obtained at low values of applied stress (345–690 kN m^{-2}, 50–100 psi) and thus follow the trends, commonly observed in published rock data, that the static values of Young's modulus are in general lower than those obtained by sonic methods and the static Poisson's ratios are typically higher. Raw oil shale tends to be a highly consolidated and not too porous material (0·5–3%), so that only low values of pressure are sufficient to provide good contact between the acoustic transducer and the shale to obtain reasonable acoustic velocity data. At temperatures where porosity in oil shale is increased by loss of water and/or kerogen, the Poisson's ratio obtained from ultrasonic measurements can become negative because of the effect of pore spaces and heterogeneities on sonic velocities and the sensitivity of the ratio to small variations in shear (s) wave or compressional (p) wave velocities.

For consolidated rocks the static and dynamic moduli tend to agree well (within 10% or less). However, for rocks with cracks, fractures and cavities, differences as large as 300% have been reported (Clark, 1966).

It is therefore difficult to overemphasize the importance of careful sample selection and well characterized experimental set-ups for measuring temperature dependent elastic properties. Most measurements are made on cylindrical cores. It is important that the cores be as crack-free and as uniform in grade (kerogen distribution) as possible.

There have been many techniques developed to measure the elastic properties of rocks. The interested reader is referred to Jaeger and Cook (1976) for a review and comparison of the primary techniques. Tables 12.1 and 12.2 provide definitions of the elastic moduli and equations for extracting the elastic moduli from ultrasonic velocities.

Oil shale is transversely isotropic. Therefore, six elastic coefficients are required to completely characterize the elastic moduli. A comprehensive study of the temperature dependent elastic properties has not yet, to our knowledge, been published. Two possible reasons for this are the relative paucity, until recently, of data on the temperature dependent elastic properties and that the complexity of applying these data to field conditions would require such a study to be part of a large

534   *Joel DuBow*

TABLE 12.1
Definitions of Elastic Moduli

$$E = \text{Young's modulus} = \frac{\text{Longitudinal stress}}{\text{Longitudinal strain}} = \frac{\text{Stress}}{l/l_0} \quad (\text{dynes} \times \text{cm}^{-2})$$

$$G = \text{Modulus of rigidity} = \frac{\text{Shearing stress}}{\text{Shear deformation}} \quad (\text{dynes} \times \text{cm}^{-2})$$

$$\nu = \text{Poisson's ratio} = \frac{E}{2G} - 1 = \left(\frac{\text{Transverse contraction}}{\text{Longitudinal extension under tensile stress}}\right)$$

$$K = \text{Bulk modulus} = \frac{1}{V_0} \times \frac{dV}{dP} = \text{Relative change of volume with applied pressure}$$

oil shale development. The current energy supply climate has delayed these large scale projects for an indeterminate time. The existing data concerning the temperature dependence of oil shale elastic properties comes from three groups. Closman and Bradley (1979) at Shell studied shales where nahcolite and dawsonite were present. Ozdemir *et al.* (1981) of the Colorado School of Mines and Colorado State University, studied Green River oil shale from Anvil Points and Debeque, Colorado which did not contain significant quantities of nahcolite or dawsonite. These two studies utilized triaxial testing apparatus to determine Young's modulus and the Brazilian technique to determine tensile strength. A third study, at Colorado State University (Petit, 1978; Mraz *et al.*, 1980), utilized ultrasonic pulse propagation techniques to determine Young's modulus, Poisson's ratio and bulk modulus.

TABLE 12.2
Elastic Moduli from Sonic Velocities

$$E = V_s^2 \left[\frac{3V_p^2 - 4V_s^2}{V_p^2 - V_s^2} \right]$$

$$G = \rho V_s^2$$

$$K = \frac{3V_p^2 - 4V_s^2}{3}$$

$$\nu = \frac{V_p^2 - V_s^2}{2(V_p^2 - V_s^2)}$$

where ρ = density, V_p = longitudinal wave velocity, V_s = shear wave velocity.

12.4. ANALYSIS OF OBSERVED PROPERTY VARIATIONS

The variation with temperature of the structural properties of oil shale was initially studied by P. R. Tisot and H. W. Sohns at the (then) US Bureau of Mines in Laramie, Wyoming (Tisot, 1975). Their work demonstrated the severe reduction in strength and subsequent plastic flow of oil shale in various configurations of significance to oil shale retorting. For example, Fig. 12.3 shows initial decrease and subsequent increase in the ability of oil shale to withstand a load.

Tisot also demonstrated that plastic flow of the rock, combined with coking of shale, could result in permeability plugging *in situ* and possibly even above ground retorts. This result was inferred from

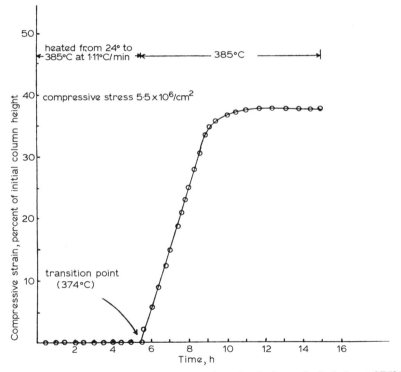

Fig. 12.3. Typical strain–time result showing plastic flow of oil shale at 375°C and subsequent stabilization of mechanical properties deriving from recementation.

536 *Joel DuBow*

observations of alterations with temperature of oil shale contained within a tightly sealed chamber containing two pieces of oil shale separated by glass beads. The beads provided permeability and represented propped fracture structures in *in situ* retorts. The resulting flow around the beads and escape of oil, gas and rock fines presaged a serious problem of maintaining permeability in *in situ* retorting, a problem which remains to be solved.

This work was performed with extreme care and provided information of importance to engineers working in the field. However, it was not a study of the fundamental mechanical constitutive properties and was thus not readily applied to numerical simulations of mechanical behavior for commercial scale configurations.

Closman and Bradley (1979) of Shell Development Corp., Houston, Texas studied the temperature dependence of the tensile strength, compressive strength and Young's modulus of oil shale. Their focus was on the leached and saline zone shales of significance to Shell's wet process recovery techniques. Most shales in this study were from nahcolite bearing intervals in the Green River Formation. However, the core samples studied included nahcolite bearing and nahcolite free samples. The cores were $12 \cdot 7$ mm ($\frac{1}{2}$ in) diameter disks, $6 \cdot 4$ mm ($\frac{1}{4}$ in) thick.

Tensile strength was measured under uniaxial stress, using the Brazilian indirect measurement technique on an Instron machine which was loaded at a constant platen displacement of $0 \cdot 051$ mm min^{-1} ($0 \cdot 002$ in min^{-1}). The Brazilian test results do not provide a measurement directly yielding Young's modulus. The qualitative variation of the Young's modulus has to be inferred from the stiffness (load–displacement) curve of the sample.

The compressive strength was measured in a triaxial test apparatus in which the load at a fixed temperature was increased to the point of failure. Measurements on both tests were made from room temperature up to a maximum of 260°C. Data were obtained both parallel and perpendicular to the bedding planes. Figure 12.4 shows reference loading and acoustic signal propagation directions for the measurements described in this work. The 'horizontal' loading direction is interchangeably used with 'parallel' loading direction in the text.

The confining pressure was varied between 0 and 609 MPa. Since only axial, and not radial, deformations were measured, no direct estimate of Young's modulus could be obtained. An estimate of the change in modulus was made by relating changes in stiffness (from

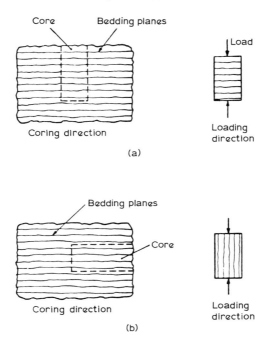

Fig. 12.4. Coring and loading directions for the vertical and horizontal cores used in the mechanical property tests for oil shale. (a) Vertical cores; (b) horizontal cores.

load–strain curves) directly to changes in Young's modulus. Young's modulus was estimated from the steeply sloping load and large displacement portions of the stress–strain curve. The inaccuracy inherent in this estimate is indicated by the absence of variation of Young's modulus with grade of shale.

In a more recent study, by Ozdemir et al. (1981), both axial and radial deformations were measured as a function of load during triaxial tests. In this work the Young's modulus and Poisson's ratio were directly measured, along with the compressive and tensile strengths. Their results did reveal a dependence of compressive strength and Young's modulus on grade. Increasing kerogen content resulted in a reduction in strength and a concomitant increase in Poisson's ratio. This study utilized samples from a different region of the Green River Formation than did the Shell group, but the results should be comparable since the Shell group studied nahcolite-free samples.

12.4.1. Tensile Strength

A comparison of these data illustrates the main features of the temperature variation of the compressive and tensile strengths of oil shale. Tensile strengths parallel and perpendicular to the bedding planes are given in Figs 12.5 and 12.6. Both the vertical and horizontal cores exhibit a decreasing tensile strength with temperature. Both sets of

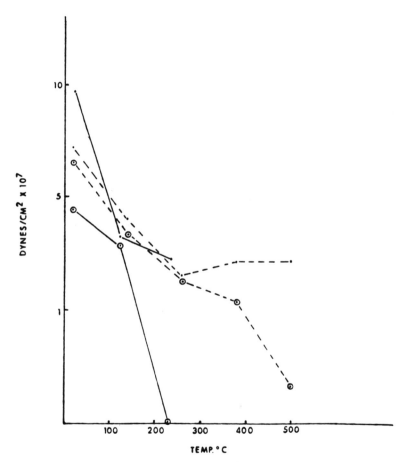

Fig. 12.5. Tensile strength measured parallel to the bedding planes as a function of temperature with grade as a parameter. (·——·) 18·1 gallons ton^{-1}, Closman and Bradley (1979); (·----·) 20 gallons ton^{-1}, Wang et al. (1979); (⊙——⊙) 42·4 gallons ton^{-1}, Closman and Bradley (1979); (⊙---⊙) 40 gallons ton^{-1}, Wang et al. (1979).

Temperature effects 539

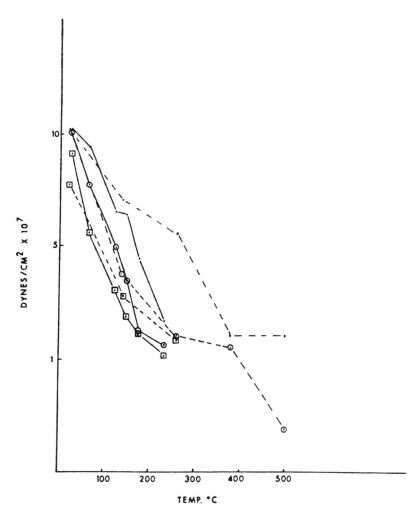

Fig. 12.6. Tensile strength measured perpendicular to the bedding planes as a function of temperature with grade as a parameter. (·——·) 15·6 gallons ton^{-1}, Closman and Bradley (1979); (·----·) 15 gallons ton^{-1}, Wang et al. (1979); (⊙——⊙) 28 gallons ton^{-1}, Closman and Bradley (1979); (⊙---⊙) 25 gallons ton^{-1}, Wang et al. (1979); (□——□) 42·4 gallons ton^{-1}, Closman and Bradley (1979); (□---□) 40 gallons ton^{-1}, Wang et al. (1979).

540 *Joel DuBow*

data exhibit a pronounced decrease in tensile strength with temperature. The major loss of strength occurs by 150°C. It is also noteworthy that the sample containing nahcolite (75·5 liters tonne^{-1}, 18·1 gallons ton^{-1}) was considerably weaker than those without.

The monotonicity of these data made them readily amenable to curve fitting. Closman and Bradley (1979) used a multiple regression technique while the Colorado School of Mines (CSM) used stepwise regression starting with single variable fits and progressing to more complicated expressions, until further terms added little to the goodness of fit relative to the increased complexity involved.

For vertical cores (failure across bedding planes), the equations are:

$$S_t = 3105 - 425 \cdot 7 \ln T - 18 \cdot 46G \quad \text{(Ozdemir } et \ al., 1981) \tag{12.6}$$

$$S_t = 288 + \frac{7 \cdot 79 \times 10^8}{T_r^2} - 19\ 300\ \frac{G}{T_r} - 0 \cdot 299G^2$$

$$\text{(Closman and Bradley, 1979)} \quad (12.7)$$

For horizontal cores (failure along the bedding plane), the equations in the two studies are:

$$S_t = 1666 - 214 \ln T - 1 \cdot 416G \quad \text{(Ozdemir } et \ al., 1981) \tag{12.8}$$

$$S_t = -25 \cdot 3 + \frac{6 \cdot 48 \times 10^8}{T_r^2} - 2 \cdot 983 \times 10^4\ \frac{G}{T_r} + 0 \cdot 386G^2$$

$$\text{(Closman and Bradley, 1979)} \quad (12.9)$$

where S_t = tensile strength (psi), G = grade (gallons ton^{-1}), T = temperature (°C), T_r = Rankine temperature.

The more complicated equations in the Shell model are a better fit to their data which, however, only extend up to 260°C. However, the CSM equations were focused on elucidating qualitative trends over the full 450–500°C temperature range. A statistical test for significance of a given variable is the F coefficient. The magnitude of this coefficient, as well as the weighted residuals test described in Ozdemir *et al.* (1981), indicate that grade is not as significant as temperature in determining the variation with temperature of the tensile strength. However, Closman and Bradley (1974) find a much more pronounced variation of the tensile strength with grade than do the CSM group, whose equations predict a linear decrease in tensile strength with grade. One reason for this less pronounced grade dependence could be sample non-uniformity. However, it is seen that higher grades exhibit a

Temperature effects

more rapid decrease of strength with temperature since macroscopic oil shale samples are almost always non-uniform to some extent and thus result in data scatter; the standard deviation about mean values obtained in the 5–10 sample replications reported in the CSM data run from 20% to as high as 50%. Because of the non-uniform distribution of kerogen in the sample the quoted values of average grade may include sections with much higher and much lower concentrations. Tensile strength tests highlight the consequences of this non-uniformity.

It is thus seen that tensile strength decreases significantly as a function of temperature. Higher grades exhibit a more rapid decrease in strength with temperature. The low temperature tensile strength is higher for loading perpendicular to the bedding planes than the tensile strength parallel to the bedding planes. At temperatures exceeding 150°C this relationship reverses, with the vertically loaded cores (perpendicular to the bedding plane) losing virtually all the tensile strength at temperatures above 380°C. Finally there was no observed effect of heating the atmosphere, either nitrogen flushed or oxygen available, on the tensile strength.

12.4.2. Compressive Strength

The Shell and CSM studies are again the principal sources of data regarding compressive strengths. The Shell data are confined to a room temperature to 260°C temperature range. The CSM data base is far more comprehensive and extends to nearly 500°C.

The compressive strength (failure stress) perpendicular to the bedding planes as a function of temperature, with grade and pressure as parameters, is shown in Fig. 12.7. Both Shell and CSM data show qualitatively similar trends. The resulting curve fits for perpendicular compressive strength are given as:

$$S_c = 3 \cdot 48 \times 10^5 T^{-0 \cdot 615} \times \exp{(3 \cdot 99 \times 10^{-4} P - 0 \cdot 0214 G)} \quad \text{(Shell)}$$

$$\tag{12.10}$$

$$S_c = 10 \cdot 75 + 1 \cdot 86 / G - 2 \cdot 40 \ln T + 0 \cdot 0040 P) \quad \text{(CSM)} \tag{12.11}$$

where S_c = compressive strength (psi), G = oil shale grade (gallons ton^{-1}), T = temperature (°C), P = confining pressure (Pa).

For convenience, the equation derived in Ozdemir *et al.* (1981), for horizontal cores comprehensive strength, is given below:

$$S_c = 12 \cdot 9 + 212 \cdot 6 / G - 2 \cdot 85 \ln T + 0 \cdot 0045 P \tag{12.12}$$

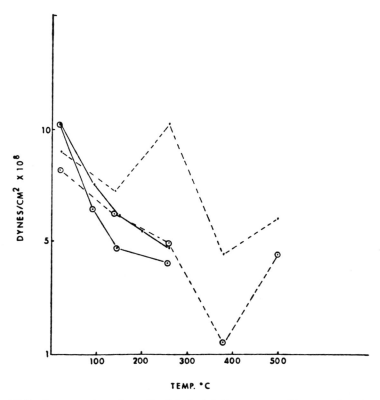

Fig. 12.7. Compressive strength (labeled failure strength) as a function of temperature for cores loaded perpendicular to the bedding planes. (·———·) 19·2 gallons ton^{-1}, 500 psi, Closman and Bradley (1979); (·— — —·) 20 gallons ton^{-1}, 500 psi, Wang et al. (1979); (⊙———⊙) 19·2 gallons ton^{-1}, 100 psi, Closman and Bradley (1979); (⊙— — —⊙) 20 gallons ton^{-1}, 50 psi, Wang et al. (1979).

The similarity of the equations for compressive strength parallel and perpendicular to the bedding planes point to similar mechanisms underlying the compressive strength behavior at elevated temperatures.

An interesting feature arises in vertical core compressive strength data. The shale initially loses strength, rapidly from room temperature to 140°C, and less rapidly thereafter up to 380°C. Above 380°C up to the onset of retorting, the shale regains some compressive strength. Some shales exhibit a local peak about 260°C.

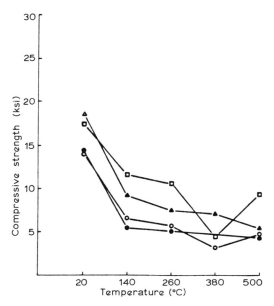

Fig. 12.8. Compressive strength as a function of temperature for 40 gallons ton^{-1} horizontal (parallel) loaded oil shale with pressure as a parameter. Confining pressures: (●) 50 psi; (○) 500 psi; (△) 1000 psi; (□) 1500 psi.

Similar, although less pronounced, manifestations of this type of behavior are exhibited in the horizontal core compressive strength depicted in Fig. 12.8. This high temperature strengthening was also observed earlier by Tisot and Sohns (1971) and DuBow (1980a, b) in a number of the constitutive properties of oil shale as part of a broad based thermophysical properties study. The amount of strength regained decreases as the grade increases.

An examination of the effects of grade on the temperature dependent compressive strength is given in Figs 12.9 and 12.10. These data indicate the monotonic decrease in compressive strength with the increasing grade. The data from this and earlier figures also demonstrate the monotonic increase in compressive strength with increasing confining pressure.

An analysis of the data and curve fit equations above shows that three variables dominate the temperature dependence of the compressive strength: temperature, confining pressure and grade. For vertical

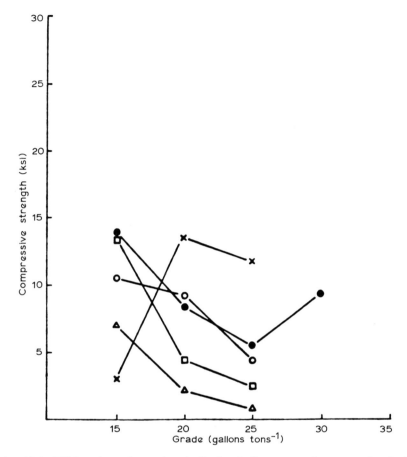

Fig. 12.9. Effect of grade on (vertically loaded) compressive strength of oil shale for various temperatures at 50 psi confining pressure. (×) 20°C; (●) 140°C; (○) 260°C; (△) 380°C; (□) 500°C.

cores the temperature variation had the largest effect on compressive strength, followed by grade and lastly by pressure.

For horizontally oriented cores the grade was the most significant variable, followed by temperature and lastly by confining pressure. This dependence may be viewed in terms of the relative effect on horizontal and vertical strength measurements of rich sections of a core. It is known that for grades greater than 146 liters tonne^{-1} (35

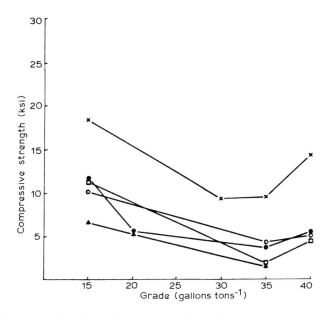

Fig. 12.10. Effect of grade on (horizontally loaded) compressive strength of oil shale for various temperatures at 50 psi confining pressure. (×) 20°C; (●) 140°C; (○) 260°C; (▲) 380°C; (□) 500°C.

gallons ton^{-1}), the kerogen forms a continuous phase and an inorganic matrix may be considered as being suspended in a matrix of the organic material. For grades less than 125 liters tonne^{-1} (30 gallons ton^{-1}) the reverse is true. Ultimate strength, limited by failure and buckling, could be more affected by buckling failure of a rich seam perpendicular to the varves than a loss of strength of sections of a core loaded parallel to the bedding plane. However, the effect on elastic properties, discussed below, provides a useful contrast since thin sections of rich material have a smaller relative impact on acoustic velocity data.

The samples tested by Ozdemir et al. (1981) did not contain any detectable nahcolite or dawsonite. These compounds weaken the material considerably, both at room temperature and elevated temperature (Closman and Bradley, 1979). As with the tensile strength, heating the atmosphere, either nitrogen flushed, or oxygen available, had no significant effect on the strength of oil shale. Further results of

546 *Joel DuBow*

this study indicate that heating time, once thermal equilibrium was attained, had little effect on strength.

While grade was the most significant variable determining compressive strength for horizontal cores, the absolute reduction in strength with increasing grade was greater in vertical cores. For both orientations, the compressive strength decreased with temperature. The major loss of strength occurred on heating from ambient to 140°C. Some strength was regained between 380 and 500°C, possibly as a result of pore pressure caused by evolving gases and fluids and the incompressibility of fluid filled pores, along with recementation of the material after kerogen removal. Consistent with this hypothesis is the observation that, while confining pressure increased strength for all temperatures and grades evaluated, the relative increase was much greater at high temperatures, where strength was quite low. Finally, it was observed that horizontally loaded cores exhibited larger triaxial compressive strengths than vertically loaded cores. This result could arise from a combination of the effects of rich varves and the definition of mechanical failure in the triaxial test sequences. However, the mechanisms underlying these observed variations may be further elucidated using the temperature variation of the elastic moduli. Young's modulus and Poisson's ratio are discussed in the next section.

12.4.3. Elastic Properties

The elastic moduli, Young's modulus and Poisson's ratio, were obtained by direct measurement in the triaxial test cell employed by Ozdemir *et al.* (1981) and Miller *et al.* (1979). However, the advantages of direct measurement are somewhat offset by the sensitivity of measured radial displacement to positioning of the radial displacement probe. Unless the sample is quite uniform, the displacement probe could easily be placed at a leaner section, and hence deform less, or alternatively a richer one, and hence deform excessively. These effects will particularly confound static Poisson's ratio measurements.

The temperature dependent dynamic properties of oil shale were also measured using ultrasonic techniques in a series of studies at Colorado State University begun by Petit (1976) and continued by Mraz *et al.* (1980). As noted earlier, the values of Young's modulus obtained using ultrasonic pulse techniques are generally higher than those obtained by static techniques. The application of these techniques would be limited by extremely non-linear stress–strain behavior, which causes the instantaneous slopes to be unrepresentative of the

Temperature effects 547

strains obtained under typical conditions in commercial retorts. Since oil shale is highly consolidated and can bear large loads, at least at low temperature, ultrasonic data should provide reasonable qualitative elastic property estimates. The experimental set-up used (Mraz, 1981) facilitates the simultaneous measurement of both p-wave (vibration parallel to propagation direction) and s-wave (vibration, or polarization, parallel to propagation direction) velocities.

The variation of acoustic velocity with temperature and grade for vertically oriented cores is given in Fig. 12.11. These data exhibit the following trends:

1. The s- and p-wave acoustic velocities decrease monotonically with temperature.
2. The s- and p-wave acoustic velocities decrease monotonically with grade.

The anisotropy of acoustic velocity parallel and perpendicular to the varves is elucidated in Figs 12.12 and 12.13, which show the variation of p- and s-wave velocities with temperature for 96 liters tonne^{-1} (23 gallons ton^{-1}) shale. These data permit an evaluation of the uniaxial anisotropy of these velocities. From these data it is seen that the p-wave velocity perpendicular to the bedding planes is lower and decreases with temperature more rapidly than the p-wave velocity parallel to the bedding planes. Just the opposite is true for the s-wave velocity. The data in Mraz (1981) further illustrates the general trend that in higher grade oil shales, in which kerogen forms the continuous phase, both s- and p-wave velocities exhibit similar temperature dependence and have a much smaller amount of anisotropy.

A clue as to the source of these variations is obtained by studying the variation of the acoustic velocities versus grade and temperature for retorted and burnt (oxidized) shale. As shown in Figs 12.14 and 12.15, the acoustic velocities of retorted shale are a function of grade prior to retorting. This is indicative of the effect of increased porosity obtained on retorting. However, when the retorted shales are heated, there is no appreciable variation of acoustic velocity with temperature. A similar situation is obtained for burnt oil shales. In burnt shale the residual carbon is removed by heating in air. The variation of acoustic velocity with prior grade and temperature is shown in Fig. 12.16. These results indicate that the kerogen is the origin of the observed temperature variation of the acoustic velocities in oil shale.

The compressive strength increases with increasing confinement

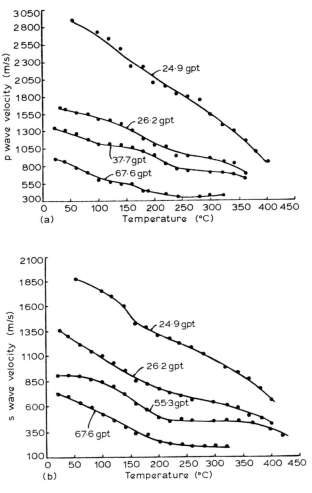

Fig. 12.11. Acoustic velocity as a function of temperature and grade. (a) Dependence of p-wave velocity on temperature for Green River oil shales; shale grade is shown as the variable parameter (see text). (b) Dependence of s-wave velocity on temperature for Green River oil shales; shale grade is shown as the variable parameter (see text).

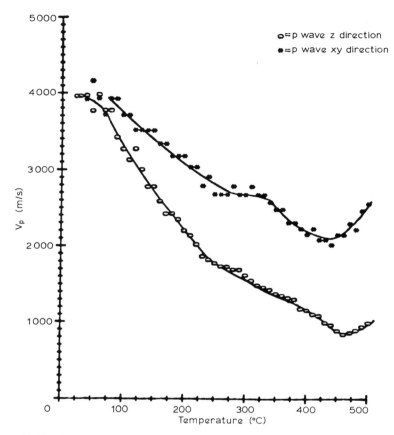

Fig. 12.12. Comparison of V_{pz} and V_{pxy} versus temperature for 23 gallons ton^{-1} oil shale. V_{pxy} exhibits less sensitivity to temperature than V_{pz}. However, differences between V_{pxy} and V_{pz} are not as large as observed in lower organic content (18 gallons ton^{-1}) shale.

pressure. It is thus expected that the acoustic velocities will increase with increased uniaxial stress. This expectation is confirmed in the data (Mraz, 1981) which exhibit a linear increase of acoustic velocity with increase in applied uniaxial stress. This variation makes pressure a significant parameter for designing ultrasonic or acoustic retort diagnostics. These velocity data permit the ready calculation of Young's modulus and Poisson's ratio. Young's modulus along the z direction,

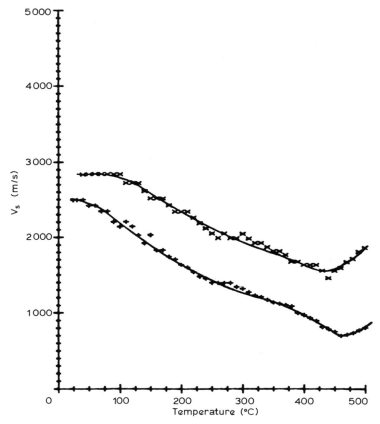

Fig. 12.13. Comparison of V_{sz} and V_{sxy} versus temperature for 23 gallons ton^{-1} oil shale. (+) s-wave, z direction; (×) s-wave, xy direction.

E_z, versus temperature for various grades of shale is given in Fig. 12.17. These data may be compared with the value of Young's modulus obtained from static test data given in Figs 12.18 and 12.19. These data exhibit the anticipated trend that the dynamic values exceed the static values. Since oil shale at room temperature is relatively non-porous and consolidated, the discrepancy between the two moduli is less than at elevated temperatures, where exfoliation and cracking, along with plastic flow, significantly impact the acoustic velocities and the accuracy of both static and dynamic measurements.

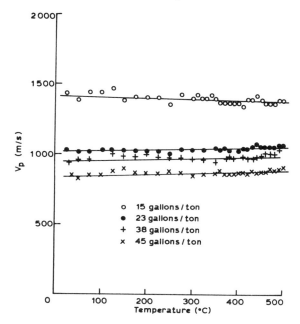

Fig. 12.14. V_{pz} versus temperature for retorted oil shale. Data is given for four different prior grades. Little temperature dependence is observed for retorted shale relative to raw oil shale. This behavior is responsible for the strong temperature dependence of V_p in raw oil shale.

The general trends in Young's modulus data are that the Young's modulus decreases with increasing grade and increasing temperature. The relative change with grade and temperature is far more pronounced in lower grade shales than in higher grades. As the grade is increased the intergrade differences and variation with temperature are increasingly similar. This is indicative of the domination of shale properties by kerogen, which forms a continuous matrix at concentrations exceeding 146 liters tonne^{-1} (35 gallons ton^{-1}). Poisson's ratio is highly sensitive to the difference in shear and transverse acoustic velocities. At elevated temperatures the combination of chemical and physical processes such as kerogen decomposition cracking caused by differential thermal expansion of various components of the rock, changes in porosity, pore pressure and delamination lead to potentially wide variability in Poisson's ratio. In static testing measurements,

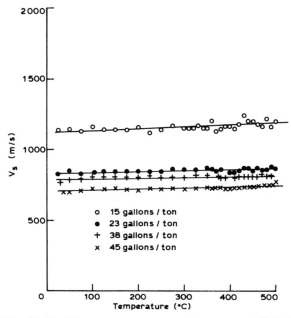

Fig. 12.15. V_{sz} versus temperature for retorted oil shale.

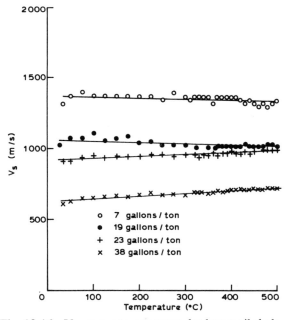

Fig. 12.16. V_{sz} versus temperature for burnt oil shale.

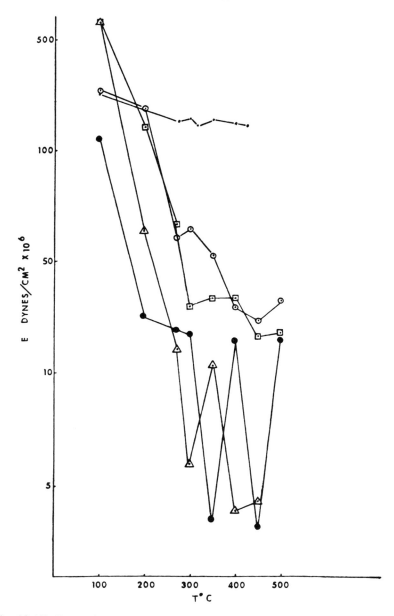

Fig. 12.17. Dynamic Young's modulus (E_z) as a function of temperature and grade derived from acoustic velocity data. (·) 7 gallons ton^{-1}; (○) 18 gallons ton^{-1}; (⊡) 23 gallons ton^{-1}; (△) 30 gallons ton^{-1}; (●) 40 gallons ton^{-1}.

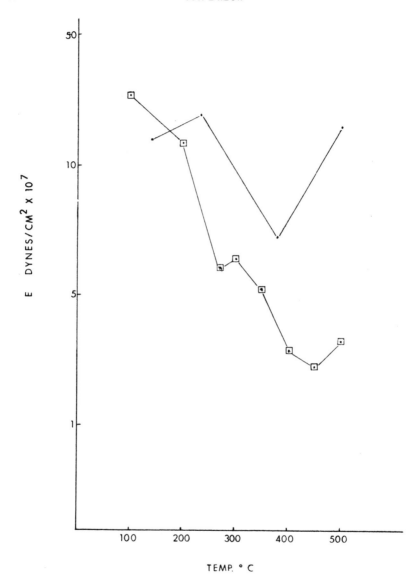

Fig. 12.18. Comparison of static and dynamic Young's modulus (E_z) versus temperature for oil shale at 50 psi confining pressure. (·) Wang's data, 15 gallons ton^{-1}; (□) DuBow's laboratory, 18 gallons ton^{-1}.

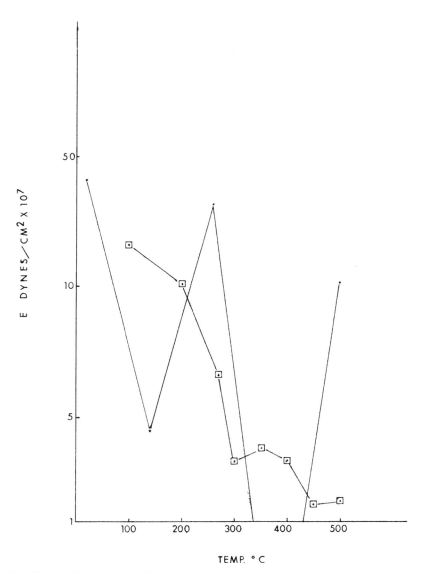

Fig. 12.19. Comparison of static and dynamic Young's modulus (E_z) versus temperature for oil shale for 50 psi confining pressure. (·) Wang's data, 25 gallons ton^{-1}; (□) DuBow's laboratory, 23 gallons ton^{-1}.

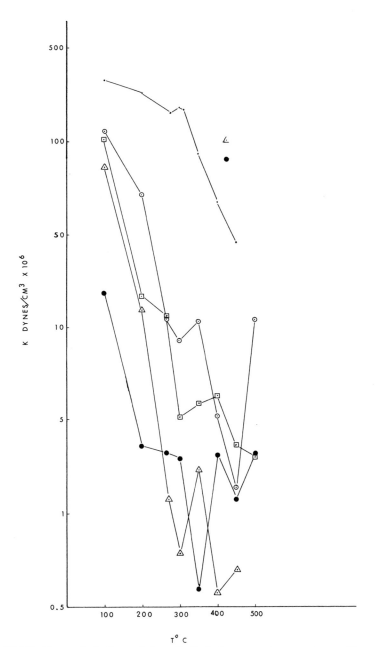

Fig. 12.20. Dynamic bulk modulus of oil shale versus temperature with grade as a parameter. (·) 7 gallons ton^{-1}; (⊙) 18 gallons ton^{-1}; (□) 23 gallons ton^{-1}; (△) 30 gallons ton^{-1}; (●) 40 gallons ton^{-1}.

Temperature effects 557

grade non-uniformity leads to a degree of uncertainty in radial deformation obtained from single point indicators.

There are complementary sources of error in static and dynamic Poisson's ratio data for oil shale. The dynamically obtained data are sensitive to microcracks and porosity but average over the volume, while static data are sensitive to non-uniform kerogen distribution and the resulting radial deformations. As a result, the values of these data fluctuated substantially and sometimes assumed negative values. Therefore, these data have more qualitative and comparative rather than quantitative significance. For example, the value of the dynamic Poisson's ratio was reasonable at low temperatures (0·25–0·35) but at elevated temperatures it approached or exceeded unity and hence became unrealistic. For triaxial tests, those values exceeding unity are possibly a result of anomalously large radial displacement of the radial sensor because of sensor placement near richer sections of the shale. Figure 12.20 shows the dynamic bulk modulus as a function of temperature with grade as parameter. It is seen that the bulk modulus decreases with increasing grade and increasing temperature. It is noteworthy that the bulk modulus exhibits the same qualitative behavior as the Young's modulus.

12.4.4. Passive Mechanical Testing
In addition to the dynamic and static mechanical tests, in which a combination of steady state and time varying mechanical loads are applied to the sample, and the resulting displacement determined, useful data may be obtained from measuring the acoustic emission from the sample under small or minimal load. Since the acoustic emission represents a time varying mechanical output with zero (or nearly zero) mechanical input, it may be classified as a passive test in which output is measured with a thermal input as opposed to a mechanical input. These data were obtained in conjunction with TGA (thermogravimetric analysis), DSC (differential scanning calorimetry) and electrical resistivity data (Mraz *et al.*, 1980). These measurements help elucidate the variations observed in the mechanical property data. In that regard, they are indicative of the extreme utility obtained by studying a complex heterogeneous material such as oil shale using a variety of complementary measurement techniques.

The complementary measurement techniques involve thermoanalytical (DSC, TGA, etc.) and constitutive property measurements (electrical, thermal and acoustical conductivity and capacities) and,

558 *Joel DuBow*

when available, analytical chemical (composition) data. The thermo-analytical data provide information on the reaction temperatures, heats of reaction, reaction rates and other chemical thermodynamic and kinetic data. The constitutive property measurements determine the wave, or energy, propagation properties in the material and the relative amount of energy stored (capacities) as well as the amount of energy dissipated (conductivity) in the material as energy propagates through the sample. Analytical chemical characterizations determine the composition of the sample before and after heating and can be used to determine the substances emitted during heating. The acoustic emission test is, in the thermal analysis nomenclature, called thermo-sonometry (TS). Measurement of the sample response to the applied acoustic field is called thermoacoustometry.

Typical data obtained from thermosonograms for shales of varying grades are shown in Fig. 12.21. Four regions of TS activity may be distinguished. One region extends from approximately ambient temperature to approximately 250°C. The second region occurs from 300 to 380°C. A third region of activity extends from approximately 380 to 470°C and a fourth, smaller region from 480 to 500°C. These four regions may be identified with four distinct processes occurring within the oil shale.

The mechanisms underlying this behavior are better understood when correlated with variations in electrical properties and the DSC. Details are described in Lonvik *et al.* (1980).

The first process, which occurs from ambient to 250°C, based on combined DSA, TS and electrical data, and work by other authors, is identified with the decomposition of light hydrocarbons from the organic matter in the shale. The second process, occurring between 300 and 380°C, and characterized by peaks in the a.c. resistivity and a peak in the thermosonograms, is a physical transition consistent with delamination of the varves and detachment of inorganic from the organic phases, perhaps coupled with structural alterations in the kerogen. This region is also characterized by an abrupt degradation of mechanical strength and a minimum in the mechanical strength versus temperature.

The third process, occurring between 380 and 472°C, represents the decomposition of kerogen and has been widely studied in the litera-ture. The final process observed, at 480–500°C, was not precisely identified because of the absence of analytical identification of the products. The results obtained are consistent with secondary decom-

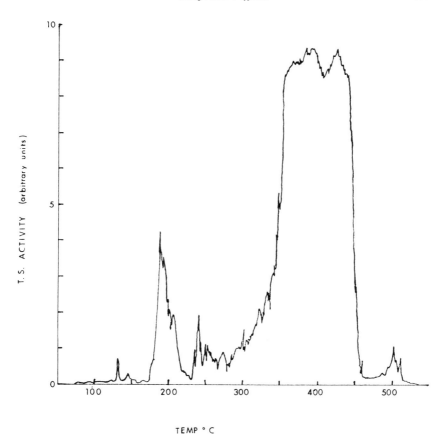

Fig. 12.21. Thermosonogram (acoustic emission) of a typical oil shale sample showing four distinct regions. Each region is indicative of a chemical or physical transition occurring within the shale.

position of kerogen with resulting carbon residue or radical species.

This study helped elucidate the chemical and physical changes occurring while oil shale is heated and explain the variations in mechanical properties. Studies of this type demonstrate the utility of multiple, complementary measurements on complex, heterogeneous materials such as oil shale and also demonstrate the differential sensitivity of these various measurements to particular processes occurring within the material.

560 *Joel DuBow*

The measurements described above were taken in a flowing inert gas such as dry nitrogen or helium to avoid oxidation of the kerogen during heating. While such oxidation does not significantly affect the mechanical properties of oil shale, it has a major qualitative and quantitative effect on thermoanalytical characterizations and electrical properties (DuBow, 1980a, b).

In conclusion, the mechanical properties vary substantially with temperature. These variations are consistent with the four mechanisms identified with the temperature regimes discussed above. Thus from room temperature to about 250°C the decomposition of light hydrocarbons leads to gradually reduced strength in the oil shale rock. Between 300 and 380°C a physical phase transition leads to significant weakening of the rock and plastic flow when there is sufficient kerogen. Between 380 and 470°C the strength remains low but increases somewhat because of recementation of the shale matrix and internal pore pressure increases. Above 470°C the mechanical properties are not well characterized. Further, structural variables such as cracks, faults and highly non-uniform kerogen concentration, can also substantially affect the measured strength of oil shale. Increasing the confining pressure increases the measured strength and rigidity of oil shale. However, heating rate and oxidizing or non-oxidizing atmospheres do not substantially affect the mechanical properties of oil shale. Oil shale retorting occurs in packed beds with 10–25% void volume. The properties of these retort beds may be modeled as two-phase media. These models and certain applications of the mechanical properties are reviewed in the next section.

12.5. APPLICATIONS OF THE TEMPERATURE DEPENDENT ELASTIC MODULUS RESULTS

The major applications of the temperature dependent mechanical properties data have been: to assess permeability plugging and the support capability of retort beds; to assess the support capability of retort walls; and to assess the feasibility of acoustical diagnostics of above ground and *in situ* retorts.

Use of the dynamic mechanical, or acoustical, properties of retort beds as a process diagnostic tool requires a model of the heterogeneous, two-phase media consisting of oil shale and void spaces. At higher temperatures, it is conceivable that the oil and escaping gas vapors could alter the properties of these two media and further complicate the problem. However, at present there is no evidence of this. The most apparent effect of elevated temperatures is the decrease in void volume arising from compaction of the shale rubble and crushing of spent shale. The structural alteration of the kerogen and plastic flow of the rock between 300 and 380°C leads to potential plugging of the retort permeability. The acoustic velocities of the oil shale rock decrease. This decrease is a function of the overburden pressure on the oil shale.

12.5.1. Acoustical Properties of Oil Shale Rubble Beds
The acoustical properties of the elastic moduli for multi-phase media in general and two-phase media in particular have been a problem of keen interest in the field of geophysical exploration. For porous sedimentary rocks, one phase is commonly the mineral matrix and the second the gas or liquid fluid that occupies the pore space in the material. Individual oil shale rocks may be considered as two-phase media wherein inorganic minerals comprise one phase while organic kerogen comprises the other. This approach has been useful for modeling the thermal transport properties in oil shale (Wang *et al.*, 1979).

A number of the models for two-phase media are reviewed in the thesis by Mraz at Colorado State University (Mraz, 1981). The models usually derive the relationships between shear and bulk moduli of saturated rock and unsaturated rock. In modeling these moduli, assumptions need to be made about the interconnection or isolation of the pores and the resulting equilibration, or lack thereof, of pore fluid pressure throughout the pore space. Other assumptions include the relationship of the wavelength of the acoustical signal in relation to pore size and the shape and orientation of the pores and rubble particles. In addition, an assumption needs to be made regarding the interaction or lack of interaction between the individual cracks or between the pores. The resulting moduli are usually denoted with an asterisk to indicate an 'effective' modulus, as opposed to a modulus for a homogeneous material. The approach that was found most useful for evaluating the feasibility of remote acoustical diagnostics for oil shale was that developed by Kuster and Toksoz (1974*a*, *b*). They employed dynamic loading and derived effective or composite elastic moduli in terms of the scattering of elastic waves. Other models typically employed static, or time invariant, loading which requires pore pressure

562 *Joel DuBow*

equilibration times faster than the propagation times of the acoustic or seismic waves.

The effective moduli and effective density may be used to derive the effective propagation velocities. Two critical limitations on most models are the assumption of a particular pore shape and the neglect of multiple scattering by pores. These assumptions restrict the model to low concentrations of non-interacting pores, or inclusions. However, oil shale rubble consists of pores with many different aspect ratios, defined as the ratio of crack width to crack length, and pore sizes. Since acoustic velocities are strongly affected by the discontinuities caused by cracks, the effective p- and s-wave velocities are more influenced by small aspect ratio pores than by larger aspect ratio pores. Moreover, if the pore is gas-filled, the p-wave velocity is more affected than the s-wave velocity, while both velocities are equally affected by water or fluid saturation. In addition, the effective p- and s-wave velocities decrease with increasing porosity. Kuster and Toksoz ($1974b$) extended their model to include media which have a discrete spectrum of different aspect ratio pores, a relationship predicting pore closure as a function of axial pressure and an extension to include multi-phase media. A significant extension was later made to include a formal inversion of the velocity versus pressure data to obtain a spectrum of pore aspect ratios. Their model is presented in a form which was more amenable to simulating acoustic diagnostics with the available data, than were other models which are usually based upon self-consistent calculations, or upon integral equations which are difficult to invert, or upon assumptions which are overly restrictive. Mraz (1981) used that model to evaluate the potential for acoustical diagnostics of oil shale retorts. This analysis was limited by an insufficiently complete data base regarding the elastic properties of individual constituents and the tensor structural properties of oil shale and oil sands. However, the qualitative results obtained with available data, as described below, indicate the potential utility of these techniques and thus warrant further study.

Experimental scale models for determining acoustic velocities for oil shale rubble were made by Mraz (1981). These models require rubble distributions scaled approximately as that obtained in a commercial scale retort. Different particle sizes were obtained by crushing oil shale and separating by size with various sieves. The average grade of the rubble was obtained by retorting the sample in a furnace and correlating the percentage weight loss with shale density and subsequently

Temperature effects

shale density to grade. From the shale density, and a careful weighing of the rubble inserted into a sample holder, along with precise knowledge of sample chamber dimensions, the total volume of shale rubble and void space could be found by subtracting the shale volume from the total volume. Figure 12.22 shows the p- and s-wave velocities as a function of average particle size, with applied uniaxial pressure as a parameter. As anticipated, both V_p and V_s increase for higher pressures, with a larger effect on V_p than V_s. This effect is more pronounced than the reduction in pore volume, which was less than an order of 10% over the entire pressure range.

The acoustic velocities of an oil shale retort are dominated by five parameters: grade, pressure, size distribution, temperature and thermal history. The accuracy of estimating a parameter value obtained from an indirect measurement is limited by the accuracy of the forward wave propagation model. Thus, the utility of acoustical diagnostics will be limited by the ability to incorporate knowledge of retort properties as constraints in the propagation model typically in the form of constitutive properties. As an example, if pressure, grade, thermal history and size distribution are known, temperature profiles will be far easier to extract than if such data were unknown or incompletely known. The preliminary model presented (Mraz, 1981) was based upon an idealized Kuster–Toksoz model of a retort as a two-phase medium which exhibits an effective velocity and an effective density.

The effective media approach requires assumptions of macrohomogeneity (although not microhomogeneity). The macrohomogeneity assumption requires a uniform average grade throughout the retort, a uniform average particle size distribution and a random particle orientation with respect to the bedding planes. Another assumption, which likely needs to be relaxed for practical retort optimization models, is that of lateral uniformity within a retort bed. This uniformity function reduces the problem to a one-dimensional model in which only z directions of variation are considered. Table 12.3 lists the parameters of the model retort that were assumed.

In scaling these retort parameters to a laboratory system the most significant scaling factor is the ratio of the characteristic medium size to the wavelength of the acoustic wave. The frequency of the acoustic wave in pulse experiments is estimated from the fundamental frequency in a Fourier transform of the received acoustic signal. For particle diameters of $0\cdot4–0\cdot6$ m, acoustic velocities from $1500\,\mathrm{m\,s^{-1}}$ to $2200\,\mathrm{m\,s^{-1}}$, signal frequencies of 600 Hz and wavelengths of $2\cdot5–3\cdot7$ m

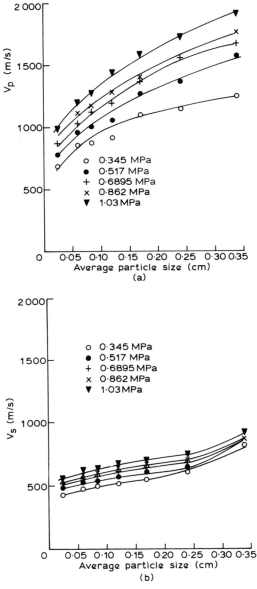

Fig. 12.22. (a) V_p versus mean particle size of 23 gallons ton^{-1} shale rubble; sensitivity of velocity to particle size is more pronounced at higher levels of pressure. (b) V_s versus mean particle size of 23 gallons ton^{-1} shale rubble; slightly different behavior than p-wave velocity is noted.

TABLE 12.3
Physical Specifications of Model *in situ* Retort Bed

Retort property	*Value*
Shale density	$2\,250\ \mathrm{kg\,m^{-3}}$
Shale grade	96 liters tonne^{-1} (23 gallons ton^{-1})
K' (related to interparticle friction)	0·22
μ' (coefficient of friction between walls and bed)	0·84
Zero depth bed porosity	0·20
Overburden pressure	$0{\cdot}0\ \mathrm{MPa}^{a}$
Bed width (square bed)	50·0 m
Bed depth	75·0 m
Particle size	0·4–0·6 m

^{a} Assumes a slight void at the top of the shale retort bed and thus no pressure exerted on the shale rubble from present overburden.

are obtained. These lead to a ratio of wavelength to particle size of 6. A similar ratio is obtained for shale with a particle size range of $+7, -5$ and a corresponding velocity of $1900\ \mathrm{m\,s^{-1}}$ at 1 MPa pressure and an acoustic frequency of 95–100 kHz. This scaling permitted laboratory scale models of a retort to be readily constructed and acoustic velocities in retort beds to be estimated.

An estimate of the effective acoustic velocity incorporating known properties of the retort bed may be simplified by calculating the velocity at room temperature and incorporating knowledge of the variation of effective velocity with temperature, pressure, grade and particle size distribution. It is assumed that average grade, particle size distribution and orientation are known. It is also assumed that velocity changes within a particular retort bed will result from changes in void volume and not total void volume. This distinction is significant in that the retort diagnostic tool will depend upon monitoring changes observed in acoustic velocity and not on absolute values of velocity, which are much more difficult to estimate accurately. In addition, the thermal history of a retort may be incorporated by denoting the different regions of the retort and their boundaries (raw shale, retorting shale, retorted shale and burnt shale with relatively abrupt transitions). For simplicity, raw shale was considered the region between 25 and 500°C, retorted shale between 500 and 600°C and burned shale regions delineated; the effective acoustical velocity will then be a function of temperature and pressure.

566 *Joel DuBow*

From the acoustic velocity experimental data, the temperature dependence may be described in an exponential form:

$$V(T_0) = V_{p0} \exp(-BT) \qquad (12.13)$$

where T_0 is the initial temperature (°C); T is the actual temperature (°C); V_{p0} is the p-wave velocity at the initial temperature; and V is a parameter obtained from the temperature variation data. The data obtained from Mraz indicated that the equation for $V_{p,\mathit{eff}}$ is as shown below:

$$V_{p,\mathit{eff}}(T) = V_0 \exp[-2\cdot12 \times 10^{-3}(T-25)] \qquad (12.14)$$

From the experimental data the acoustic velocity will increase linearly with increasing pressure. The pressure dependence (empirically derived from the $+7$, -5), particle size distribution and shale rubble can be fitted to the equation given below:

$$V_{p0}(P) = 1040 + 865P \qquad (12.15)$$

where P equals pressure in MPa.

The pressure and temperature dependence of the p-wave acoustic velocity may be combined into a single equation, valid over the temperature interval $25 < T < 500°C$, as given below:

$$V_{p,\mathit{eff}}(T, P) = (1040 + 865P) \exp(-2\cdot12 \times 10^{-3}(T-25)) \qquad (12.16)$$

Figure 12.23 shows the predicted V_p versus temperature for 96 liters tonne^{-1} (23 gallons ton^{-1}) rubble. These data may be understood in terms of an exponential temperature dependence and the effect of a relatively small pressure closing easily closed micro-cracks and compacting the rubble. However, as the pressure is increased, a relatively small effect on packing and void closing can be obtained. The contrast between solid shale and rubble acoustic velocities is shown in Fig. 12.24. Curve A shows the p-wave velocity for a solid shale sample. It is noteworthy that shale rubble exhibits a lower acoustic velocity compared to a solid shale sample. The major differences arise from the existence of voids in the shale which would be equivalent to cracks and porosity in a solid two-phase sample. It was also assumed that the variation of velocity for individual shale particles is the same as for a solid sample, but it is seen that the model predicts less relative influence of temperature on the effective acoustic velocity for the rubble.

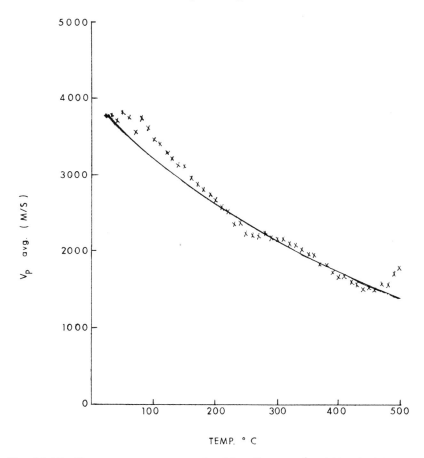

Fig. 12.23. V_p versus temperature for 23 gallons ton^{-1} rubble. Solid line is predicted variation using single exponential and the crosses are experimental data.

For retorted oil shale the lack of temperature dependence makes the effective p-wave velocity a function only of pressure. Applied pressure often crushes retorted and burnt rubble particles. To compensate for this effect, the data used in the model were scaled down (by 72%), velocities extrapolated from raw oil shale data and summarized in the equation given below:

$$V_{p,eff}(P) = 283 + 865P \qquad (12.17)$$

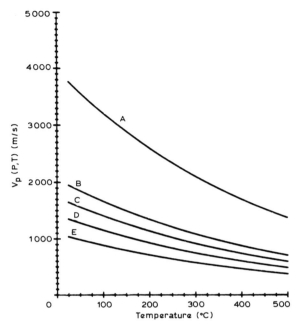

Fig. 12.24. Predicted V_p versus temperature behavior in shale rubble at different pressures. Curve A represents solid shale velocity. Curves B–E are for the following values of pressure: B, $P = 1.052$ MPa; C, $P = 0.705$ MPa; D, $P = 0.358$ MPa; E, $P = 0.0$ MPa. The curves demonstrate the decreasing influence of oil shale temperature dependence as media rubble consolidation decreases.

For combusted or burnt shale, the comparable equation is:

$$V_{p,\text{eff}}\, c(T, P) = 94.2 + 0.256T + 865P \quad (600 < T < 950°C) \quad (12.18)$$

Another important variable in the Kuster and Toksoz model is the effective density. The effective density is a function of the original shale matrix density, the pressure, the void volume, the density of fluid in the voids and thermal history. For raw, retorted and burnt shale, the effective density is a linearly increasing function of temperature. For retorting shale the effective density is a complex function of increasing temperature. Given these variations, it is necessary to incorporate models for temperature profiles and pressure profiles within the retort. Typical temperature profiles for an *in situ* retort with air injection are described in Jones (1980). Pressure variations within a retort have been treated by Carley (1976). Pressure and temperature variations and changes in the effective density are the implicit equations which can be combined into a single equation or, more simply, the values of the profiles can be computed, a table developed and an empirical table look-up procedure used to compute the resulting density. This is also the likely manner in which field data would be processed, i.e. empirically derived pressure and temperature data utilized to compute effective density in a particular region.

From temperature and pressure profiles, the acoustic velocity profile may be found. In principle, given pressure profiles, the temperature profiles may then be derived from acoustic velocity data. Moreover, the pressure variation of acoustic velocities, for typical retort pressures, is less than the temperature variation, so that anticipated lateral pressure gradients will not substantially limit system accuracy.

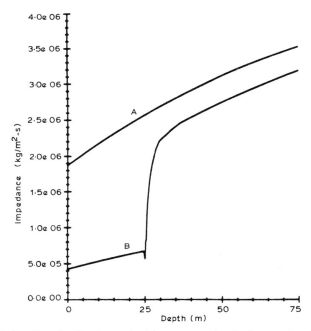

Fig. 12.25. Predicted effective velocities versus depth for an *in situ* retort. Curve A is for before-processing conditions and curve B represents velocity during processing. Note the large discontinuity (curve B) coinciding with the retorting front position in the velocity profile.

Conversely, given the temperature and pressure profiles, a resulting profile of acoustic velocities may be derived. It is important to include the change in density during retorting in the pressure calculation in order to reduce a significant overestimation of the pressure versus depth. Figure 12.25 shows the predicted acoustic velocity profile both prior to and during retorting for the *in situ* retort defined in Table 12.3.

The variation of p-wave velocity with depth in raw shale arises from the variation of pressure with depth in the retort bed. Superimposed on this variation, which may be considered as a base line, are temperature- and process-induced variations. As is seen in the Fig. 12.25, these are sharp and detectable variations in acoustic velocity with depth. In a practical retort it is likely that the width of zones 2 and 3 will be smeared over a greater distance than that obtained in an idealized model, with a resulting broadening of the region in which the velocity exhibits a minimum.

The foregoing calculations show that the acoustic velocity variation with temperature and pressure in oil shale beds yields velocity profiles that show the potential of acoustical diagnostics for delineating zones of operation in a retort and, in an imaging configuration, detecting lateral inhomogeneities. Indeed, the works of Fert (1976), Elkington (1978), Bartel (1979) and Fitzpatrick (1979) indicate a growing practical interest in these techniques.

12.5.2. Pillar Strength

Both *in situ* and above ground processing require a knowledge of the mechanical properties of oil shale. While temperature dependent effects are more significant for *in situ* processing, safety considerations make them important for surface process mining operations also. Underground pillars separating rooms being processed or mined require sufficient strength at predicted temperature extremes to support overburden pressures. A knowledge of minimum pillar size permits an evaluation of optimum resource extraction ratios.

Determining minimum pillar size requires a knowledge of several thermophysical properties. These include thermal conductivity, thermal diffusivity, Young's modulus, Poisson's ratio, compressive strength and tensile strength. Thermal conductivity and thermal diffusivity are reviewed by DuBow (1980a). The static and dynamic mechanical properties have been reviewed earlier in this chapter.

Based on a data base and an understanding of the thermophysical properties, heat flow into pillar walls surrounding a retort bed may be computed. Heat flow calculations have been reported in two recent studies. Ozdemir *et al.* (1981) used finite difference and finite element techniques to compute temperature profiles in pillar walls. Wilhelm *et al.* (1979) derived a closed form analytical solution to the problem of heat flow exterior to a right circular cylinder with a right circular cylindrical disk (representing the retorting zone) propagating down its axis. The result was an estimate of the temperature distribution exterior to the retort walls.

From these estimates of temperature profiles and a knowledge of the temperature dependent static and dynamic mechanical properties, a profile of the mechanical strength may be determined. This profile permits the estimation of minimum pillar sizes and design of retort and mine structures. Agapito and Hardy (1982) have recently reported a pillar design technique incorporating induced horizontal stresses and Ozdemir *et al.* (1981) a numerical model based on thermally induced compressive and shear stresses.

Since oil shale is thermally insulating, regions of elevated temperature tend to be concentrated near the retort wall. Exceptions to this occur when cracks exist in the wall as a result of stress propagation during fracturing, thermomechanical stresses occurring during retorting or natural fracture resulting from leaching of saline minerals or other geological conditions. In these instances heat can flow to the interior of the pillar through convection as well as the conduction that occurs in an intact, consolidated pillar.

The decrease of strength of the oil shale below 125°C to 150°C limits the employment of conventional mine and pillar design techniques to low temperatures. At higher temperatures structural design considerations are much more complicated.

12.5.3. Other Applications

Another related application of the temperature dependent mechanical properties is in the design of rubble piles of spent shale and in the use of spent shale as a structural material for road beds. Carley (1976) has reviewed pressure distribution in oil shale beds and Plancher *et al.* (1980) have reviewed pavement applications of spent shale. Heistand and Holtz (1980) have reviewed other applications.

In addition to active seismic and acoustical diagnostics of oil shale, passive acoustical diagnostics can also be useful probes. Strahle (1978) reviews combustion noise, a turbulence (from gas flows)–combustion

572 *Joel DuBow*

(from liberated heat) interaction noise. The combined geophysical acoustic emission and noise from a retort caused by burning and rock motion, along with the large size of process beds, have made data interpretation in acoustic emission diagnostic tests difficult. However, a systematic study of acoustic emission from oil shale cores and modeling of rubble bed emissions using techniques similar to those employed by Mraz (1981) has yet to be reported.

Since the failure of an *in situ* pillar could have such massive consequences for underground workers and surface structures (because of subsidence), a pillar strength diagnostic tool would be of importance. In addition, since permeability plugging can significantly reduce the efficiency of a retort process bed, an indirect diagnostic of mechanical strength and rubble distribution would be a valuable process tool. DuBow (1980*a*, *b*) discusses tools of this type.

These types of diagnostic tool could be based upon a correlation of electrical and acoustical properties with mechanical properties, temperature or chemical–physical transitions occurring in the shale. Since, as shown in DuBow (1980*b*), the electrical properties are not single valued functions of temperature, an estimate of the relative effect of various transitions and reactions occurring in oil shale upon the electrical or acoustical properties is significant for the design of retort diagnostics. Hong and DuBow (1980) have developed a model of a radio-frequency diagnostic for *in situ* retorts. Fitzpatrick (1979) indicates how acoustical holography could be useful for *in situ* processing. However, the accuracy of imaging diagnostics is limited by the accuracy of the acoustic or electrical constitutive properties. Finally, at high power levels, the same types of radio-frequency and acoustical imaging systems used to detect the state of various regions of an oil shale retort could be used to alter the state of a region of a retort through the selective deposition of energy. In this manner the temperature dependent acoustical (and electrical) properties could be utilized in both a process diagnostic and a process control system.

12.6. CONCLUSIONS

The following conclusions can be drawn from the data on the dependence of the mechanical properties of oil shale on temperature and on application considerations:

1. The static and dynamic mechanical properties of oil shale

Temperature effects 573

exhibit considerable variation with temperature. The rock weakens substantially as temperature is increased.

2. Higher grades exhibit a smaller relative decrease, with temperature, in the static and dynamic mechanical properties.
3. The values of the static and dynamic mechanical properties decrease with increasing grade of shale.
4. Heating rate does not significantly affect the mechanical strength of oil shale.
5. Ambient gas atmosphere does not significantly affect the mechanical strength of oil shale.
6. The values of the static and dynamic mechanical properties increase with increasing confining pressure.
7. The acoustical velocities exhibit the same qualitative temperature and pressure variation as the dynamic mechanical properties.
8. At elevated temperatures, and for sufficient time to result in the complete conversion of kerogen to oil, gas and residual carbon, the static and dynamic mechanical properties of oil shale increase slightly in magnitude, but not to a level approaching those expected at room temperature.
9. The observed grade and temperature variation of the mechanical properties of oil shale arises from the presence of kerogen and its transformation during heating. The most significant transformations are:

   (i) the loss of light and volatile compounds between 100°C and 150°C;
   (ii) the physical separation of organic and inorganic phases and structural alteration of kerogen in the temperature range 200–350°C;
   (iii) the pyrolytic decomposition of kerogen into oil, gas and residual carbon between 350 and 400°C;
   (iv) the recementation of the remaining material between 450 and 550°C; however, the presence of significant quantities of inorganic compounds, such as nahcolite and dawsonite, which decompose at temperatures below 500°C, will add additional structure to the temperature dependence of the mechanical properties; in addition, structural flaws and macro-inhomogeneity will further complicate (and likely reduce) the values of the static and dynamic mechanical properties.

10. Triaxial test data show that compressive strength with stress applied parallel to the varves was higher than the compressive strength with stress applied perpendicular to the varves. This result is attributed to the non-uniform distribution of kerogen, and the influence of a band of kerogen-rich varves which exhibit lower mechanical strength than leaner varves. The premature failure of this rich band of kerogen will appear sooner in perpendicular loading than in loading parallel to the varves.
11. The tensile strength for perpendicularly loaded cores is higher than for parallel loaded cores below 150°C. Above 150°C the situation reverses.
12. An F-value statistical regression analysis of tensile and compressive strength led to the following conclusions:

   (i) for compressive strength temperature was the most statistically significant variable for perpendicular loading while grade was the most statistically significant variable for loading parallel to the varves;
   (ii) for tensile strength, temperature was the most statistically significant variable for both directions of loading.

13. The dynamic Young's and dynamic bulk moduli exhibit qualitatively identical temperature and grade dependence. Both decrease with increasing grade and temperature, with a slight increase at temperatures over 400°C.
14. Acoustic velocities in oil shale may be predicted using effective two-phase media models. Application of these models to typical retort configurations demonstrates the potential for useful retort process diagnostic tools.
15. The combination of conduction heat transfer and temperature dependent mechanical properties permit a prediction of pillar strength in underground retorts.
16. Correlation of acoustical properties with static mechanical properties, electrical properties and chemical and physical transitions, occurring within the oil shale, is useful for increasing understanding of the behavior of the material and for the process and retort diagnostic tools.
17. Since the major decrease in mechanical strength occurs between room temperature and 150°C, predicting the mechanical strength, and hence pillar and mine design, of structures where a significant portion of the volume equals or exceeds this tempera-

Temperature effects 575

ture is difficult. However, the thermally insulating nature of oil shale results in limited pillar heating for practical retort temperatures and pillar sizes.

REFERENCES

Agapito, J. and Hardy, M. (1982). Induced horizontal stress method of pillar design in oil shale, *Proc. 15th Oil Shale Symposium*, Colorado School of Mines, Golden, pp. 191–7.

Bartel, L. (1979). Use of electrical geophysical techniques to monitor *in situ* oil shale recovery processes, *Proc. 12th Oil Shale Symposium*, Colorado School of Mines, Golden.

Baughman, G. and Hedrickson, T. (1978). *Synthetic Fuels Data Handbook*, 2nd edn, Cameron Engineers, Denver.

Bridges, J., Erik, J., Snow, R. and Sresty, G. (1982). Physical and electrical properties of oil shale, *Proc. 15th Oil Shale Symposium*, Colorado School of Mines, Golden.

Carley, J. F. (1976). Pressure distribution in beds of oil shale rubble, *Lawrence Livermore Lab. UCRL-51957*.

Clark, G. (1966). Deformation moduli of rocks. *Testing Techniques for Rock Mechanics*, ASTM, Philadelphia.

Closman, P. and Bradley, W. (1979). The effect of temperature on tensile and compressive strength and Young's modulus of oil shale, *J. Soc. Det. Eng.*, 301–12.

Dimitryev, A., Kuzyayev, Y., Protasev, Y. and Yamshehikov, V. S. (1972). Physical properties of rocks at high temperatures, *NASA Technical Translation F-824*.

DuBow, J. (1980*a*). Indirect evaluation of fossil fuel processes, *Proc. 8th Symposium on Fossil Fuel Processing*.

DuBow, J. (1980*b*). Thermophysical properties of oil shale, *US Department of Energy Report EF-77-S-03-1584*.

Elkington, W. (1978). Sound monitoring of oil shale retorts, *US Patent 4,082,145*.

Evans, W. (1963). A system for combined determination of dynamic and static elastic properties, permeability, porosity and resistivity of rocks. PhD Dissertation, University of Texas, Austin.

Eyring, H., Lin, S. H. and Lin, S. M. (1980). *Basic Chemical Kinetics*, Wiley Interscience, New York.

Fenix and Scisson Inc. (1976). Phase I report on technical and economic feasibility of *in situ* shale oil recovery, *US Bureau of Mines Contract 5024173*. Tulsa, Oklahoma.

Fert, J. W. (1976). Evaluation of oil shales using geophysical well-logging techniques. In: *Oil Shale*, eds T. F. Yen and G. V. Chilingar, Elsevier, New York.

Fitzpatrick, G. (1979). Seismic imaging by holography, *Proc. IEEE*, **67**(4), 536–53.

576 *Joel DuBow*

Heistand, R. and Holtz, W. (1980). Retorted shale research, *Proc. 13th Oil Shale Symposium*, Colorado School of Mines, Golden.

Hong, S. and DuBow, J. (1980). Monitoring *in situ* retorting processes in oil shale with reflected and transmitted electromagnetic waves, *J. Appl. Phys.*, **51**(7), 3953–6.

Jaeger, J. and Cook, N. (1976). *Fundamentals of Rock Mechanics*, 2nd edn, Chapman and Hall, London.

Jet Propulsion Laboratory (1980). Instrumentation and control needs for fossil energy processes, *US Department of Energy Report 5030-437 Rev A*.

Jones, D. (1980). Sensitivity study of oil shale retorting, MS Thesis, Colorado State University, Fort Collins.

Kuster, G. and Toksoz, M. (1974a). Velocity and attenuation of seismic waves in two phase media: Part I, Theoretical formulations, *Geophysics*, **39**, 587–606.

Kuster, G. and Toksoz, M. (1974b). Velocity and attenuation on seismic waves in two phase media: Part II, Experimental results, *Geophysics*, **39**, 607–18.

Logan, J. (1979). Deformational behavior of Colorado oil shale, *Int. J. Rock Mechanics and Mineral Science*, **7**.

Lonvik, K., Rajeshwar, K. and DuBow, J. (1980). New observations on chemical and structural transformations in Green River oil shales. *Thermochimica Acta*, **42**, 11–19.

Miller, R., Wang, F. D. and DuBow, J. (1979). Mechanical and thermal properties of oil shale at elevated temperatures, *Proc. 11th Oil Shale Symposium*, Colorado School of Mines, Golden, pp. 135–46.

Mraz, T. (1981). Acoustic properties of oil shale and tar sand for retort diagnostics, MS Thesis, Colorado State University, Fort Collins.

Mraz, T., Rajeshwar, K. and DuBow, J. (1980). An automated technique for thermoacoustiometry of solids, *Thermochimica Acta*, **38**, 211–23.

Ozdemir, L., Ropchan, D., Miller, T., Wang, F., Sladek, T., Young, C., DuBow, J. and Fausett, D. (1981). Effect of *in situ* retorting on oil shale pillars, *US Bureau of Mines, Final Report 40262031*.

Petit, C. (1976). Ultrasonic and thermal properties of oil shale during transient heating, MS Thesis, Colorado State University, Fort Collins.

Plancher, H., Miyake, G. and Peterson, J. (1980). Shale oil products as replacements for petroleum counterparts in pavement applications, *Proc. 13th Oil Shale Symposium*, Colorado School of Mines, Golden.

Rzhevsky, V. and Novik, G. (1981). *The Physics of Rocks*, Mir Publishers, Moscow; John Wiley, New York.

Sladek, T. (1980). *An Assessment of Oil Shale Technologies*, US Office of Technology Assessment, Washington DC.

Spencer, J. and Nur, A. (1976). The effects of pressure, temperature and pore water on velocities in westerly granite, *J. Geophys. Res.*, **81**(5), 899–904.

Stagg, K. and Zenkiewicz, W. (1968). *Rock Mechanics in Engineering Practice*, J. Wiley, New York.

Strahle, W. (1978). Combustion noise, *Prog. in Energy Combust. Sci.*, **4**, 157–76.

Terzaghi, K. (1945). Stress conditions for the failure of saturated concrete and rock, *Proc. Am. Soc. Test. and Mat.*, **45**, 777–801.

Timur, T. (1977). Temperature dependence of compressional and shear wave velocities in rocks, *Geophysics*, **42,** 950–6.

Tisot, P. R. (1975). Structural response of propped fractures in Green River oil shale as it relates to underground retorting, *US Bureau of Mines Report of Investigations 8021.*

Tisot, P. R. and Sohns, H. W. (1971). Structural deformation of Green River oil shale as it relates to *in situ* retorting, *US Bureau of Mines Report of Investigations 7576.*

Walsh, J. (1965). The effects of cracks on compressibility of rocks, *Geophysical Research*, **70,** 381–9.

Wang, Y., Rajeshwar, K. and DuBow, J. (1979). The dependence of the thermal transport parameters on organic content for Green River oil shales, *J. Appl. Phys.*, **50**(4), 2776–81.

Wilhelm, H., DuBow, J. and Hong, S. (1979). Analysis of transient temperature distribution exterior to an in site oil shale retort, *In-situ*, **3**(3), 227–45.

Index

Acceleration, surface uplift blasting, 403–4

Access, required for stress measurement techniques, 154, 156, 158

Acoustic emission testing, 557–9
noise in, 571–2

Acoustic properties, rubble beds, 561–70

Acoustic test methods, engineering properties obtained by, 266, 267, 270, 276, 278, 546–57

Acoustic wave(s)
anisotropy of, 547–8
velocities
density effects on, 568–9
organic volume effects on, 547–8
particle size effects on, 563–4,
pressure effects on, 563–4, 566–8
temperature effects on, 548–52, 566–8
uniaxial stress effects on, 549

Air
blast data, Occidental process rubblization, 512
comparison pycnometers, 25, 29

Air/foam cuttings, samples obtained from, 128, 129–30

Albite, 21, 233

Aluminized explosives, 420, 466, 515
TNT slurry, 488, 497, 505

Analcime, 20

Analytical methods
fracture mechanics, 217
oil shale, 7, 147–9

ANFO (ammonium nitrate/fuel oil), 51, 53, 54, 325, 450, 464, 465, 470, 474, 481, 483
characteristics of, 327, 420, 492

Anisotropic mechanical properties, 251–3

Anisotropy
experimental observation of, 231, 236–7, 251–3
strain-induced, 102, 114
stress-induced, 105, 112

Anticlastic plate bending, shear modulus determined by, 166, 169, 178

Anvil Points Mine, 47–51, 469–70
blasting experiments at, 51, 53–4
crushing experiments at, 51

Index

580

Anvil Points Mine—*contd.*
fragmentation tests at, 139–40
location of, 45, 232
mine design at, 47–50, 53
Mobil operations at, 52–4, 469–70,
474–6
Paraho operations at, 60–1, 470,
474–6
shale samples
elastodynamic fracture studies
of, 426
engineering properties of, 234,
237
mechanical properties of, 263,
265
non-representative nature of, 230
Ashfalls, volcanic, 20
Assay, *see* Fischer assay
Atlantic Richfield Co. (ARCO), 46,
54, 61, 470
Atlas Copco Boomer hydraulic drill
rigs, 480, 481
Australia, shale oil production, 5, 9
Averaging intervals, effect of, 313,
314
Axisymmetric grid, explosive fragmentation model,
323–5

Backfilling methods, *in situ* retort
chambers, 58
Barrel (petroleum unit), xix, 8
Barrier branching, explosive frag-
mentation crack, 347,
356
Basin, angle of dip of, 139
Bedding planes, 18, 19
fractures along, 136–7
Bingham substances, 528
Bings (spent shale dumps), 15, 571
Birefringent models, fragmentation
studies in, 338–73
Bitumen deposits, defined, 17
Bivariate regression analysis, 272,
276–7, 280
correlation coefficients for, 272,
276–7

Black Sulfur Creek, 46, 68
Blast hole
diameters, Occidental process,
486, 488, 489, 490
expansion, 396, 403
Blasting
parameters, Occidental process,
490–1
tests, 51, 53–4, 62
Brazilian surface mining, 466–8
burial depth effect, 446–8, 449,
489, 490′
charge shape effect, 448, 450
explosive fragmentation,
mine simulation, 450–3
rock samples, 377–86, 446
see also Fast Leader. . . ,
Lofreco. . .
Borehole
cracks
creation of, 343–5
reflected stress waves effect on,
345–8
deformation gauges, 155, 156
stresses, 340–1
Bragel CB (IRECO) aluminized
slurry explosives,
466, 468
Branching, *see* Cracks, explosive fragmentation, branching
Brazil, shale oil production in, 465–8
Brazilian test, 186
tensile strengths determined by,
186–95, 266, 267, 534,
536
organic volume effect on, 192–5,
235
statistical analysis of, 270, 276,
278
Brittle failure, criteria for, 96–7,
196–7
Brittle–ductile transition, triaxial
testing, 237
Bureau of Mines (USBM)
cooperative agreements, 46, 53, 54,
56, 61, 66
experimental mining and retorting
operations, 45

Index

Bureau of Mines (USBM)—*contd.*
Horse Draw shaft facility, 46, 66–8
liquid explosive fracturing tests by, 459–60
location of Piceance Creek Basin tract, 232
overcoring techniques developed by, 154–6
statistical studies by, 263–5, 266
Burgers (viscoelastic) models, 259, 528
Burial depths, explosive charge, 446–9, 490
Butterfly-shaped fringe loops, dynamic photoelasticity, 363
BX *in situ* retorting project, 68

C-4 explosive, 446
Caliper logs, 146
Cannel coals, 17
Carbon analysis, organic matter determined from, 26, 27
Carbonate
minerals, oil shale, 20–1
shales, examples of, 183, 185, 332
Cartesian coordinate system, 74
Catch boxes, mud cuttings, 130
Cathedral Bluffs Project, 46, 58, 484
Cauchy elastic materials, non-linear approach, 100–1
Cauchy's formulas
strain, 84, 85
stress, 76, 79–81
Caustics, photoelasticity, 365
CAVS (Crack And Void Strain) fracture model, 318–19
calculational zones for, 319–20
Characteristic equation, stress analysis, 81
Charge separators, surface uplift blasting, 402, 415, 416
Charges, *see* Columnar charges, Explosive(s)
Chert nodules, samples blemished by, 23, 25

Chong–Kuruppu precracked specimens, fracture testing, 218–19
Classical breakage models, 396–8
Classical theory (of elasticity),
compared with other methods, 189–90
tensile stresses in, 187, 340
ultimate tensile strength computed using, 191
Clay, oil shale structure affected by, 20, 21, 31
Clay-rich shales, examples of, 183–4, 332
Closure, mine, temperature-dependent creep as cause of, 257
Coal measure rocks
rheological model for, 259
triaxial testing of, 243
Coals, defined, 17
Coefficient of variation (statistical), 268
values quoted, 270–1
Cohesion, 95
bedding plane orientation effect on, 237
measurement of, 215, 231, 238, 239, 250, 266, 267
regression analysis of, 277
statistical characteristics of, 271, 281
Coke, formation of, 30
Colony Development Operation, 46, 54–5, 61–2
Colony Mine, 54–5, 61–2, 470, 476–9
joints and fractures in, 140–3
location of, 54, 232, 470
shale samples
blast-induced fragmentation, 450–1
creep testing of, 260
engineering properties of, 235, 237, 238
non-representative nature of, 230
triaxial testing of, 238, 241–2
Colorado
first oil shale retort in, 45

Index

582

Colorado—*contd.*
see also Federal Lease Tracts C-a,
C-b
Colorado School of Mines (CSM)
Anvil Points leased by, 52
temperature-dependent mechanical
properties study by,
540
Columnar charges, 391–2, 412
bottom detonation effects, 412–13,
450
length(s)
effects of, 452, 488, 489
used by Occidental Oil Shale
Co., 488, 489
Compatibility conditions, strain–displacement, 87–8
Compliance
coefficients, three-dimensional, 237
matrix, 99
tensors, plastic tangential, 111
Composition, uniformity in, 20
Compressive strengths
bedding plane orientation effect
on, 237, 281
confining pressure effect on, 541,
543
organic content effects, 65, 92–3,
178, 181, 184, 200–1,
541, 543–6
organic volume effect on, 541, 543–
6
regression analysis of, 277, 278
specimen size effect on, 159–60
strain rate effects on, 65, 200–1
temperature effect on, 541–6
testing of, 22, 172–85
values quoted, 47, 53, 54, 67, 92,
542–5
Compressive stresses, defined, 74, 77
Computer(s)
aided design (CAD), blasting techniques modeled by,
417
programs
explosive simulation, 294, 295–6,
318
regression analysis, 176, 200

Computer(s)—*contd.*
programs—*contd.*
statistical analysis, 268, 313
surface uplift blasting, 408
Concrete, triaxial testing of, 243
Confidence interval (CI), 268
values quoted, 270–1
Confining pressures
compressive strengths affected by,
541, 543
elastic moduli affected by, 529,
531
triaxial testing, values quoted, 216,
249, 250, 252, 536
Constituent minerals, absolute
density affected by,
26, 28–9
Constitutive modeling, 73, 90–2, 122
Constitutive relationships
material behaviour, 72, 73, 88–90,
123
rock and shale
elastic
linear, 97–100
non-linear, 102–4
three-dimensional stress/strain,
166–7, 169, 171–2
viscoelastic, 116, 118
Contact widths, finite element model,
188–9
Continuity condition, neutral loading,
108, 109
Continuous Failure State Triaxial
Tests, 243
Continuum
damage concepts, 435–41
theory, rock mechanics, 71–124
Conversion factors
general units, xix
oil yield/organic matter, 27, 28
Convexity postulate, rock strength
models, 93–4
Cooperative agreements
Bureau of Mines, 46, 53, 54, 56,
61, 66
Department of Energy, 62, 63, 68,
484, 487, 501
management of, 64–5

Index

Cord detonation systems, 416, 505
Cordin Dynafax 350 framing camera, 374–5, 380
Coring sampling methods, 129, 133–6
  fracture records in, 133–4
  handling techniques in, 134–6
  identification techniques in, 135
  storage conditions in, 136
Correlation coefficients, statistical analysis, 269, 272
  fracture property data, 311–17
  oil yields, 273, 311–12, 314–17
  values quoted, 273–5
Costs
  development, 12
  materials handling, 14–15
Coulomb criterion, 94
Coulomb–Navier criterion, 94–6, 97
Coursen breakage theory, 399, 400, 407, 412
Cowboy Canyon
  creep testing of samples from, 206, 210–11
  mechanical testing of samples from, 167, 179–81, 185, 193
Crack(s)
  explosive fragmentation
    branching, 344–5
      causes of, 348
      rock specimens, in, 383, 385
    circumferential patterns, 345–8, 397
    dominant, 344–5
    initiation, 343, 349
      zones, 356–7
    radial patterns, 328, 330, 344–5, 397
    velocities
      maximum, 343–4
      rock, in, 380, 383, 385
      values quoted, 338, 339, 344, 351, 366
  lengths, stress intensity factors dependent on, 220–1
  propagation, explosive fragmentation testing, 343
  shape, fracture stress affected by, 428–9, 431, 432

Crack(s)—*contd.*
  size, fracture stress affected by, 428–9, 431, 432
  velocities, value quoted, 338, 339, 344, 351, 366
Cranz–Schardin multiple spark gap camera, 340, 375–6, 379–80
Cratering experiments, 450–3, 491
Creep
  behavior, 124, 205–13
    experimental determination of, 209–10
    models for, 115–16, 206–9, 257–60, 262
    organic volume effect on, 211–12, 213
    rheological models for, 206–9, 259–60, 262
    statistical analysis of, 210–13
    stress levels effect on, 211–12, 213
    temperature effects on, 238, 253–6, 260–3
    three-parameter rheological model for, 259–60, 262
  functions, 257–60
Critical slope, load–deformation curves, triaxial testing, 244–5
Critical stress intensity factors, 343, 353, 355, 427
  values quoted, 339, 428
Cross-sectional areas, change of, during axial loading, 245
Crude oil reserves, oil shale resources compared with, 7
Crushed zones, borehole detonation, 343, 396
Crushing processes
  Anvil Points Mine experiments, 51
  energy requirements for, 15
Cubic triaxial testing, anisotropy determined by, 231, 237–8, 251–3
Curve-fitting
  strength/yield data, 315–17

584 Index

Curve-fitting—*contd.*
temperature-dependent mechanical
properties, 540, 541
tensile strength data, 313–15
Cutoff, maximum tensile strength, 96
Cyanoacrylate adhesives, 298, 357,
359

Damage
activation and growth, fragmentation model, 436–7, 439
parameters, 435, 436
contour plots of, 448, 449, 450,
451, 452
Dashpot (viscous) models, 114, 203,
258, 528
Dawsonite
temperature-dependent properties
affected by, 534, 545
thermal decomposition of, 527, 575
Deformation plasticity theory, 106–10
variable moduli models, 109–10
Deformational plastic models, 108–9
Delamination, 30, 399, 558
Delay detonators, inaccuracy of, 418
Delayed detonation
fragmentation particle size effect
on, 369–73
Occidental process, 494
surface uplift blasting affected by,
406–8, 463, 464
Density
difference, stratification caused by,
18
logs, 146, 147
measurement of, 29
oil yield related to, 28, 321
organic volume related to, 25, 26,
28, 233, 321–2
plasticity affected by, 323
regression analysis of, 276, 278
rubblized shale, organic volume related to, 562–3
values quoted, 25–6
Department of Energy (DOE), US
cooperative agreements with, 62,
63, 68, 484, 487, 501

Department of Energy (DOE),
US—*contd.*
Laramie Energy Technology
Center, 45–6, 63–5
purchase agreements with, 66
Department of Interior, US
attitude to oil shale development,
11
leasing by, 46, 52, 56, 58, 59, 60
see also Bureau of Mines
Depth, confining stresses affected by,
161–2, 215
Depth of burial, explosive charge,
446–8, 489, 490
Desiccation cracking, 23, 136, 302
Detonation
caps, selection of, 418
pressures, values quoted, 389, 403
velocities, values quoted, 459
Development
examples of, 9
factors affecting, 8–9
reasons for, 2–5
Deviatoric strain tensors, 85
Deviatoric stress
dilation–initiation, 247, 248, 267
statistical characteristics of, 271
tensors, 82–3
versus longitudinal strain curves,
triaxial testing, 244–7
Devonian black shales, 6, 31
density–yield relationship for, 28
organic matter, composition of, 27
tensile strength/organic volume relationship for, 194
Diagnostic tools, retorting system,
525–6, 561–70, 571–2,
574
Diagonal joints and fractures, 136
Differential scanning calorimetry
(DSC), 557, 558
Dilatation, defined, 85
Dilatational waves, 340
Direct-pull tensile tests, 297, 298–300
results from, 304, 307–8, 310–11
statistical correlations of, 311–18
Discontinuities, ignored in continuum
theory, 90, 122

Index 585

Distribution, oil shale, 5–6, 9–11
Dolomite
  oil shale structure affected by,
        20–1, 184, 194, 233
  rheological model for, 259
DOSCO Twin Boom mining
        machine, 477, 478
Dowelltown Member, 214
Drilling
  costs, break-even point, 4
  logs, 144–5
  muds, sample contamination by,
        132
  parameters, underground, 53, 470,
        489–90, 497
  techniques, 50, 53, 62
Dummy index, stress analysis, 78, 79
Dynamic bulk modulus
  shear modulus related to, 561
  temperature and organic volume
        effects on, 556–7
  see also Shear modulus
Dynamic elastic coefficients
  acoustic methods to determine,
        178, 183, 531, 532,
        533, 535, 546–52
  regression analysis of, 276, 278
  statistical characteristics of, 270
Dynamic fracture stress, defined, 439
Dynamic photoelasticity techniques,
        339–40
Dynamic stress intensity factors, 343,
        349, 425, 426–8
Dynamic Young's modulus, tempera-
        ture and organic
        volume effects on,
        553–5

Echelon fractures, 136
Economic factors, site development,
        4–5
Edge restraint effects, uplift blasting,
        402, 408, 411, 415, 419
Effective (acoustically coupled)
        modulus, 561–2
EGDN (ethylene glycol dinitrate)
        explosive, 459

Elastic moduli
  definition of, 247, 248, 531
  organic volume effect on, 167, 176,
        178, 179, 196, 199,
        234–7, 537, 551–7
  explosive fragmentation model,
        322–3, 326
  static versus ultrasonic test
        methods, 532, 533
Elastic–plastic yield data, explosive
        fragmentation model,
        323, 326–7
Elasticity theories
  linear, 97–100, 122, 123
  non-linear, 100–5, 123
Elastodynamic fracture mechanics,
        425–33
Elasto-plastic behavior, kerogen,
        527–8
Elastoviscoplasticity, 117–19
Electrical analogue models, elasto-
        plastic behavior, 528
Electrolinking, 460
Endochronic plasticity, 119–20
Endochronic viscoplasticity, 120–1, 124
Energy
  considerations, strain-rate de-
        pendency models, 204
  functions, strain, 102, 172
  requirements, shale crushing, 15
Engineering properties
  classification of, 263
  examples of, 247, 248
  factors affecting, 234–8
Enrichment processes, 16
Environmentalist groups, develop-
        ments affected by,
        11–12
Epoxy adhesives, tensile test samples
        prepared with, 298–9
Equilibrium, equations of, 86–7
Equity Oil Co., 46, 68
Estonia SSR
  enrichment processes used in, 16
  shale oil production, 9
Explicit finite-difference (EFD)
        methods, 295
  calculational sequence for, 296–7

Index

Explicit finite-difference (EFD)
  methods—*contd.*
  input data requirements for, 295–7
Exploration, oil shale, 7
Explosive(s)
  characteristics of, 491, 492
  failures, mechanical testing, 198
  fragmentation
    general principles of, 318–19,
      394–402
    models, 318–20, 338–86
      baseline calculations for, 327–8
      computational zones for,
        319–20
      data input to, 323–7, 444–5
      hard rock, 396–8
      oil shale, 398–402
      quasi-static analysis of, 329–32,
        340
    tests
      Brazilian surface mining,
        466–8
      field testing, 51, 53–4, 62, 384,
        459–61
      model studies, 339–73
        enhancement effect, 368–73
        experimental procedure for,
          339–40
        flaw effects in, 348–57
        homogeneous media, 343–8
        joint effect, 357–68
        particle sizes from, 368–73
        rock plate, 373–86
      wave systems produced in,
        340–2
  strain-rates, 195, 205, 424, 435
  types of
    aluminized, 420, 466, 515
    aluminized TNT slurry, 488, 497,
      505
    ANFO (ammonium nitrate/fuel
      oil), 51, 53, 54, 325,
      450, 464, 465, 470, 474
    characteristics of, 327, 420, 492
    Bragel CB, 466, 468
    C-4, 446
    effect on crack patterns of, 346

Explosive(s)—*contd.*
  types of—*contd.*
    EGDN (ethylene glycol
      dinitrate), 459
    IRECO, 403, 466
      characteristics of, 491, 492
    NG (nitroglycerine), 459, 460
    PETN (pentaerythritoltetrani-
      trate), 339, 346, 359,
      374, 377
    semi-gelatin dynamite, 51, 474
    TNT (trinitrotoluene), 460, 461,
      488, 505
    used by Occidental Oil Shale
      Co., 488, 497, 505
Extraction ratios, values quoted, 48,
  53, 62, 65, 476, 483
Exxon Co. (USA), 62, 470

F-value tests, factors affecting
    temperature-dependent
    mechanical properties,
    540, 574
Failure
  criteria, 92–7, 123
    one-parameter, 93, 96–7
    three-parameter, 96
    two-parameter, 93, 94–6
    see also Coulomb. . . ,
      Griffith. . . ,
      Lankford. . . ,
      Mohr. . .
  mechanisms, strain-rate de-
    pendency, 202–5
  modes
    detonation-induced, 398
    explosive (during testing), 198
    strain-rate effect on, 196
  surface
    expression of, 409, 412
    uplift blasting, 409
Fast Leader blasting technique, 413,
  416–18
Faults
  orientation of, 162, 357
  study of, 136–43
  types of, 151

Index

Feasibility studies
material properties for, 8, 230, 232
regression equations used in, 263, 282

Federal Lease Tracts
C-a, 56–8, 512
engineering properties of samples from, 235
location of, 56, 232
C-b, 58–9, 484
location of, 58, 232
strength parameters of shale from, 95–6
U-a and U-b, 59–60, 482
engineering properties of samples from, 235

Finite material approaches, constitutive modelling, 91
Finite-difference techniques, 295–7
Finite-element methods (FEMs), 295
high-temperature creep behavior, 255
input data requirements for, 295
modified split cylinder test, 217, 218, 220–1
split cylinder test, 188–9

Firing precision, surface uplift blasting, 406

Fischer assay, *see* Yield
procedure, 7, 26, 27, 147–8

Flat-jack stress measurement techniques, 152–3

Flaw
activation, fragmentation model, 348–57, 424, 433–4
size
fragmentation particle size affected by, 371
shale tensile strengths affected by, 332

Flexural strengths, statistical characteristics of, 270

Flow
plasticity
generalization of, 118–19
hypoelasticity compared with, 112

Flow—*contd.*
plasticity—*contd.*
rock mechanics, 110–12
rules, plasticity, 111, 118, 123

Folds, 151
Formation joints, 36
Fourier transforms, acoustic models, rubble beds, 563
Fracture(s)
energies
experimental determination of, 301–2
values of, 307, 309
mechanics, 217–23
modulus, values quoted, 47
records, core sampling, 133–4
stresses
crack size effect on, 428–9, 431, 432
strain-rate dependent, 424–5
strain-rate effect on, 429, 430, 432
study of, 136–9
toughness, 301
experimental determination of, 23, 217–23, 301–2
stress analysis of, 220–3

Fracturing behavior, strain-rate effects on, 197–9

Fragment
shapes, explosive fracturing, 398; *see also* Pie-shaped. . .
size
calculation of, 444
experimental versus predicted, 438
strain-rate effect on, 425
surface area, 437–8, 440

Fragmentation
mechanisms
Coursen breakage theory, 399
not considered in classical hard-rock breakage theory, 396
models, 318–20
computational models, 441–5
data input for, 442, 444–5

Index

Fragmentation—*contd.*
models—*contd.*
computational models—*contd.*
fragment size calculation for, 444
stress-wave propagation in, 338, 340–86, 441–3
development of, 433–45
in situ retorting calculations using, 453
numerical simulation, 445–53
strain-rate sensitive, 423–54
processes, control of, 292
tests
fracture studies in, 139–40
Geokinetics Inc., 140
Occidental Oil Shale Co., 56
see also Explosive fragmentation tests
Fredholm integral, 428
Free index, stress analysis, 78
Free-face effects, blasting, 343–8, 424, 435, 446, 449

Gamma-ray logs, 147
Gas
penetration, hydrofracture effects on, 391
pressurization, explosive fragmentation, 327, 331–2, 333, 338, 386, 397–8, 399
Gasoline consumption, 3
Gassaway Member, 214
Gassy mines, MSHA classification, 57, 59, 66
Geokinetics Inc. (GKI)
Fast Leader blasting technique, 413, 416–18
in-place retorting process, 16, 62–3, 394, 461–5
efficiency of, 63, 465
Lofreco blasting technique, 392–420
test site
explosive fragmentation model for, 318–20
location of, 62, 293, 461

Geokinetics Inc. (GKI)—*contd.*
test site—*contd.*
tensile strengths of shale from, 310–11
Geolograph drilling-depth recording instruments, 134
Geometry conditions, strain–displacement, 87–8
Geophysical logs, 145–7
Government
interference with development, 11–12
ownership of Green River Formation (USA), 11
support for development, 9
surface uplift blasting tests by, 390, 391
see also Bureau of Mines, Department of Energy, Department of Interior
Grade, *see* Organic volume. . . , Yield
Grain size, oil shales, 18
Granite
blast-induced fragmentation studies on, 373–4, 377, 378, 380, 382–3
triaxial testing of, 243
Great-depth stress measurement techniques, 156–8
advantages/disadvantages of, 158
Green River Basin
core samples, fractures in, 134
hydraulic fracturing in, 37
Green River Formation
description of, 9–10, 44, 292–3
factors affecting development of, 10–12
ownership of, 11
typical sample from, 19, 168
see also Laney Member, Mahogany Zone, Piceance Creek Basin, Tipton Member, Uinta Basin, Wilkins Peak Member
Griffith criteria, 96–7, 187, 436
Ground motion, Occidental process rubblization, 507–11
Gulf Oil Corp., 56, 512

Index

Hardness, oil shale, 270, 477
Haystack Mountain, *in situ* retorting tests on, 46
Heading rounds, 51, 53, 62, 67, 471–5
Heat-carrier materials, retorting 13
Heaviside loading, 426, 430
High-speed photography, 339, 340, 374–7, 404
High-temperature
creep testing, 260–3
effects, 30–1, 231–2
Historical perspective, 43–68
Holography
acoustic, 572
blast-induced fragmentation testing, 376–7, 382–3
Homalite 100, fragmentation studies in, 338–73
Hondros modified (elasticity) theory, 187, 188, 189, 191
Hookean (elastic) model, 114, 259, 528
Hooke's law, 97, 123
linear generalized expression of, 102, 103, 171, 172
Horizontal stress, vertical stress, effect on, 151
Horse Draw shaft, 46, 66–8
Hot-film flow logs, 147
Hydraulic fracturing, 36–8, 390–1
in situ stress measurement by, 156–8
orientation of, 157, 162
tests in Piceance Creek Basin, 161–2
Hydraulic rotary drilling, samples obtained from, 130–3
Hydrogen content, 21
analysed by neutron logs, 146
thermal behavior affected by, 29–30
Hydro-stone (commercial gypsum product), 379
Hyperbolic equations, yield/organic matter conversion, 27, 28
Hyperelasticity theory, 100, 102–3, 176

Hypoelasticity theory, 103–5
flow plasticity compared with, 112

Idealized models, continuum theory, 97
practical conditions for, 91
IJUMP computer code, 408
Illite, 20, 192
Impulsively overloaded laminate failure, 398, 404
In situ retorting methods
permeability required for, 185, 195, 292, 391, 461, 462–3
see also Geokinetics. . . , Modified *in situ* (MIS). . . Occidental. . . , Rio Blanco. . . , True *in situ* (TIS). . .
In situ stresses, 149–52
creep caused by, 254–5
measurement of, 152–9
mine stability affected by, 159–61, 231–2
Piceance Creek Basin, 161–2
Incremental approaches
constitutive modelling, 92
flow plasticity, 110–11
hypoelasticity modeling, 103–5
Incremental theory of plasticity, 108–9
Index notation, stress analysis, 75, 76, 78
Indirect tensile tests
results from, 186–95, 235, 267, 305, 308–9
statistical analysis of, 270, 276, 278, 311–18
see also Brazilian. . . , Modified split cylinder. . . , Split cylinder. . .
Inert decks, surface uplift blasting, 402, 415, 416
Inertial effects
fracture behavior affected by, 424
ignored in strain-rate failure model, 202

590 Index

Instantaneous shut-in pressure (ISIP), 156–8
measurement of, 156
Instrumentation, mechanical testing, 174–5, 191, 197, 209–10
Internal friction angles, 95
measurement of, 95, 215, 231, 238, 239, 250
orientation effect on, 237
regression analysis of, 277, 279
statistical characteristics of, 271, 281
Intrinsic time measure, 119, 124
Invariants
deviatoric stress tensor, 83
stress tensor, 81, 95
Irati (Brazil), prototype mine and plant, 465
IRECO explosives, 403, 466
characteristics of, 491, 492
Isodynes method, stress analysis, 188, 189
Isotropy
assumed in strength models, 93
lateral, 18–20, 22, 169
see also Transversely isotropic material

Janach unloading model, 202
Joint-initiated (JI) fracture
model material, 361, 366
rock specimens, 380, 382, 384, 385
Jointed systems
fragmentation
rock affected by, of, 386, 401, 402
studies on, 357–68, 380–2
practical examples of, 36, 49, 357, 358
Joints
open, effects in blast-induced fragmentation, 384
study of, 136–43

Kamp Kerogen, Geokinetics Inc. test site at, 399, 419, 461

Kelvin–Voight (viscoelastic) model, 115–16, 259, 262
creep functions for, 116, 259–60
Kernel functions, creep model, 208, 209
Kerogen
content of, Brazilian oil shale, 465–6
decomposition cracking of, 551
description of, 44, 251
elemental composition of, 27
high-temperature
decomposition of, 30, 524, 558, 573
softening of, 255, 527, 535
physical characteristics of, 21, 31
plastic flow of, 255, 257
transition temperature, 535
properties affected by oxidation of, 560
rock structure affected by, 21, 545
strength, 323
loss, 30, 560
volatile hydrocarbons lost from, 558, 560, 573
Kronecker delta functions, 80
Kumar brittle-fracture model, 203
Kurtosis, 269, 281
values quoted, 270–1
Kuster–Toksoz model
acoustic velocities in, 563–4
density effects in, 568–9
rubble bed analysed using, 561, 562, 563

Lagrangian methods, 295–6, 445
Laminations, 18, 19, 169
in situ retorting affected by, 401–2
see also Delamination
Laney Member, vertical joints and fractures in, 137, 138, 139
Lankford failure criterion, 196
Laplace transform, elastodynamic equation, 427–8
Laramie Energy Technology Center (LETC), 45–6, 63–5
Rock Springs test site, 64, 307

Index

Lateral extension, surface uplift blasting, 393, 415–18
Lateral uniformity, 18–20, 167
Leached Zone, 34
  characteristics of, 32
  oil extraction from, 68, 536
Limestone
  bedding and joint sets in, 357, 358
  blast-induced fragmentation studies on, 373–4, 377–81
  composition of, 374
  rheological model for, 259
  triaxial testing of, 242–3
Linear elasticity, 97–100, 123, 206
Linear viscoelasticity, 114–17, 124
Liquid explosive fracturing tests, 459–61
Lithologic logs, 144–5
Loading–unloading
  cycles, multiple-stage triaxial testing, 244–6
  definition of, 107–8
  reversibility of, in hypoelastic materials, 104–5
  treatment contrasted in elasticity and plasticity models, 106
Locations of oil shale development, 5–13, 232, 465
Lofreco blasting technique, 392, 394
  detonators and explosives for, 418–20
  general principles of, 394–402
  lateral extension by, 393, 414–18
  overburden raising step in, 392–3, 402–15
  random nature of retort beds, 400–1
  retort beds resulting
    ordered, 415–16
    randomized-to-ordered ratios, 400–1, 415
Logan Wash Mine, 46, 55–6, 484
  mechanical properties of shale from, 263, 265
Logging, types of, 144–7
Low (power) explosives, used in surface uplift blasting, 419–20, 464

Lurgi-type pilot retorts, 58

Macro fractures
  origin of, 138–9
  study of, 139–43
Mahogany Zone, 34
  characteristics of, 32, 35, 47
  in situ retorting experiments in, 62, 293, 461
  laminations in, 18, 19
  location of, 32
  minerals in, 35, 192
  non-representative nature of samples from, 230–1
  shale samples
    conversion factors for, 27
    fracture energies of, 306–7
    mechanical testing of, 167, 172, 179–81, 193
    tensile strengths of, 304-6, 310–11
Marble, triaxial testing of, 243
Marlstone, oil shale classified as, 251
Materials handling
  costs of, 14–15
  equipment for, 66, 466, 477, 478, 479, 480, 481, 483
Matrix notations, 74
Maximum shear
  strain, determination of, 86
  stress(es)
    criterion of failure at, 93, 94
    determination of, 82
Maxwell (viscoelastic) model, 114–15, 117, 259
Mechanical characterization, 165–224
Mechanical logging instruments, 144
Mechanical properties, 31–8
  application of, 223–4
  mineralogical effects on, 184–5, 193–4, 233
  organic volume, effect on, 23, 167, 178, 181, 184, 200–1, 224, 314–15, 317–18, 322, 326, 538–41, 543–6
  statistical characteristics of, 270–1
  strain-rate dependence of, 196–7, 199–202

592   *Index*

Mechanical properties—*contd.*
  temperature dependence of, 526, 528, 538–46
  typical values quoted, 67
Methanol, as alternative fuel, 3
Micro fractures, core and outcrop sample, 138
Microcracking mechanisms, 113, 203
  triaxial testing, in, 243
Mine
  opening, orientation of, 160
  simulation, blasting tests, 450–3
  stability, *in situ* stresses, effect of, 159–61
Mine Safety and Health Administration (MSHA),
  classifications, 57, 59, 66
Mineralogical logs, 145
Mineralogy, mechanical properties affected by, 183–5, 192–4, 233, 236, 332
Minerals, discovery of new types of, 145, 148–9
Mining
  research, 457–517
    Horse Draw shaft, 67
  types of, 15
Mixed mode loading, explosive fragmentation, 350
Mobil Oil Corp., experiments by, 52–4, 469–70, 471–3, 474–6
Modelling methods, 294–7
Moderate depth stress measurement techniques, 154–6
  advantages/disadvantages of, 155–6
Modified *in situ* (MIS) retorting, 230, 253–5, 292, 483–516
  creep effects in, 254–5
  high-temperature effects in, 231–2
  pillar temperature profiles in, 524
  true *in situ* retorting compared with, 394
  see also Occidental Oil Shale Inc.
Modified split cylinder tests, 218–19, 236

Mohr–Coulomb failure
  criteria, 95–6, 214, 239
  law, 214, 215, 239
Mohr–Coulomb strength parameters
  determination of, 214, 215, 231, 239, 282
  regression analysis of, 277, 279
  statistical characteristics of, 271
  values quoted, 250
Mohr's circles, 75, 78, 82, 214, 238–9 242
Moiré method, split cylinder test loading analyzed by, 187–8
Moisture, conservation of, in samples, 135, 298, 302
Morocco, oil shale deposits in, 5
Motion, equations of, 86–7
Mud cuttings, samples obtained from, 130–3
Multi-acoustical (physical property) logs, 146
Multi-deck charges, damage regions calculated for, 453–4
Multi-hole blasting, 318
Multi Mineral Corp., cooperative research agreement with, 66–7
Multiple Failure State Triaxial Tests, 243
Multiple regression analysis, 279, 280
  correlation coefficients for, 278–9, 280
  temperature-dependent mechanical properties, 540
  thermal behavior, 528
Multiple-stage triaxial (MST) testing, 242–51
  mechanical properties from, 247–8, 266, 267, 271, 277, 279
  possible reason for non-success of, 247
  procedures for, 244–7
  strength parameters obtained from, *see* Mohr–Coulomb. . .

Nahcolite, 23, 32, 35
  mechanical properties of, 67

Index

Nahcolite—*contd.*
  shale mechanical properties
    affected by, 540, 545
  temperature-dependent elastic
    properties affected by,
    534
  thermal decomposition of, 527, 573
Natural parting planes (NPPs), 134,
    308, 311
Naval Oil Shale (NOS) Reserve
  engineering properties of shale
    from, 237, 249
  location of, 45, 232
  tensile strengths of shale from,
    304–6
  statistical analysis of, 311–15
Neutral loading, continuity condition
    at, 108, 109
Neutron (hydrogen content) logs, 146
Newtonian (viscous) models, 114,
    259, 528
Nitroglycerine (NG) explosive, 459,
    460
Non-linear elasticity theory, 100–5,
    123
Non-linear pseudo three-dimensional
    creep model, 207–9
Non-linear three-dimensional stress–
    strain relationships,
    166–72
  experimental investigation of,
    172–85
Non-linear viscoelastic model, strain-
    rate dependency,
    202–5
Notations
  continuum mechanics, 74
  strain analysis, 84–5
  stress analysis, 75–8
Nuclear plants, retorting heat from,
    16, 46
Numerical methods, 217, 255, 294–7

Occidental Oil Shale Inc. (OOSI), 55
  Cathedral Bluffs Project, 58, 484
  design equations (for retorts 6–8),
    488–91

Occidental Oil Shale Inc.
    (OOSI)—*contd.*
  engineering properties of shale
    from, 249
  explosives used by, 488, 497, 505
  ground motion studies, 507–12
  in-place retorting process, 16, 55–6,
    58, 484–516
  retort
    1E, 56, 484–5
    2E, 56, 485
    3E, 56, 485–6
    4, 56, 486–7, 510
    5, 56, 491–5, 509, 510
    6, 56, 495–501, 504, 510
    7/8/8x, 56, 501–12
Octahedral normal and shear stress–
    strain curves, 101
Octahedral plane, defined, 82
Octahedral strains, 85
Off-tract disposal, 57, 58
Oil shales
  characteristics of, 17, 251
  definition of, 2, 16–17, 44
  typical mechanical properties of,
    67, 270–1
  typical physical properties of, 20–1,
    48
  typical samples of, 19, 168
Opencast mining operations
  Brazil, 465–8
  USA, 56, 58
Organic matter, oil shale, *see*
    Kerogen
Organic volume
  acoustic velocities affected by, 547,
    548
  Brazilian oil shale, value quoted,
    465–6
  compressive strength dependent
    on, 65, 92–3, 178,
    181, 184, 200–1, 541,
    543–6
  compressive strengths affected by,
    541, 543–6
  creep behavior affected by, 211–12,
    213
  density relationships, 25, 26, 28,
    233, 322

594   *Index*

Organic volume—*contd.*
  elastic coefficients dependent on,
      167, 176, 178, 179,
      196, 199, 234–7, 537,
      551–7
  explosive fragmentation model,
      322–3, 326
  elastic/plastic yield affected by,
      235, 323
  tensile strengths affected by,
      192–5, 322, 326, 537,
      538–40
  Mahogany Zone, 314–15
  Tipton Member shale, 317–18
  ultimate strain to failure affected
      by, 200, 205
  weight conversion for, 26–7
  yield related to, 27, 93, 147–8, 173,
      210
Orientation, sample, mechanical
      properties affected by,
      237, 280–1, 536, 541,
      545
Outcrop(s)
  oil shale, 7
  vertical fractures and joints in,
      137
  sampling, 128
Overburden
  off-site disposal of, 57, 58
  raising, 392–3, 402–14, 461
    ballistic motion in, 410, 414
    blast hole orientation effect dur-
      ing, 395
    edge pre-splitting used in, 402,
      411
    factors affecting, 395
    fallback in, 413–14
    lifting efficiency of, 395
    movements in, 408
    vertical velocities of, 402, 404–5
  stresses due to, 150, 159, 160,
      161–2
  creep caused by, 254–5
  laboratory simulation of, 213,
      215
  thickness of, 461
  surface uplift blasting affected
      by, 395, 416, 419

Overburden—*contd.*
  thickness-to-retort-depth ratio, 402,
      408, 409, 463
Overcoring techniques, 154–6
  advantages/disadvantages of, 155–6

p–q diagrams, 214, 215, 216, 239,
      241, 245, 246, 247, 251
P-waves, ultrasonic, 547
P-waves (primary dynamic pressure
      waves), 340–1
  typical velocities of, 341
Parachute Creek Member, 31
  Colony Development operation on,
      46, 54–5
  modified *in situ* retorting experi-
      ments in, 293
  shale samples
    creep testing of, 206, 210–11
    mechanical testing of, 167,
      179–81, 185, 193, 264
    statistical analysis of, 231
  Union Oil Company production
      from, 45, 51–2, 65–6
Paraho Development Corp., 60–1,
      470, 474–6
Particle(s)
  size distribution, 18
  velocity measurements, *in situ* re-
      tort blasting, 402, 404,
      405, 410, 509–10
Passive mechanical testing, 557–60
Path-independent behavior models,
      101
Peak strengths, triaxial testing, 242,
      248–50
Peeling-type failures, mechanical test-
      ing, 199
Penny-shaped cracks, factors affect-
      ing fracture, 428–31,
      432–3
Permeability
  hydrofracture effects on, 391
  physical properties affected by, 29,
      524, 560
  processes to create, 16, 185, 195,
      292, 461, 462–3
  see also Void volume ratios

Index

PETN (pentaerythritoltetranitrate) explosive, 339, 346, 359, 374, 377
Petrobras (Petroleo Brasileiros SA), 465
Petroleum
applications, 3
US reserves, 4
world reserves, 4
Phillips Petroleum Co., 60
Phosphoria Formation, oil deposits in, 29
Photoelasticity studies, dynamic observation, 339–40
Physical properties, 20–1, 48, 265, 266, 267
regression analysis of, 276, 278
statistical characteristics of, 270
Piceance Creek Basin
hydraulic fracturing tests in, 36, 161–2
in situ stresses in, 161–2
in-place retorting methods in, 56–7, 58
joint systems in, 36
location of, 10, 32
mechanical characterization of samples from, 172, 185, 230–1
mining map of, 232
modified *in situ* retorting experiments in, 293
Pie-shaped segments, explosive cracking, 344, 357, 398
break up of, 347, 348, 355
Pillar(s)
creep of, 257
deformation measurements, 68
failure, 55, 238
heat flows in, 571
sizes, 52, 53, 54, 62, 469, 476, 481
selection of, 159–60, 476, 570–1
strength, *in situ* retort, 570–1, 572, 574–5
stresses, 53
temperature effects on, 524
Pinch tests, 300–1
experimental results from, 305

Plane cracks, factors affecting fracture, 431–3
Planning, shale properties required to enable, 8, 230, 232, 282
Plasticity
density dependence of, 235, 323
theories
deformation, 106–10, 123
flow, 110–12, 123
Pneumatic drilling, samples obtained from, 128, 129–30
Poisson's ratios, 97, 169
dynamic versus static measurement, 551, 557
mineralogical effects on, 184
organic volume effect on, 167, 176, 178, 180
regression analysis of, 277, 279
static versus ultrasonic test methods, 533
statistical characteristics of, 270, 271
temperature and organic volume effects on, 551
values quoted, 98, 151, 180, 444
Political factors, oil shale development affected by, 11–12
Pore(s)
aspect ratios, acoustic velocities affected by, 562
pressure effects, thermal behavior, 529–31
Porosity
mechanical properties affected by, 29, 529
statistical characteristics of, 270
Post-mortem assembly, rock fracture testing, 377–8, 383
Potential resource rocks, defined, 2
Powder factors
fragmentation tests, 446, 449
in situ retorting, 392, 402, 412, 465, 497, 504
normal underground mining, 51, 53, 54, 67, 470, 474
surface mining, 468

596   *Index*

Precracked split cylinder test specimens, 218–19
Pressure decay, surface uplift blasting, 397, 399, 403
Pressure-containment devices, blast-induced fragmentation testing, 339, 374
Pressurization model, crack internal, 319–20
Price differential, shale oil vs. natural oil, 12
Primary creep, 254–5, 258
Primary (*P*) stress waves, 340–1
Principal axes, stress, 80–2, 150
Principal shear
  strains, 85–6
  stresses, defined, 82
Production methods, 13–16, 230
  see also Geokinetics. . . , *In situ* retorting, Modified *in situ* (MIS) retorting, Retorting, Occidental. . . , True *in situ* (TIS) retorting
Progressively fracturing theory, 113–14, 123
Propping mechanisms, crack-opening, 320, 402
Prototype Oil Shale Leasing Program, 46, 56, 59–60
Purchase agreements, Department of Energy (USA), 66
Pyrite nodules, samples blemished by, 23, 25

Quarry blasting, fragmentation model of, 367, 368
Quartz, oil shale, 20
Quasi-static analysis, explosive fragmentation model, 329–32

Radioactive pulse logs, 147
Radio-frequency diagnostic, *in situ* retorting, 572

Rankine criterion, 96
Rate-dependence, viscoelastic, 117
Rayleigh wave velocity, 343–4
Recementation, high temperature, 531, 560, 573
Reflected stress wave systems
Reflected wave systems
  borehole cracks affected by, 345–8
  crack initiation by, 352–5, 446
  notation of, 342
Regression analysis
  computer program for, 176
  creep behavior, 211–13
  engineering properties, 272–81
  mechanical properties, 167, 171
  strain-rate effects, 200–2
  temperature-dependent mechanical properties, 540, 541
  see also Bivariate. . . , Multiple regression. . .
Relaxation tensor, viscoelastic, 116
Relaxed areas (tectonically), 37
Resistivity measurements, 146, 557, 558
Resource evaluation techniques, 8
Retardation times, creep testing, 262–3
Retorting
  recovery efficiency
    Geokinetics process, 63, 465
    Occidental process, 485, 486, 495, 500
    Rio Blanco process, 514
  systems
    described, 13–14, 44
    process diagnostics in, 525–6
    throughputs quoted, 52
    see also In situ retorting. . . , Geokinetics. . . , Modified *in situ*. . . , Occidental. . . , Rio Blanco. . . , True *in situ*. . . ,
Retorts, surface
  TOSCO, 54, 58, 60, 61, 62
  Union B, 60, 65
Reworked zones, borehole detonation, 343, 396–7

Index

Rheological models
  creep behavior, 206–9, 238, 258–60
  three-parameter, 259–60, 262
Rich shale, 7
  behavior in sorting of mud cuttings, 131–2
  estimates of, 7
Rio Blanco Oil Shale Co., 46, 56–8, 512–16
Rock
  engineering, 72
  fracture model, 318–19
  mechanics
    continuum theory of, 71–124
    introduction to, 71–3
    oil shale, 16–31, 45, 67–8, 71–3
    scope of, 73–4
  mechanics research, Horse Draw shaft, 67–8
Rock Springs
  fracturing experiments at, 37, 46, 64, 460
  LETC test site at, 64, 307
  location of, 10
  tensile strengths of samples from, 307–10
    statistical analysis of, 316–18
  true *in situ* retorting experiments at, 64, 293
  vertical joints and fractures at, 137
Roof
  falls, Anvil Points, 48, 49–50
  sag
    measurements of, 48, 53, 54
    tensile stresses induced by, 160
  stability, stress factors influencing, 160
Room-and-pillar mine design, 47–8, 51, 52, 54, 58, 60, 61, 476, 478, 479, 482
  creep effects in, 206
Rotary drilling
  development tests for, 50, 53, 476
  samples obtained by, 128, 129
Rotary–percussive drilling
  techniques, 488
Rotating drum camera, 374–5

Rubble beds, acoustic properties of, 561–70
Rubblization
  Geokinetics/Lofreco process, 62–3, 393, 414–18, 461, 463
  Occidental process, 55, 483–4, 505–6
  blast damage caused during, 493, 495, 506
  rock fragmentation tests for, 56, 140; see also Explosive fragmentation tests
Rulison Mine, location of, 232
Running averages, fracture properties, 305, 306, 307, 308, 309
  statistical analysis of, 311
Rupture pressure, hydraulic fracture stress measurement, 157

S-waves, secondary or shear dynamic pressure waves, 341
  typical velocities of, 342, 428
S-waves, ultrasonic, 547
Saline Zone
  characteristics of, 32
  mechanical properties of samples from, 236
  oil extraction from, 67, 536
Sample preparation
  blast-induced fragmentation testing, 376, 377
  creep testing, 209
  fracture testing, 302–3
  mechanical testing, 170, 173, 191, 197, 209, 214–15, 218
Sampling, 21–2, 23, 127–33
  subsurface, 129–33
  surface, 128
Sandstone
  rheological model for, 259
  triaxial testing of, 243
Scarp-like failure, surface uplift blasting, 409, 411
Scotland
  oil shale mining in, 15
  shale oil production in, 5, 9

598 *Index*

Secant
stress–strain models, 91, 101–3
(Young's) modulus, 247, 248, 531
Secondary (steady state) creep,
254–5, 258
strain rates, factors affecting, 260,
262
Sedimentary conditions, oil shale
affected by, 17–18
Sedimentary rocks, oil content of, 7
Seismic studies, 403–4, 508–10
Seismic vibration, surface uplift blasting, 419
Semi-gelatin dynamite, 51, 474
Shadow zones, explosive fragmentation, 342, 356
Shale
classified as compressed clay, 90
shakers, mud cuttings sampling,
130, 131–2
Shallow water deposition, 36
Shear
failure, 82, 93, 94–5, 199
modulus
calculated from compressive
strength, 98
dynamic bulk modulus related
to, 561
experimental determination of,
166, 169
organic volume, effect on, 169,
171
regression analysis of, 276, 278
statistical characteristics of, 270;
see also Dynamic bulk
modulus
strains
engineering, 84–5
tensorial, 84
(S) stress waves, 341, 342, 428
stresses, maximum, determination
of, 82
Shell Oil Company
hydraulic fracturing experiments
by, 36–7
in-place retorting tests by, 46, 536
Shock wave, *see* Stress wave. . .
Shore hardness, statistical characteristics of, 270

Silver Point Quad, triaxial testing of
samples from, 213,
214–15
Sinclair Oil and Gas Co., 46
Single-stage triaxial (SST) testing,
213–16, 239–42
Skewness, 268–9
values quoted, 270–1, 281
Slump faults, samples blemished by,
23, 24
Soil mechanics
plasticity theory in, 106
rock behavior expressed by, 72
Solid mechanics equations, 72, 86–90
Sonic, *see also* Acoustic. . . , Ultrasonic
Sonic (density determination) logs,
146
Sound velocity determinations, 234,
236, 237
Source rocks, defined, 2
South Africa, shale oil production, 5,
9
Space-heating applications, 3
Spalling fracture behavior
explosive fragmentation, 342,
353–5, 382, 398, 493
in situ retorting, 500
Specific charges, *see* Powder factors
Specific gravity, regression analysis
of, 276
Spent shale
acoustic velocities in, 547
applications for, 61, 571
piles, 15, 571
Split barrel core sampling, 133
Split cylinder test, 166, 186–95
experimental details for, 191–2
precracked specimen version of,
218–19
theory of, 186–91
Spring (elastic) models, 114, 258, 259,
528
Spring-and-dashpot models, 114–16,
117, 203, 258–60, 528
Standard Oil Company, 54, 56, 60, 512
State of strain, defined, 85
State of stress, 76
stress wave propagation, 362

Index

Statistical analysis
  computer programs for, 268, 313
  fracture properties, 311–17
  temperature-dependent mechanical
    properties, 540
Statistical analysis and modelling,
    263–81
  bivariate regression analysis, 272,
    276–7, 280
  correlation coefficients used in,
    269, 272, 273–5
  data sources for, 263–5
  methods and procedures, 266–8
  multiple regression analysis, 279,
    280
  properties analyzed by, 265–6
  statistical and distributional charac-
    teristics for, 268–9
Statistical characteristics, 268–9
  mechanical properties, 270–1
Statistical Package for the Social Sci-
    ences (SPSS) compu-
    ter programs, 268
STEALTH computer programs,
    295–6, 318
Steam injection methods, 68
Stepwise multiple regression analysis,
    280
Stepwise regression analysis, temper-
    ature-dependent
  mechanical properties, 540
Stiffness matrices, 171
  coefficients for, 171, 178, 182
  cubic triaxial testing, 252
Strain(s)
  analysis, definitions and notations
    for, 84–5
  hardening, 106
  rosette devices, 153, 155
  softening, 106, 113, 123
  three dimensions, in, 83–6
Strain-rate effects, 195–205
  compressive strengths affected by,
    200–2, 205
  dynamic fracture affected by, 423–54
  elastic moduli dependent on, 196,
    199, 200, 205, 532
  experimental determination of,
    197–8

Strain-rate effects—*contd.*
  failure mechanisms for, 202–5
Stratification, depositional, 18
Stratigraphic logs, 145, 303
Stratigraphic variations
  fracture properties, 304–11
  fragmentation model materials, 325
  oil yield, 34, 306–7
Strengths, temperature effects on,
    256–7
Stress(es)
  analysis
    definitions and notations, 75–8
    fracture toughness testing, 220–3
    modified split cylinder test, 220–3
    split cylinder test, 187–91
    two-dimensional, 75, 78
  distributions, normalized
    modified split cylinder test, 221,
      223
    split cylinder test, 189–90
  in situ, see In situ stresses
  intensity factors
    Chong–Kuruppu specimen
      center hole radius effect on,
      221–2
      crack length effect on, 220–1
    elastodynamic fracture, 425,
      426–8, 430
    explosive fragmentation, 343
    factors affecting, 351, 355
    fracture mechanics, 217–18
  levels
    creep behavior affected by,
      211–12, 213
    defined, 175–6
    elastic coefficients dependent on,
      176–7, 323
  relaxation, 115, 124
  tensors, components of, 76–7
  thermal expansion, 529
  three-dimensional, 74–83
  vectors
    components of, 79, 80
    defined, 75–6, 84
  wave propagation
    explosive fragmentation model,
      338, 340–2, 396–8,
      441, 443, 445

600 *Index*

Stress(es)—*contd.*
  wave propagation—*contd.*
    explosive fragmentation
      model—*contd.*
        enhancement effects in, 368–73
        flaw effects on, 348–57
        homogeneous media affected
          by, 343–8
        jointed material affected by,
          357–68
        rock sample, 377–86, 446, 448
Stress–strain relationships
  non-linear three-dimensional,
    166–72
  transversely isotropic elastic mate-
    rial, 98–100
Stress/strain relief techniques, stress
  measurement, 153,
  154
Strike faults, 151
Subsidence theory, 409
Summation convention, stress analysis, 79
Sunedco (Sun Oil Co.), 60
Supply and demand, oil shale development affected
  by, 12
Surface
  mining operations, 56, 58, 465–8
  sampling, 128
  stress measurement techniques,
    152–4
    advantages/disadvantages of,
      153–4
  uplift blasting
    general principles of, 394–402
    overburden raising by, 392–3,
      402–14
  uplift blasting methods, 140, 318,
    390–420
Surface-drilled *in situ* retorting
  methods, 62–3, 64,
  465
  see also Lofreco blasting technique
Sweden, shale oil production, 9
Synthetic fuel, definition of shale oil, 2
Synthetic Liquid Fuels Act (USA,
  1944), 45

Talley Energy Systems Inc., 38, 293
Tangent (Young's) modulus, 247,
  248, 531
Tangential stress–strain models, 92,
  101, 103–4
Tangential stresses, borehole, 340
Tar oil, coal, 17
Tar sands, 5, 17
Temperature
  acoustic velocities affected by, 547,
    548
  creep behavior affected by, 236,
    253–6, 260–3
  effects, 523–75
    analysis of, 535–60
    applications of, 560–72
    measurement of, 531–4
    mechanisms underlying, 526–31
  elastic moduli affected by, 256–7,
    546–57
  mechanical properties affected by,
    256–7, 538–46
Tenneco, 58
Tennessee oil shales, 6, 27, 28, 31,
  194–5
Tensile strengths, 185–95
  experimental determination of,
    22–3, 191–2, 297–301,
    534, 536
  flaw size effect on, 332
  mineralogical effects on, 184–5,
    193–4, 233, 332
  organic volume effect on, 192–5,
    322, 326
    Mahogany Zone, 314–15
    Tipton Member, 317–18
  possible experimental methods for,
    293–4, 297–8
  statistical characteristics of, 270
Tensile stresses, defined, 74, 77
Tensile tests
  direct-pull, 297, 298–300
  indirect, 166, 186–95, 298, 300–1
Tension
  cracks, samples blemished by, 23,
    24
  failure surface, 96
Tensors, second-order, 78, 85

Index

Tertiary creep, 254–5, 260
  rupture times for, 254, 260–2
Testing
  machines, stiffness of, 198
  modes, oil shale, 22
Texas oil field, development of, 11
Thermal analysis techniques, 557–60
Thermal behavior effects, 19–31
Thermal decomposition
  kerogen, 30, 524
  rocks, 527
    Arrhenius-type equation for, 527
Thermal expansion, stresses due to, 529
Thermal mechanical properties, 253–63
Thermogravimetric analysis (TGA), 557, 558
Thermophysical characterizations, 525–6
Thermosonometry (TS), 558–60
Three-point-bend tests, fracture toughness determined by, 301–2
Thrust faults, 151
Time
  effects, 206–8, 212–13
  factors, continuum theory, 91
  limitations, resource development, 4–5
Time-dependent flow models, 114, 124, 254–6, 258–60
Time-step scaling techniques, explosive fragmentation model, 297, 329
Timing effects, surface uplift blasting, 406–7, 415, 464
Timoshenko–Goodier elasticity theory, 187, 188, 340
Tipton Member, 31
  hydraulic fracturing in, 37, 38
  in situ retorting experiments on, 64
  minerals in, 192
  shale samples
    fracture energies of, 309
    mechanical testing of, 167, 172, 179–81, 185, 193
    strain-rate effects, 200–1

Tipton Member—*contd.*
  shale samples—*contd.*
    tensile strengths of, 307–10
    statistical analysis, 316–18
TOODY-IV stress wave propagation code, 445
TOSCO (The Oil Shale Company)
  Colony Mine development, 61–2, 470
  retorts, 54, 58, 60, 61, 62
Transportation, fuel demands for, 3
Transversely isotropic material
  elastic model for, 98–100, 171–2
  oil shale represented as, 169, 217, 238, 253, 533
Trenching, sampling method, 128
Tresca criterion, 94
Triaxial testing
  cubic, 231, 251–3
  elastic moduli obtained by, 271, 277, 279, 534, 537, 546
  mechanical properties obtained by, 266, 267, 271, 277, 279, 536
  multi-stage, 242–51
  single-stage, 213–16, 239–42
Trinitrotoluene (TNT) explosives, 460, 461
True *in situ* (TIS) retorting, 62–3, 230, 292, 458–65
  experiments on, 63, 64, 293
  modified *in situ* retorting compared with, 394
  vertical control of retort front in, 394–5
  see also Geokinetics. . .
Tuffs, 20, 136, 138
Two-dimensional stress analysis, 75, 78
Two-phase models, acoustic properties of, 561–70

Uinta Basin, 10, 35, 60, 62–3
  fragmentation tests at, 140
  true *in situ* retorting experiments in, 62, 293

Index

Ultimate compressive stress
experimental values of, 182
prediction of, 178, 181
Ultimate strain to failure, organic
volume effect on, 200, 205
Ultrasonic methods, elastic moduli
determined by, 178, 183, 531, 532, 533, 535, 546–52
Underground mining operations, 468–83
see also Anvil Points Mine, Colony
Mine, Union Mine
Uniaxial compressive testing, 174–5
mechanical properties obtained
from, 267, 270, 276, 278
Uniaxial strengths, 92–3
Union B retorts, 60, 65
Union Mine
location of, 51–2, 232
shale samples
engineering properties of, 235, 249
mechanical properties of, 263, 265
Union Oil Co., 11, 45, 51–2, 65–6, 479–81
Unit conversions, xix
Universal-joint arrangements, direct-
pull tensile testing, 298
University of Colorado at Denver
(UCD),
statistical studies by, 263–5, 268
USSR, shale oil production in, 9, 16
Utah
oil shale deposits in, 6, 10, 35
shale oil production in, 59–60, 62–3
see also Federal Tracts U-a and
U-b

V-cut detonation arrangements, 51, 53, 67, 392, 412, 463, 468

V-notched specimens, fracture tough-
ness testing, 301–2
Variability, importance of in *in situ*
retorting, 333
Variable moduli models, deformation
theory of plasticity, 109–10
Varves, 18, 20, 169, 527, 574
Vectors, tensor definition of, 78
Viscoelastic models
rock behavior expressed by, 72, 528
strain dependency of, 203–5
Viscoelasticity, linear, 114–17, 124
Viscoplastic potential function, 118
Viscoplasticity, endochronic, 120–1, 124
Void strain, defined, 320
Void volume ratios, 524, 560
Geokinetics process, 63, 394, 400, 401
Occidental process, 484, 485, 486, 492, 497, 504
Rio Blanco process, 516
Volcanic ash falls, 20
Volterra–Frichet relation, 260
von Karman's notation, 78
Vugs, 140, 143

Walking W design blasting arrange-
ments, 393, 403
Wave propagation models
blast-induced, 338–73
ultrasonic, 561–70
see also Stress wave propagation. . .
Weathering effects, samples affected
by, 128
Wedging action, tensile failure, 190
Weibull functions, 436
Westergaard's solution, crack tip
stress, 221
White River Shale Project, 46, 60, 482–3
Wilkins Peak Member, hydraulic
fracturing in, 38
Wire saws, 173, 174, 197, 302

Index 603

Withdrawal Order, oil shale, 11
World Energy Conference, barrel of
   oil defined by, 8
World reserves, petroleum, 4
World resources, shale oil, 2
Wrapping, core sample, 135, 144,
   302

X-ray diffraction techniques, 148–9,
   236

Yellow Creek, 56
Yield, 7
  analysis of, 7, 26, 27, 147–8
  density relationships, 28, 321
  organic content calculated from,
    27, 93, 147–8, 173, 210
  regression analysis of, 276, 278
  statistical characteristics of, 270
  stratigraphic distribution of, 34,
    306–7
  stress, 235, 323
Young's moduli
  calculated from compressive
    strength, 98

Young's moduli—*contd.*
  experimental value of, 177
  maximum value of, 176
  mineralogical effects on, 183
  organic volume effects on, 167,
    178, 179, 196, 199,
    234–7, 322–3, 326,
    551, 553–7
  Poisson's ratios calculated from,
    97, 169
  prediction of, 178, 179
  regression analysis of, 276, 277,
    278, 279
  static versus ultrasonic test methods,
    532, 533, 546
  statistical characteristics of, 270,
    271
  strain-rate effects on, 196, 199, 200,
    205
  temperature effects on, 256–7
  types of, 247, 248, 531, 532
  value used in explosive fragmenta-
    tion model, 444

Zones, retorting, delineated, 565
  acoustic velocity profiles in, 567–9